International Conference on Frontier of Science and Society (ICFSS-2017)

IOP Conference Series: Earth and Environmental Science
Volume 156

Makassar, Indonesia
3 – 5 November 2017

ISBN: 978-1-5108-6396-5
ISSN: 1755-1307

Printed from e-media with permission by:

Curran Associates, Inc.
57 Morehouse Lane
Red Hook, NY 12571

Some format issues inherent in the e-media version may also appear in this print version.

Printed by Curran Associates, Inc. (2018)

For permission requests, please contact the Institute of Physics
at the address below.

Institute of Physics
Dirac House, Temple Back
Bristol BS1 6BE UK

Phone: 44 1 17 929 7481
Fax: 44 1 17 920 0979

techtracking@iop.org

Additional copies of this publication are available from:

Curran Associates, Inc.
57 Morehouse Lane
Red Hook, NY 12571 USA
Phone: 845-758-0400
Fax: 845-758-2633
Email: curran@proceedings.com
Web: www.proceedings.com

Table of contents

Volume 156

International Conference on Frontier of Science and Society (ICFSS-2017): Maritime Perspectives to Public Discourse

3–5 November 2017, Makassar, Indonesia

Accepted papers received: 14 May 2018
Published online: 1 June 2018

Preface

Papers

IOP Conf. Series: Earth and Environmental Science 156 (2018) 011001 doi:10.1088/1755-1315/156/1/011001

International Conference on Frontiers of Sciences and Society (ICFSS-2017)

PROFILE OFINDONESIAN LECTURER FORUM (ILF)

Indonesian Lecturer Forum (ILF) has been established for about five years after long discussion among Indonesian lecturers across the nation in a facebook group, known as Dosen Indonesia. It is interesting that a cyberspace discussion group turned into a real time working group uniting thousands of lecturers across the nation having one vision to promote Indonesian education. Officially the ILF was declared on August, 24th, 2013 in Bandung, West Java and was notarized on January 8th, 2014.

The ILF is independent organization having no direct association with the office of its member, as any member of this association represent him/herself. The main objective of this organization is to improve the quality of Indonesian lecturers and that of the higher education nationwide. In order to achieve the objective, the ILF has carried out a wide range of scientific activities, such as seminar, workshop, and conference. In addition, the ILF has also encouraged her members to work together conducting community services at the regional, national, as well as international level. At this point, the ILF has successfully publish scientific journals for various field of studies. Since the member of the ILF are coming from a very diverse field of study, geography, and background because of which it becomes possible to have multidisciplinary partnership and collaboration in every activity being held.

As an independent organization, the ILF has the freedom to use and manage the fund collected either from an individual member or from non-committed sponsorship. The ILF headquarter is in Bandung and right now it has 20 branches across the nation, with nearly 1000 active members. Anyone who wish to know more about the ILF feel free to visit www.fdi.or.id or contact the ILF at this email address: sekretariat@fdi.or.id.

About Conference

International Conference on Frontiers of Sciences and Society (ICFSS-2017)

Indonesia Lecturer Forum

Thema : Empowering Nation's Maritime Uniqueness to Enrich Global Future

Venue : Makassar, Indonesia, November, 3-5, 2017

Background

Indonesian maritime perspectives came up into public discourse again after the presidential election of 2014 ended. The newly installed President, His Ecxellency Joko Widodo (known as Jokowi) has stated in his vision statement submitted prior to election, then adopted as his presidential political manifesto, to: "(1) focus on strengthening Indonesia's maritime security, (2) expand the canvas of regional diplomacy to cover the entire region of the Indo-Pacific, and (3) project the Indonesian navy as a respected regional maritime power in East Asia" (Liow & Shekhar, 2014). During presidential campaign, Jokowi repeatedly announced his willingness to transform Indonesia into a "global maritime axis" (*poros maritim dunia*). In a minute after his inauguration as the 7th President of Republic of Indonesia, Jokowi clearly instructed his newly establish cabinet to adopt the maritime vision into policy.

IOP Conf. Series: Earth and Environmental Science 156 (2018) 011001 doi:10.1088/1755-1315/156/1/011001

Indonesian Lecturer Forum organized the International Conference on Frontiers of Sciences and Society 2017. This conference is a moment as the Indonesia's visions to focus on strengthening Indonesia's maritime security simultaneously to broaden the Uniqueness of region. The conference take place in Makassar, Indonesia, November, 3-5, 2017. The conference is hosted by Indonesian Lecturer Forum (ILF) Makassar chapter in collaboration with Universitas Lancang Kuning, Indonesia, and Universiti Kebangsaan Malaysia.

The mission of ICFSS 2017 conference is to provide a forum for collaboration and experience exchange as well as invites academia, researchers, professionals and decision makers from all parts of society and all regions of the world to share their knowledge, experiences, concepts, examples of good practice, and critical analysis with their international peers.

Keynote Speaker

Zulfikar Mukhtar, M.Sc (Ministry of Fisheries and Marine, Republic of Indonesia)
Dr. Phil. Timo Marcus Dulle, Bonn University, Germany
Prof. Dr. Jamaluddin Jompa, Universitas Hasanuddin, Indonesia
Prof. Dr. Ekwan Toriman, Universiti Kebangsaan Malaysia, Malaysia

INTERNATIONAL ADVISORY BOARD

Prof. Jamaluddin Jompa, Universitas Hasanuddin, Indonesia
Prof. Dr. Masrurah Mochtar, Universitas Muslim Indonesia
Assoc. Prof. Azrin Adnan, Universiti Sultan Zainal Abidin, Malaysia
Prof. Dr. Mazlin Mokhtar, Universiti Kebangsaan Malaysia
Prof. Dr. Ekhwan Toriman, Universiti Kebangsaan Malaysia
Prof. Indah, Universitas Diponegoro, Indonesia
Assoc. Prof. Dr. Mukhtasar Syamsuddin, Universitas Gadjah Mada, Indonesia
Prof. Dr. Mudjia Rahardjo, Universitas Islam Negeri Maulana Malik Ibrahim Malang, Indonesia
Assoc. Prof. Dr. Kamarul Shukri Mat Teh, Universiti Sultan Zainal Abidin, Malaysia
Assoc. Prof. Iqbal Jawad, Universitas Hasanuddin, Indonesia

SCIENTIFIC COMMITTEE

Ismail Suardi Wekke, Ph.D., Sekolah Tinggi Agama Islam Negeri Sorong, Indonesia
Marhamah Nadir, Ph. D. Universitas Hasanuddin, Indonesia
Dr. Syamsia, Universitas Muhammadiyah Makassar, Indonesia
Irmawati Sagala, MA. M.Sc, Universitas Islam Negeri Sultan Thaha Saifuddin, Indonesia

ORGANIZING COMMITEE:

Amril Arifin, M.Si, Ak. CA (Universitas Muhammadiyah Makassar, Indonesia)
Rayudiaswati Budi, M.Si (Universitas Muslim Indonesia, Indonesia)
Fivi Elvira, MA (Universitas Bosowa, Indonesia)
Irfan Andi Pallaloi, M.Kom (Universitas Sulawesi Barat, Indonesia)
Azwar Anwar, M.Si. Ak, CA (Universitas Negeri Makassar, Indonesia)

IOP Conf. Series: Earth and Environmental Science **156** (2018) 011001 doi:10.1088/1755-1315/156/1/011001

Dien Triana, M.Si (Politenik Negeri Ujungpandang, Indonesia)
Kardina, MA (Universitas Fajar, Indonesia)
Mirfan, S. Kom (Sekolah Tinggi Manajemen Informatika Handayani, Indonesia)
Muhammad Haris Ibrahim (Universitas Hasanuddin, Indonesia)
Basir Basir (Universitas Hasanuddin, Indonesia)
Purnama Isti Khaerani (Universitas Hasanuddin, Indonesia)
Anny Lamya Munasirah (Universitas Hasanuddin, Indonesia)

IOP Conf. Series: Earth and Environmental Science **156** (2018) 011002 doi:10.1088/1755-1315/156/1/011002

Peer review statement

All papers published in this volume of *IOP Conference Series: Earth and Environmental Science* have been peer reviewed through processes administered by the proceedings Editors. Reviews were conducted by expert referees to the professional and scientific standards expected of a proceedings journal published by IOP Publishing.

Fisherman's Wife Role in Extending Household Income in Palu Gulf

Haslinda B. Anriani[1], Harifuddin Halim[2], Rasyidah Zainuddin[3], Ismail Suardi Wekke[4], Anzar Abdullah[5]

[1]Tadulako University, Jl. Soekarno Hatta Km. 9 Tondo Palu Sulawesi Tengah 94118, Indonesia
Email: haslinda.tadulako@gmail.com
[2]Universitas Pejuang RI, Jl. Baruga Raya Antang Makassar 90244. Indonesia
Email: herijuddinhalim@upri.ac.id
[3] Universitas Bosowa, Jl. Urip Sumoharjo Km. 4 Makassar 90232, Indonesia
Email: Rasyidahzainuddin@bosowa.ac.id
[4] Sekolah Tinggi Agama Islam Negeri (STAIN) Sorong, Jl. Sorong-Klamono Klablim Km. 17 Sorong 98917, Indonesia
Email: ismail@stain-sorong.ac.id
[5] Universitas Pejuang RI, Jl. Baruga Raya Antang Makassar 90244. Indonesia
Email: Anzarabdullah@upri.ac.id

Abstract. This paper aims to reveal the productive role of the fisherman's wife in improving her household's economy. Qualitative research with case study approach to five multi-profession fishermen's wives as fish seller, cake seller, daily needs seller, and others. The main data collection uses structured interviews. The focus of the interview is their role on improving the household economy. The data obtained found the equation and interpreted as a conclusion. The results of this study show that the role of fishermen's wives in improving their household economy include: selling (traditional) cakes, selling daily necessities, selling furniture, selling fish by peddling in the market, and sending it to other areas through distributors

Keywords : Fisherman's wife , income, household

1. Introduction

The involvement of the Fishermen's wife in assisting her household economy has always been an important part in the realities of Fishermen's life. In fact, their existence is more often the determinant of the sustainability of these households when they are in economic pressure. Various activities that are generally Fishermen's wives do related to the economic activities of the household, among others open the shop daily needs, cleaning trawls from sea plants, and so forth. In addition to the above activities, other activities of value productive and not less important is related to fisheries. In the context of this activity, the Fisherman's wife indirectly also contributes to the work of her husband. Some of the productive activities include sorting fish by type, preserving fish, making fish jerky, and selling fish to the market or peddling around the village.

Considering the social reality of the Fisherman's wife, Kusnadi asserted that "the economic role of coastal women is strong and dominating, both at the level of household and community level. At the household level, coastal women become "one of the pillars of the economy" of households, in tandem with their husbands. The functional relationship of husband and wife in meeting the economic needs of households is complementary or assist "household economic needs, but also to determine the availability of economic resources for household needs [1]. Similar perspectives are also stated by Rauf about ideally every member of the household to have their respective duties in order to establish a good cooperation [2]. Who does what is one of the patterns of division of labor in the household so that it does not have to be done by a certain wife or member only. The division of domestic work can also be distinguished by physical form and occupation. So term "Who" here mean family members of men or women. The perspective expressed both by Kusnadi and Rauf implied a construction of Gender Equality. As it is known that equality is the equality of the conditions of men and women to obtain opportunities and rights as human beings in order to be able to participate in political, economic, socio-cultural, defense and national security activities and equality in enjoying the results development [3]. Gender equality is the absence of discrimination against women and men in opportunities, allocation of resources, benefits and access to health services. Equality of men and women is achieved when there is equality in power and influence, equality in opportunity and freedom to work or endeavor, equality in education level, ambition, internet, talents and abilities, equality in sharing domestic responsibilities and caring for children, equality in freedom from pressure, intimidation, violence against women at home and at work.

The experts suggested division of roles in the household that men and women were divided into, among others, through or based on their "nature" values. Men through nature (earning a living), earning "dignity" while women only struggle with their nature [4,5,6]. But these circumstances must be maintained in order not to cause competition among them, for the sake of social order. Indeed, reality has shown similar situations and conditions to take place safely and in an orderly fashion, when the family at the core of the society is in a stable state of non-volatility and its role in the ladder because of internal and external issues. This theory appreciates that the balance of roles in the household is very important in order to maintain the existence of harmony in the household both in performing the functions and duties as husband and children each perform its functions and role in the household.

Research on the role of women in satisfying the needs of families, among others research in 2009 on the Role of Female Head of Households in the Village In an Effort to satisfy Family Living Needs; furthermore research in 2005 on the role of *Passambu* in Farmers Community in Kambiolangi Sub-District, Alla District, Enrekang District; and research on Women of Inter-Area Traders in Enrekang Regency. All of these studies focused on the economic role of women [7,8,9]. Regarding with the above descriptions and the study, clearly illustrated the classification of tasks that occur naturally among families of fishing communities. Therefore, this paper reveals one of the role of Fisherman's wife in based on local social reality.

2. Method

This paper used a case study approach of five Fishermen's wives that work as fish sellers. They held in-depth interviews about their family profiles, about the reasons they sell fish, the strategies they use in selling fish. In addition, their actions are also observed to synchronize with their statements. Both types of data are matched and analyzed.

3. Result

In this study, the main subjects chosen were Fishermen's wives that work as fish sellers as many as five people, as follows: Wahyuni's involvement in helping her husband is based on fulfilling the family's economic needs in her Fisherman's household as a fish seller. When the west season took place, her husband did a job as a carpenter while Wahyuni also switched professions as a traditional cake seller in the morning. The work is done as a temporary job. In normal weather and Fishermen activity again improved then Wahyuni's family back to do the job that is as a fish seller.

Wahyuni has been working on this for about 10 years and also assisted by his son. If the amount of fish supply brought by her husband in large quantities, then Wahyuni contact the fish retailer to come to his house to take fish for sale. Conversely, if the amount of fish supply is less. Then Wahyuni bring it to market. In addition to pursue the job, she also sometimes look for other additional income is selling cakes at home. In carrying out their work, when they want to catch fish, they must first complete themselves with the provision of their lives to go to sea to fulfill their needs at sea. Apart from that at the time of taking the fish should prepare capital of approximately one million five hundred thousand with the time used is also uncertain sometimes 24 hours sometimes 48 hours and sometimes use 12 hours depending on their condition at sea.

Fishermen's wives have a wide variety of joys and sorrows depending on their luck condition. This is important because their earnings cannot be predicted in every day if that day is their lucky day in getting results that are very exciting for them. Luck in question, for example sometimes earn income of 1.5 million from once to sea but vice versa. On the contrary, when they are on a bad day or a disadvantaged day their earnings sometimes do not match their expectations or have luck on that day they spend capital of one million five hundred thousand rupiah. Similarly to those whose income is not fixed. When it comes to good days then the money earned can be up to 3 million rupiah from once to sea.

The activities of the Fishermen's wife within the scope of the Fishermen's family that in carrying out their activities, the actual income earned does not remain dependent on the weather and market conditions of the day. The large number of fish from the catch of fishermen is very influenced income of wife on that day. Wahyuni's husband has a position as a boat owner. In that status he must bear all operational costs when going to go to sea. The amount of capital they use is relative, according to the distance they take for seven days or ten days and sometimes even twenty days at sea. This depends on the location of their fish *rumpon*.

In addition, the amount of capital used varies depending on the capacity of the boat used whether the boat is using a car engine (consisting of two engine units) or with a small capacity engine. The cost that they spend on high-capacity vessels also requires substantial operational costs. Their role as Fisherman's wife is to help her husband sell the fish outside the area for example to Poso, Parigi, and Biromaru. Their income is tentative, depending on their situation at the time. For example, when the west season fish prices are usually categorized as expensive because the Fishermen less than the maximum activity. As a result the fish supply is reduced, causing fish prices to rise or in the language of the "more expensive fish" market. Indeed the activity of the Fisherman's wife is clearly different from the Fisherman's wife who plays as a *Sawi*. It is each of them move in different situations and conditions depending on the sustenance of the day. Actually the Fishermen have different activities and each Fisherman has different boats and boats also depending on which boat and boat they occupied to join. Likewise, the amount and value of their income also vary. It is unique from the community of Fishermen's wives that in order to obtain high sales revenue other than the large amount of fish supply is also determined by the area where they sell the fish whether in their own area or outside the city as mentioned above.

Mantasia's involvement as a fish seller is motivated by cultural factors of hereditary work done by family environment. Since long time, her family has been selling fish and supplying fish to various areas around him. In carrying out the process of buying and selling fish, Mantasia do it by using a large scale tool that the size of the fish ranging from 7 kg-20 kg range by using pick up vehicles. In the neighborhood, Mantasia's husband is known as a skipper of a ship. Mantasia with her husband did the work at home selling fish in pretty good amounts in kilograms. Target catch of her husband is a fish that manifold big fish such as snapper, Tuna fish, and the like. in addition, it also sells to other areas by using pickup vehicles. Through this way he earns 5-10 million rupiahs per three days and even up to 50 million. Ships used also diverse. 1 boat with large capacity, 4 small capacity vessels.

Mantasia activities as a fish seller while her husband as a Fisherman in the sea is very influential on the number of wife's income. It is determined by the catch of fish that brought her husband from the sea. Similarly, the position of the husband on the ship is very influential to the family's income every day whether the Fisherman once a day or in three days or in a week or a day. The number of fish that her husband earns each at sea is also determined by the distance the *rumpon* that he has.

Faridah's activity is to sell fish and fish providers to other areas such as Sigi, Donggala, Poso, and Tentena. She used pickup trucks loaded with 25-30 boxes of fish made from cork. She did it with her husband who works as a financier. Income earned ranged from 5-10 million rupiah per three days and even more.

From the daily life of socio-economic life, Faridah's family is well established because the Fisherman is a capitalist. Each fishing activity as a guide in carrying out its activities as a sailor. In addition to spending capital to go to sea in the amount of high enough that sometimes resulting from the sea and even then uncertain, sometimes the income is profitable sometimes loss. This is due to weather and luck factors that day. But even so the experience and knowledge of the sea has come to the level of proficiency and expertise in performing routine activities is to go to sea.

Likewise Faridah as a housewife in this case as a woman who succeeds in wading his life working together with her husband in order to achieve a success and become a meaningful human figure in his life to his own household and the environment. Thanks to the success she gained, then how to sell his fish was different from other fish sellers. Members and partners engaged in activities as Fishermen's wives who sell fish, generally sell fish in the market or bring them to fish auctions. While the woman is no longer carry out activities like that, but the Fishermen's wife only ordered its members to sell it outside the area or other areas. By using your own car that has been prepared to transport the fish. In her activities, Rosnawati sometimes works as a fish seller and also works as a manager of various needs needed by the local people, for example the needs of kitchen appliances, furniture needs and other consumptive needs. It can serve in accordance with consumer demand by way of credit starting demand in installments for 3 months to 10 months. Generally they pay according to the agreement eg paying on a weekly. Rosnawati pursue her work motivated by the work hereditary done by her families so that the job as a fish seller is a cultural heritage inherent family of their family, in the sense that this worker is a mandatory job they do. In running it, Rosnawati also as a trader of nine basic materials. Another activity is to form an *arisan* group.

As a Fisherman's wife, Rosnawati can be judged as a wife whose income is adequate. Her husband's position on the ship as a crew can contribute enough to his wife who works as a fish seller, so the income of this fisherman's wife can be quite a lot. The average income from selling the fish they earn once sailing between 3-5 million.

Sahriani as a Fisherman's wife pursue her job as a fish seller caused by habit factor. In addition, the large number of family members also contributed to their daily needs in their households. Furthermore, Sahriani states her involvement as a fish seller is a job that he persistently hereditary done by their families. However, this activity is not carried out continuously caused by bad weather, so sometimes he does other activities selling traditional cakes that are typical of the region in the environment. In carrying out her work, he sells fish in the market from the catch of her husband who is positioned as a *Sawi* (labor). Sahriani's income is not fixed sometimes 500 thousand dollars to 1 million rupiahs per seven days. Sahriani whose job status of her husband as a *Sawi* in the ship greatly affects her activities as a fish seller. The amount of fish that her husband earns is very small, so he sells fish in the market, from the catch of her husband's partner who gets the supply of more fish then he sells it to the nearest market. Looking at the descriptions of all the informants above, the authors conclude in the form of the following table.

Tabel 1. Fishermen's wife role on fisher houshold in coastel of Palu Gulf.

No.	Informants	Husband role	Income (Million)	Other works	Reason to work
1.	Wahyuni (27 years old)	Ship owner	1,5-3	Selling Cake	Family needs
2.	Mantasia (34 years old)	Ship owner	5-50	-	Hereditary habits
3.	Faridah (25 years old)	Ship boss	5-10	-	Helping the household economy
4.	Rosnawati (30 years old)	Sawi or Labour	3-5	Selling daily life need, goods credit	Hereditary work
5.	Sahriani (31 years old)	Sawi or Labour	0,5 -1	Selling traditional cake	Family heritage

Source: primary data.

Based on the above table illustrated the activities of Fishermen's wives viewed from the structure of their work as a Fishermen's wife who has a dual role in the household, namely as domestic workers and also as a public worker. The women's dual burden serves as a household manager and has full responsibility for fulfilling and serving the needs of the husband as a housewife and on the other hand the women are each responsible for the fulfillment of the household's economic needs. The Fisherman's wife does this activity as a fish seller that holistically the Fisherman's wife works as a fish seller, because it is dominated by activities as a wife who helps the husband in terms of fulfilling the family's economic needs. This they do because the desire factor and willingness as a wife who wants to help the husband in terms of economic fulfillment of Fishermen's household.

4. Discussion

Based on the above descriptions, basically the Fishermen's wife who is involved in the job as a fish seller is basically caused by the factors of fulfilling the family's economic needs. They do so by helping their husbands to earn a living in order to achieve economic needs in their households, at the level of need that is in the household environment of the fisherman. The existence of economic functions in the family is a factor in family relationships not only based on the interest to continue offspring but also view the family as a working relationship. husband not only as head of household but also as head of work [10]. By him it is in the social life of Fishermen households, actually a wife is very varied in terms of activities and functions as a housewife who works as a breadwinner. In carrying out its activities were also varied this can be seen in terms of the actions underlying them in carrying out its activities as a Fishermen's wife who works as a breadwinner. It can be seen from the background of socio-economic life of their household, both from the husband's family and from the wife's family it is also very affected, so that the actions of the wife of the Fisherman in doing his activities sometimes also influenced by cultural values factor is inherent in hereditary is done by their families, so the Fisherman's wife is involved in the job as a fish seller.

Observing the reality of the Fishermen's wife, it calls as the term 'partnership' and 'alignment' husband and wife, where both are in the same status. Pandu provides an example in terms of decision making in terms of household, both in bargaining status equal equivalent weight. The same status carries the same human dignity and dignity. Both will respect each other, mutual help, mutual support, filling each other's flaws, sharing each other's flavors, share the fate [11].

If the above description is related to the concept of 'gender' then the role of fisherman wife in the household economic activity shows the existence of gender equality in the field of domestic role [12]. This condition can be a cause for widespread opportunities for women, including mothers to improve themselves by continuing higher education, which in turn will lead them to enter both formal and informal employment [13]. The implied part of the mother's time for educational activities, economic and leisure time causes the mother's function as a primary educator to start shifting.

At the opposite angle, the realization of equality makes the creation of balanced relationships between husband and wife. There is a shift in the form of: (1) the wife serves while the husband is serviced, (2) the wife serves while the husband is served, (3) the husband of the ruler while the wife of the executive, (4) independent husband while the wife depends, (5) husband is venerated while wife venerate.

5. Conclusion

The involvement of the wife as a fish seller in a Fisherman's household is based on (a) the fulfillment of household economic needs, and (b) hereditary inheritance. In this case, the role of the wife becomes more than the husband who only catch fish in the sea, while the wife in addition to performing the duties of housewives also play a role in the economic field. Nevertheless, the Fishermen's wife considers all these roles to be part of his participation in the family.

6. Acknowledgements

The authors would like acknowledge to the dean of the social and political science faculty in the official permitting to do this research. Also thank you to the all informants for the interview and other data supports.

7. References

[1] Kusnadi. (2009). Keberadaan Nelayan dan Dinamika Ekonomi Pesisir, Yogyakarta: Ar-Ruzz Media.

[2] Rauf, (2008). Angkatan Kerja Wanita, Kasus tiga Kota Di Sulawesi Selatan. Perpustakaan Nasional: Katalog Dalam Terbitan.

[3] Kristina Ni Nyoman. (2013). *Referensi: Modul Pelatihan Pengarasutamaan Gender Bidang Kesehatan (PUG-BK) Bagi Tenaga Kesehatan, Kemenkes RI,* Badan Pengembangan dan Pemberdayaan SDM Kesehatan Balai Besar Pelatihan Kesehatan Jakarta.

[4] Pandu.M.E.dkk. (2001). *Perempuan di Sulawesi Selatan* (Ikhtisar Regional) CIDAP17.18.

[5] Megawangi, Ratna. (1999). Membiarkan Berbeda: Sudut Pandang Baru Tentang Relasi Gender. Cet. I. Bandung: Mizan.

[6] Mosse, J.C. (1993). *Gender dan Pembangunan.* Terjemahan oleh Hartian Silawati. Yogyakarta: Pustaka Pelajar.

[7] Musrifah. (2009). Peranan Kepala Rumah Tangga Wanita Di Pedesaan Dalam Upaya Memenuhi Kebutuhan Hidup Keluarga (Kasus 5 Janda Cerai Desa Sidorejo Kecamatan Pulokulon, Grobogan). *Thesis*. Universitas Negeri Semarang.

[8] Mahmud, Abdul Rahman, (2005). Peranan Passambu dalam Masyarakat Petani di Kelurahan Kambiolangi Kecamatan Alla Kabupaten Enrekang. *Tesis* Program Pascasarjana Universitas Hasanuddin, Makassar.

[9] Herawati, Netty. (2015). Perempuan Pedagang Antar Wilayah (Studi Kasus Pada Empat Keluarga Pedagang Kecamatan Alla di Kabupaten Enrekang). *Disertation*. PPS Universitas Negeri Makassar.

[10] Narwoko, (2004). *Sosiologi Teks Pengantar dan Terapan*, Jakarta: Penerbit Kencana.

[11] Pandu.M.E.dkk. (2001). *Perempuan di Sulawesi Selatan* (Ikhtisar Regional) CIDAP17.18, p.45.

[12] Budiman, Arief. (1985). *Pembagian Kerja Secara Seksual, Sebuah Pembahasan Sosiologis Tentang Peran Wanita di Dalam Masyarakat*. Jakarta: Penerbit Gramedia.

[13] Fakih, Mansour. (1999). *Analisis Gender & dan Tranformasi Sosial.* Yogyakarta: Pustaka Pelajar.

IOP Conf. Series: Earth and Environmental Science **156** (2018) 012002 doi:10.1088/1755-1315/156/1/012002

Social Conflict in Malifut Sub-district of North Halmahera Regency

Ebin E Danius
Philosophy of Divinity Department, University of Hein Namotemo, Jl. Kawasan Pemerintahan, Vila Vak I, North Halmahera Regency, North Molucas Provinces, Indonesia 97762.
E-mail: ebin@unhena.ac.id

Abstract. Social conflict is always understood as a social phenomenon that occurs in a plural society. The assumptions about conflict are always understood as a conflict of interest for limited resources. Conflict groups are formed as a continuation of the common interest in dealing with other groups. Such assumptions eliminate the location as a source of conflict. This paper wishes to show that to understand a conflict is not enough with social analysis but also needs location analysis. The location of the residence of a community participates in determining the pattern of life and behavior patterns that impact psychologically on the way a society sees the conflict. Specifically, the location taken for observation is Malifut District, North Halmahera district where the place had been a conflict between citizens. The author wants to show that the coastal area as a social environment Malifut community has a share in determining the occurrence of conflict. In this case, the authors will discuss the Malifut community's understanding of the conflict and show the influence of the coastal environment on the understanding of it.

Keywords: Social Conflict, Malifut, Makian, Pagu, North Halmahera, North Maluku

1. Introduction

Malifut is one of the sub-districts in North Halmahera Regency of North Maluku Province. In December 1999 inter-community conflict occurred in the region. The conflict then spread to various regions in North Maluku province. There are several analyses of the conflict. Chris Wilson [1] argues that the conflict occurred as a result of the influence of the conflict that occurred in Ambon. The main contributing factor to the conflict, according to Wilson, is the lack of public confidence in the security apparatus that has led people from two religious communities to choose to secure themselves through groupings based on religious identity. In addition, according to Wilson this conflict is also caused by the formation of sub-district Malifut separate from Kao District. Some villages inhabited by indigenous people refuse to be part of the newly established sub-district.

Klinken [2] argues that conflicts in Indonesia are strongly linked to communal identities. In such conflicts, the opposite sides use a certain identity as a unifying feature. In addition, identity is also used to distinguish between groups in conflict. Identity factors are important when natives dealing with newcomers who differ in religion and culture. Another important factor is politics. Economic disparity, cultural differences, religion, and ethnicity are not strong enough to generate a conflict. Political interests provide direction toward achieving the expected goals of the group. In this case, the political factor can be considered as one that unifies the differences to then generate an open conflict. Related to the conflict in North Maluku Klinken views that the conflict in North Maluku is a direct result of the fall of the New Order regime. Such a view is attributed to the efforts of local politicians to build influence in order to gain a share within the new province to be established.

Bertrand [3] sees on a wider scale by linking conflicts that occur in areas including the conflict in Malifut as a result of the collapse of the New Order's hegemonic power which with the collapse led to the seizure of influence from interest groups to influence the course of government. The collapse of the New Order regime prompted the seizure of influence from interest groups to influence the way of government. Thus it can be understood that the occurrence of conflicts in certain areas including in Malifut is the result of elite opposition in the central government. Such an understanding also implies that the problems occurring at the local level have no power to trigger conflict.

IOP Conf. Series: Earth and Environmental Science **156** (2018) 012002 doi:10.1088/1755-1315/156/1/012002

2. Method

All analyzes of the Malifut conflict use a social and political approach as a starting point for understanding the conflict. To understand social conflicts, social and political discourse in people's life is an important factor that can not be ruled out. However, all of the analyzes described above show that the conflict is understood to be limited to a sociological perspective with a primary consideration of external factors which are then linked to the internal conditions of society. Such an analysis seems to exclude the internal conditions of society. Although of course the internal aspects of society are also related to the social situation but the psychological aspects and the influence of the location of the constituent elements of society become important things that also need to be considered. To that end, the writer approached the conflict from the perspective of the Malifut community itself. Are there certain influences from the location of their residence with their perspective on the conflict? Does the coastal area as a residence produce different perspectives on how they perceive the conflict? In this case the authors believe that people's understanding of the conflict and the way they live it have little influence in the conflict.

3. Results

3.1. The Origin of Malifut Conflict

The Malifut region before 1972 was an area populated by ethnic Pagu. There are several villages in the region, namely Sosol, Wangeotak and Tomabaru villages. In 1972 the local government at that time was the government of North Maluku Regency seeking to the displacement of people on the Makian island. The reason for the move was in a study showing the potential danger of an eruption of the Kie Besi volcano which is expected to drown the island of Makian. The reason for the displacement of the people was in a study showing the potential danger of an eruption of the Kie Besi volcano which is expected to drown the Makian island [4]. The displacement program to moving out people from Makian island to other areas considered more secure is not only a program of the regency government but also a central government program. The displacement of the population began officially in 1975 and from then on gradually until 1982 the people of the island of Makian were transferred to the territory of Halmahera Island.

One of the locations of the displaced people of the Makian island is the Malifut region which was then part of the Kao district. The local people welcomed the inhabitants of the Makian island. Local people even provide their residence as a temporary residence for migrants from the island of Makian. Friendship is also quite good where it is seen with the provision of land from indigenous to immigrants. Giving the land according to the native population to support the early life of the immigrants.

Some sources discussing the Malifut conflict argued that the main factor causing the Malifut conflict was the creation of the Makian Malifut sub-district [5] [6]. The Makian ethnic who are immigrants in Malifut are delighted to the establishment of the sub-district. The Makian ethnic who are immigrants in Malifut are delighted at the establishment of the sub-district.The forming of the new sub-district gives the Makian ethnic authority in Malifut to take care of their own sub-districts. The forming of the new sub-district gives the Makian ethnic authority in Malifut to take care of their own sub-districts. Prior to the establishment of the subdistrict, the population of the Makian ethnic group still relied on the sub-district government on the Makian island. This situation is of course difficult for them because the location of the island of Makian far from Malifut. Note that the Makian ethnic migration process to the Malifut area was not accompanied by the displacement of the sub-distric administration. This has an impact on government administration that is still controlled by the sub-district on the island of Makian. The establishment of Makian Malifut sub-district based on Government Regulation number 42 year 1999 where one of the articles regulates the removal of Makian Subdistrict located on the island of Makian. This abolition for Makian ethnic groups has been relocated to victory because they think in reality many residents on the island of Makian are refusing to move. With the removal of Makian sub-district on the island of Makian making the inhabitants still surviving on the island of Makian will itself out of the island.

For the indigenous people, the establishment of Makian Malifut sub-district raises its own problems. As indigenous people they refused to join the new district. In addition to the issue of the name of the sub-district as well as ethnic issues become an important factor. In addition to this, according to Judo Poerwowidagdo [7] in addition to ethnic issues, minor conflicts between indigenous peoples and Makian ethnic communities have never been resolved completely. In this case people are not quite sure that the government can solve their problems. As a result people choose to solve their own problems in ways that are deemed appropriate by them.

3.2. Conflict in Perspective of ethnic Pagu

The indigenous people who inhabit Malifut are the ethnic Pagu community. Ethnic Pagu is one of the original ethnic groups in Halmahera Island. This ethnicity is spread over several villages in Kao District. Specifically in the discussion of the Malifut conflict, this Pagu ethnic inhabit some villages in Malifut namely Sosol, Wangeotak and Tomabaru villages. Although living on the coast, the main livelihood of ethnic Pagu is on forest products and gardening. The hallmark of this ethnicity is their calm nature with a softer intonation of language than the other ethnic Halmahera.

For indigenous peoples, the coming of Makian ethnic to their territory is not a big deal. The socialization done by the government and their friendly attitude made the acceptance process work well. It should also be pointed out that the place of the people of the Makian ethnic who is now the first village of the Sosol villagers. The name of their village was originally Malifut. Malifut itself the name of a tree that grows coastal beach. However, because of the frequency of their village hit by the tide, they decided to move their village to a safer place. The initial location of their village was then used as a settlement location for migrant residents of Makian ethnicity. This placement is according to citizens Sosol as an act that does not respect them as indigenous people. But their anger was not directed at the Makian ethnic but to the government that placed the migrants in their territory.

In addition, such rapid development with all the supporting facilities provided to the immigrants evokes a certain jealousy. Their silence is more on their efforts not to take the issue to a position that will confront them with Makian settlers. The most obvious thing in their view is that the state has enough money given to migrants. As indigenous people, they have never felt the attention of the government where it can be seen from the facilities provided to them. The very rapid development that is intended for the migrants raises a certain jealousy for the indigenous people.

Observing the development developments given to migrants also led to a sense of ownership of the land they owned. For indigenous people the government should appreciate their existence and the land they have. The central government in this case seemed to ignore the ownership of the land and assume that the land given to migrants from the island of Makian is unprocessed land so that it is a state land that can be given to anyone who is desired by the state. This fact also generated suspicion from the indigenous population of the managers governing the removal of the Makian population into their territory.

Land issues are a major issue for indigenous Malifut people [8]. Governments that do not pay attention to their complaints are then added to the attitude of ethnic Makian who seem to want to control their land makes the indigenous people's discontent increasingly accumulate. The result is that they take back the land left to the migrants. In addition they began to draw clear boundaries of their land with the land granted by the government to the Makian ethnic.

3.3. Conflict within Makian Ethnic Perspective

Ethnic Makian is one of the ethnic groups in North Maluku and inhabits the island of Makian. Makian ethnic community though live on the island surrounded by sea but the main livelihood of society is farmer. The inadequate marine environment makes farming an important thing to do.

The removal of Makian ethnic from the island of Makian to Malifut is not without problems. The main rejection of Makian ethnicity towards this transfer process was their ownership of their land on the island of Makian. Approaches and guarantees provided by the government make Makian citizens willing to move to the Malifut region.

In view of the Makian ethnic community, the conflicts are caused more by misunderstandings between Makian ethnic and indigenous people. They recognize that the indigenous people's welcome to them is very good. Their arrival was welcomed by the natives and even some of them raised their brothers with indigenous Malifut.

The main misconception is the way the Malifut people see the lives of migrants from Makian ethnicity. The rapid economic development is not due to government grants but because of their hard work in managing agricultural land. In this case, the Makian ethnic seeks to maximize every land they earn to make money. Makian ethnicity in managing the land is also based on their life experiences on the island of Makian. According to them the island of Makian in addition to very limited land also because the land on the island of Makian is not fertile enough so that their agricultural produce is not enough to help to improve the economy. This is also the cause of the many ethnic Makian who migrated to various areas in North Maluku. In this case, what they are doing is the utmost effort to harness the potential of nature without intending to make the indigenous Malifut marginalized. In such a way, Makian ethnic realizes also that the conflict is somewhat due to their lifestyles that always make the most effort in their economic improvement [9].

4. Discussion

The pattern of relations between migrants and indigenous peoples is a factor that needs to be taken into account in understanding the Malifut conflict. The coming of Makian ethnic occupation into the life of the indigenous people of Malifut changed the way the indigenous peoples perceived the value of the land they possessed. Of course this is not a major issue but different lifestyles with different characters and religions are factors to consider in understanding conflict.

As has been pointed out that the purpose of this paper is to show the influence of the location in this case the coastal region to the conflict that occurred. The research shows the different ways in which both communities see conflict. The ethnic Pagu people who are indigenous Malifut even live in coastal areas but their lifestyle is the lifestyle of rural people. An important feature in the life of the inland Halmahera people is in their more subtle language speech and a slightly shy lifestyle. Another important feature is their view of the land. For them the land is part of the living area used for survival. Land boundaries do not become important because basically the utilization of land is carried out for the sake of survival. In such a way of thinking the land is not viewed in an economic perspective but rather on a functional perspective. With such a perspective, land granting to migrants from Makian ethnic groups is regarded as a sign of friendship where the gift is also embodied in the sense of preservation of the indigenous population towards migrants.

In contrast, migrant settlers from Makian ethnic who have been accustomed to the limitations of land in their place of origin use the land grant to the maximum. The migrants immediately demonstrated progress in the economic field. Of course in addition to their own efforts, the government's attention in providing their life support infrastructure is a factor that can not be ruled out in favor of their lives in a new place.

From here clearly visible meeting of two poles of life. Although both communities rely on agricultural produce, their location determines their perspective on the land they own. Indigenous peoples who initially lived in Halmahera were familiar with the abundance of forest products where this formed their mindset in a way of looking at unlimited natural resources for their lives. The vastness of the land makes the indigenous people less concerned with maximizing their land for economic development. This is fundamentally different from the Makian settlers. Living on an island surrounded by sea makes them realize the importance of land use maximally. Especially with the sea conditions that do not always support make them lend their livelihood on agricultural products. Land granting by the government makes them maximally utilize the land. Even with no hesitation they began to penetrate new forest areas that will be used as agricultural land. This action then also triggered the indignation of the indigenous people because in their understanding all existing land belonged to them by custom.

It is also clear that the Malifut conflict is not a fully politically charged conflict and not a purely natural resource overlap in the region. Communities of the two ethnic groups do not see it as an important issue because basically the main issues that emerge are land ownership rights and unfair government treatment [10].

5. Conclusion

Coastal residential areas with major searches not on fishery produce make two groups of people fight over land for use. In this case the conflict is the result of differences in perception of the land and the difference in the treatment of the land itself. Maximum utilization of existing land by migrants makes their economically visible lives quickly improve. This result also changed the way the indigenous people view their land. This change makes the land no longer seen in the functional aspect but at the economic value of any land they have. Thus changes in the values of life become more materialistic. This change also gave rise to a new awareness that all this time they are only used by migrants. Such an assumption triggers them to take back all the land they once gave to migrants when they arrived for the first time. Thus the conflict that occurred in Malifut from the perspective of indigenous people and migrants is a land-grabbing conflict.

6. References

[1]. Wilson C 2008, *Etno Religious Violence in Indonesia; From Soil to God* (New York: Routledge) p 97
[2]. Klinken G van, 2007, *Perang Kota Kecil, Kekerasan Komunal dan Demokratisasi di Indonesia* (Jakarta : Yayasan Obor) pp 1-29 (in bahasa Indonesia)
[3]. Bertrand J 2012 *Nasionalisme dan Konflik Etnis di Indonesia* (Yogyakarta:Ombak) pp 18-19.
[4]. Braithwaite, J and Dunn, Leah, 2010, *Maluku and North Maluku*, (ANU Press), p. 198
[5]. Duncan, Christopher R., 2005, The Other Maluku : Chronologies of Conflict in North Maluku, *Indonesia,* No.80 (Oct.,2005), pp.53-80
[6]. Wilson, C., 2005, The Ethnic Origins of Religious Conflict in North Maluku Province, Indonesia, 1999-2000, *Indonesia,* No.79 (Apr.,2005), pp 69-91
[7]. Poerwowidagdo J 2003*Menuju Rekonsiliasi di Halmahera*(Jakarta:PPRP) pp 106-108.
[8]. Smith, Claire Querida, 2009, The Contested State and Politics of Elite Continuity in North Maluku, Indonesia (1998-2008), *Dissertation*, (The London School of Economics and Political Science) pp 121-122
[9]. Umar, Rivai, Menelusuri Jejak Konflik Halmahera Utara, in Kasman, Hi Ahmad and Oesman, Herman, 2000, *Damai yang Terkoyak : Catatan Kelam dari Bumi Halmahera*, (Kerjasama Kelompok Studi Podium, LPAM Pemuda Muhammadiyah Maluku Utara dan Madani Press) p 88 (in bahasa Indonesia)
[10]. Hermenkens, Anna-Karina, and Timmer Jaap, 2011, Conflicting States: Violent Politics in North Maluku, Indonesia, *Etnofoor,* Vol.23, No. 2, State/Violence (2011), pp.56-77.

Controlling DC-DC Buck Converter Using Fuzzy-PID with DC motor load

Jumiyatun Jumiyatun,[1] Mustofa Mustofa[2]

[1]Electrical Engineering Department, Tadulako University, Palu, Indonesia.
Email: jum@untad.ac.id
[2]Mechanical Engineering Department, Tadulako University, Palu, Indonesia.
Email: mustofa@untad.ac.id

Abstract. Circuit of DC-DC converter is a device converting from remains DC input voltage to variable output voltage. The components used for switching converter can be MOSFET, IGBT, Thiristor, GTO and others. This paper discusses DC-DC buck converter to reduce the voltage by using MOSFET switching that is used to control the output voltage of buck converter remain constant with DC motor load. To control the output voltage the buck converter remains constant and stable, it is required proper control to regulate the duty cyclenya. The methods are combination of conventional control methods namely Proportional Integral Derivatives (PID) and intelligent control of Fuzzy logic. It was simulated by using MATLAB software. The simulation results show the combination of PID control and Fuzzy Logic can improve the performance. Small error is almost zero and the output voltage of the buck converter remains constant even though the motor load is changed.

Keywords: Buck Converter, fuzzy PID kontroller, Motor DC

1. Introduction

To get an output voltage that has a high quality and efficiency, it usually uses a switching system rather than linear systems that have power losses. This paper discusses the control of output voltage DC-DC buck converter which serves to decrease the voltage till remain constant, when DC motor load is altered using MOSFET switching. Combination of PID control and Fuzzy logic are used that regulates their cycle duty. On switching converter,MOSFET, IGBT, Thiristor, GTO and others components are applied. PID method is a conventional control that has been widely used both for research and applications in the industry. While *Fuzzy like PD* is a fuzzy control intelligent whose behavior resembles PD.

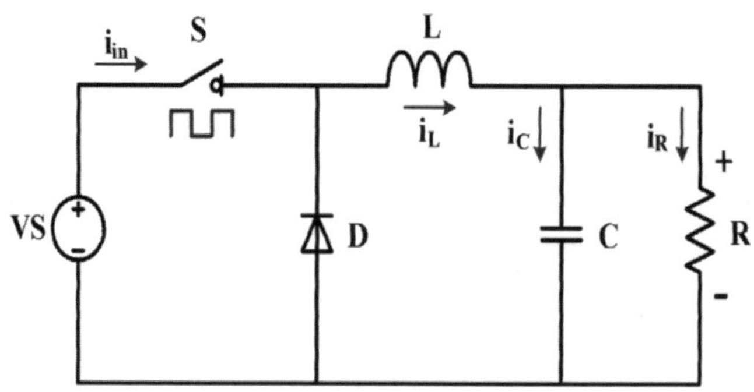

Fig 1. Buck converter circuit

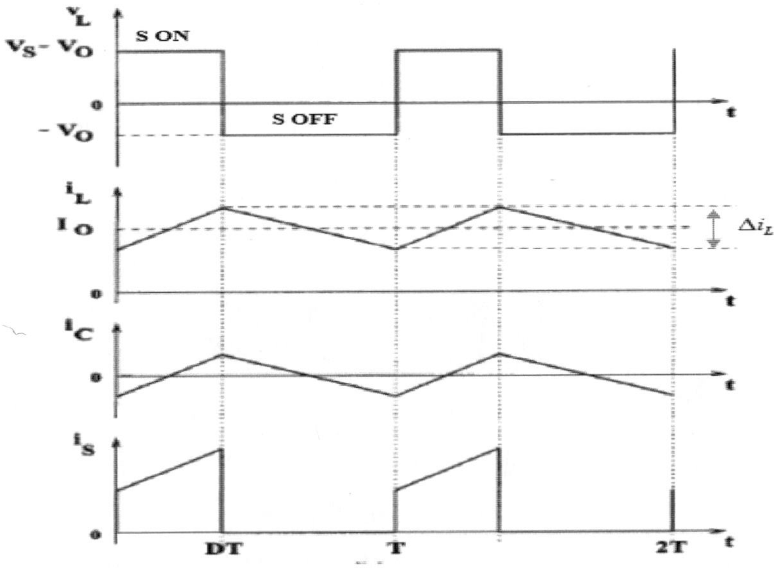

Fig 2. Wave response of buck converter

Some studies that have been done related to buck converter control.Amit, et al. [1] discusses speed control driven by FOPID DC motor to see performance and robustness orvarying loads. Bhagyashri et al. [2] figure out load changes and input voltage reducing efficiency and power of the converter. The function of fuzzy control is to maintain the output voltage constantly even though the input and load voltages are changed in fast.Sutedjo et al. design construct DC-DC converter module with PI controller. It can control the output voltage of buck and boost converter circuit to be stabler even with the condition of input voltage is not balanced, when the circuit works [3]. Jumiyatun dkk., compare two methods; conventional PI method and Fuzzy-PI Logic to control the output voltage of buck converter remaining constant. Controlling using Fuzzy-PI controller is better performance and faster response than conventional method [4].

2. Method

In this study, DC-DC converter is designed by determining the correct parameters value of the main components of buck converter, namely inductor (L), Capacitor (C), Resistor (R), duty cycle, and DC motor instead of R load. During simulation, combination of Fuzzy logic control and PID performs better output steady state response.

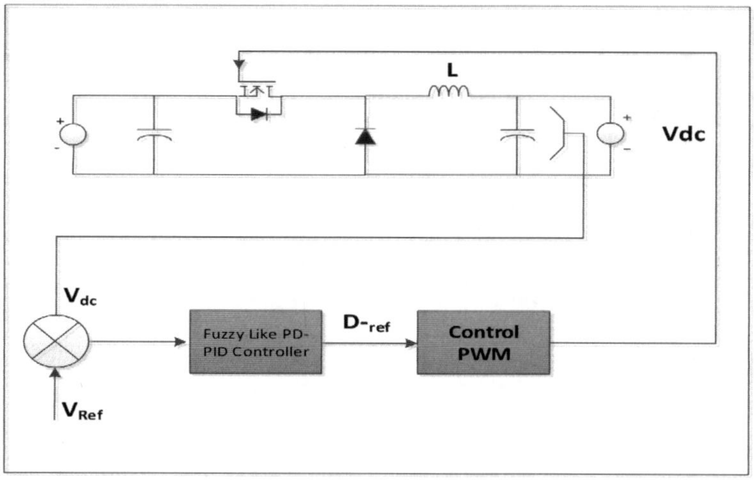

Fig 3. System diagram

As can be seen in Fig 3, the system diagram above the buck converter receives the input and current voltage from the DC source, then converts the output voltage lower than the input voltage and the result is fed to the load. Thus, the voltage that flows into the load can be valued according to the set point that has been determined and then output from buck converter in feedback to control.

a. Buck converter [5]

designing the system, the parameter values of each component of the buck converter as seen in Table must be set so that the whole system runs well.

Table 1. Parameters of buck Conventer

Parameter	**Value**
Input voltage (Vin)	500 Volt
Output voltage (Vo)	200 Volt
Ripple (ΔV)	5%
Arus Output (Io)	12 A
Ripple Arus (ΔI)	10%
Frekuensi Switching	10 KHz

- Determine of duty cycle values (1)

$$Son = Soff$$

$$(\text{Vin} - \text{Vo})\text{ton} = \text{Vo.toff}$$

$$\text{Vin.ton} - \text{Vo.ton} = \text{Vo.T} - \text{Von.ton}$$

$$\text{Vin.ton} = \text{Vo.T}$$

$$\frac{\text{ton}}{\text{T}} = \frac{\text{Vo}}{\text{Vin}}$$

$$\text{D} = \frac{\text{Vo}}{\text{Vin}} = \frac{200\text{V}}{500\text{V}} = 0{,}4$$

- Determine resistor value (2)

$$R = \frac{Vo}{Io} = \frac{200V}{12A} = 16{,}667\Omega$$

- Determine inductor value (3)

$$\Delta IL = 10\% x12A$$
$$\Delta IL = 1{,}2A$$

$$L = \frac{Vin(1-D)D}{\Delta ILxfs}$$

$$L = \frac{500V(1-0{,}4).0{,}4}{1{,}2Ax10Khz}$$
$$L = 0{,}01H$$

- Determine capacitor value (4)

$$\Delta V = 5\% x200V \quad R = \frac{Vo}{Io} = \frac{200V}{12A} = 16{,}667\Omega$$
$$\Delta V = 10V$$
$$C = (D.Vin + \Delta V)\frac{(1-D)}{fs.R.\Delta V}$$
$$C = (0{,}4.500V + 10)\frac{(1-0{,}4)}{10000x16{,}667x10}$$
$$C = 0{,}0000756\ \mu F$$

Table 2. Parameters of Motor DC

Parameter	**Unit(SI)**
Mechanical power($P_{mekanik}$)	5 HP
Anchor voltage (E_s)	240 V
Nominal speed Nominal ($\omega_{nominal}$)	1220 Rpm
Anchor resistance (R_a)	0,6 Ω
Anchor current (I_a)	16,2 A
Anchor inductance (L_a)	0,012 H
Inertia (J)	I Kg.m^2

2.1. Designing controller method

To get the output voltage constant buck converter used to supply DC motor load using combination of Fuzzy like PD and PID control, the steps are as follows:

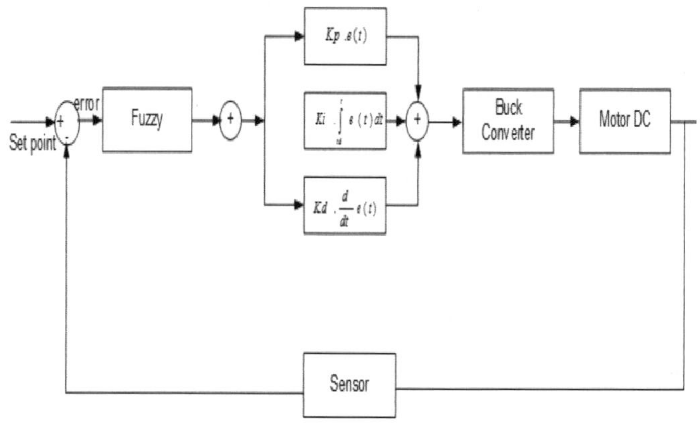

Fig 4. Block diagram of combination Fuzzy & PID Logic Control at Buck Converter with DC Motor Load

A. PID (Proportional Integral Derivative)[6]

Proportional Integral Derivative (PID) is a controller to determine the precision of an instrumentation system with the characteristics of feedback on the system. The PID control component consists of 3 types: Proportional, Integral and Derivative. All three can be used simultaneously or independently depending on the response we want to a plant. In the PID control, the error signal e (t) is the control input, while the control output is the control signal u (t). The relationship between the control input e (t) and the control output u (t) is:

$$u(t) = K_p \left\{ e(t) + \frac{1}{\tau_i} \int_0^t e(t)dt + \tau_d \frac{de(t)}{dt} \right\}$$

Or in form of Laplace Transformation :

$$U(s) = K_p \left(1 + \frac{1}{\tau_i s} + \tau_d s\right) E(s)$$

Thus, transfer function of PID controller is :

$$\frac{U(s)}{E(s)} = K_p \left(1 + \frac{1}{\tau_i s} + \tau_d s\right)$$

Block diagram of PID controller, namely :

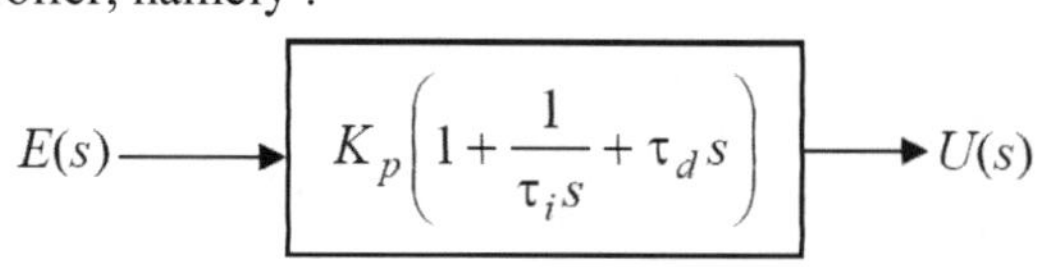

which :

K_p = Proportional reinforcement
τ_i = Integral time
τ_d = Differential time

IOP Conf. Series: Earth and Environmental Science **156** (2018) 012003 doi:10.1088/1755-1315/156/1/012003

The function of the PID component is as follows:

1. Proportional control functions to accelerate the rise time.
2. Integral control functions to eliminate steady state errors.
3. Derivative control functions to improve system stability and reduce overshoot.

B. Fuzzy like PD controller [7].

Fuzzy like PD controller is a fuzzy controller that has a behavior resembling a PD controller. Design consists of the process of quantization, fuzzifikasi, inference, and defuzzifikasi. In the Fuzzy control type used in this design, the formation of Ke and Kde is used to normalize the boundary E and ΔE into the limit of -3 to +3. I used to denormalize the control signal limits from 0 s / d 3 to 0 s / d U_{max} or -3 s / d 3 to U_{min} s / d U_{max}. The design of the fuzzy control system is shown in Figure 5.

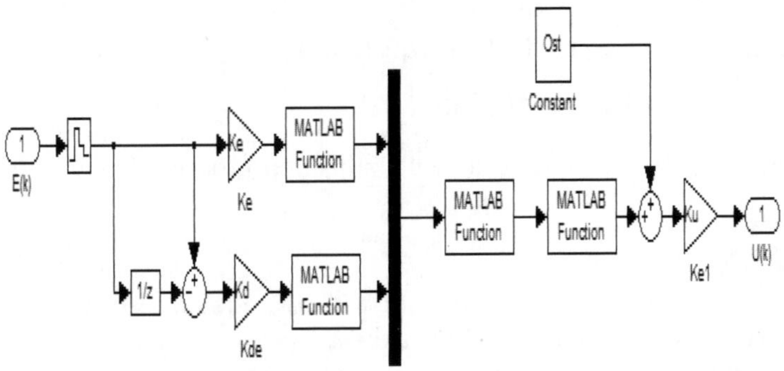

Fig 5. Block diagram of logic control on buck converter

Figure 6. A linear representation, membership level mapping that defines membership functions for E, ΔE and control signals in a normalized form on triangular shapes with 7 support sets.

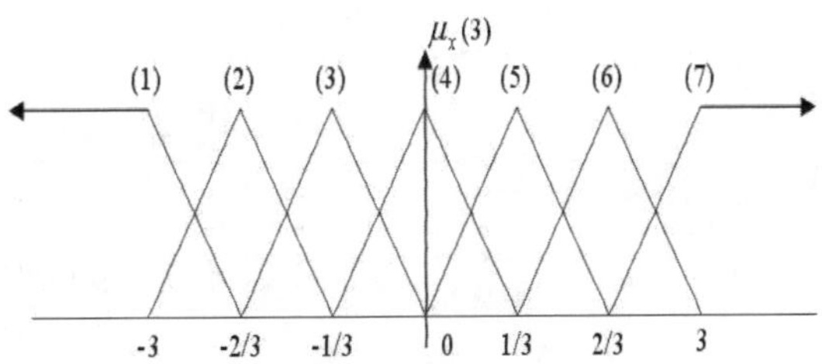

Fig6. Triangle function

Create a basic fuzzy rule that is used to determine the control signal with the help of inference. The preparation of fuzzy control rules with the membership set for each of the seven input entries is shown in Table 3.

Table 3. Preparation of Fuzzy Rules

E/dE	1	2	3	4	5	6	7
1	1	1	2	2	3	3	4
2	1	2	2	3	3	4	5
3	2	2	3	3	4	5	5
4	2	3	3	4	5	5	6
5	3	3	4	5	5	6	6
6	3	4	5	5	6	6	7
7	4	5	5	6	6	7	7

Inference here is Mamdani methods in which the equation is:

$$\mu_y(k) = \max\left[\min\left\{\mu_u(k), \mu_R(E(i), \Delta E(j))\right\}\right]$$

To get values of action control (u) , it needs defuzzification process that is calledCOA (*Center Of Area*).

3. Results and Discussion

The simulation test of DC-DC buck converter with DC motor load with load variation using Fuzzy-PID control is shown on the response graph below:

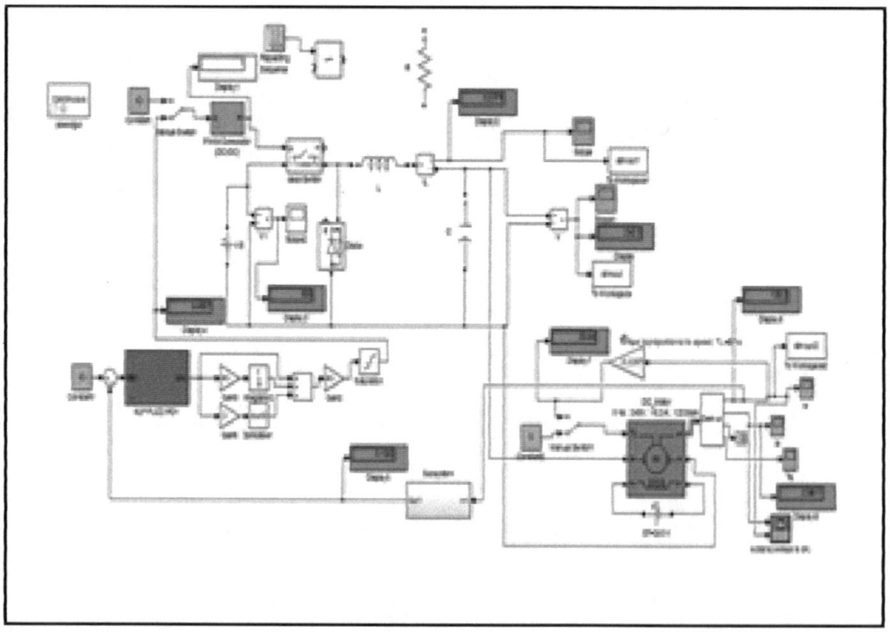

Figure 7. Image Simulation of whole Buck Converter with DC Motor Load Using Fuzzy Control - PID.

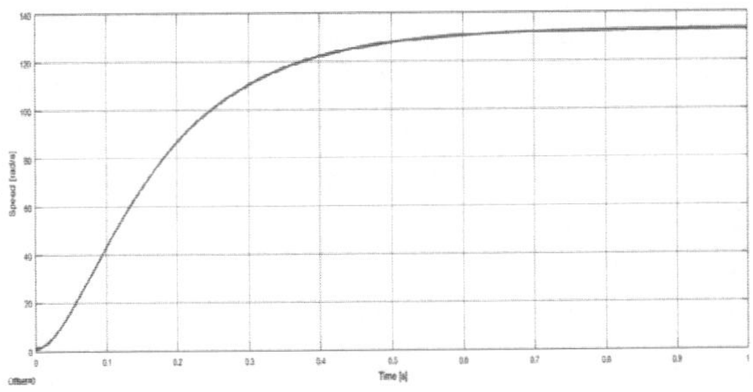

Figure 8.Response on zero load and nominal speed (ωnom) = 133.2 rad / s

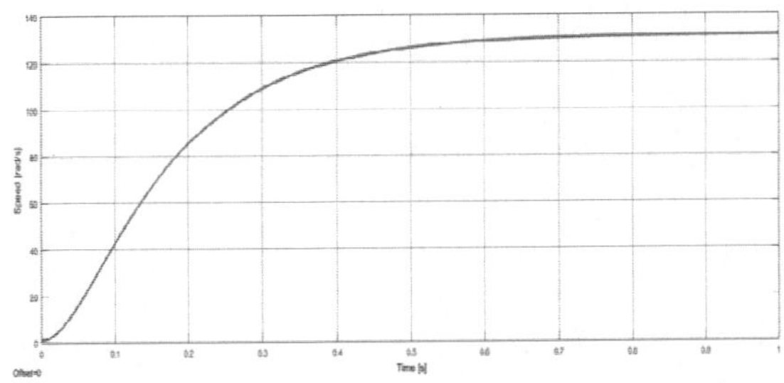

Figure 9.Response on Load 10 and nominal speed (ωnom) = 131.3 rad / s

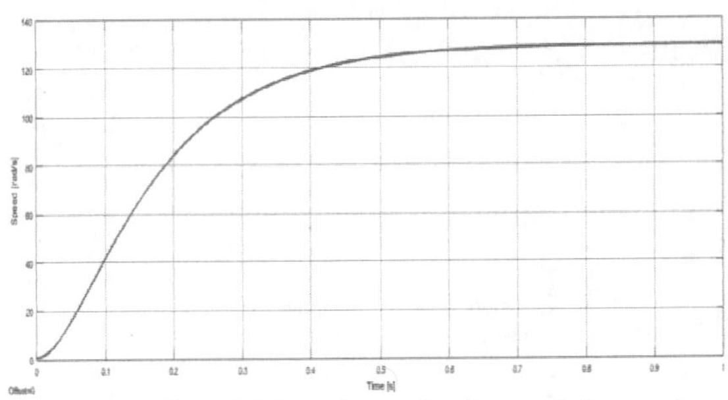

Figure10.Response on Load 20 and nominal speed (ωnom) = 129,4 rad / s

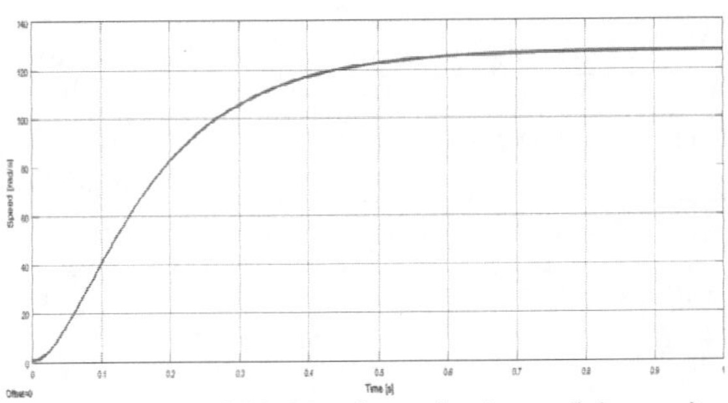

Figure 11.Response on Load 29,2 and nominal speed (ωnom) = 127,7 rad / s

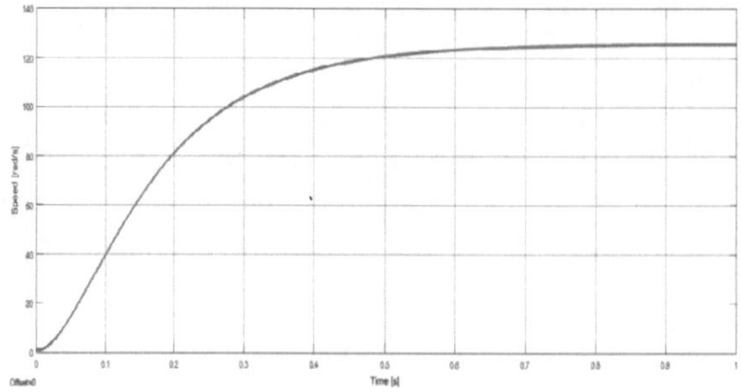

Figure 12.Response on Load 40 and nominal speed (ωnom) = 125,7 rad / s

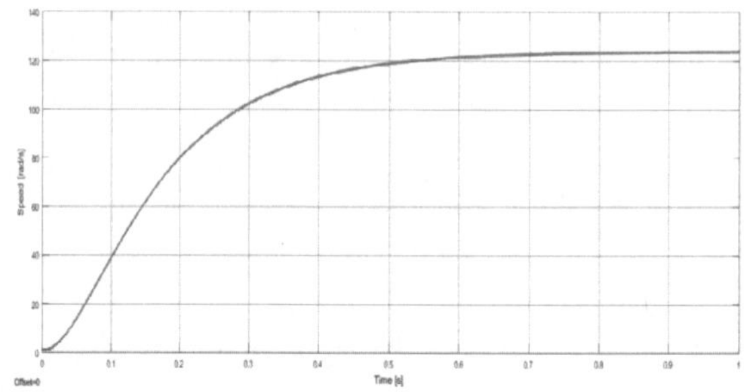

Figure 13.Response on Load 50 and nominal speed (ωnom) = 123,9 rad / s

Table 4. Testing on Open Loop

V.Input (v)	**V.Output (v)**	**Torse(Load)**	**Nominal Speed (rad/s)**
500	240,2	0	133,1
500	239,9	10	131,3
500	239,9	20	129,4
500	239,8	29,2	127,7
500	239,8	40	125,7
500	239,8	50	123,8

Table 5. Testing of suing Fuzzy-PID controller

V.Input (v)	**V.Output (v)**	**Torse(Load)**	**Nominal Speed (rad/s)**
500	240,3	0	133,2
500	240	10	131,3
500	240	20	129,4
500	240	29,2	127,7
500	240	40	125,7
500	240	50	123,9

IOP Conf. Series: Earth and Environmental Science **156** (2018) 012003 doi:10.1088/1755-1315/156/1/012003

4. Conclusion

The simulation result of controlling buck converter with DC motor load using Fuzzy-PID control with load variation shows better performance than when open loop. The output voltage of the buck converter remains constant at 240 volts although the load is variable, in which the nominal speed of the motor follows the given load.

5. Refferences

[1] Amit S. Chopade, Swapnil W.Khulbalkar, A.S. Junghare and M.V.Aware," Fractional Order Speed Controller For Buck – Converter Fed DC Motor,"International Conference on Control, Measurement and Instrumentation, IEEE. 2016
[2] Bhagyashri U, Patil, Satyawan R.jagtap. International Conference on Computation of Power Energy, Information and Communication. 2015
[3] Sutedjo, Zaenal Efendi, Dina Mursyida. Politeknik Elektronika Negeri Surabaya-ITS
[4] Jumiyatun, Andi Fatmawati, "Komparasi Metode Pi Dan Fuzzy-Pi Pada Pengendalian Tegangan Output DC-DC Buck Converter", Prosiding SNTE Vol 2. 2017
[5] Rashid, Muhammad. Power Electronic Handbook .Akademik Press.Canada.2001
[6] Gamayanti, Nurlita.Kontrol PID. Institut Teknologi Sepuluh Nopember. Surabaya.2007
[7] Ana Ratna Wati, Dwi. Sistem Kendali Cerdas. Cetakan Pertama.Graha Ilmu.2011
[8] Ogata, Katsuhiko. Teknik Kontrol Automatik. Jakarta, Erlangga. 1996
[9] Rashid, Muhammad H.Terjemahan Elektronika Daya.PT prehallindo.Yogyakarta.1993
[10] Mohan, Ned. Power Electronic Handbook : Converters, Aplication, and design, Second Edition. 1995

IOP Conf. Series: Earth and Environmental Science **156** (2018) 012004 doi:10.1088/1755-1315/156/1/012004

Attoriolong Tradition of Lake Preservation

Kamaruddin Mustamin[1], Rahmawati Rahmawati[2], Abbas Abbas[3] Akbar Akbar[4]

[1]Institut Agama Islam Negeri Gorontalo, Jl. Gelatik, Heledulaa, Kota Tim., Kota Gorontalo, Gorontalo, 96135, Indonesia.
Email: kmaustamin@iainkendari.ac.id
[2]Institut Agama Islam Negeri Kendari, Jl. Sultan Qaimuddin No 17 Baruga, Kendari, Sulawesi Tenggara, 93563, Indonesia
Email: rrahmawati@iainkendari.ac.id
[1]Institut Agama Islam Negeri Gorontalo, Jl. Gelatik, Heledulaa, Kota Tim., Kota Gorontalo, Gorontalo, 96135, Indonesia
Email: aabbas@iainkendari.ac.id
[1]Institut Agama Islam Negeri Gorontalo, Jl. Gelatik, Heledulaa, Kota Tim., Kota Gorontalo, Gorontalo, 96135, Indonesia.
Email: akbar@iainkendari.ac.id

Abstract: Lake as a source of life for the fishermen has a very significant potential in influencing the views and behaviour of the fishing community's culture. Events such as tidal lake erratic, flooding, high hardness water waves, even the lack of fish catches and the death of sailors in the lake, are symptoms that serve as the base outlook for the fishermen errors that they do against the unseen force which they consider to be the ruler of the lake. The fishing community of Lake Tempe has run *attoriolong* and established rules related to the management of the lake since the pre-Islamic period (before 1610 AD), led by a *Macua Tappareng*. *Macua Tappareng* is considered understand the ancestor details and rules in controlling the use of the lake, including the mystical of the lake. This study explores the local wisdom of coastal fishing communities of Lake Tempe in interacting with the lake led by *Macua Tappareng*. *Macua Tappareng* hereditary ensures that *attoriolong* tradition with all the rules and policies related to the lake is applied correctly. This research is a descriptive-analytic study that intends to describe as well as analyze the regulations and procedures on how *Macua Tappareng* in preserving the lake. The research shows that *Macua Tappareng* has the legitimacy and strategic position in protecting the lake, as evidenced by the continued preservation of some rules of fishing, such as a ban on fishing in particular place and time. Besides, the tradition of *Maccera' Tappareng* is still considered liabilities fishing communities in honour of the ruler of the lake even become additional tools of the governments in disseminating and implementing local regulations.

1. Introduction

The system of *attoriolong* in the Bugis society contains the various rules concerning human relations with God, human beings, human beings with ancestral spirits, and human beings with all nature and its surroundings [1]. The religious system of Bugis ethnic group is almost inseparable from their daily cultural and social economic values. This means that the personality, behaviour, attitudes, deeds, arts, health, and everyday social activities are guided, supported and linked not only with the belief system or the teachings of religion and customs or customary law but also with values of culture and ethnicity

[2]. Thus, their response to external stimuli and pressure are often based on the complexity of these elements.

Buginese who follows *attoriolong* believes in the existence of a fixed rule, which overcomes everything that happens like the world done by humans. Supracosmic rules are stable, harmonious and eternal. The rule is the source of all human glory and happiness. Based on this rule, the human life becomes intact. It contains a fixed and definite archetype that gives meaning to everything that is not fixed and uncertain. The meaning of each human act is self-relative and temporary, but it needs to be adapted to the absolute order of nature. What is appropriate or consistent in human life against that entire background is all right. What is deviant, incompatible or against it is dysfunctional, false, misguided, and sinful. The participation of human behaviour in the rules raises human life to be authentic, meaningful and valuable. *Attoriolong* in the context of the coastal community of Lake Tempe presents a central figure as a stalwart and implementer of some *attoriolong* rules called *Macua Tappareng* [3]. *Macua Tappareng* is the leader of customary law chosen by the indigenous community in a specified period. *Macua Tappareng* is voted based on its ability to preserve the traditions and culture of the coastal community of Lake Tempe.

2. Method

This research is a qualitative research with social anthropology approach [4] which is observing the life of the fisherman in coastal of Tempe lake of Sengkang City of Wajo Regency, primarily related to the procedures and rules they hold in carrying out fishing activities in the lake. This research was conducted by in-depth interviews with *Macua Tappareng* as the central figure who leads the fishermen. In this case, there are three *Macua Tappareng*, one of them is the current one, while the others two are the former ones. In addition, other vital informants randomly selected are fishermen, community leaders, religious leaders, youth leaders, including local government from RT / RW, sub-district heads, district heads, and other stakeholders such as Tourism Office staff of Wajo.

3. Findings and Discussion

Geographically, Lake Tempe is located between 11.90.53'-12.00.04 ' in East Longitude and 40.03'-40.09' South Latitude. Tempe Lake is associated with two other lakes: Sidenreng Lake in Sidrap Regency and Buaya Lake in Tanasitolo sub-district, Wajo. These three lakes unite to form a vast water area in the rainy season and can cover the settlements of communities in three districts. Presented in Fishery Report FAO of UN (1995) that Tempe lake is a system of three natural lakes, Tempe Lake, Sidenreng lake, and Buaya lake [5].

Tempe Lake is better known located in Wajo because the most important area is this region, mainly Tempe district where the capital of Wajo is located, and the other three communities are Belawa, Tanasitolo, and Sabbangparu. Other areas of Lake Tempe are situated in Soppeng and Sidrap districts. It can be seen from Bappeda data that Tempe Lake occupies three regions with seven districts. The most significant part of the lake is located in Wajo Regency which consists of four areas of Tempe, Sabbangparu, Tanasitolo, and Belawa. Soppeng district consists of Marioriawa and Donri-donri sub-districts, and the narrowest part is Sidrap district with Panca Lautang sub-district. Lake Tempe with an area of about 13,000 ha is a potential water location in South Sulawesi. In addition to being one of the largest freshwater fish producers, the lake is also used as a place to live on the water with floating home systems, as a place of growing floating vegetation, a place to live various species of rare birds and as the second largest tourist destination in South Sulawesi other than Toraja. The potential of the lake with multiple types of fish that live in it and the function of the lake that can be utilized to support the survival of humans and several other ecosystems, and make Lake Tempe as a significant destination for the surrounding community to make a living.

Coastal communities of Lake Tempe are predominantly Muslim. The data shows that the percentage of converts to Islam reaches 99.5% of the population that is married on the coast of Lake Tempe. The influence of Islam in their lives is powerful in the tradition that they carry out so far,

IOP Conf. Series: Earth and Environmental Science **156** (2018) 012004 doi:10.1088/1755-1315/156/1/012004

primarily related to the life cycle. The entry of Islam can influence the existing culture, but the influence does not necessarily eliminate the identity of tradition which leads to the presence of a dialectic of the two characters that then creates new values and standards. The coastal fishermen of Lake Tempe annually carry out the ritual *Maccera' Tappareng*, the ceremony as the start of the fishing season which is interpreted as a ceremony for the alms of the earth or reject the reinforcements. This ritual aims to prevent fishers from disasters during fishing activities in the lake, and they can catch more fish, so that fishermen can be more prosperous. This *Maccera' Tappareng* ritual can only be carried out under the leadership of *Macua Tappareng. Maccera' Tappareng* is part of the mysticism surrounding the lake that has been done for generations, so their needs for knowledge and exceptional skills belonging to *Macua Tappareng* is inevitable. This is where the vital role of a *Macua Tappareng* in fishermen community as a central figure of the implementation of the sacred ritual of the fishermen life takes place.

While *Macua Tappareng* has some rules that he must address to all fishermen in the form of specific restrictions to be obeyed by the fishermen. This prohibition, for example, is in the form of a fishing ban on the lake on Thursday night to Friday afternoon after the *zuhur* praying. This is intended to allow the fish to breed and give the fishermen a chance to rest once a week. It is also at once as part of the integration of Islamic values that is for Muslim fishermen to perform the Friday prayers. Other restrictions are they are not allowed to cross the corpses in the lake, should not wash the mosquito nets in the lake, should not catch fish without the use of head coverings, should not sing in the lake except the song associated with the song calling the fish, should not date someone in the lake, especially between two young people. All these restrictions are intended to honour the fish so that the fish will not disappear and respect the lord of the lake. The lord of the lake according to the belief of society is a creature appointed/authorized by Allah SWT. As his representative in guarding Tempe lake, so it needs to be respected by preventing the lake lord to be furious. If the ban is violated, then the ruler of the lake is wrathful, and it is believed that the fish in the lake will disappear. So, the local people consider the fish is *walli* (guardian of God in the lake).

If the ban is violated, *Macua Tappareng* will punish the fisherman/ community with the term *dosa* (sanctioned) by prohibiting them to fishing for three days and are required to perform ritual *maccera' tappareng* at his own expense without the help of other community members. This ritual is intended as an apology to the lord of the lake for the mistakes that have made. All members of the community in the lake environment served as a supervisor in charge of supervising people who violate customary rules and then reported to *Macua Tappareng*. The prohibition and custom rules in Tempe Lake are traditional local wisdom which aims to maintain the survival of all existing ecosystems in the lake, thus creating a balance of treatment between humans and the lake environment. In such a vital position, *Macua Tappareng* is also a government agent in applying government regulations in managing lakes, such as the prohibition of using bombs, stuns, and other tools that cause damage and threaten the sustainability of the ecosystem.

Local wisdom has various forms, sometimes in the form of values and ethics, sometimes also in the form of works [6]. Local wisdom in the form of works, for example, is shown in the study of Fadly Usman et al that these artificial hollows are arrayed on the beach as one of the local wisdom in Lampon village to reduce the intensity of inundated tsunami flow. This study evaluates the efficiency of a hollow topography which can be seen on the beach along Lampon village in Indonesia. These artificial hollows are arrayed on the beach as one of the local wisdoms in Lampon village to reduce the intensity of inundated tsunami flow [7].

A study in the United States by Daniel W. Schneider entitled Local Knowledge, Environmental Politics, and the Founding of Ecology in the United States: Stephen Forbes and "The Lake as a Microcosm" (1887) indicate how the birth of ecology is tied to local knowledge and the domestic politics of environmental transformation. The study is conducted by tracing the connections between scientists and local fisherman underlying the research on floodplain lakes presented in 'Lake as a Microcosm'[8]. Another case proves that local wisdom plays a vital role in ecological, economic, and social systems of society. The study of S.Bendem-Ahlee, et.al is to explore traditional knowledge in the ecological use of resources in the case of honey hunting in a Melaleuca forest. It shows the s local wisdom that reflects the complex interrelationship between knowledge, belief, and practice covering the knowledge level and management system and taking the next generation into account [9]. In Thailand, local wisdom becomes the basis for the development of community culture and production processes. Utit Sungkharat et al. reveal that the concept of community culture is developed through the local community leaders through the production process, which is based on the idea of self-sufficiency and self-reliance. They point out that the development has led to the decrease in residents' use of technology and return to the use of labour. Finally, the local wisdom helps people save on production costs and reduce pollution [10]. In this way, local wisdom positively influences the environment and people's lives. Some studies have shown that local wisdom is a significant basis for protecting and maintaining the environment from damage.

4. Conclusion

Macua Tappareng as a stakeholder in the culture of fishing communities of Tempe Lake has proven the responsibility in maintaining *attoriolong* (local wisdom). *Macua Tappareng* thus becomes a personal symbol of lake guarding as a source of life for fishermen. Various rules of management/utilization of the lake continue to be respected and adhered to the fishermen because a *Macua Tappareng* always exists to escort. In fact, this position is often used by the local government to socialize and implement local regulations.

5. References

[1] Abidin AZ 1971 Notes on the Lontara'as historical sources. *Indonesia.* **12** 159-172.

[2] Hamonic G 1991 God, Divinities and Ancestors: For the Positive Representation of a" Religious Plurality" in Bugis Society, South Sulawesi, Indonesia.

[3] Mustamin K 2016 *Maccéraq Tappareng (Kajian Mistisisme terhadap Tradisi Danau dalam Attoriolong Masyarakat Bugis Kabupaten Wajo)* (Doctoral dissertation, Universitas Islam Negeri Alauddin Makassar).

[4] Bernard HR 2017 *Research methods in anthropology: Qualitative and quantitative approaches.* Rowman & Littlefield.

[5] Naing N, Santosa HR, Soemarno I 2009 Kearifan Lokal Tradisional Masyarakat Nelayan pada Permukiman Mengapung di Danau Tempe Sulawesi Selatan. *Local Wisdom.* **1**(1) 19-26.

[6] Meliono I 2016 Understanding the Nusantara Thought and local wisdom as an aspect of the Indonesian Education. *Tawarikh.* **2**(2).

[7] Usman F, Murakami K, Kurniawan EB 2014 Study on Reducing Tsunami Inundation Energy by the Modification of Topography based on Local Wisdom. *Procedia Environ. Sci.* **20** 642–650.

[8] Schneider DW 2000 Local Knowledge, Environmental Politics, and the Founding of Ecology in the United States. Stephen Forbes and "The Lake as a Microcosm" (1887). *Isis J. Hist. Sci. Soc.* **91**(4) 681–705.

[9] Bendem-Ahlee S, Kittitornkool J, Thungwa S, Parinyasutinun U 2014 Bang Kad: A Reflection of Local Wisdom to find Wild Honey and Ecological Use of Resources in Melaleuca Forest in the Songkhla Lake Basin. *Silpakorn Univ. J. Soc. Sci. Humanit. Arts.* **14**(3) 77–100.

[10] Sungkharat U, Doungchan P, Tongchiou C, Tinpang-nga B 2010 Local wisdom: The development of community culture and production processes in Thailand. *Int. Bus. Econ. Res. J.* **9**(11) 115.

Using an electrochemical assay to determine the biofilm elasticity change as a response to toxicant exposure

Retno Novvitasari Hery Daryono[1], **Agnes Feurtet-Mazel**[2], **Stephanie Bouletreau,**[3] **Frederic Garabetian**[4]

[1] State Islamic University of Maulana Malik Ibrahim Malang, JawaTimur 65144, Indonesia. Email: retno@uin-malang.ac.id
[2] CNRS, EPOC, UMR 5805, Station Marine d'Arcachon, France, 33120 Arcachon, France. Email: agnes.feurtet-mazel@u-bordeaux.fr
[3] Univ. Paul Sabatier (Toulouse 3), EcoLab UMR 5145, Toulouse, France. Email: stephanie.bouletreau@univ-tlse3.fr
[4] University of Bordeaux, EPOC, UMR 5805, Station Marine d'Arcachon, 33120 Arcachon, France . Email: frederic.garabetian@u-bordeaux.fr

Abstract: Elasticity is a trait of biofilm physiognomy which relates to cell clustering and can be measured by means of an electrochemical assay based on rotating disc electrode (RDE). This study aimed at testing the hypothesis according to which exposure of phototrophic biofilm to toxicant could reduce its elasticity. We compared biofilms developed for 21 days, in four sets of 6 replicated experimental units, in absence and presence of isoproturon at two concentrations of the inoculating suspension of biofilm, 103 and 104 diatom cell mL-1. Biofilm thickness and elasticity were measured based on RDE assay, bacterial and diatom density were measured by microscope-based numerations.Very thin biofilms (< 10 µm) were obtained as compared with a previous study. This might be linked with the way we selected the initial biofilm providing the suspension and the way we developed its growth. The biofilm elasticity mean values in the presence of isoproturon was quasi twice lower (60 ± 10 and 60 ± 41 µm rpm0.5) than the treatment without isoproturon (138 ± 93 and 115 ± 104 µm rpm0.5), for initial biofilm concentration of 103 and 104 respectively, but there was no significant difference between the mean values of each treatment. Nevertheless, the present preliminary study demonstrated the feasibility of an experiment dedicated to assessing biofilm elasticity changes as a response to toxicant exposure.

Keywords: electrochemical assay, biofilm elasticity, toxic

1. Introduction

Physiognomy is a term broadly used in plant ecology to describe the combination of the external appearance of vegetation, its vertical structure, and the growth forms of the dominant taxa e.g. coniferous forest or grassland [1]. Biofilm assembled microbial communities also have three dimensional architectures [2]. Biofilm thickness, porosity, viscoelasticity, size of cell clusters are some traits of biofilm physiognomy. Allowing the assessment of biofilm thickness and elasticity, the recent application of electrochemical method using Rotating Disc Electrode (RDE) provided the investigations in biofilm physiognomy with new insights [3].

Previous studies suggested the effect of toxicants on phototrophic biofilm physiognomy. One study found that the adpressed and stuck diatoms *Achnanthinidumminutissimum*, *Encyonemaminutum* and *Navicula minima* characterized the phototrophic biofilm in cadmium and zinc contaminated conditions[4]. Conversely, in uncontaminated conditions the phototrophic biofilm was characterized by the filamentous diatom *Melosiravarians*. Changes in the physiognomy of a *M. varians* dominated phototrophic biofilm were also observed after exposure to copper [5]. The authors reported no taxa replacement but shortening of the *M. varians* filaments to short tufts. Consistently, exposure to cerium oxide nanoparticles caused the flocculation in exopolymeric substances of a non-axenic strain of *Chlamydomonasrheinardti*[6]. Together the results of these studies suggest that cell clustering as a mean to attenuating individual cells exposure to a toxicant, could be an adaptive response of (some) microbial populations or communities to chemical stress.

In biofilms, cell clustering would be the reduction of the extracellular volume of the biofilm that could be measured by a loss of biofilm elasticity. We thus hypothesized that exposure to a toxicant would reduce phototrophic biofilm elasticity as a trait of biofilm physiognomy. To address this question we aimed to compare the biofilm thickness and elasticity of phototrophic biofilms exposed, or not, to a well-defined biocide. Thickness and elasticity were measured by means of the RDE electrochemical method on lentic phototrophic biofilms developed under controlled temperature, light and nutrient conditions in absence and presence of the phenylurea herbicide isoproturon (20 μg L^{-1}) at two concentrations of the inoculating biofilm suspension (10^3 and 10^4 diatom cell mL^{-1}).

2. Methodology

2.1. Experimental Design

Four sets of 6 replicated experimental units were prepared (Table 1). Isoproturon half live concentration is reached after 71 days of exposure[7]. Hence, during the present 21-day colonization experiment, isoproturon concentration was assumed to remain constant in the experimental units. Two different diatom concentrations of the inoculating suspension of biofilm, 10^3 and 10^4 cell mL^{-1}, were tested to ensure optimal biofilm colonization. The inoculating suspension of biofilm was prepared by scrapping and homogenizing two biofilm colonized glass slides (kindly provided by B. Chaumet and S. Morin, IRSTEA) in 150 mL of Dauta medium [8]modified by the addition of siliceous (10 mg L^{-1}), at pH ~ 7.7. In March 2016, clean glass slides had been immersed for 15 days in a pound (Gazinet-Cestas, France) then placed for 3 weeks, at 20°C, in a continuously illuminated aquaria filled with Dauta medium before use in this experiment.

Table **1.** - Experimental setup (RDE 1-24: identification number of the electrodes)

Isoproturon	Diatom cell concentration of the inoculating suspension of biofilm	
	10^3 cell mL^{-1}	10^4 cell mL^{-1}
0 μg L^{-1}	6 replicates RDE 1-6	6 replicates RDE 13-18
20 μg L^{-1}	6 replicates RDE 7-12	6 replicates RDE 19-24

The assay was conducted by developing phototrophic biofilm onto RDE as artificial substrate in a 100 mL propylene vials experimental unit. Identified RDE were immersed in 40 mL Dauta medium inoculated with 0.2 or 2 mL the biofilm suspension (10^5 cell mL^{-1}), with or without addition of 100 μL of a 50 μg L^{-1}isoproturon solution (stock solution: 50 mg L^{-1}). The 24 experimental units (4 treatments x 6 replicates) were incubated for 21 days, at 20°C, by moderate shacking (80 rpm), exposed to a 45 μmol s^{-1} m^{-2} radiation (neon light) with a 12:12 photoperiod. The RDE was vertically suspended to the vial cap to prevent biofilm accrual by particle sedimentation onto the working surface of RDE. In such an assembly, however, the RDE working surface was not straight exposed to direct neon light radiation.

2.2. Biofilm thickness and elasticity measurements

The method is based on a basic experiment in electrochemistry, the RDE assay. A 3-electrode-system is immersed in an electrochemical cell filled with an electrolytic solution and connected to a potentiostat. The 3 electrodes are: (i) the working metallic electrode, RDE, on which biofilm develops;(ii) the counter electrode closing the electrical circuit and (iii) the reference electrode monitoring the potential of the electrolyte. Intensity, I, of the steady-state diffusion current on the RDE working surface is controlled by the RDE rotation speed, □, and by the thickness of the biofilm colonising the RDE working surface (Figure 1).

For a given value of □□(rpm), the difference between the steady state current intensity measured on the naked RDE $I_{(t0)}$ and on the colonised RDE $I_{(t21)}$ permit to assess the biofilm thickness □ (μm) using Equation 1.

Equation 1: $\delta = n\,F\,D\,C^{*}\,S\,[i(t_{21j})^{-1} - i(t_0)^{-1}] \times 10{,}000$

Where n is the number of electrons, F the Faraday constant (96485 C mol^{-1}), D the diffusion coefficient in both water and biofilm set to 6.8 x 10-6 cm^2 s^{-1} at 20°C, C* the electroactive species concentration in the bulk solution (0.00001 mol cm^{-3}), and S the active RDE area (0.196 cm^2). The value measured at 300 rpm was retained as the estimate of biofilm thickness ($□_{300rpm}$).

The biofilm elasticity is determined by relating the biofilm thickness and rotation speed, considering the following law:

Equation 2: $\delta = 1 / (\delta_0^{-1} + K\Omega^{0,5})$

Where δ_0 (μm) is biofilm thickness at zero rpm, and K (μm^{-1} $rpm^{-0,5}$) the coefficient used to parametrise biofilm elasticity as 1/K (μm $rpm^{0,5}$).

The 24 identified RDE were 5-mm diameter platinum cylinder (electrical conductor) coated with a teflon® cylinder (electrical insulator)[3]. The reference electrode was a saturated calomel electrode (SCE) (REF 421, Radiometer Analytical, France). The counter electrode was a cylindrical grid of platinum immersed into the electrolyte solution that surrounded the working electrode. A 0.01M potassium ferrocyanide $[Fe(CN)_6]^{2-}$ and ferricyanide $[Fe(CN)_6]^{3-}$ solution was used as a electrolyte in 1M KCl. Ferrocyanide oxidation current intensity was measured at 0 V/SCE at which no water electrolysis and no oxygen reduction occur. Measurements were performed at 20°C.

In the laboratory, RDE was mounted on a motor axis plugged using mercury contacts and was rotated by a direct current motor system. The motor speed was controlled and measured using a tachymeter. Diffusion current was then measured at the potential 0V for each RDE rotation speed between 100 and 1200 by steps of 100 ppm. Rotation speed was limited to 1200 rpm to prevent biofilm erosion. Before t_0 measurements, every RDE were polished using sandpaper and cleaned with Milli-Q water. At the end of the incubation, RDE were sampled and their intensity were measured. Then, RDE working surface was immersed in 0.5 mL of Dauta media and 70 μL of neutralized formaldehyde and preserved for storage at 4°C until further analysis.

2.3. Numeration of diatom and bacteria colonizing the RDE working surface

The biofilm colonizing the working surface of the RDE was recovered by scrapping with a sterile plastic tool and sonication in an US bath at 37 kHz (30 s). After sonication the RDE was rinsed with 0.5 mL of Milli-Q water and removed. The resulting 1070 μL-biofilm suspension was vortexed (5 minutes) before storage at 4°C and prior to aliquoting for bacterial and diatom numerations within a few days.

For determination of bacterial density, 100 μL aliquot of the appropriate cell suspension dilution was stained with 200 μL 4'6-diamidino-2-phenylindole (DAPI, 0.01 mg mL^{-1}) and collected by filtration on 0.2 μm pore-size black polycarbonate filter. Counts were carried out on a OLYMPUS BH2 microscope fitted for epifluorescence at 1250x magnification and results were expressed as cell per cm^2 using equation 3.

Equation 3: Bacterial density (cell cm^{-2}) = NC x (SF/SC) x (VS/VF) x (1/SRDE)

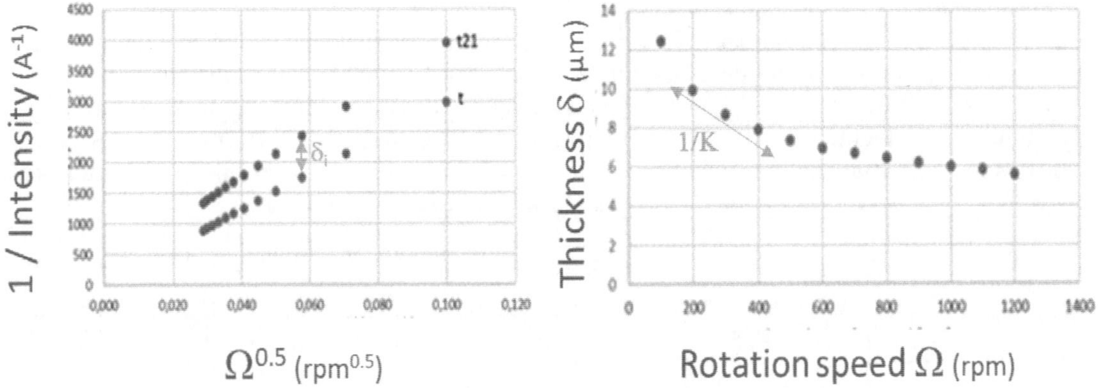

Figure 1. Inverse current intensity evolution with the electrode rotation speed measured on RDE #1 at t0 (blue symbols) and 21 days (red symbols). Thickness evolution with the electrode rotation speed measured on RDE #1 after 21 days of colonization.

Where NC (cell) is the mean number of bacteria on the calibrated counting grid, SF (mm^2) the filtration area on the filter, SC (mm^2) the area of the counting grid at the 1250x magnification, VS (mL) the total volume of biofilm suspension, VF (mL) the volume of filtered biofilm suspension and SRDE (cm^2) the RDE colonising surface.

Diatom density were determined directly or after 5-fold dilution. Counts were carried out on an OLYMPUS BH2 microscope at 200x magnification using a Nageotte counting chamber and results were expressed as cell per cm^2 using equation 4.

Equation 4: Diatom density (cell cm^{-2}) = (NC x VS) / (VN x SRDE)

Where NC (cell) is the mean number of diatom on the counting grid (5 to 10 fields counted, 1.25 µL each), VS (mL) the total volume of biofilm suspension, VN (mL) the Nageotte counting chamber volume, and SRDE (cm^2) the RDE colonising surface.

2. 4. Statistics

Biofilm elasticity was deduced by fitting experimental values to Equation 2 by means of non-linear least squares fits using Jkp Ads Excel add-in. Only significant fits (p-value $\leq$ 5%) were considered. At least, 3 out of 6 replicates were significant for each treatment. We thus assumed to consider only 3 replicates per treatment for further analyses.The non-parametric Mann-Whitney U-test (Merlin Excel Add-in) was used to check for differences in biofilm thickness and elasticity, diatom and bacterial cell numbers between the different treatments.

3. Results

Mean (± SD) biofilm thicknesses measured at 300 rpm were slightly lower in presence of isoproturon, 2 ± 1 and 3 ± 1 µm than in absence of isoproturon, 5 ± 3 and 4 ± 2 µm, for inoculating concentration of 10^3 and 10^4 diatom cell mL^{-1}, respectively (Table II and Figure 2.A). Differences between isoproturon treatments, however, were not significant (p-value > 0.05). The inoculating concentration of diatom has had no significant effect on biofilm thickness, either.

There were no significant differences between mean (± SD) values of diatom numbers in the biofilms developed in absence of isoproturon with an inoculum of 10^3 diatom cell mL^{-1} ($1.1 \pm 0.5\ 10^5$ cell cm^{-2}) as compared to other treatments including biofilms developed in absence of isoproturon with an inoculum of 10^4 diatom cell mL^{-1} ($2{,}2 \pm 0.8\ 10^4$ cell cm^{-2}) and those developed in presence of isoproturon (1.3 ± 1.0 10^4 and $1.0 \pm 0{,}3$ 10^4 cell cm^{-2}, with inoculating concentration of biofilm suspension of 10^3 and 10^4 diatom cell mL^{-1}, respectively) (Table II and Figure 2.B).

Table 2. Raw data of replicate sets of measurements of biofilm thickness, elasticity, diatom and µbacteria densities of biofilms developed in May 2016 for 21 days in Dauta medium at 20°C in relation with different concentrations of inoculating biofilm suspension and/or isoproturon.

Inoculum concentration *cell ml^{-1}*	**Biocide treatment** *µg l^{-1}*	**RDE** *n°*	**Thickness** *µm*	**Elasticity** *1/K as µm rpm*	**Diatom numbers** *cell cm^{-2}*	**Bacteria numbers** *cell cm^{2}*
10^3 cell ml^{-1}	0	1	8,8	243	1,6 x 10^5	3,8 x 10^6
		2	4,8	93	1 x 10^5	4,2 x 10^6
		3	2,4	79	6,6 x 10^5	3,2 x 10^6
	20	1	2,7	55	2,4x 10^5	2,9 x 10^6
		2	1,7	55	9,6x 10^5	1,5 x 10^6
		3	3,3	72	4,8 x 10^5	1,8 x 10^6
10^4 cell ml^{-1}	0	1	4	68	3 x 10^5	3,7 x 10^6
		2	2,2	42	1,6 x 10^5	4,2 x 10^6
		3	6,4	235	1,9 x 10^5	6,3 x 10^6
	20	1	1,1	27	9,9 x 10^5	4,6 x 10^6
		2	3,8	105	1,2 x 10^5	5,6 x 10^6
		3	2,5	46	6,7 x 10^5	2,5 x 10^6

Among treatments, mean (± SD) bacterial densities in the biofilm were in the same order of magnitude ranging from 1.6 to 6.3 10^6 cell cm^{-2} (Table II). No significant differences in bacterial density means were found between biofilms developed with or without isoproturon and/or with inoculating concentration of biofilm suspension of 10^3 or 10^4 diatom cell mL^{-1} (Figure 2.C).

The mean (+/-SD) values of biofilm elasticity in the treatment without isoproturon were quasi twice higher (138 ± 93 and 115 ± 104 µm rpm$^{0.5}$, for inoculating concentration of biofilm suspension of 10^3 or 10^4 diatom cell mL^{-1} respectively) than values found for biofilms developed in presence of isoproturon (60 ± 10 and 60 ± 41 µm rpm$^{0.5}$, for inoculating concentration of biofilm suspension of 10^3 or 10^4 diatom cell mL^{-1} respectively) (Figure 2. D). Differences, however, were not significant (p-value > 5%).

4. Discussion

In the present study, diatom and bacteria densities onto the RDE working surface confirmed the occurrence of a biofilm colonising the RDE working surface after 21 days of incubation in controlled conditions. Measured thicknesses therefore may be related to biofilm occurrence on the RDE working surface. However, we found very thin biofilm (< 10 µm) as compared to values previously reported in the River Garonne for the same colonization duration (71 – 343 µm) [3]. In agreement, lower bacterial densities were found in the present study as compared to the River Garonne experiment. Diatom density, on the other hand, were not that low in the present experiment. This could be explained by the use of Dautamedium which is a culture medium dedicated to diatom culture. We thus hypothesize that conditions in the river such as current velocity or solar radiation were more likely to promote the development of complex and thick biofilm, except for the diatom community.

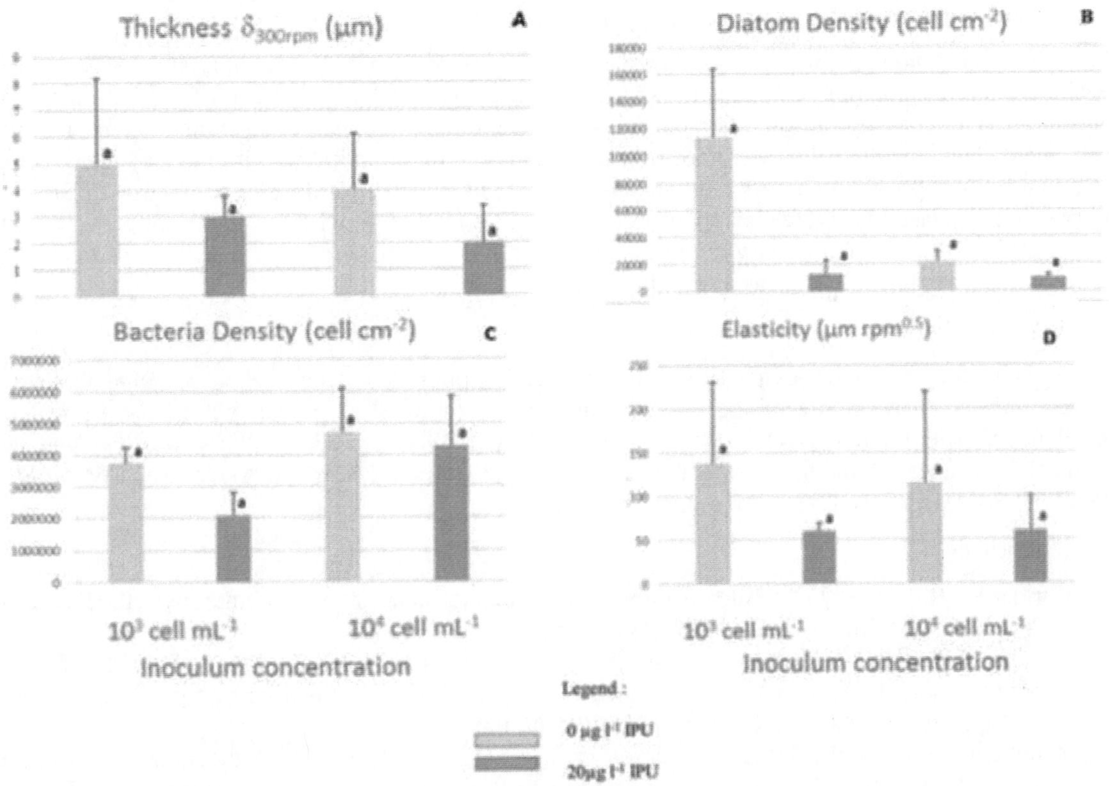

Figure 2. Biofilm thickness (a), elasticity (d), diatom (b) and bacteria (c) densities of biofilms developed for 21 days in Dauta medium at 20°C, in May 2016, with different concentrations of inoculating biofilm suspension and/or isoproturon. Bars represent the mean of 3 replicates and error bars represent the standard deviation. Distinct letters on bars indicate significantly different values at the 5% level.

Whether there were not significantly different values between the treatments, we obtained the lowest thickness values (ca. 2µm) when biofilms were developed in presence of isoproturon. An initial concentration of 5 µg L^{-1} of isoproturon was shown to reduce the rate of diatom colonization after 34 days of exposure [7]. The authors reported that small diatoms with high growth rate, *Naviculaparvulum* and *Sellaphoraseminulum,* dominated the isoproturon contaminated biofilms and could be indicators of isoproturon contamination. Diatom species replacement could explain biofilm development in spite of isoproturon addition. However, in the present experiment we did not perform diatom species identification to confirm it.

Biofilm elasticity ranged between 60 and 138 µm $rpm^{0.5}$. These values were not very important compared with the elasticity of River Garonne 21-day old biofilms (790 - 1300 µm $rpm^{0.5}$) [3]. This was nevertheless in agreement with the thickness data as elasticity is a function of the biofilm thickness. Again, there was a trend that isoproturon has a tendency to reduce by ca. two-fold the elasticity of biofilm. However, this result should not be considered as conclusive because (i) differences were not significant and (ii) the elasticity is a function of the biofilm thickness. In order to compare the elasticity, biofilm of similar thicknesses should be compared. To fulfill this condition, a different experiment designed with longer colonization duration in the higher contaminated treatment is needed to better characterize this contaminant effects on this physiognomic variables which are biofilms thickness and elasticity.

5. Conclusion

Biofilm colonisation was successfully measured in this study as diatoms and bacteria well developed on the rotating disc electrode, nevertheless there was no effect of initial inoculum concentration and/or isoproturon treatment on bacterial and diatom densities.As compared to previous experiment, developed biofilms were not very thick and this might be linked with the way we selected the initial biofilm providing the inoculating suspension and the way we developed its growth.Biofilm thickness and elasticity were lower in presence of isoproturon whereas differences were not significant.The present preliminary study demonstrated the feasibility of an experiment dedicated to assess biofilm elasticity changes as a response to toxicant exposure.

6. Acknowledgments

This study has been carried out with financial support from the French National Research Agency (ANR) in the frame of the Investments for the future Programme, within the Cluster of Excellence COTE (ANR-10-LABX-45) as part of PhytoCOTE project. Authors acknowledge the technical assistance members of the technical staff of Station Marine d'Arcachon (UMR 5805 EPOC):L. Bourasseau, B. Etcheverria and C. Portier.

7. References

[1] Whittaker, R. H. (1973). Editor. Ordination and Classification of Communities.Handbook of Vegetation Science 5. Junk, (The Hague, The Netherlands)

[2] Stoodley, P., Hall-Stoodley, L., Boyle, J. D., Jorgensen, F., Lappin-Scott, M. (20000. Environmental and genetic factors influencing biofilm structure. Community Structure and Co-operation in Biofilm. Cambridge University Press.

[3] Boulêtreau, S., Charchosset, J. Y., Gamby, J., Lyautey, E., Mastrorillo, S., Azémar, F., Moulin, F., Tribollet, B., Garabetian, F. (2011). Rotating disk electrodes to assess river biofilm thickness and elasticity. Water Research45, 1347 – 1357

[4] Feurtet-Mazel, A., Gold, C., Coste, M., Boudou, A. (2003). Study of periphytic diatoms communities exposed to metallic contamination through complementary field and laboratory experiments. Journal de Physique IV107, 467-470

[5] Barranguet, C., Plans, M., van der Grinten, E., Sinke, J.J., Admiraal, W. (2002). Development of photosynthetic biofilms affected by dissolved and sorbed copper in a eutrophic river. Environmental Toxicology and Chemistry 21 (9), 1955 – 1965

[6] Rohder, L. A., Brandt, T., Sig, L., Behra, R. (2014). Influence of agglomeration of cerium oxide nanoparticles and speciation of cerium (III) on short termeffects to green algae Chlamydomonasreinhardtii. Aquatic toxicology152, 121 – 130

[7] Pérès, F., Florin, D., Grollier, T., Feurtet-Mazel, A., Coste, M., Ribeyre, F., Ricard, M., Boudou, A. (1996). Effects of the phenylurea herbicide isoproturon on peryphytic diatom communities in freshwater indoor microcosms.Environmental Pollution94 (2), 141-152

[8] Dauta, A. (1982). Conditions de développement du phytoplancton.Annales de Limnologie18, 217- 268

Analyze the effect of Molybdenum on Heat Resistance Stainless Steel Casting SCH 22

Achmad Sambas,[1] Gamawan Ananto[2]

[1] Bandung State Polytechnic for Manufacturing, Bandung, Indonesia
Email: sam@polman-bandung.ac.id,
[2] Bandung State Polytechnic for Manufacturing, Bandung, Indonesia
Email: waloeyo_gamawan@yahoo.co.uk

Abstract. There was SCH 22 component named hanger that is produced by casting, where it is one of the components applied at 1100°C work temperature. Problems on the product is the occurrence of *scaling*, causing the the function failure and shortening its lifetime. Generally, this *scaling* occurs when the material is exposed to high temperature, then kinds of iron oxide layers are arisen such as FeO, Fe_3O_4, and others. Based on the literature, Molybdenum can be added to the SCH 22 with maximum limit of 0.5%. The purpose of this study is to know the influence of Molybdenum to the mechanical properties and analyze the phenomenon of *scaling* on Heat Resistant Stainless steel Casting (SCH 22). It is expected that molybdenum could increase the *scaling* resistance in the high temperature. The method is making specimens through metal casting process where the molybdenum content are made to 0.3%, 0.7%, 1.1%. Each specimen should be heated at 1150°C and held for 4 hours. Then the mechanical strength and micro structure are observed to see the effect such molybdenum addition. The result is obtained that the 0.7% addition of molybdenum element can reduce the degradation due to *scaling* on Heat Resistance Stainless Steel Casting SCH 22.

Keywords: scaling, molybden, stainless steel casting, heat resistance.

1. Introduction

The SCH 22 is a Heat Resistant Steel Casting, that is generally applied to working temperature between 650°C and 1315°C. The material is selected because of its resistance to corrosion, it is stable at high temperatures, and its resistant to creeps or heat-breaking/ cracking. The material is applied to heat-resistant components in the cement processing industry, fertilizer industry, oil/ gas industry and in the oven industry heating. It is expected to use local content of the component in product supply, and promotes the development of this material in terms of quality and quantity.

Hanger is one of the components applied at 1100°C. Problems that occur on the product is the occurrence of *scaling*, that is causing the function failure and shortening its life time. Generally *scaling* occurs because the material is exposed to high temperatures, then kinds of iron oxide layers are arisen such as FeO, Fe_3O_4, and others. There are various ways to increase oxidation resistance at high temperatures, such as adding 25-30% Cr, <2% Si, <4% Al, Yttrium and Cerium as well as by addition of Ni [1][2]. Based on the literature can also be given Molybdenum to the limit of 0.5%.[1]. On other hand, Ferro-Molybdenum addition has influence on impact toughness values of the sintered low alloy steel [3]. The influence of the addition of Molybdenum element to the Heat Resistant Steel of SCH 22, so far has not been studied by practitioners and scientists on a national scale.

The purpose of this study is exploring the influence of Molybdenum to the mechanical properties and analysing the phenomenon of scaling on Heat Resistant Stainless Steel Casting/ SCH 22. Figure-1 shows the hanger components that have been heated at high temperatures. The component is damaged by degradation.

Figure 1. Scaling on Hanger products

2. Method

The casting process is designed to obtain the test sample through the manufacture of mold, melting and followed by modification by varying the amount of molybdenum from 0.3% to 1.1%. The specimen pouring scheme can be seen in following figure 2.

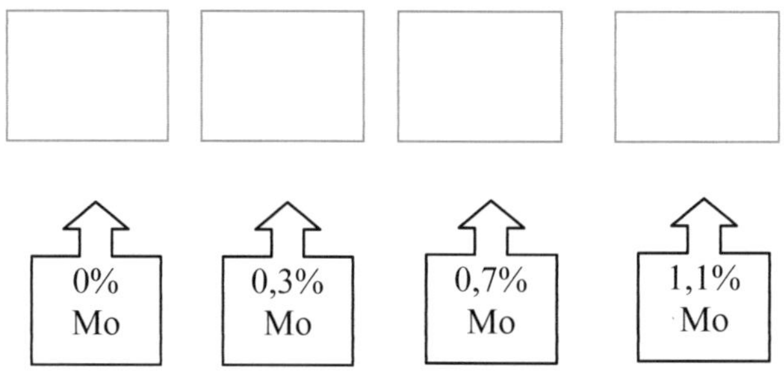

Figure 2. Schemes of molybdenum modification

Specimen test preparation by metal casting process had the following stages; making 4 pieces of mold that is containing of each 3 test specimens in the form of Y-block, following with mold casting with varying amount of Molybdenum.[5][7] The process of alloying and pouring can be seen in Figure-3. When the Y blocks test sample are already frozen and cold, then it should be removed from the mold. The next process is to clean the Y block test sample from the attached sand. Figure-4 shows a sample of Y blocks that have been cleared.

Preparation of the test sample for tensile test specimen by cutting Y block test sample, then grinding with cylindrical shape. The cylindrical form is then obtained by lathe machine in accordance with the standard size of the tensile test. Figure 5 Shows the sample preparation process for tensile test specimens.

IOP Conf. Series: Earth and Environmental Science **156** (2018) 012006 doi:10.1088/1755-1315/156/1/012006

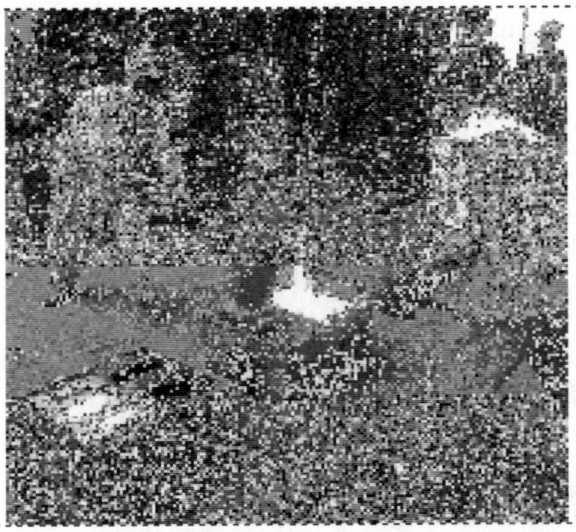

Figure 3. Alloying and pouring process.into the molding.

Figure 4. preparation for Y block specimens

Figure 5. preparation for tensile test specimens.

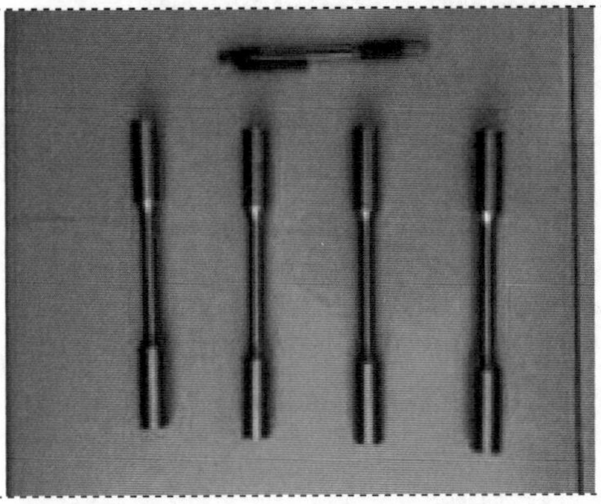

Figure 6. spesiment test for tensile trength

Hard test specimens by cutting the Y block test sample then grinding process with a cylindrical shape. The cylindrical form is by lathe in accordance with the standard size of the hard test. Figure 7 shows the sample preparation process for hardness test specimen.

Figure 7. preparation for hardness test specimens.

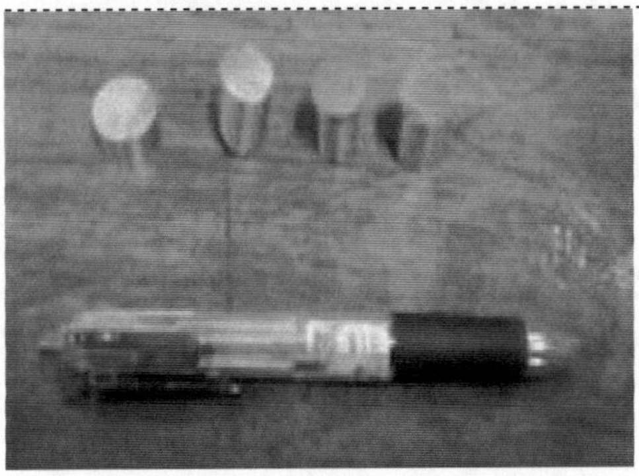

Figure 8. spesiment test for hardness

The next step is heating process on each specimen with the temperature of 1150°C and 4 hours holding time, using a heating stove.

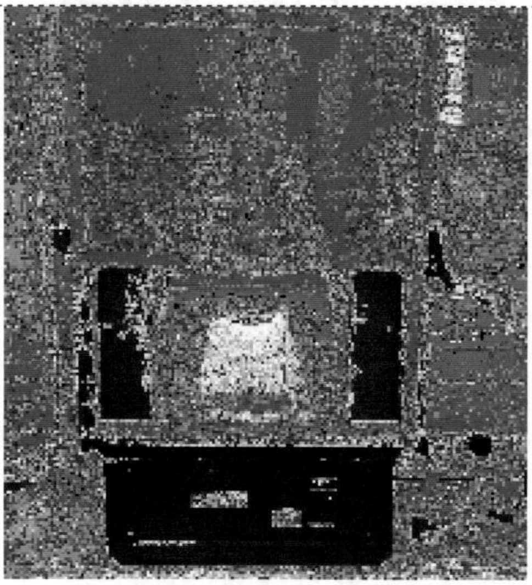

Figure 9. Heating of specimen with heating stove

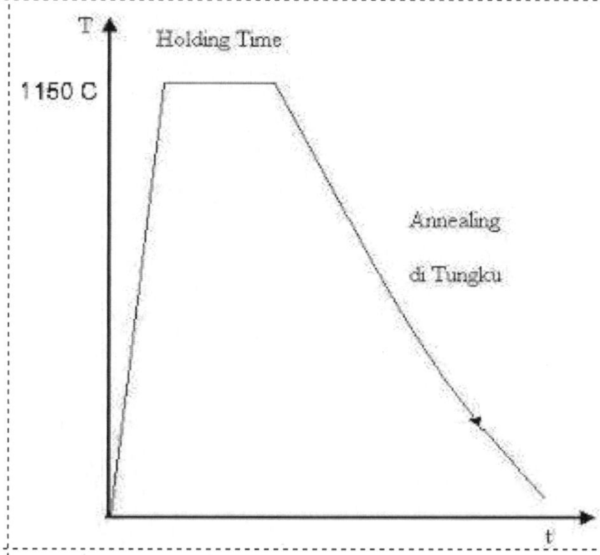

Figure 10. The cycle of heatreatment

Preparation for micro structure test specimen is cutting Y block test sample, then grinding process with cylindrical shape. The cylindrical form is then processed in the lathe according to the different diameter. The cross section to be observed, sanded gradually from 240 to 1000 mesh. Then, it is polished by using alumina powder and etching process afterwards, by etching *vilella* for 18 minutes.

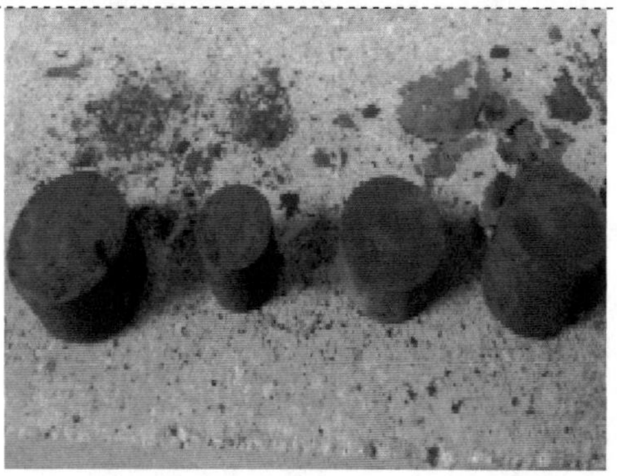

Figure 11. preparation for tensile test specimens.

Figure 12. spesiment test for tensile trength

Figure 11. shows a specimen that has been heated at a temperature of 1050°C, and has been held for 4 hours. Figure 12 mentions the observation with an optical microscope for viewing and determined the *scaling* mechanism on Heat Resistant Steel. Microscopic observations are performed by optical microscope with magnification of 100x and 500x, performed at the center and edge of the specimen. Such testing is performed to determine the phase configuration that occurs.

3. Results and Discussion

3.1. Results of Chemical Composition Characterization

The carbon content exceeds the target where the target at JIS standard is max 0.45 % and actual 0.50 % but it can still be tolerated because it is still close to the DIN standard which has a maximum carbon content of 0.5 %. For the composition of chemical elements such as Si, Mn, Ni and Cr are according to the standard.

Table 1. Chemical composition

Standard	Chemical composition (%)						
	C	Si	Mn	P	S	Ni	Cr
JIS	0,35-0,45	≤ 1.75	≤ 1.5	≤ 0,04	≤ 0,04	19-22	23-27
Foundry results	0,50	1,1	1,2	0,03	0,00	19,8	25,4

All specimens have the same composition as those listed in Table 1. While the difference is its molybdenum due to the modification. The results of the composition of the molybdenum modification are shown in Table 2.

Table 2. The results of the composition of the molybdenum modification

Specimens	I	II	III	IV
Target % Molybdenum	0	0,3	0,7	1,1
% Molybdenum actual	0	0,15	0,5	0,7

It is noticed that the target did not match with actual result. This was analyzed in addition to improper ladle volume calculation and incomplete fluid stirring process in the ladle, which caused the liquid became to the non homogeneous. But the data can still be used anyway because the Mo content tends to increase.

3.2. Results of Mechanical Testing Characterization

The tensile testing process is based on the JIS Z 2201 standard.[8] Figure 11 shows the curve of tensile test results.

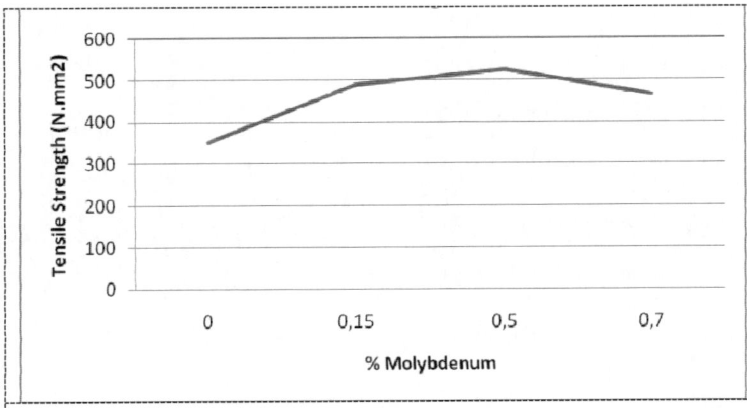

Figure 13. Tensile test curve.

Tensile strength values of about 400 to 500 N/mm2 are obtained. The value of the tensile strength standard value for SCH 22 material is 440 up to 640 N/mm^2. The highest tensile strength is in the amount of molybdenum of 0.5 %. The above curve shows that the influence of molybdenum increases its tensile strength but not in the very significant value.

3.3. Results of Micro structure Characterization

Observation of micro structure on all specimens that have been heated at 1150°C with 4 hours holding time, is done with 100x magnification, on the side or edges of objects, with following results.

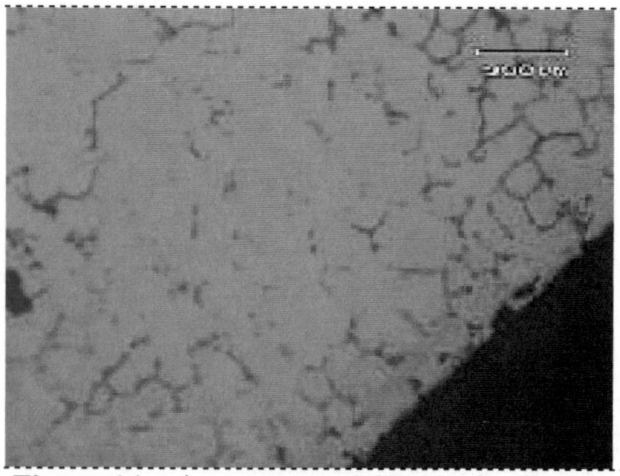

Figure 14. microstructure SCH 22 after heated 1150°C with held for 4 hours , without modification Mo.

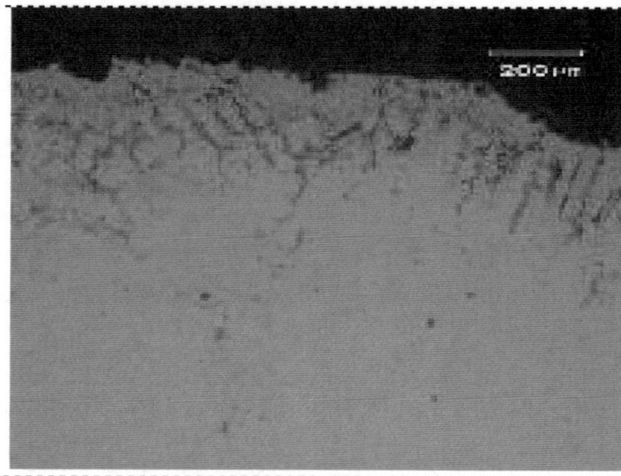

Figure 15. Microstructure SCH 22 after heated 1150°C, held for 4 hours with modification 0.15% Mo

Figure-14 shows the micro structure of stainless steel SCH22 without molybdenum modification. There is degradation at the grain boundary up to 2 mm deep from the surface. Scalling begins from the grain boundary of the carbide that surrounds the austenite phase. Further, Figure-15 shows the micro structure with 0.15 %, in which the degradation at the grain boundary is approximately 400 mm deep from the surface.

Figure 16. Microstructure SCH 22 after heated 1150°C, held for 4 hours with modification 0.5% Mo

Figure 17. Microstructure SCH 22 after heated 1150°C, held for 4 hours with modification 0.7% Mo

Meanwhile, Figure-16 shows the micro structure with 0.5 % molybdenum modification, with the same heating at 1150°C in 4 hours holding time where the degradation at the grain boundary is approximately 200 µm depth. For these additions of molybdenum element, the degraded areas of samples are dwindling. In Figure-17, modification of 0.7% molybdenum shows the degradation at the grain boundary with the depth of approximately 50µm, and, the degraded areas are more dwindling.

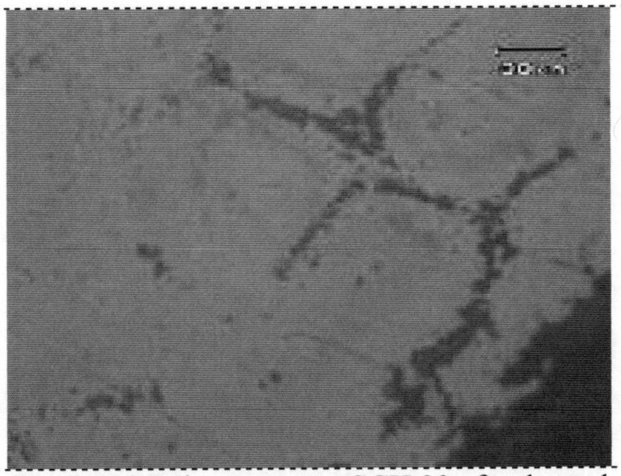

Figure 18. Microstructure SCH 22 after heated 1150°C, held for 4 hours without modification Mo, 500x magnification.

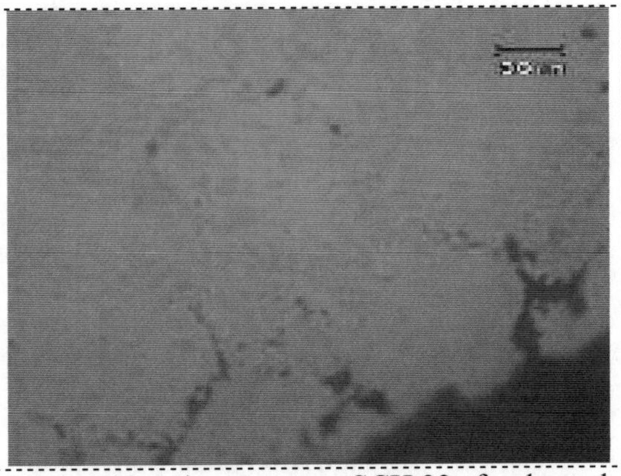

Figure 19. Microstructure SCH 22 after heated 1150°C, held for 4 hours with modification 0.7% Mo, 500x magnification.

Figure-18 and Figure-19 compare the SCH 22 micro structure between 0% molybdenum with a 0.7% molybdenum. With 500x magnification it appears more clearly morphologically of the grain boundaries that are affected by degradation or scalling. From such observation of micro structure, with the higher percentage of molybdenum up to 0.7% it is shown that the degradation due to oxidation or scaling on carbide grain boundaries is getting thinner. It indicates that the molybdenum element can strengthen the bond at the carbide grain boundary, thereby it is increasing the heat resistance in the area as well.

4. Conclusion

After the number of processes, the addition of molybdenum element up to 0.5% into Heat Resistant Stainless Steel Casting /SCH 22 material can increase its tensile strength, although not so significantly. However, the addition of such element up to to 0.7% has no effect on the value of hardness. Moreover, this 0.7% addition of molybdenum element can reduce the degradation due to scaling on SCH22 stainless steel as well.

5. References

[1] Davis, J.,R. (1997). ASM Speciality Handbook: Heat Resistant Material ASM Handbook.

[2] Yudi Prasetyo (2013), "Heat Resistance Material Selection", retrieved from https://yudiprasetyo53. wordpress.com. December 25th 2016.

[3] R. Yilmaz. A. Gokce (2007). "The Effect of ferro-Molybdenum addition on the Microstructure and Mechanical Properties of Sintered steel". Advanced Material research vol 23, pp 71-74. Trans Tech Publications, Switzerland.

[4] Aziz Abdul, Bambang Suharno. (2007). "Karakterisasi Kekuatan Tarik dan Struktur Mikro Baja Cor Tahan Panas HK 40 Berbahan Baku Ferro Nikel Lokal", Seminar Nasional Metalurgi dan Material (SENAMM) 1 2007 Universitas Indonesia.

[5] Beeley, P. (2001). Foundry Technology, Butterworth Heineman, Oxford.

[6] Dwi Mulyo, Agung. (2014). Pengaruh Deoksidasi Aluminium Terhadap Porositas Gas Pada Baja Sch 22. Tugas Akhir Politeknik Manufaktur Bandung

[7] Heine Richard W. (1967). Principles of Metal Casting, Mc Graw Hills.

[8] JIS. (1992). Ferrous Material and Metalurgi. Tokyo: JIS International.

[9] Muhamad Daud Pinem. (2005). Jurnal Teknik SIMETRIKA Vol. 4 No. 1–April 2005:301–306. Sambas,

[10] Achmad. (2014). "Analisa kegagalan dan pengembangan material baja cor tahan panas sch 22 pada kasus *lip replaceable*" , Prosiding Seminar Nasional Teknologi Terapan SV UGM.

IOP Conf. Series: Earth and Environmental Science **156** (2018) 012007 doi:10.1088/1755-1315/156/1/012007

Role of Bajo Women in Wakatobi Island

Fatirahwahidah[1] and Mansur Mansur [2]

[1]Institut Agama Islam Negeri Kendari, Jl. Sultan Qaimuddin No 17 Baruga, Kendari, Southeast Sulawesi, 93563, Indonesia
Email: tira-idha@yahoo.co.id
[2]Institut Agama Islam Negeri Kendari, Jl. Sultan Qaimuddin No 17 Baruga, Kendari, Southeast Sulawesi, 93563, Indonesia
Email: mmansur@yahoo.co.id

Abstract: This paper investigates the unique side of Bajo women at Pulau Pandai Besi in Wakatobi. Bajo women are able to take over their husband's role. Bajo fishermen's households present different conditions from the common fishermen community in which the man and woman can share roles in the family. Although naturally, men and women are created differently which leads to the role difference, the local wisdom of Bajo indicates that women are very dominant in their households. The husband only works in the sea to earn a fortune. Meanwhile, Bajo women, in addition to taking domestic roles such as cooking, taking care of children and cleaning the house, are also in control of the development and survival of their families when their husbands go sailing in the sea. This research is qualitative research. This research was conducted in Peka Besi region of Wakatobi in 2017. The data of this research were collected through observation and interview. The study reveals that women are the leaders and managers of their households. Regarding economic needs, Bajo women process and market the fish caught by their husbands, open a small shop of basic necessities, and sell a variety of traditional food.

1. Introduction

In many ways, every culture has a different conception and outlook towards women. There is a culture that positions a woman with a very respectable. Women seat in a special position. However, there is also a culture that makes women meaningless, like slaves, and does not have any bargaining power in social life. Besides, in the division of labour, people still use gender as a criterion in determining the type of work that is considered appropriate for men and women.

Sanderson (2001) suggests that all humans use gender as a major criterion in the social division of labour [1]. Meanwhile, people usually have a suitable role for both genders, and they also describe the specific roles either only for men or women. The fact is supported because men are considered more powerful regarding a woman's physical appearance, which in reality is weaker than men.

Gender roles always exist in the community from the beginning until now, often called the concept of patriarchy. Therefore, the gender roles are biased and unbalanced. As a result, unbalance roles often lead to injustice and discriminatory according to women, as well as when taking part in workforce, women often get more work in the factory or the office, with low wages. After that, they have to do their household tasks like as cooking, washing, and caring for children [2].

Patriarchal society is a society that has a referral system based on the agreement of men. In the patriarchal system, the conditions are very marginal. It makes the women do all the domestic works. The exclusion of women in a patriarchal society in terms of the pattern of the division of labour between men and women is manifested very clearly, where more men dominate the public sector, women are placed in the domestic sector.

In a patriarchal society, the labour division does not show a balance pattern. In work, men are more valued than women's work. Women's works tend to receive very little appreciation. This is due to the social construction based on body/physical of women and men. Thus, the women will be intimidated by the patriarchal system, while men control the work of the public sector even more. This condition leads to the patriarchal culture. This patriarchal culture will remain alive and well preserved in the lives of gender-based society. With these conditions, it will display a rule that insists women to be a nurse and guardian based on men' thoughts, so women are often called living creatures, who are in second phase of life. Under these circumstances, the position and status of women are increasingly cornered and marginalized in life.

In addition to the marginalized position, second-class status given by the patriarchal culture also affects the performance and the presence of women in a patriarchal society. Thus, the women's works are considered insufficient to be classified as high quality work. The emergence of the notion that the quality of women's work is not caused by the values in the patriarchal society that considers women cannot work. Women can only receive and enjoy the results of the work carried out by men. It is born of a patriarchal culture that positions men as the leader and breadwinner for women. Thus, the position of women is only considered as a nurse and guardian who merely serve the interests of workers.

There are several reasons why many people discredit the women's works. These are: (1) physical body of men and women in which the women are considered weak while men is physically strong; (2) the women's characters are sentimental, gentle, and like tidying up and managing things. Many people in this patriarchal system really believe these factors. As a result, the women will only do light work, such as serving and taking care of fewer works. Nonetheless, serve and care are the evidence of women's existence in domestic life.

In addition, serving and caring work in the domestic sector led to an idea that fits the women appropriate work that is in a mattress, a well, and a kitchen. With the existence of these images, born unequal division of labour in the concept of gender equality will only be a dream of a gender-sensitive society. Unequal division of labour between men and women in the domestic and public sector will bear a double workload for women. However, such expenses are considered as a supporting role in the man's work, not as a woman who can work in spite of all the myths and gender bias. Likewise, the domestic roles of men and women in Wakatobi Bajo are separated.

2. Research Methods

This research is qualitative research. This research was conducted in Peka Besi region of Wakatobi in 2017. The data of this research are collected through observation and interview.

3. Findings and Discussion

There are several theories on the division of labour are:

Naturalistic theory assumes that the role differences between men and women are natural. The difference of women and men anatomy is a major factor to determine their social roles. Men have a major role in society because they are considered more powerful, potential, and productive. Because women have reproductive organs, their movement is limited. These differences lead to the separation of functions and responsibilities between men and women, where men play a role in the public sphere, and women in the domestic field [3]; [4].

Nurture theory considers the differences in gender relations between men and women are not determined by biological factors but by the construction community. In other words, that the social role is considered raw and understood as a religious doctrine, in fact not the fate from God, and also as a product of biological determination but as a product of social construction. Therefore, understanding one of the gender values caused by natural factor is prevalent in the society [4].

According to Budiman, the labour division based on gender (sex) is the work division based on the biological and socio-cultural differences, where women are gentle, dependent, emotional, and cannot work hard, while men are leader, independent, secure, and rational. Therefore, men and women have their respective roles [5]. It will be entrenched in the community and is considered as something natural. According to Skolnick (1974) the labour division based on gender leads to the psychological differences caused by biological and socio-cultural to assign the role of men and women [6]. Nevertheless, Budiman explains some things related to the labour division based on gender as follows [5]:

1) The factors are based on the socio-economic needs of the community. This requirement is based on the real needs of the community system.
2) The factors which are based on a psycho-cultural system with social institutions that dissemination and breeding system of labour division. The system is based on the gender labour division which becomes patriarchal system. This system is not solely about abstract belief system, but is supported by social institutions.

According to Saptari, gender refers to the circumstances in which individuals are born biologically as men and women acquiring the social characterization as men and women through the attributes of masculinity and feminists which are often supported by values or symbol systems of society [7]. Fakih states that gender is an inherent nature of men and women who constitute the social category or characterization (femininity and masculinity) that is socially and culturally reflected in behaviour, personality, attitudes, beliefs, appearance, work, sexuality, family responsibilities, and others [8].

The dichotomy is not based on biological, but rather on socio-cultural relations of men and women who are affected by the values of society or system of symbols and structures of the community. The hallmark of the fungible nature can change over time, differ from place to place, and even vary from class to class based on the gender concept. For example, it is known that a woman is beautiful, emotional or motherhood while a man is considered reliable, rational, manly, and brave and others. Fakih presents five forms of injustice, as follows [8]:

1) Stereotype
The stereotype refers to the labeling of one gender is negative which leads to inappropriate judgment. In a society, many stereotypes attached to women which affect the women negatively. For example, because women are considered friendly, gentle, neat, it is more appropriate to work as a secretary, a teacher of kindergarten and many other stereotypes in society.

2) Subordinate
Subordination is the notion that one gender is considered inferior or subordinated position compared to the other sex. For example, a woman takes care of domestic work because she is considered as "home" or "friend to return".

3) Exclusion (marginalization)
Marginalization is the condition or process of marginalization of one sex about the current/main job which resulted in poverty. For example, a technological development led to what was previously done manually by women taken over by the machine operated by men.

4) Double load (Double burden)
Double Burden is the treatment of one gender where you were working far more than the other sex. As it happens in the society that women are managing households, many women bear the burden of domestic work more and longer.

5) Violence
Violence often happens toward the specific genders. Generally, women are abused because of gender generally women because of gender differences. The violence includes rape, beatings, harassment and creating dependency.

Finally, this study indicates that the women are the leaders and managers of their households. Regarding economic needs, Bajo women process and market fish catches of their husbands, open a small shop to sell basic necessities, and sell a variety of traditional food.

4. Conclusion

Naturally, men and women are created differently. This difference leads to the role differences in many ethnicities, inlcluding Bajo ethnic in Wakatobi. A husband as the leader of the household, protector and educator families cannot contribute fully because they need to go for work to earn living. Therefore, Bajo's women in Wakatobi, in addition to doing household jobs like cooking, taking care of children and cleaning the house, are in charge of domestic role with regard to the development of family life after the death of her husband at sea. The role of women in Wakatobi Bajo appears to be very dominant in that women are leader and housekeeper. Regarding economic needs, women Bajo process and sell the fish caught by their husband, open a small shop, and sell a variety of traditional food.

5. References

[1] Sanderson SK 2001 The evolution of human sociality: A Darwinian conflict perspective. Rowman & Littlefield.

[2] Setiadi EM, Kolip U 2011 Pengantar sosiologi: pemahaman fakta dan gejala permasalahan sosial: teori, aplikasi dan pemecahannya. Kencana.

[3] Nugroho R 2008 Gender dan Strategi. Yogyakarta: Pustaka Pelajar.

[4] Nugroho R 2008 Gender dan Administrasi Publik. *Yogyakarta: Pustaka Pelajar.*

[5] Budiman K 2000 Feminis Laki-Laki dan Dikurusus Gender. Magelang: IndonesiaTera.

[6] Skolnick AS, Skolnick JH 1974 Intimacy, family, and society. Little, Brown.

[7] Saptari R, Holzner BM 1997 Perempuan, kerja, dan perubahan sosial: sebuah pengantar studi perempuan. Pustaka Utama Grafiti.

[8] Fakih M 1996 Membincang Feminisme: diskursus gender perspektif Islam. Risalah Gusti.

Information technology Governance standards on mobile applications for fishing zone based onCobIT 5 Framework in Majene

Irfan Andi Palalloi[1]; Azwar Anwar[2]; Syarifuddin[3]
[1]Universitas Sulawesi Barat. Indonesia
Email: irfan_ap@unsulbar.ac.id
[2]Universitas negeri Makassar;
Email: azwaranwar@fdi.or.id
[3]STIE YPUP Makassar
Email:Syarif070707@gmail.com

Abstract: Most of the activities of Majene regency society dominant as a fisherman, by and large, they work based on the hereditary experiences of their ancestors. This is proven by fishery industry statistic that highest from other industry with 18,30 % in the distribution of the gross regional domestic product. In each specific case, utilization of technology becomes a necessity that plays a key role. Adoption of technology for fishermen groups in use of GPSequipment has frequently committed by the government also non-profit organization go through training and mentoring. Nowadays there are some modern mobile applications has been developed by government agency assist the group of fishermen handy on managing their fishing activity. Such us ZPPI data row from Lapan Satelite, nelpinwas also known as smart fisheries, infrastructure development for space oceanography (indeso). However, all of them carry out of the risk and problems on the user side. One of them related toaccuracy and reliability. In this research, we elaborate technical factor, governance, through Cobit framework and analyze the best practice standardfor implementing the technology. All in all, the result presented of the governance standard on control and implementing technology under customer dimension in information technology governance on the standard process to ensure benefit delivery for implementing mobile application fishery in DKP Majene regency.

Keyword: IT, Governance, CobIT 5, fishery, framework.

1. Introduction

Indonesia is the largest archipelagic country in the world and has coastal communities that run daily activities as fishermen in the traditional way. There have been many studies and concepts was offering for the community to improve methods in fishing activities so as to improve the standard of living of the people[1]. The activities of these fishermen are generally classified into small-scale fisheries groups[2][3][4], the use of technology is inevitable by the spread of various tools and information systems targeting this group [3][5][6]. Similarly, the fishermen community in West Sulawesi, especially in this research in the region of Majene Regency, based on data from the Ministry of Marine and Fisheries (DKP) Majene District there are some technical equipment and information systems that have been used or mentoringfor the community or group of fishermen by the government. Utilization of this technology includes fish tracking equipment, GPS equipment, potential fishing zone information (ZPPI), nelpin applications. The diversity of this technology has two sides, on the one hand, the benefits but also bring new problems on the other side. This is the responsibility from DKP Majene districts to think about the strategy and regulation of utilizing and managing the utilization of Information Technology optimally, measurable and effective. This is where IT governance is needed that can be a reference for the government in formulating policies, procedures, and controls to manage information risk from the use of information technology itself [7]. In this study focusing on the utilization of information systems in helping improve the productivity of fishing communities is the ZPPI Lapan information system and smart fisherman application (nelpin) developed by balitbang KKPP. Further aligning the objectives of DKP as well as objectives related to information technology. So it can be formulated activities as a reference for IT governance that can be applied in DKP Kabupaten Majene [8].

2. Methods

The first stage of this activity is to know the stakeholder needs of the district government, in this case, DKP Majene in accordance with its mission vision, so it can be known the added value that will be generated in DKP policy related to Information Technology.

Fig. 1. The Governance Objective; Value Creation[8]

Furthermore, alignment of institutional objectives with the objectives relating to Information Technology produces activities that correspond to the best practices in IT Governance[7][8][9]. This best practice is the basis for developing IT Governance standards related to the utilization of online-based information system application so that all evaluation, control,and development can be done continuously to protect fisherman society in their daily activities. The testing accuracy of fisherman supporting information system is done in some water points in Majene, as data to confirm the accuracy of much-needed information, on ZPPI and Nelpin applications.

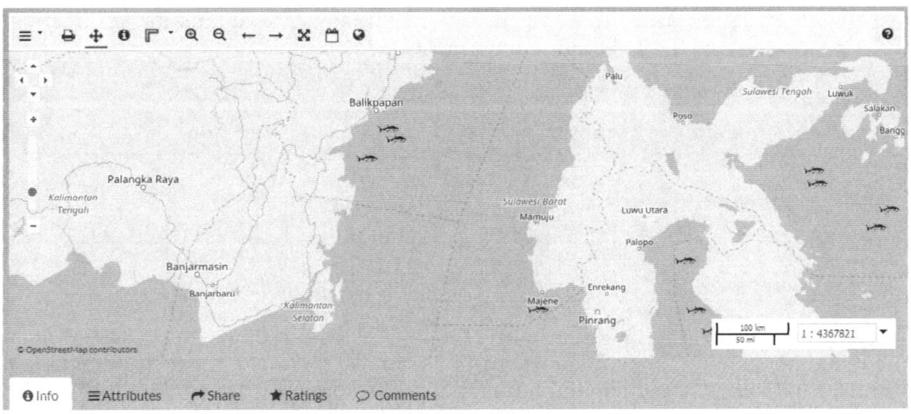

Fig. 2. ZPPI interface view

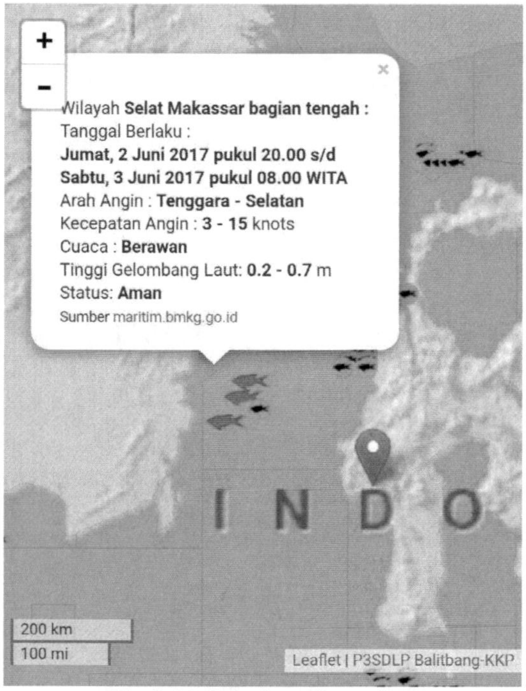

Fig 3. Nelpin interface view

All previous processes form the basis for formulating appropriate IT Governance Standards, as well as measures for evaluation of their implementation.

3. Results and Discussion

3.1. Stakeholder needs

The needs of DKP Majene stakeholders relate to the utilization of technologies such as the expected benefits from the use of technology, this can be seen in the following table.

Table 1. Stakeholder Need DKP Majene -Related Technology

	Fisherman	Industry
Benefit Realisation	High	High
Risk Optimisation	Neutral	High
Resources Optimisation	Low	High

3.2. Alignment Process

According to the vision of mission DKP, found targets to be achieved are:

Vision:Majene Regency as Aquaculture Production Center.

One of the corresponding missions is "Increasing Aquaculture Production".

From the Cobit 5 matrix, we can align to Enterprise Goals (EG), Optimization of business process functionality. And in accordance with IT Related Goals (ITG), Knowledge, expertise,and initiatives for business innovation[8][9]. One of the corresponding missions is "Increasing Aquaculture Production".

IOP Conf. Series: Earth and Environmental Science **156** (2018) 012008 doi:10.1088/1755-1315/156/1/012008

From the Cobit 5 matrix, we can align to Enterprise Goals (EG), Optimization of business process functionality. And in accordance with IT Related Goals (ITG), Knowledge, expertise,and initiatives for business innovation [8] [9]. The EG and ITG alignment bring us to the Process Reference Model In the Evaluate, Direct, and Monitor (EDM) process which is the Cobit 5 focus on the Governance side. The process that best suits the conditions and data profile DKP Majene is in the process of EDM 02 Ensure benefits delivery.

3.3. Metrics Assessment

Measurements of IT governance evaluation onEG-11[8][10][11]which can be formulated under DKP conditions include::

- Frequencyofbusinessprocesscapabilitymaturityassessments
- Trendofassessmentresults
- Satisfactionlevelsofboardandexecutiveswithbusinessprocesscapabilities

In IT-related activities (ITG-17) the evaluation matrix refers to:

- Level of business executive awareness and understanding of IT innovation possibilities
- Levelofstakeholdersatisfactionwithlevelsof IT innovationexpertiseandideas.
- A numberofapprovedinitiativesresultingfrominnovative IT idea.

When these metrics are applied to the technology used by the community we can see the following results.

Table 2. Result of application observers

EG	ZPPI	Nelpin
Frequency of business process capability maturity assessment	High	Neutral
Trend of Assesment result	Neutral	High
Satisfaction levels of board and executive with business process capabilities	Low	High
ITG		
Levelofbusinessexecutiveawarenessandunderstandingof IT innovationpossibilities	Low	Neutral
Levelofstakeholdersatisfactionwithlevelsof IT innovationexpertiseandideas	High	Neutral
A numberofapprovedinitiativesresultingfrominnovative IT idea	Neutral	High

3.4. IT governance process practice accordant with DKP Majene

The best practices that can be applied in accordance with the conditions of DKP Majene relating to information technology on EDM-02 [8]comprise:

Table3. Governance Practice EDM-02 for DKP Majene

EDM02.01 Evaluate value optimization	**Activities**
Process Goals - The enterprise is securing optimal value from its portfolio of approvedIT-enabled initiatives, services,and assets.	- Understand stakeholder requirements. - Understand the key elements of governance - Understand and regularly discuss the opportunities - Understand what constitutes value for the enterprise - Evaluate how effectively the enterprise and IT strategies - Understand and consider how effective current roles - Consider how well the management of IT-enabled investments - Evaluatetheportfolioofinvestments
EDM02.01 Evaluate value optimization	
Process Goals - Optimal value is derived from IT investment through effective valuemanagement practices in the enterprise.	- Define and communicate portfolio and investment. - Define requirements for stage-gates - Define and communicate enterprise
EDM02.03 Monitor value optimization.	
Process Goals - Individual IT-enabled investments contribute optimal value	- Define A Balanced Set Of Performance - Obtain Regular And Relevant Portfolio - Collect relevant, timely, complete, credible and accurate data

4. Conclusions

From the data and the result of the alignment of vision and mission DKP Majene knew that the best practices of IT governance in accordance with activities to manage, resulting in policies related to the use of technology in the routine fishing communities can be run in a measurable and developed as needed. So the constraints and problems that arise in the future already have standard procedures and become the reference for all parties.

IOP Conf. Series: Earth and Environmental Science **156** (2018) 012008 doi:10.1088/1755-1315/156/1/012008

5. References

[1] D. R. A. Samantha M. Berdej, "Bridging Organizations Drive Effective Governance Outcomes for Conservation of Indonesia's Marine Systems," The PLoS ONE, vol. 11, no. 1, 2016.
[2] FAO, "Voluntary Guidelines for Securing Sustainable Small-Scale Fisheries in the Context of Food Security and Poverty Eradication, Food and Agriculture Organization of the United Nations," FAO, Rome, Italy, 2015.
[3] D. R. D. B. M. H. R. G. M. R. L. R. D. Ratana Chuenpagdeea, "The global information system on small-scale fisheries (ISSF): A crowdsourced knowledge platform," Marine Policy, 2017.
[4] X. B. M. S. Emilie Lindkvist, "Micro-level explanations for emergent patterns of self-governance arrangements in small-scale fisheries—A modeling approach," The PLOS ONE, vol. 12, no. 4, 2017.
[5] W. D. H. Liam M. Carr, "Testing fisher-developed alternatives to fishery management tools for community support and regulatory effectiveness," Marine Policy, vol. 67, pp. 40-53, 2016.
[6] G. M. E. J. L. H. Elianny Domínguez-Tejo, "Marine Spatial Planning advancing the Ecosystem-Based Approach to coastal zone management: A review," Marine Policy, vol. 72, pp. 115-130, 2016.
[7] S. B. D. D. Sana Bent Aboulkacem Guetata, "The Architecture Facet of Information Governance: The Case of Urbanized Information Systems," Procedia Computer Science, vol. 64, pp. 1088-1098, 2015.
[8] ISACA, "Enabling Process - Cobit 5," Isaca, USA, 2012.
[9] M. J. Yiwei Gong, "An interoperable architecture and principles for implementing strategy and policy in operational processes," Computer in Industry, 2013.
[10]C. K. Rachel J. Shucksmith, "Data collection and mapping – Principles, processes, and application in marine spatial planning," Marine Policy, vol. 50, pp. 27 - 33, 2014.
[11] J. Strickland-Munro, H. Kobryn, G. Brown and S. A. Moore, "Marine spatial planning for the future: Using Public Participation GIS (PPGIS) to inform the human dimension for large marine parks," Marine Policy, vol. 73, pp. 15 - 26, 2016.

Extension of Religious Ritual Functions in Development Process: Study of *Mandi Safar* in Air Hitam Ocean Indonesia

Irmawati Sagala

SulthanThaha Saifuddin State Islamic University (UIN) Jambi, Indonesia;
Indonesian Lecturer Forum (FDI)
Email: irma.sagala@fdi.or.id

Abstract. The ritual of *Mandi Safar*, bathing together on the beach using the prayer rajahs performed by the community on every Wednesday of the last week of Safar (second month of Hijri), has been designed by the local government as one of the supporting programs for economic development in Air HitamLaut Village, Jambi. The tradition that derived from religious rituals in the Malay society to avoid catastrophes and calamities along Safar faced controversy at first because it was considered as heresy by some Muslims. Nevertheless, the ritual evolved simultaneously into a popular cross-ethnic tradition and had been planned as one of the national tourism icons of Jambi Province since 2017. Through political sociology and psychoanalysis approaches in social change, this paper analyzes the programming processes of this agenda and its impact. The result shows that marine tourism program of *Mandi Safar* provides a positive basis for economic development in Air HitamLaut. The program has spurred the development of local infrastructure on the one hand and has led to social change in society on the other side. Previously, the process of desacralization of the *Mandi Safar* ritual as well as the role of local customary-religious elites as a mobilizer became the key variables to empowerment process.

Keywords: *Mandi Safar*, religious rituals, economic development, desacralization, customary-religious elites.

1. Introduction

People's traditions along coastal areas are one of the important marineworld's resourcesthat often associate with sacred beliefs or values. In the coastal Malay Muslim society, *Mandi Safar* has long been known as a religious ritual which is rooted in the view of some Muslims that Allah decreases many catastrophes and calamities throughout Safar month especially every last Wednesday, so thatpeople need to perform the ritual of rejecting. There are several opinions related to the number of disasters that Allah revealed during this month, among others mentioning twelve thousand calamities[10; 12]. This ritual is known in most parts of Malaysia such as Malacca, Negeri Sembilan, Sabah and Sarawak and also in Malay regions in Indonesia, like Riau, Jambi, Kalimantan and in Bugis society of South Sulawesi.

Mandi Safar Ritual is performed by taking a bath either individually or together by using water which has been mixed with rajah or recited with medication verses or seven verses which begins with the word "*sala>mun*" (peace) in al-Quran [12]. The procession and the use of rajah causesome other Muslims considered this ritual as *bid'ah*(heresy) and *khurafat* (superstition)and opposed it; even the Malaysian government prohibits this ritual. However, some groups of Malays continue to practice it with various modifications. The kingdom of Malacca for example, although the royal considered *Mandi Safar* as*bid'ah* but the King does not abandonbutmodifies it into a cultural tradition with the name of *PestaAdatNegeri* (the CulturalCelebration of the Country)[1].

Air HitamLaut Village, in Jambi Province-Indonesia, is one of the areas that still preserves*Mandi Safar* ritual and manages it as one of the regional development assets. Air HitamLaut Settlement was formed by Bugis nomads since 1964 and officially became a village in 1973. The people performed this ritual since the beginning of the formation of settlements until present. Since 2003, TanjungJabungTimurRegency government has taken part in the *Mandi Safar* ritual then managesit to be a regional tourism activity as one of the economic development programs. Despite the controversy, the program was widely accepted by the society and continuing by local government, and even since 2017, it was planned as a national tourism event of Jambi Province.

This paper discusses the transformation process of *Mandi Safar* ritual into a regional tourism event and its contribution to the economic development of society in the region. I haddone an indirect studyon this subjectsincefive years ago when I had contact with my colleagues from Air HitamLaut Village. However, I have only conducted the structured research since mid-2017 through library study, observation and interviews of both local people and participants from outside the village.

2. Result and Discussion

In the last few decades, the potential of religious tourism has been increasingly overshadowed by both the private sector and the government of the countries as an asset of development. In fact, religious tourism, especially the pilgrimages, is currently at the first rank in world tourism activities. For example, the holy trip to Kumbha Mela in India earns 70 million visitors, and Muslims pilgrimage to Saudi Arabia reaches 2 million people per year [19].The United Nations World Tourism Organization (UNWTO) estimates that about 300-330 million people visited the world's major pilgrimage sites [6]. The development of religious tourism is also a potential opportunity for the development of tourism areas in ASEAN, a region that has many sites, celebrations and other religious attractions from various major world religions [6].

Refer to the fact above; various religious rituals are developed into religious tourism activities to encourage the economic development. *Mandi Safar* ritual in Air HitamLaut Village also experienced an extension of the religious-cultural ritual functions into a supporting element of economic development. Air HitamLaut Village is one of the nine villages in SaduDistrict with an area of 206.7 Ha (11.35% of district area) and a population of 2,033 people (16.64%, the third densely populated region). Like the typical character of Buginese immigrants known as hard-working, Air HitamLaut Village people work as farmers, fishers, and traders [see, for example, 2; 21]. Trade activity is less prominent due to limited transportation facilities that connect with other regions. In 2016, the Air HitamLautVillage has a Village Fund Allocation (ADD) Rp. 538,378,162, - which is equivalent to 13.34% of the total budget of all villages in the district [4; 5]. Despite its abundant natural potential [10], the poverty and limitations of transport infrastructure are still themain problems faced by Air HitamLautvillagers especially before the last decade (interview, Arsyad, 2017).

In the beginning, this program of extension faced the debate about the status of *Mandi Safar* ritual itself in Islamic law. In other cases, religious groups tend to be resistant to religious tourism activities, at least in the early days of introducing of this concept, for fear that the labeling of such tourism would undermine the sanctity of religious activities[19].But in this case, *Mandi Safar* ritual faces a challenge from the internal Muslim community itself in which some scholars declare this event as *bid'ah* and should be abolished. Consequently, the government and the clerical group of ritual supporters must be able to redefine and modify the ritual of *Mandi Safar* so that the society would widely accept it.

Furthermore, it is also challenged to prove the proclamation by the government that the religious tourism activities of *Mandi Safar* contribute to the economic development of the people and the region inTanjungJabungTimur Regency, especially in Air HitamLautVillage. Instantly, the impacts of economic development activities are measured through the direct increase of local and regional income from *Mandi Safar* event. That the achievement of development is not determined by a single factor, the impact measurement requires then a holistic approach. It is in this context we need to examine the development supporting factors throughout the whole process.

To explain the transformation process of *Mandi Safar* ritual into a regional tourism activity as well as its impact on economic development, we need to discuss at first the status of *Mandi Safar*in religious and cultural views by explaining the dynamic of the procession and its meaning over time. Understanding of ritual position becomes the foundation for understanding the desacralization efforts undertaken by the local government as a first step in transforming *Mandi Safar*to a religious tourism program. The process lasting several years has periodically brought some changes in the community.

2.1 Procession and the Meaning of Mandi Safar in Air Hitam Laut.

Although rooted in the same belief and performed at the same time, *Mandi Safar*'s ritual procession differs in different regions. But in general, this ritual procession consists of joint bath activities, accompanied by prayers/rajahs and musical attraction. This particular character shows the aspect of the sacredness of the ritual as well as the form of a celebration side that features festivity.

We can divide*Mandi Safar* ritual into three stages of development. At first, *Mandi Safar* is a local community ritual performed in each home individually andvery concise. In this period, people understood *Mandi Safar* as part of Islamic rituals. Although writing rajah on media is not mandatory, but it can be read directly before bathing, people preferred to use rajah written on leaves. At the day, people brought leaves that had been written by religious leaders to their homes for then soaked into the water for bathing (interview, Arsyad, 2017).

In the second stage, *Mandi Safar*was performed together by bathing on the beach, but still within the local scope. This change had made since the 1970s on the initiative of M. ArsyadSitte as the first Head of Air HitamLaut Village who reflected uponthe need to strengthen social ties and Islamic brotherhoods in society through joint activities. Also, this eventwas intended as a community vacation (interview, Arsyad, 2017). In this stage, the people ate together after bathing. However, this tradition is not practiced anymore in recent years. The organizing committee only prepares meals for guests from local government and mass media. Finally, *Mandi Safar* enters the third stage where this ritual expands its function from local tradition into tourism activities managed by the government of TanjungJabungTimur Regency which until 2017 also involves the government of Jambi Province.

In this stage, the ritual process undergoes a meaningful development in which the profane aspects are increasing. PesantrenWaliPeetu(Islamic boarding school) community plays a vital role in preparing the core activities. Every year, the *santri-s*(students of Islamic boarding school) are in charge of collecting mango leaves and making the ornamental eggs, while the women prepare the food. The *santri-s* and *ustadz-s* (students and teachers) write then the verses on mango leaves which will be tied to the body on the day of the ritual. The people go to the beach by *pompong* (boat) on the day of ritualbringing the goods. They also became the principal officers throughout the peak of the event such as lifting the tower, entertaining guests and leading the customary ceremony. The committee also emerges a new tradition: presenting a miniature ofthe minaret, decorated with boiled eggs in ornamental wraps, built on a raft which is then floated on the beach in the peak of ritual.In 2017, the miniature of the minaret has a height of about 6 m and the size of the raft 3 x 3 m.Moreover, the committee organizes modern music performances and some competitions such as kite competitions, *mallanjong*, ornamental boats and others throughout the event. Merchants open up market around the beach for a few days

The core event begins with an official ceremony just like any other government event. Some prominent figures give the speech containing exposure to programs and development targets. At the end of the ceremony, the participants pray together for the goodness of the future. Before bathing ritual, the regional leaders (local government and customary leaders) throw boiled eggs from the tower rafts to people while the people water them. This attraction symbolizes cooperation between leaders and the people. Because the number of eggs is not sufficient for the number of attendees, the people scramble to get it and creating an atmosphere of festivity. Finally, the people bath in the beach where some of them pray to God for their hopes.

Since *Mandi Safar* was performed together on the beach, almost residents participatedin ritual as their engagement in community togetherness. This view correlates with the Costly Signaling Theory of Ritual-CSTR [17], that religious ritual hasa meaning of communication among followers in which they establish common identity, build mutual trust and affirm the commitment to shared goals. In CSTRs view, followers are even more solid with more demands and rules are given by their religion although it will spend a lot of costs. At this stage, *Mandi Safar* ritual begins to shift the meaning of purely religious rituals into cult cultures. Divine things start to mix with the profane ones.

Besides several processional changes in the second and third stages, there is also an increase in the number of participants from other areas that some of them cometo the village before the ritual day. Participants or tourists are not only from the surrounding regions but various others in Jambi Province even from surrounding provinces. Residents interviewed said they were happy to welcome tourists as guests in their homes without collecting fees. Otherwise, guests confess warmly welcomed like relatives by the local people. This welcoming pattern of guests often becomes a starting point of closer friendship between them.

2.2 Mandi Safar in Review of Religion and Culture.

Ritual is an ordinary action in human life which, like belief, is a form of religiosity [7]. It has a distinctive character and earns the respect of its followers as a sacred experience [13]. All religions from the most primitive one, have their religious rituals which consist of obligations and restrictions (taboo) and can be distinguished from other profane magical rites [7]. Although both domains are different, at the practical level there is often a syncretism between sacred rituals and profane, especially in societies with an active cultural system. As a result, profane things are sometimes regarded as sacred, and ultimately become the object of debate in religious life, as does the *Mandi Safar* ritual.

Functionally, although it always has a relationship with the belief system, the ritual may have other functions or purposes such as in the formation of social cohesion [16]. As an element of religion, rituals should be viewed as collective phenomena although they are sometimes performed individually [7]. *Mandi Safar* Ritual for example, although initially done independently in the home of each citizen, this ritual is still seen as a collective action because the people practice it simultaneously with consciousness and the corresponding procession in a region.

Referring to al-Quran and hadith as the primary source of Islamic law, *Mandi Safar* has no proposition in Islam. In fact, the roots of trust and some parts of the procession can be judged as *bid'ah*and *khurafat*, so it becomes *haram*(forbidden). The growing debate surrounding *Mandi Safar's*status warmed up since the ritual became increasingly popular after being a regional tourism event. Some religious figures from TanjungJabungTimurtriedthen to explain that *Mandi Safar* is not a ritual determined in the teachings of Islam, but only a tradition that developed in the Muslim community. Furthermore, some prominent scholars of Jambi then approved the *Mandi Safar* ritual, like KH. M. Said Magwie and KH. Prof. Dr. Sulaiman Abdullah, by considering the function of activities as a tradition that has positive values [12].

Ritual processions are gradually modifiedby the development of situations and conditions in society. Modifications that occurred in the implementation of *Mandi Safar* also indicate that this activity is a mere tradition, not a ritual derived from the teachings of Islam (interview, Arsyad, 2017). But in practice, people connect with the need for the meaning of rituals personally and simultaneously while they can also accept ceremonial processions determined by tradition without disturbing their conceptions of the ritual's significance. In fact, it sometimes happens that intuitively held assumptions and domain-specific expectations about the causal properties of the entities involved are disconnected[16].

However, in interaction over two years with several people from Air HitamLautin Jambi city, I found that the meaning of *Mandi Safar*for the people is not as simple as choosing whether the activity is just a community tradition or Islamic ritual. My colleagues born in the 1980s received *Mandi Safar* as part of their traditional hereditary cultures as it was socialized to them.Still, they regarded it as an activity with sacred value. The prayer procession is one of the primary lures that encourages them to always go home in the moment of *Mandi Safar*.Regardless of how many this ritual influences their lives, people inherit the tradition from previous generations even though they do not fully understand the underlying rationale and meaning of each procession. Although in different social settings, the symbolic meaning of Mandi Safar ritual for Air HitamLaut society has similarities with the Kwaio tribe community, in the Solomon Islands, that most people understand rituals more as a hereditary tradition, already done and inherited from the predecessors [9].

2.3 The Programming of Tourism and the Social Transformation.

To discuss the programming of *Mandi Safar* ritual into a tourism event, I like to highlight from the standpoint of the elite contestation in government programs. Foucault in I NyomanWijaya[22] said that there are relations among issues, knowledge, and power in every discourse. Analyzing the development of tourism in Bali, Wijaya showed thatsome interest groups manage their insight on a subject to achieve their interests. A contestation occurs on how each teamseeks to influence the government in determining the form of Bali's tourism being developed.But it must be realized that power relations are not necessarily oppressive or destructivebut can also trigger new breakthroughs for future progress.

Like the case in Bali, local elites in Air HitamLaut Village caught the government programs on tourism development for empowering their region. *Mandi Safar* ritual was then developed into a tourism event because it is considered potentially improve the economy of the community, as well as to preserve local culture.This study shows that the programming process of the *Mandi Safar* ritual as a tourism event more as the elite action than the public one. Therefore, the analysis of the interests and identity discourse of elites are very strategic for understanding the future of the program [8].The main players in this process are Sitte family who keeps *adat* (customary) and religious leadership through PesantrenWaliPeetu, and at the same time holds political power as village head until 2014. The gathering of leadership positions in this family gives them the legitimacy and opportunity to articulate and aggregate public aspirations.

Economic activities through the *Mandi Safar* event are not the primary objectives of the programming this ritual. Furthermore, the elites are targeting the development of tourism supporting infrastructure in the region such as the construction of access roads for villages, installation of electricity, bridges, beach embankments and others. This infrastructure development is ultimately expected to boost the village economy. Personally, Arsyad also admitted that his family and *pesantren*also have an interest concerning infrastructure development. The development of *pesantren* depends on the realization of road construction adjacent to the *pesantren's*development land. Furthermore, this tourism programming will also preserve, within certain limits, local traditions. The preservation of this culture even naturally affirms the identity of the Sitte family as a "founder and keeper" of tradition. It is in this context that power relations play their part.

On the side of local government, although *Mandi Safar* event ritual is rooted in religious beliefs, its tourism potential is feasible to be developed as a regional program. Government involvement with religious activities is also not surprising considering Tanjung Jabung Timur Regency is a predominantly Muslim region with a strong Malay-Bugis character. However, the inclusion of religious tradition activities into government programs does not necessarily indicate the strengthening of religion's influence in the public sphere. This relationship can be considered as a partnership mutualism, or vice versa, it is the government that uses religion to implement its program. In a slightly similar case, the latter model of religious-state partnership for example also becomes a pattern in the United Kingdom's policy of providing spiritual service in prison [3]. But in this case, I argue that *Mandi Safar* religious traditions have a mutual relationship with government programs, especially in developing tourism. Department of Culture, Tourism, Youth and Sports (Disbudparpora) Tanjung Jabung Timur Regency is the unit responsible for the implementation of *Mandi Safar* tourism event today. Although this is an institutional agenda, the involvement of personnel is different from everyone. Referring to the recognition of Arsyad, the most intense officer of Disbudparpora in several years is Mahasin who takes care of Tourism Development and Product Quality Standard Section of Disbudparpora. Mahasin's mastery of the development and management of the event is seen in its performance in the field, which tends to be the primary reference of strategic information related to *Mandi Safar* especially about the detail of the event's budget.

Although *Mandi Safar* has been accepted as a cultural tradition by the Muslim community in Jambi province, the efforts of desacralization continue to be carried out by the government with local elites. While it connects solidly with the teaching of Islam, Mandi Safar ritual will continue to reap the controversy and will ultimately be difficult for the government to promote it as a governmental program. Therefore, the government should at first minimize the roots of religiosity*Mandi Safar* ritual. In plural countries such as Indonesia, *the* transformation of religious values intopublic policies in the public sphere needs the process of desacralization of spiritual values and transform it into shared values of the inhabitants [18]. It means that all stakeholders should declare that *Mandi Safar* is a tradition or cultural activity, although it contains Islamic rites. Thus, this desacralization will also broaden the acceptance of the *Mandi Safar* tourism program in other communities outside of Islam.

For this purpose, the elites play a critical role again. Many theorists explain the importance of the elite in the process of social change, both in a positive and a negative sense [11]. The type of elite that influences social change also varies depending on the culture within the local society. Muhammad ArsyadSitte played the key roles in modifying the tradition of Mandi Safar ritual. ArsuatmanArsyad, Sitte's son, continues then his position. In his capacity as a customary-religious elite, Arsyad's role is crucial in transforming the people's beliefs of *Mandi Safar* ritual on the one hand, and establishing strategic agreements with other clerics in Jambi on the other hand. Through the discussions with the Muslim scholars and socialization through Islamic lectures, the community accepts the practice of Mandi Safar as a cultural tradition. The addition of various community activities during the event also supports the strengthening of the ritual desacralization. I found in the observation that Chinese citizens also participated in events such as helping with the boat transportation and managing sound systems in matches. This transformation would expand the opportunity to anyone, regardless of their religion and ethnicity,to participate in a series of activities, as like tourism activities in Thailand inspired by Buddhist teachings in which people with the various background can join it [6].

The problem then, is the religious meaning of *Mandi Safar* in society changed? As discussed above, the religious roots of Mandi Safar are still inherent in the meaning system of some communities. Some people interviewed stated that the meaning of the Mandi Safar event is a chance to pray for rejecting the catastrophes and calamities and, at the same time, to ask for the fortunes. Also, the interviewed tourists acknowledged that their interest in attending Mandi Safar was stronger because of a prayer procession that they believed to be useful in future life. Therefore, it is almost impossible to separate the ritual from its spiritual roots. This point of view is also related to the fact that majority of the population of Air HitamLaut, also the inhabitants of the surrounding villages who usually participate in the *Mandi Safar* event, are Bugis descendants. For Bugis, religion is an essential thing in life, so they always connect all their life activities with the motives of worship by preserving customs that are not against Islam [21].

Nevertheless, the community finds out the other functions of the ritual that are social and economic functions. Socially, people interact annually with unknown heterogeneous societies. This interaction creates the exchange of information and motivation about various things. This tourism activity thus impacts the modernization of the Air HitamLaut society, which becomes an essential foundation for social change that maybe at individuals level, groups and social structures [11]. In an economic perspective, people need a change of personality towards the entrepreneur's spirit for economic development. According to McClelland in Lauer [11], the entrepreneurial spirit emerges as a motivation for achievement, not on the economic motives per se. The description of the Mandi Safar procession above shows that all elements of society are undergoing a meaningful transformation of the Mandi Safar ritual, as well as facing the change of values and personality. Mandi Safar tourism Activity has driven the people of Air HitamLaut to become a more open society and better able to explore the potential of their region.

IOP Conf. Series: Earth and Environmental Science **156** (2018) 012009 doi:10.1088/1755-1315/156/1/012009

2.4 Economic Impact of Mandi SafarTourism Event

The *Mandi Safar* tourism event, which lasts only one day or several days with support events, does not significantly expand the economic activities. Although business development efforts have been undertaken, for example, the production of local resource-based snack products by the Family Skills Education (PKK) group, the work has not been seriously and sustained as it has not had market prosperity including during the *Mandi Safar* event. This tourism activity has no impact in increasing the volume of trade, as typical in tourist destination areas that operate throughout the year. As an example, religious-based tourism development in Wonosari Village of Malang City, Indonesia, has changed the pattern of people's income from farmers into traders, or farmers-traders, which ultimately affects social community structures [15]. Besides trading and some boat rentals on event day, there are no other economic activities related to this tourism event. Unfortunately, there are no detailed calculations associated with the velocity of money in the beach market throughout the event.

However, this calculation model on the impact of events is too simplistic. The effect of the programming should be measured in a long-term calculation and consider non-material implications such as cultural and local wisdom values preservation. As outlined earlier, the transformation of this event has another target that a long-term one such as roads, power grids, and bridges. Road construction will support the economic activities of the community, mainly to transport agricultural products and plantations which are the primary income of the region. There are 12.47 km^2 of farmland and plantations in Air HitamLaut Village (60% of the total area) which is mostly planted with coconut, areca nut and later oil palm. Currently, the transport of plantation products relies on boat transportation which spends more expensive financing and is sometimes blocked by ocean waves. In practical terms, the mobility of villagers in and out of villages for various activities such as trade, education, and even tourist arrivals at Mandi Safar event as well as other tourism destinations including as an alternative road to Berbak National Park. The operation of the power grid from the State Electricity Company (PLN) will also drive the micro, small and medium enterprises (UMKM) of the village community.

The development of this infrastructure is a vital aspect of tourism development. The Thai's experience of developing religious tourism shows that their success is supported by the availability of physical infrastructure, social services and education, community stability and tolerance and great promotion that reaches all target markets [6]. In support of this tourism event, regency government has provided the event's budgets and made infrastructure development plans by seeking funding support from the provincial government. Although I did not get official figures on the budget, also never published in the mass media, through calculations in fieldwork I estimate operational and publication costs to reach more than 300 million rupiahs. Unfortunately, most of the values are spinning out of the village which means it is not part of the economic activities of the local community. Furthermore, the construction of roads and bridges, as well as encouraging PLN to install electricity grid in the village are the three primary programs of regency. Every year, the district government and even the provincial government also attend the event. The presence of local leaders in the village is also used by the community to convey their aspirations. Besides, Disbudparpora even made the online promotion of the potential of Air HitamLaut Village area and Mandi Safar event.

The village secretary, Muhammad Tang, also recognized the government's commitment. For example, he explained that the village apparatus canceled their plan to repair the hanging bridge connecting the village center to the coast by road using ADD 2017 due to a regency government's proposal that would soon build a permanent bridge. The streets have also been done by the government, but still muddy and hollow soil, especially if the rainy season. The budget for the construction of permanent roads is under discussion of regency and provincial governments. The provincial government's budgetary assistance for road construction and the addition of coastal embankment building is still one of the leading hopes conveyed by local leaders to Jambi's deputy governor during the ceremony of Mandi Safar event in 2017. The operation of PLN's power grid will also commence in early 2018 while the installation of poles has begun in late 2017.

Although this process is not as fast as community expectations, especially the elite, infrastructure development began to progress. The launching of *Mandi Safar* as a national tourism event is an effort by local government and local elites to encourage the acceleration of development with higher hierarchical government involvement. It has become a common phenomenon in Indonesia that infrastructure development in the region will accelerate if there will be government visits from the higher hierarchy, especially if it is a national level. From this point of view, the government and local elites should make innovations to *Mandi Safar*ritual,so it is worthy of being a national event.

3. Conclusion

Religious rituals, like any other form of worship, should be understood as human efforts that tend to seek happiness through something divine [14]. Therefore, the orientation of religious rituals is not easily changed, although the ritual procession has transformed into a profane activity such as tourism. This adjustment is still strongly found in participants of Mandi Safar. Although there has been a shift of meaning initially as a ritual of rejection of catastrophes and calamities, Mandi Safar remains of religious value to society as a means of praying to God for their hopes. Although the implementation of Mandi Safar in recent years has more cultural significance than before, the core elements considered religious like the use of the al-Quran verses in rajah also the intent of bath to seek Allah's pleasure, are still maintained until present. The crossing of these two domains, though opposite, may eventually lead to mutually reinforcing symbiosis. The spiritual aspect makes tourism more desirable while the extension of religious ritual function into tourism activities will enliven ritual facilitation by the government.

The programming of Mandi Safar ritual into a tourism event is a mutual collaboration of the common interests between the government and the local elites, especially customary-religious elites. The two actors have the same interest in the development of tourism supporting infrastructure in the region as one of the bases for economic growth. Also, this programming event will preserve the culture and its local wisdom values. As an essential step for this programming, they do actively and sustainably the desacralization of Mandi Safar ritual. Local people are only passive players in the programming and management of this tourism event. They just participate in competitions, and a few of them join the event committee. However, they get the impact of the program concerning social change. Although direct economic results through the Mandi Safar tourism event are not significant enough, at least an increased interaction with outsider has led to a broader change in their psychology of being more open-minded, well-informed, and motivated to develop themselves.

Although the direct economic impact of this event is not significant enough, the programming of the Mandi Safar ritual into a tourism event will have a long-term effect on the economy of the community and region. This objective will occur if the local government realizes the development of village infrastructure. The TanjungJabungTimur Regency certainly needs support from the Jambi Province government, especially regarding financial aspect. But the more important thing than the realization of these development plans is the people's readiness to participate in the development itself while maintaining their local wisdom.

4. References

[1] Abdullah, A. R. H. (1997). *Pemikiran Islam di Malaysia*. Jakarta: GIP.

[2] Ammarell, G. (2002). Bugis Migration and Modes of Adaptation to Local Situations. *Ethnology, 41*(1), 51-67. doi:10.2307/4153020.

[3] Beckford, J. A. (2011). Religion in Prisons and in Partnership with the State, in *Religion and the State, A Comparative Sociology*. (eds. Jack Barbalet, Adam Possamai, and Bryan S. Turner). New York: Anthem Press. http://www.jstor.org/stable/j.ctt1wn0r14.8.

[4] BPS Kabupaten Tanjung Jabung Timur. (2017). *Kecamatan Sadu dalam Angka*. BPS Kabupaten Tanjung Jabung Timur.

[5] BPS Kabupaten Tanjung Jabung Timur. (2017). *Kabupaten Tanjung Jabung Timur dalam Angka*. BPS Kabupaten Tanjung Jabung Timur.

[6] Choe, J. and O'Regan, M. (2015). Religious Tourism Experiences in South East Asia, in *Religious Tourism and Pilgrimage Management: An International Perspective*, 2nd Edition. (eds. R. Raj and K. Griffin). CAB International.

[7] Durkheim, E. (2017). *The Elementary Forms of The Religious Life (Bentuk-bentuk Dasar Kehidupan Beragama).* (Inyiak Ridwan Muzir, trans.). Yogyakarta: IRCiSoD.

[8] Hassenteufel, P. and Genieys, W. (2011). Combiner la sociologie des élites et de l'action publiquepour analyser le changement. La démarche programmatique. http://ses.ens-lyon.fr/combiner-la-sociologie-des-elites-et-de-l-action-publique-pour-analyser-le-changement-la-demarche-programmatique-127989.kjsp.

[9] Keesing, R. M. (2012). On not understanding symbols; Toward an anthropology of incomprehension. *Journal of Ethnographic Theory, 2*(2), 406–30.

[10] L. Bakhtiar. (2015).*Ritual Mandi Safar; Praktik dan Fungsinya dalam Masyarakat.*Yogyakarta: Pustaka Pelajar.

[11] Lauer, R. H. (1997). *Perspektif tentang Perubahan Sosial.* (Alimandan S.U, trans.). Jakarta: Bina Aksara.

[12] Mursalin, A. et al. (Desember 2008). Ritual Mandi Shafar; Akulturasi Islam dan Tradisi Lokal: Studi Kasus di Desa Air Hitam Laut Kecamatan Sadu Kabupaten Tanjung Jabung Timur.*Kontekstualita, 24*(2), 84-109.

[13] O'Dea, T. F. (1995). *Sosiologi Agama: Suatu Pengenalan Awal.* (Yasogama, trans.). Jakarta: PT Raja Grafindo Persada.

[14] Oberle, J. P. (2008). Understanding Prayer, Worship, and Religious Ritual. *Generations, XXXII*(2), 55-56. http://www.watsoncaringscience.org/images/features/library/Article-Understanding%20Prayer.pdf.

[15] Rozuli, A. I. (2012). Keragaman Pranata Agama dan Budaya serta Implikasinya bagi Penguatan Kegiatan Ekonomi Masyarakat Desa. *Jurnal Interaktif, 1*(2).

[16] Sørensen, J. (2014). The question of ritual: a cognitive approach. *Scripta Instituti Donneriani Aboensis, 18*, 207-220. https://ojs.abo.fi/ojs/index.php/scripta/article/view/441

[17] Sosis, R. (Mar/Apr 2004). The Adaptive Value of Religious Ritual. *American Scientist; Research Triangle Park, 92*(2), 166-172. Accessed from ProQuest.

[18] Sukayat, T. (2011). *Internalisasi Nilai Agama Melalui Kebijakan Publik; Dakwah Struktural Bandung Agamis*. Bandung: Rieksa Utama Jaya.

[19] T{ala>, M. L and Pa>durean, A. M. (2008). Dimensions of Religious Tourism. *Towards Business Excellence, Numa>r special.*

[20] Wekke, I. S. (2013). Islam Dan Adat: Tinjauan Akulturasi Budaya Dan Agama Dalam Masyarakat Bugis. *Analisis: Jurnal Studi Keislaman, 13*(1), 27–56. https://doi.org/10.42042/ANALISIS.V13I1.641

[21] Wekke, I. S. (2017). Migrasi Bugis dan Madura di Selatan Papua Barat: Perjumpaan Etnis dan Agama di Minoritas Muslim. *Intelektualita, 6*(2), 163-180. http://jurnal.radenfatah.ac.id/index.php/intelektualita/article/view/1603.

[22] Wijaya, I. N. (2012). Relasi-relasi Kekuasaan di balik Pengelolaan Industri Pariwisata Bali. *Jurnal Humaniora, 24(2),* 141-155.

Livelihood and Rural Tourism Development in Coastal Area North Maluku Province Indonesia

Yerik Afrianto Singgalen[1], Silvanus Maxwell Simange[2]

[1]Politeknik Perdamaian Halmahera, 97762 Kabupaten Halmahera Utara, Indonesia
Email: singgalen.yerik@gmail.com
[2]Politeknik Perdamaian Halmahera, 97762 Kabupaten Halmahera Utara, Indonesia
Email: simax69@yahoo.co.id

Abstract: The livelihood of the people who live in the villages in coastal and remote areas can be developed into tourism area based on local wisdom to achieve sustainable tourism. Tunuo village is a village that has a wealth of Natural Resources and the Cultural Resources to support tourism. Nevertheless, Tunuo Village has problems related to Human Resources to support tourism. Meanwhile, the livelihood of Tunuo Village as producers of copra have constraints, namely access to economic capital. The gain from copra sale is erratic and prone to change, it encourages people to seek income from other options to meet their needs. These economic pressures encourage the development of environmentally damaging activities, such as fishing using homemade bombs so that reefs may be damaged. There are also other activities, such as the production, distribution, and consumption of traditional liquor made from palm tree sap, locally known as *Saguer*. Therefore, the present tourism as a sector that could provide opportunities for the villagers of Tunuo to earn extra income as well as efforts to preserve the environment, and culture. This research was conducted in the village of Tunuo, North Kao District, North Halmahera Regency, North Maluku province, Indonesia. The method is qualitative Participatory Action Research (PAR). The results indicate that people have a desire and a strong motivation to develop, so that the process of forming the Tourism Awareness Group (Pokdarwis) and its seven charms *(saptapesona)* can take place properly. As of the socialization of tourism awareness, as well as capacity building seminar for Pokdarwiswould be able to mobilize the community to clean up trash on the beach and to encourage communities to develop local knowledge into a leading tourism product.
Keywords: Livelihood, Rural Tourism Development, Coastal Area, Local Knowledge, Sustainable Tourism

1. Introduction

The tourism issues in North Halmahera is interesting to look at. Some previous researchers conduct studies related to policy, management and community development in North Halmahera [9;10]. Moreover, some researchers also designed models of development, decision-making system and geographic information system tourism in North Halmahera [21;24;26]. Furthermore, the government's performance in terms of governance and the promotion serves as good benchmark for the success or failure of tourism development [15; 19). Based on this, the researcher is interested in conducting further studies on the livelihood and the development of community-based tourism in North Halmahera specifically in the Tunuovillage.The geographical position of Tunuo village is very strategic to be developed as a tourist village. The natural beauty around Tunuo village, cultural diversity and typical food in the village can be developed into tourism superior product. However, people's lives in Tunuo village in relation to livelihood could hamper access to economic capital. This causes people in the village Tunuo, who live under economic pressure, to engage in activities that undermine social cohesion and the environment, such as fishing activity using homemade bombs and produce, distribute and consume traditional liquor called 'saguer' in uncontrolled basis so prone to conflict.

Tourism has become one sector which can mobilize the community to actively participate in the development. Through tourism, society is slowly accompanied to form new habits associated with tourism awareness and *saptapesona*. On the other hand, tourism could provide opportunities for local people to improve the economy and create jobs. Through the development of community-based tourism, Tunuo village will be accustomed to change from environmentally damaging activities to environmentally friendly practices. Moreover, sideline fishermen who used to catch fish using homemade bombs are accompanied to engage actively in activities that are environmentally friendly yet provide additional revenue. In addition, the communities are assisted to get accustomed to working together to clean tourism area. This study used a qualitative method with participation action research.

2.Livelihood and Tourism

Tourism affects the livelihoods of individuals and groups who live around the tourism object. Sidarta [25] in his research shows that the development of tourism affects the lives of people around attractions including aspects of livelihood and income. In the context of livelihood, Allison and Horemans (2006) describe the theoretical framework of livelihood which shows that livelihood is determined by the individual or group access to capital. Capital means that the capital of Human Resourcesor human capital, Natural Resources or natural capital, Infrastructure or physical capital, economic or financial capital and social capital. Each capital accessible by individuals is a contributing factor to access more capital, so the capital can be accumulated or converted. In livelihood context, capitals affect the sustainability of people's livelihood, as described by the following chart.

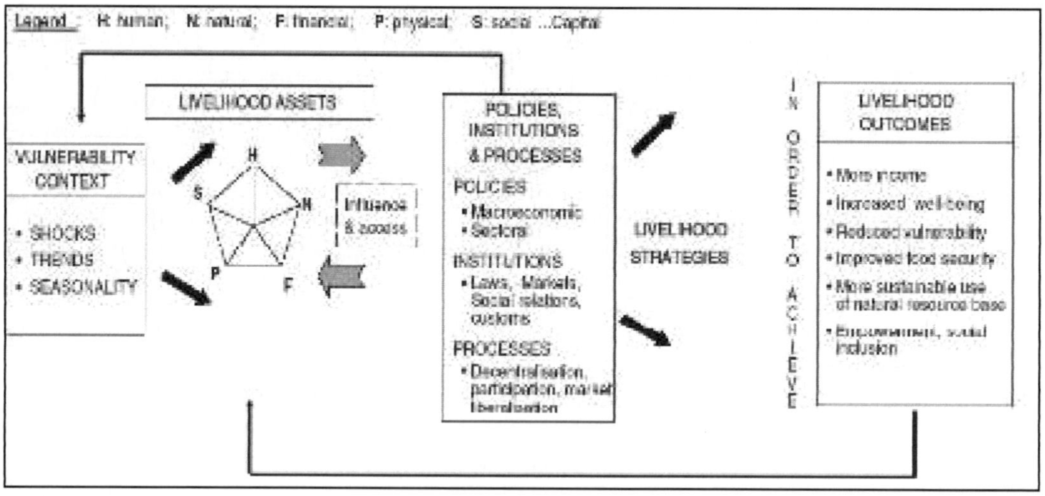

Chart 1 : Livelihood Framework (Allison danHoremans, 2006)

Chart 1 shows that the resilience of livelihood is determined by the individual or group access to capital and is influenced by the trend, seasonality and unexpected things or shocks. Those three elements can be viewed from the aspects, namely policies, institutions and processes. Policy aspects are related to sectoral policies and macro economic policies. Institutional aspects include the legal, market, social and cultural relations. While process aspects include decentralization, participation and the free trade. Changes in the aspect of policies, institutions and processes could affect the shocks, trends, seasonality and the influence of individuals or groups access to capital. Therefore, to achieve the objective and maintain livelihood, strategies are needed. The purpose of livelihood is additional revenue, improved welfare, food security, sustainable use of resources, empowerment and reduction of aspects that are particularly vulnerable to cause harm. Tourism brings about changes in livelihood systems and livelihood in the countryside [5]. People living in the vicinity of the tourism object, use natural resources, culture and attractions as a source of income [1]. Based on this, it can be seen that each of the tourism sector has an influence on people's lives around the attraction.

3. Method

This research used qualitative method with Participatory Action Research (PAR). The decision to use qualitative method was driven by the consideration that the research results will be described in narrative form. Data collection techniques were observation, interview and literature study. Observation was used to observe the lives of the people in the village Tunuo, in-depth interviews were used to obtain specific information related to the life of local communities, while the literature was used to identify studies that documented changes from year to year. After obtaining the information relating to community life in the village Tunuo, the researcher identified the potential of village tourism that can be developed to overcome the problems associated with local livelihoods.

The initial stage of the identification process was done by mapping tourism potentials manually using GPS (Global Positioning System) and digitized maps using the application Quantum GIS (Geographic Information System). After that, the researcher devised the concept of community-based tourism development and local wisdom. Furthermore, the researcher formed a Tourism Awareness Group (Pokdarwis) Village Tunuo and dissemination of tourism awareness and stepping. After conducting socialization, the researcher conducted an evaluation and organized a capacity building seminar for Pokdarwis Tunuo village. Strengthening Pokdarwis was the process of strengthening local communities' understanding of tourism, especially in terms of engineering and technical travel guide market tourism products. Meanwhile, the number of informants involved in this study were 48 people each with a background job as village government apparatus, the coconut farmers, fishermen additional sideline, artisan, grocery store merchants, and service providers working coconut. A member of the Group of Rural Tourism Awareness Tunuo participating actively as a board and members totaled 37. This research was carried out from December 2014 to October 2016. The research process took quite a long, expensive process of in-depth interviews, data collection point coordinates (location Attraction , limit poultice villages and houses, as well as water sources), the process of making a transcript, the digitization process map, the clarification process, the process of socialization, and training seminars. The main focus of this research was the problem of livelihood (livelihood) and the development of rural tourism in coastal areas.

4. Result And Disscussion

People living in Tunuo village generally work as cultivators of coconut land in that familiarly known as the coconut farmers. However, farming activities do not become the only livelihood, but there are other economic activities such as service providers for cleaning gardens, harvesting coconuts, managing coconuts into copra, planting fruits, making sago, hunting, builders, copra traders, groceries traders, andadditional sideline activities as fishermen. People in Tunuovillage do not work primarily as fishermen where most of the time used is to catch and sell fish, but only do fishing to meet the needs of daily meals. In the wind and the rainy season, the fishing activities cannot be done so the livelihood is very dependent on the results of cultivation of rice, beans, sweet potatoes, yams, bananas, and other crops. People who work as producers of copra can access social capital to sustain their livelihoods. It can be seen from the process of producing copra, which is where the process of working on plantations into palm plantations is done by mutual cooperation. The local term used is *higaro,* meaning that inviting each other to perform an activity or work. This work patternapplies in the process of cultivating the land, planting palm, coconut fruit picking, cleaning oil plantations, collecting and splitting coconuts, until the process of gutting and funk coconuts. This shows that access to social capital is able to sustain local livelihoods as farmers through networking, beliefs and values of the local culture.

The results of the sales of copra will be distributed to individuals who are involved in the work. For individuals who clean the plantations, the amount of wages is determined by the plantation area that has been cleaned. For individuals who pluck coconuts, wages are determined based on the number of coconut trees. Similarly, individuals who collect coconuts, their wages are determined by the amount of activity of a pick-up vehicle from the fumigation facility (rack) to the location for coconut collection. On the other hand, people who scrape the coconut meat, the amount of wages is determined by the number of containers (sacks). As producers of copra, in the village Tunuo there has been a change in the utilization of tools to the maintenance process of copra production process. In the past, the process of cleaning the plantations was made together by public by mutual cooperation by cutting the weeds under a palm tree using machetes. In addition, efforts to keep the coconut plantation is based on the teachings of local religion known as *sasi* (local wisdom), which means that coconuts should not be taken in a certain period. If these provisions are violated, the violator will receive punishment in the form of a disease or a curse. Meanwhile, the process of collecting coconuts is done manually, by placing a coconut over the head or by ox cart to collect coconuts. Currently, the process of cleaning the plantations can be done using automatic crop tools, while efforts to maintain coconut groves using *sasi*is done using pest poison that could prevent farmers for their losses from pests and prevent criminal acts of man to steal the palm fruit. Moreover, the process of collecting coconust is no longer carried out individually or by ox cart, since farmers choose to using land transportation such as motor vehicles or cars that can carry coconuts in large quantities as an effective and efficient way in the process of collecting coconuts in the fumigation. This shows that access to the natural capital or Natural Resources, which is the capacity to adapt, knowledge and skill of coconut farmers are able to maintain their livelihood.

Access to Tunuovillage, in this case the roads, has been updated by the regional government in 2016, causing changes in the activity of most of the coconut farmers. Formerly, the coconut farmers use the river as the main access to the location of palm plantations, considering the poor road condistionso it would take a long time and the absence of the bridge because the plantations are separated by the river. These conditions encourage coconut farmers to make a private boat and use it as transportation and conveyance of copra Tunuovillage. When the government builds bridges and renew the roads, most farmers are no longer use the river to palm plantations consider the risk of losses faced with the possibility of an upturned boat, and the limited capacity of the boat transporting copra. This shows that access to physical capital such as transportation, roads, and infrastructure for copra production process, helping farmers sustain their livelihood.

As producers of copra, the harvest time is very limited, namely three months of harvest. In addition, the selling price of copra is always changing. Sometimes, the selling price of copra fell to Rp.3,000, even increased by Rp.9.000,00. Farmers cannot control the price so that the primary focus only on production of copra. When the holidays arrived, the needs of people in the village Tunuo increase. The increase is influenced by the lifestyle that has been entrenched in the community, the shopping habits for clothing, as well as ingredients to make a cake. If copra prices decline while demand increases, copra farmers tended to owe money. Not only at the moment of holidays, but it can also occur when confronted with urgent needs such as children's education and health. This shows that, the people in the village Tunuo having problems accessing economic capital, when the change in the selling price of copra (trend) which coincides with the holidays (season).

The difficulty of accessing economic capital, does not mean that people in Tunuo village is shortage of food, clothing and shelter. In an effort to obtain side dishes, they work on plantations of fruits, vegetables, and even rice. In addition, coconut farmers also use personal boats to sea to catch fish around the Tunuo bay during waves season and catch shrimp and look for shells *(biakoli)* in the river when the waves come. Some of them hunt wild animals to be sold and consumed. This suggests that local wisdom of Tunuo village is able to sustain livelihood.

The majority of the people in Tunuo village have formal educational of Elementary School (SD), some of them do not attend school. Nevertheless, some have completed education at universities. There are several problems that currently threaten the sustainability of the environment, and social lige. The problem in question is the existence of fishing activity using a raft bomb (explosive) that damage coral reefs. Another problem is the activity of production, distribution and consumption of liquor made from palm tree sap known as *'cap tikus'*, thus making the atmosphere not conducive. Both of these activities develop due to the rise in economic needs. Considering this, tourism becomes a source of income as well as a strategy to overcome both of these issues through community empowerment as *Pokdarwis,*thatis able to raise awareness about environmental conservation and awareness to maintain the security of an effort to create a village atmosphere in accordance to the seven charms *(saptapesona)*.

Tunuo village has the potential of Natural Resources, which can be developed into tourism attractions. The natural resource potentialsofTunuo that could attract tourism is the white sand beach in Bailo bay, a rock region called by the local communities as Bakoroko, the cape of Tunuo village called the Cape of Good Hope of Tunuo Village, estuary of Tunuo,Tunuo Village River and Black Sand Beach of Gulo Village. The geographic location of Tunuo village is well placed to be developed as a tourism village, as shown in Figure 1, below.

In addition, the cultureof the people of Tunuo village, as part of the Boeng tribe is still maintained until today. Some of those cultural attractions are tide-tide dance, cakalele dance, and wedding ceremonies. Some cultural objects are typical crafts of Tunuo village, such as the accessories made from marine plants (*bahar*), believed to give the wearer strength in the form of protection from the evil. In addition, there are also traditional foods such as scallops, shrimp, sea worms, fish, yams, sweet potatoes, bananas, sago and so forth. The many types of traditional foods of Tunuo village increase the diversity of tourism potentials to get to know the lives of the rural communities. Here is a map of tourism potentials of Tunuo village, as in Figure 2 below.

Figure 1: Culture of Tunuo Village (Empirical Data,2016)

Tunuo village has the potential of natural tourism, culture and special interests. Natural attractions consist of beaches, marine and river. Beach tourism can be developed as an attraction for tourists to enjoy the scenery at sunrise and sunset. While marine tourism can be developed into attraction for surfing, scuba diving and surface diving (snorkeling). On the other hand, river tourism can be developed into a tourist river boat paddle traditional activity along the Tunuo River (kayaking / canoeing). Furthermore, cultural tourism which can be classified into two parts, namely the traditional dance and music *(cakalele, tide-tide, yangere)* and traditional ceremony (batombo). The special interest tours that can be developed is fishing excursions, seasonal tourist village (marine worm / laor), recreational tourism and culinary tourism.

The diversity of activities that can be done by travelers in Tunuo village. Tunuo village location is very strategic, because access from Tunuo village to tourist attractions can be reached on foot with a range of five to ten minutes. Transportation in Tunuo village has been updated so that the access of transportation such as cars and buses is possible. Tunuo village is still in the development phase, so that the tourism amenitas are not optimal. So far tourist can only stay in people's residence, categorized as homestay houses.

One form of community participation in tourism development is characterized by the formation of Travel Awareness Group (*Pokdarwis*) of Tunuo village. Pokdarwis ofTunuo village was established on May 3, 2016 in Tunuovillage, and was inaugurated on October 5, 2016 by the village government of Tunuo. Members of TunuoPokdarwisare documented as 35 people and led by NurmincePipidor, S.ApwithDjorsen secretary Hadi, S.Ap. The Pokdarwis goal is to increase public awareness about the importance of keeping the village Tunuo environment and preserve culture through tourism. On the other hand, PokdarwisofTunuoVillage attempt to plan governance system attractions, tourism product development and marketing of attractions. Community education related to tourism is carried out through the socialization of the seven charms and tourism awareness in the Village Hall of Tunuo, as in Figure 2 below.

It is also serves a dissemination of tourism awareness and the seven charms held in the village of Tunuo, Wednesday, such as recently held in October 5, 2016 at 09:00 WIT at the village office. The number of participants who attended this event amounted to 47 people including the

IOP Conf. Series: Earth and Environmental Science **156** (2018) 012010 doi:10.1088/1755-1315/156/1/012010

Tunuo village apparatus. Village Chief Justus Bitikal welcomed Tunuo dissemination activities in the hope that through these activities people can know the importance of protecting the marine environment around Tunuo Beach so that fishing activity using bombs can be stopped. In addition, the village leader also invited the entire community of Tunuo village to support these activities so that through the tourism sector, the public can see it as an opportunity to earn extra income. This shows that the villagers of Tunuo have a desire and a strong motivation to develop. Thus, community support could be conducted through cleaning the shoreline of a pile of garbage, as in Figure 4 below.

The beach cleaning activity also a tourism interesting activity such as held on Thursday and Friday 06 and October 7, 2016. The activity was conducted by the Pokdarwis of TunuoVillage, followed by groups of youths in the Tunuo village. This shows that people respond positively to the development of tourism in Tunuo village. Therefore the presence of tourism could bring awareness to preserve the environment and maintain the security in the Village Tunuo. Meanwhile, awareness, spirit and motivation is not enough to build tourism, but it also requires knowledge of tourism in terms of technique and travel guiding attraction marketing.

It is a marketing seminar for promotioning tourism attractions based on technology, as a measure to supplement the knowledge of tourism. The researcher developed a website desatunuo.com as a means to market the tourism products in Tunuo village. Meanwhile, the institutional administration of TunuoPokdarwis in the forms of bylaws, logo and stamp, Main Duties and Functions (TUPOKSI) and other administrative needs are accompanied by the researcher. Thus, Pokdarwis can perform its role in accordance with the vision and mission that has been set. On the other hand, the training given to strengthen the capacity of Pokdarwis is travel guiding. The researcher provided training on hospitality to the villagers of Tunuo so TunuoPokdarwiscould understand the values that must be maintained when hosting travelers. Pokdarwis activity continues to run according to the target, the researchers helped marketing tourism attraction by publishing a number of news in the local newspaper so that tourism development efforts in TunuoVillage can be known by the people in North Halmahera.

There is a publication of the activities in Tunuo village in order to market its tourism attractions. Publication on website and local media (Radar Halmahera) remain to be done so that people in North Halmahera can understand the development of tourism in Tunuo village. Additionally, Pokdarwis, in an effort to introduce tourism potential Tunuo village, conduct various efforts based on local wisdom to be developed into tourism products. One is to capture the activity of marine worm (laor) which was conducted in May-April each year, as well as the cooking process to become a special dish called *laor*. Pokdarwis, in product development, also utilize local wisdom in the form of how to produce traditional crafts of a machetes and *salawaku*, how to weave mats or *sosiru*, *bahar*bracelets, *gata-gata* and how to make traditionaldisheslike *halua*, *nasijaha*, *lalampa*, *panada*, *kokodo*, and other dishes as tourism products. Travelers can interact directly with the people in Tunuovillage to learn the process of making art crafts and local specialties as part of the culture and local wisdom.

Figure 2 : Cultural Objects : *Sosiru,Soloi,SapuLidi,Tikar,Procosigi* (Empirical Data, 2016)

Figure 2 is the craft products by the communities in North Halmahera. People in Tunuovillage has the ability to produce art or craft as an identity symbol of local livelihoods. *Sosiru* is a filtering tool made of bamboo, *saloi*is a backpack made of rattan skin, *sapulidi is* made of coconut leaves and shaped into painted ornaments, *tikar* is a mat shelter made of pandanus leaves, and *porocosigi*ismade of woven pandanleaves for rice container. In general, the art and craft is the creation of the women in Tunuovillage who use their spare time to produce the craft based on the buyer's request or hobby. The craft art buyers come from inside or outside Tunuovillage. Meanwhile, the colors and motifs of each craft has its own identity and meaning as a reflection of the lives of individuals and the Boengethnic groups (Hibualamo). The Tunuovillage Pokdarwis develop tourism products based on local wisdom to preserve the local culture.

The livelihood of the people of Tunuo Village as producers of copra shows that there are of barriers of access to economic capital. Therefore, the tourism potential of Tunuo village will allow the Tunuo villagers access to economic capital. In the early stages (startup) of tourism development, the researcher established and accompanying the Awareness Group Travel *(Pokdarwis)* of Tunuo village. In the process of Tunuo village tourism development, there are several obstacles including a lack of understanding of tourism. The same problem is also raised by Aref [3] that the problems in the process of strengthening the capacity of the tourism community is access to decision-making, lack of resources, knowledge about tourism, skills and education, and a sense of belonging to tourism. A further problem is that the direction and development policies is mainly focused on the government resulting in inequality in every tourism community. In the context of Tunuo village, the statement is very relevant. As an emerging area, Pokdarwis of Tunuo Village will always be accompanied by experts and they will provide an understanding of tourism as well as the bottom-up decision-making process, as viewed by Sakata [20]. Additionally, Nault, et al [16] found community participation in tourism development is one way to optimize the control function of the community and the benefits would be the achievement of sustainable tourism. This is in line with the community-based ecotourism development applied in northern Mongolia. According to him, the development of ecotourism communities in remote areas requires the collaboration and support of trusted community leaders and from stakeholders. In the context of Tunuo village, the research is very relevant. The Tunuo village Pokdarwis should cooperate with the village government and local government. In addition, good cooperation also take place between Tunuo village Pokdarwis with traditional leaders and public figures. This facilitates the process of tourism development in Tunuo village.

The decision making processes at the community level done collectively will minimize the potential for internal conflict. Research that outlines some of the problems in the process of decision-making at the community level [6]. According to him, the impetus given to the Tunuo Village Pokdarwis should not be focused on results alone but on the process. His research shows that local concepts need attention, as well as an understanding of the way of life according to local culture, social relationships based on understanding of the local culture, and spiritual elements. It will facilitate the decision making process so as to avoid the controversy that led to the conflict. In the context of Tunuo Village, the concept of local culture is to know the decision making process. The issues raised by community leadersareconsidered as commands based on input from members. This is done by village elders Tunuo named Umeye Gabijin (1920-1928), Dudutu Kinitaroka (1928-1935) and Frans Singgalen (1950-1973) to protect the villagers ofTunuofrom threat of Japanese soldiers during World War II. The decision making process as it persists for generations to form a habit and culture. Nonetheless, the current Tunuovillage Pokdarwis adopt cultural values decision limited to the values of trust based on the understanding the ethnic Boeng communities.

Cultural objects of Tunuo communities can be optimized as tourism products. As the research conducted that they emphasize the utilization of Cultural Resources as a tourism product [17]. The tourism products including cultural heritage sites, cultural landscapes, events or festivals. Research results show that the development of culture as a tourism product can be categorized as a niche market, and on the other hand sustainability of culture-based tourism development must gain control over strategy, the perspective of policy makers, standards, and orientation. This shows that the development of cultural tourism products are not only limited to sales, but also the orientation towards sustainable development. Therefore, the development strategy, the standards set in the development process and the decisions taken by the leadership should also be taken into account. In the context of Tunuo village, product development decisions based on local wisdom is done collectively and coordinated by the head of the Tunuo village Pokdarwis, and supervised by the village government and traditional leaders. In addition, research conducted on the involvement of local communities in the development of cultural tourism show that education and training, interpretation and orientation of tourism management to sustainability, integrated planning, cooperation, control the development of cultural tourism, government participation and stakeholder, diversification of tourism products and markets as well as the cultural capital approach (cultural capital) to support sustainable tourism [11]. On the other hand, Salazar showed that the discourse of community-based tourism in the development of cultural tourismis linked to the conflict between local tour guides with tourists that could affect the profile of the local attractions [22]. Based on this, it can be seen that in the development of tourism products based on local wisdom in the village Tunuo, orientation towards sustainable tourism is important to be balanced with efforts to prevent conflict between tour guides and tourists.

Change of mindset from farming to service industry requires a long process. Similarly, research by Schweinsberg showed an understanding of rural communities to nature in the transition from loggers into protectors of the forest [23]. In the context of Tunuo village, tourism is a new atmosphere in public life. It is characterized by standards set to adapt with the *seven charms*concept with the emphasis on hospitality, while the work as copra producers and sideline fishermen have formed a strong characteristic. In addition, the presence of tourism as an economic growth opportunit, distract people who are involved in fish bombing activities to get involved in preserving the environment and maintaining the security of the waters from fish bombing activities.

It certainly should get the attention from the government as proposed in the National Park of Uganda, that there must be communication between the government and the people living around the conservation area [2]. This would make it is easier for process monitoring and control of environmental safety. Furthermore, they mention that the benefits of community involvement to the conservation area is the ease of process control [28]. Similarly Somarriba and their friends said that community participation in the conservation process enables process control, though government support is needed to support the infrastructure and the needs of the community in the development of ecotourism [27]. Thus, it can be seen that the shift in the mindset of an agricultural society to a service industry requires monitoring and control at every process of development to achieve sustainable tourism.

Communities need to have educational support on entrepreneurship, so that they can be optimal in utilizing tourism sector to earn profit. Hallak et. al argues that local communities need to be supported by the entrepreneurial ability to support the needs of travellers [8]. On the other hand, they found community based tourism development is an instrument to build cooperation [7]. In the context of Tunuo village, tourism is a new thing and therefore the entrepreneurship education is merely on a way of making and marketing of local art crafts. In addition, the Tunuo village Pokdarwis is still in the process of development strategy that anticipates things that are detrimental to the environment due to tourist behavior through the policies on tourism attractions development. As research of showed, the presence of environmental damage due to the behavior of tourists who visit were not well controlled [4]. Therefore it is necessary to prepare preventive action through regulation or policy [13].

Tunuo village community participation can accelerate the development of tourism and village development. Like the research conducted on public participation in the governance Beraban Village attractions Tanah Lot, it showed the presence of accelerated regional development surrounding the tourism locations because of local community participation [18]. Furthermore, research on community-based ecotourism development in Central Kalimantan, shows that community participation and the synergy between the community and attraction managers can simplify the process of tourism development control [14]. Thus, it can be seen that community involvement or community development is one strategy to accelerate the development process, especially in terms of poverty alleviation [29].

IOP Conf. Series: Earth and Environmental Science **156** (2018) 012010 doi:10.1088/1755-1315/156/1/012010

5. Conclusion

The development of tourism in the Tunuo village would be able to encourage local communities to face the problems of livelihood in relation to access to economic capital. People in the village along with the Tunuo village government support the dissemination activities, seminars and trainings on tourism. Moreover, the response of the local community was embodied in *gotong-royong*(collective) activities to clean up the shoreline from trash and invitations *(higaro)* to quit the activity of catching fish using homemade bombs (explosives). On the other hand, the implementation of the concept of seven charms *(saptapesona)* is able to suppress the growth of alcohol consumption to minimize the potential for conflict due to alcohol. This shows that tourism is able to empower the community in order to achieve sustainable development of social, economic and environmental aspects.

6. References

[1] Abdilah, dkk. 2016. Dampak Pengembangan Pariwisata Terhadap Kehidupan Masyarakat Lokal di Kawasan Wisata (Studi Pada Masyarakat Sekitar Wisata Wendit, Kabupaten Malang). Jurnal Administrasi Bisnis Vol 30.

[2] Ahebwa, Wilber Manyisa., et al. 2011. Tourism revenue sharing policy at Bwindi Impenetrable National Park, Uganda: a policy arrangements approach. Journal of Sustainable Tourism, Vol 20 (3) : 377-394.

[3] Aref, Fariborz. 2010. Barriers to community capacity building for tourism development in communities in Shiraz, Iran. Journal of Sustainable Tourism, Vol 19 (3) : 347-359.

[4] Cheng, Tien-Ming., et al. 2012.The influence of place attachment on the relationship between destination attractiveness and environmentally responsible behavior for island tourism in Penghu, Taiwan. Journal of Sustainable Tourism, Vol21(8): 1166-1187.

[5] Dharmawan, A. 2007. Sistem Penghidupan dan Nafkah Pedesaan :Pandangan Sosiologi Nafkah (Livelihood Sociology) Mazhab Barat dan Mazhab Bogor. Jurnal Sosiologi Pedesaan, Vol 1 (2).

[6] Farrelly, Trisia Angela. 2011.Indigenous and democratic decisionmaking: issues from community-based ecotourism in the BoumāNational Heritage Park, Fiji. Journal of Sustainable Tourism, Vol19(7) : 817-835.

[7] Gascón, Jordi. 2012. The limitations of communitybased tourism as an instrument of development cooperation: the value of the Social Vocation of the Territory concept. Journal of Sustainable Tourism, Vol 21(5) : 716-731.

[8] Hallak, Rob., et al. 2012. Examining tourism SME owners'place attachment, support for community and business performance: the role of the enlightened self-interest model. Journal of Sustainable Tourism, Vol21(5) :658-678.

[9] Hangewa, A. 2010. Pengelolaan Wilayah Pesisir Pulau-Pulau Kecil untuk Pengembangan Pariwisata Bahari di Kabupaten Halmahera Utara. [Tesis]. Magister Geografi Universitas Gadjah Mada.

[10] Lahamadi, dkk. 2016. Implementasi Kebijakan DinasPariwisata dan Kebudayaan dalam Pengembangan Objek Wisata di Pulau Kumo (Studi Kasus Pulau Kumo Kecamatan Tobelo Kabupaten Halmahera Utara). Jurnal Administrasi Publik Vol 3 (400).

[11] Loulanski, Tolina & Loulanski, Vesselin. 2011. The sustainable integration of cultural heritage and tourism: a meta-study. Journal of Sustainable Tourism, Vol19(7) : 837-862.

[12] Mahura, dkk. 2010. Analisis Kebijakan Pengembangan Wisata Bahari (Kasus Pulau Tagalaya dan Pulau Kumo di Kabuapten Halmahera Utara). [Tesis]. Program Studi Teknik Perikanan Institut Pertanian Bogor.

[13] Mair, Judith & Laing, Jennifer H. 2013. Encouraging pro-environmental behaviour: the role of sustainability focused events . Journal of Sustainable Tourism, Vol 21(8): 113-1128.

[14] Moch, S &Mialana, P. 2016. Community Based Ecotourism Development of Sebangau National Park in Central Kalimantan Indonesia. Hildiktipari , International Journal of Tourism and Hospitality Study, Vol1 (1): 93-105.

[15] Mokola, dkk. 2015. Optimalisasi Tugas dan Fungsi Bidang Pemasaran dalam Mempromosikan Pariwisata di Kabupaten Halmahera Utara (Studi tentang Komunikasi Pemasaran di Dinas Pariwisata dan Kebudayaan). Jurnal ACTA DIURNA Vol 4 (3).

[16] Nault, Sebastien & Stapleton Paul. 2011.The community participation process in ecotourism development: a case study of the community of Sogoog, BayanUlgii, Mongolia. Journal of Sustainable Tourism, Vol19(6): 695-712.

[17] Peters, Mike., et al. 2011. The resource-based and the marketbased approaches to cultural tourism in alpine destinations. Journal of Sustainable Tourism, Vol19(7) : 877-893.

[18] Rastitiati, J. 2016. The Participation of Beraban Village Community in The Operational Management of Tanah Lot Tourist Destination. Hildiktipari , International Journal of Tourism and Hospitality Study, Vol1 (1): 85-92.

[19] Reihara, Y. 2014. Kinerja Pemerintah Kabupaten Halmahera Utara dalam Mengembangkan Objek Wisata Guna Menunjang Pendapatan Asli Daerah. Vol 1 (3).

[20] Sakata, Hana & Prideaux, Bruce. 2013. An alternative approach to communitybased ecotourism: a bottom-up locally initiated non-monetised project in Papua New Guinea. Journal of Sustainable Tourism, Vol21(6): 880-899.

[21] Salamena, A. 2013. Perancangan dan Implementasi Sistem Informasi Geografis Objek Wisata di Kabupaten Halmahera Utara. [Skripsi]. Teknik Informasi Fakultas Teknologi Informasi Universitas Kristen SatyaWacana.

[22] Salazar, Noel B. 2011. Community-based cultural tourism: issues, threats and opportunities. Journal of Sustainable Tourism, Vol 20(1): 9-22.

[23] Schweinsberg, Stephen C., et al. 2011. Understanding communities'views of nature in rural industry renewal: the transition from forestry to naturebased tourism in Eden, Australia. Journal of Sustainable Tourism, Vol 20(2): 195-213.

[24] Sero, A. 2012. Model PengembanganPariwisataBahariBerbasisMasyarakat di Kabupaten Halmahera Utara. Jurnal Nasional Pariwisata Vol 4 (1).

[25] Sidarta, T. 2002. Dampak Perkembangan Pariwisata terhadap Kondisi Lingkungan, Sosial dan Ekonomi Masyarakat (Studi Kasus Kawasan Pariwisata Sanur, Denpasar-Bali). [Tesis]. Program Pascasarjana Universitas Diponegoro.

[26] Singgalen, dkk. 2013. Analisis Prioritas Pengembangan Pariwisata di Kabupaten Halmahera Utara. [Skripsi]. Program Studi Sistem Informasi Fakultas Teknologi Informasi, Universitas Kristen Satya Wacana.

[27] Somarriba-Chang, Matilde de los Angeles&Gunnarsdotter, Yvonne. (2012).Local community participation in ecotourism and conservation issues in two nature reserves in Nicaragua . Journal of Sustainable Tourism, Vol20(8): 1025-1043.

[28] Strickland-Munro, Jennifer& Moore, Susan. (2012).Indigenous involvement and benefits from tourism in protected areas: a study of Purnululu National Park and Warmun Community, Australia . Journal of Sustainable Tourism, Vol21(1): 26-41.

[29] Yuwanti, S. 2016. Designing Tourism Based Activities for Poor Regions Case of Brebes, Central Java. Hildiktipari , International Journal of Tourism and Hospitality Study, Vol1 (1): 106-116

IOP Conf. Series: Earth and Environmental Science **156** (2018) 012011 doi:10.1088/1755-1315/156/1/012011

Cleanliness Policy Implementation: Evaluating Retribution Model to Rise Public Satisfaction

Surya Dailiati[1], Hernimawati Hernimawati[2], Prihati Prihati[3], Bunga Chintia Utami[4]

[1,2,3,4] Universitas Lancang Kuning, Pekanbaru, 28265, Indonesia
Telp: +620761 53108, Fax: +620761 52248
Email: bunga@unilak.ac.id

Abstract: This research is based on the principal issues concerning the evaluation of cleanliness retribution policy which has not been optimally be able to improve the Local Revenue of Pekanbaru City and has not improved the cleanliness of Pekanbaru City. It was estimated to be caused by the performance of Garden and Sanitation Department are not in accordance with the requirement of society of Pekanbaru City. The research method used in this study is a mixed method with sequential exploratory strategy. The data collection used are observation, interview and documentation for qualitative research as well as questionnaires for quantitative research. The collected data were analyzed with interactive model of Miles and Huberman for qualitative research and multiple regression analysis for quantitative research. The research result indicated that the model of cleanliness policy implementation that can increase of PAD Pekanbaru City and be able to improve people's satisfaction divided into two (2) which are the evaluation model and the society satisfaction model. The evaluation model influence by criteria/variable of effectiveness, efficiency, adequacy, equity, responsiveness, and appropriateness, while the society satisfaction model influence by variables of society satisfaction, intentions, goals, plans, programs, and appropriateness of cleanliness retribution collection policy.

Keywords: Evaluation, Policy, Cleanliness/Solid Waste.

1. Introduction

Policy of cleanliness retribution in Pekanbaru City regulated in Pekanbaru City Regional Regulation No. 10 of 2012 on Service Retribution Garbage/Cleanliness (hereinafter referred to as Regulation of Cleanliness of Pekanbaru City). Regulation of Cleanlinessof Pekanbaru City regulates on objects and retribution, imposition and retribution tariff, collection procedures and cleanliness retribution calculation, provision of inspection and sanctions both administrative sanctions and criminal sanctions. Although the provisions of cleanliness in the Pekanbaru City is already regulated in Regulation of Cleanliness of Pekanbaru City mentioned above, but the realization of this cleanliness retribution does not meet with the target which is expected by the Regional Government of Pekanbaru City. Local Revenue (PAD) of Pekanbaru City that has not reached 100% of the target one of them due to still not optimal of tax retribution receipts of Pekanbaru City. This is due to the lack of evaluation of policy implementation of cleanliness retribution especially implementations of Regulation of Cleanliness of Pekanbaru City. This Regulation of Cleanliness of Pekanbaru City has not been regulated this evaluation issue. According to Dunn as actualized by Fischer (2014: 229) "one of success factor of public policy implementation is the evaluation of public policy itself".

This study was inspired by research such as research by Raffel *et. al.*, (2015) which indicates in general handling and garbage management in London City conducted by way of collecting, transportation, and disposal to the final disposal (TPA) but have not yet completed the existing problem, and research by Muller (2013) which showed that garbage management policy with recycle model can become another alternative to increase local revenue in Berlin City. Additionally, this study also referred to research by Supardi (2010) which concluded that in the implementation of the retribution charged for garbage has a contribution and influence on the effectiveness to increase garbage retribution towards the achievement of regional revenue at Cleanliness and Cemeteries

IOP Conf. Series: Earth and Environmental Science **156** (2018) 012011 doi:10.1088/1755-1315/156/1/012011

Department of Palembang City, and research by Rochdianingrum (2011) with topic potential of cleanliness retribution of household garbage in the Local Revenue in Semarang City have not been explored optimally. Dunn as actualized by Fischer (2014: 229) stated "public policy can be analyzed through a series of intellectual activity conducted in activities process which are politically favored. The political activity appears in a series of activities which includes the agenda preparation, policy formulation, policy adoption, policy implementation, and policy evaluation. While the activity of problem formulation, forecasting, policy recommendations, monitoring, and policy evaluation are activities that more intellectually". Associated with this Dunn's statement, he lack of regulation concerning the evaluation in Regulation of Cleanliness of Pekanbaru City regulating the cleanliness retribution resulting the ineffectiveness of the cleanliness retribution collection in Pekanbaru City.

This research is based on the principal issues concerning the evaluation of cleanliness retribution policy which has not been optimally be able to improve the Local Revenue of Pekanbaru City and has not improved the cleanliness of Pekanbaru City. It was estimated to be caused by the performance of Garden and Sanitation Department are not in accordance with the requirement of society of Pekanbaru City..

2. Research Method

This study used research *mixed method* (research methods mix). Hybrid approach *(mixed method)* or a combination or hybrid and the like. Initially characterized by a variety of definitions, currently developing a definition aimed to unify the various viewpoints ever. The definition is based on *"core characteristic of mixed method research"* is a method that combines a variety of methods, philosophy and orientation of the study design (Indrawan and Yaniawati, 2014: 76).

A mix of research methods approaches that combine qualitative research with quantitative research to be used together in this research activity, in order to obtain more comprehensive data, valid, reliable and objective (Creswell in Yandra, 2017). Reasons for usingresearch methods *mixed methods* considering this study intends to evaluate the cleanliness retribution policies in improving people's satisfaction Pekanbaru city, so that the necessary qualitative and quantitative research as well. Qualitative research is used to elaborate the opinion of the framers and implementers in government Pekanbaru City through interviews, whereas qualitative research is used to get input from people in Pekanbaru by distributing questionnaires as well as to formulate a model of policy evaluation cleanliness Pekanbaru City and models of people's satisfaction policy cleanliness Pekanbaru city.

This study uses a strategy *of concurrent*embedded.This strategy is similar to triangulation models concurrent, because both collect qualitative and quantitative data at the same time. What distinguishes it is this model has a primary method that guide the project and secondary data which has a supporting role in any study procedures. Secondary methods are less dominant / role (either qualitative or quantitative) is plugged in*(embedded)*into a more dominant method (qualitative) (Cresweil, 2010: 8). The reason this study usingstrategy *concurrent embedded* for research is more focused evaluation of the framers and implementers are done qualitatively through interviews, while the evaluation of Pekanbaru City community complements done quantitatively through questionnaires.

This study begins with the efforts of researchers collecting data by interview or observation to understand the various issues in accordance with the object and context to be studied. The search results were analyzed thoroughly in ways that are usually done with a qualitative approach. Full understanding of the research object as a strong foundation to formulate a survey instrument that will be used. In the form of a questionnaire survey instrument as in quantitative methods deployed for further processed to obtain quantitative conclusions for further interpretation.

IOP Conf. Series: Earth and Environmental Science **156** (2018) 012011 doi:10.1088/1755-1315/156/1/012011

2.1 Qualitative Data Analysis

Data analysis is the process of arranging and combining the data into a pattern, theme, category, whereas interpretation is to give meaning to the analysis, explaining the pattern or category, and find the relationship between some of the concepts. Researchers describe the perspective of interpretation is not the truth. Analysis and interpretation of data in qualitative research is basically not the things that go together, both carried out since the beginning of the study. Data analysis was performed so that the data have been obtained will be more meaningful. Data analysis is the process of simplification of data into a form that is easier to read and interpret.

Miles and Huberman (1992: 20) explains the data analysis is a systematic process of searching for and set the record interviews, field notes, and *theriders* othergathered for herding understanding. The analysis involves working with the data, set it up, separating into units that can be managed, combine them, searching for patterns meet important matters and what was known and decide what will be passed on to others.

Analysis of the data used in the research is iterative models of Miles Huberman (1992: 20) that is carried out through three phases, which include the following:

a. data reduction
   data reduction performed in the electoral process, concentration and simplification of data. The data obtained from the field may be the fairly extensive and complex. That's necessary for data reduction. In this case only the data relevant to the study objectives are being taken, while irrelevant discarded. In this data reduction, data is selected only data relating to hygiene policy evaluation in Pekanbaru, while data that is not relevant to the research problems discarded / not used. Data are summarized and been things that are principal only, focused on important data, then look for themes and patterns. Thus the reduced data will provide a clearer picture, making it easier for researchers to gather and add the relevant data further. Data reduction performed in conjunction with the data collection process until completed.
b. Presentation of Data
   After the data is reduced, then the next step is the presentation of data. Through the presentation of the data, the data will be organized, arranged in a pattern of relationships that are easy to understand. Presentation of data presented in the form of narrative text. Presentation of text and narrative form is also one of the characteristics of qualitative research.
c. Conclusion manufacture, Verification and Reflection
   qualitative research method does not use a statistical model (model correlations between variables) as yng used in quantitative research as an instrument to produce the output of the data analyzed. In this process the researcher sebgai key instrument to interpret rationally research on empirical data collected systematically categorized. So as a publisher of research instruments; Liti itself then penje; weld proposed by researchers as the output data is processed in the head (mind) researchers must be rational (reasonable).

2.2 Quantitative Data Analysis (Regression Analysis)

Analysis of the data used in the study is the method of regression analysis. In regression analysis will be developed an *Estimating equation* (regression equation) are mathematical formulas that find the value of the dependent variable of independent value known. Regression analysis was used mainly for forecasting, which in this model there is a dependent variable and independent variables. In practice, the method of regression analysis is often divided between *simple regression* and *multiple regression* (have now, 2012: 405-406). Called *simple regression* if there is only one independent variable, while the so-called *multiple*regression,if there is more than one independent variable.

In this study there is 1 (one) dependent variable, namely the policy evaluation Pekanbaru City cleanliness, and six (6) independent variables, namely effectiveness, efficiency, adequacy, flattening or similarity, responsiveness and accuracy. Under these conditions, the methods of analysis used is multiple regression, the equation can be expressed as follows:

$$Y = \alpha + \beta_1 X_1 + \beta_2 X_2 + \beta_3 X_3 + \beta_4 X_4 + \beta_5 X_5 + \beta_6 X_6 + e$$

Description:

Y = policy evaluation cleanliness Pekanbaru City
X_1 = Effectiveness
X_2 = Efficiency
X_3 = Sufficiency
X_4 = Flattening or similarity
X_5 = responsiveness
X_6 = precision
α = Konstantan Regression
$_{1,}\beta_{2,}\beta_{3,}\beta_{4,}\beta_{5,}\beta_6$ = Regression coefficients
e = *Random Error*

Later in the regression analysis coefficient of determination. The coefficient of determination ($R^{2)}$essentially measures how far the model's ability to explain variations in the dependent variable. Coefficient of determination is between zero and one.R value^2 smallmeans the ability of independent variables in explaining the variation is very limited dependent variables. A value close to the mean of independent variables provide almost all the information needed to predict the variation of the dependent variable (Ghozali, 2006: 83).

2.3 Community Satisfaction Model Regression Analysis

Analysis of the data used in the study is the method of regression analysis. In regression analysis will be developed an *Estimating equation* (regression equation) are mathematical formulas that find the value of the dependent variable of independent value known. Regression analysis was used mainly for forecasting, which in this model there is a dependent variable and independent variables. In practice, the method of regression analysis is often divided between *simple regression* and *multiple regression* (have now, 2012: 405-406). Called *simple regression* if there is only one independent variable, while the so-called *multipleregression,* if there is more than one independent variable.

In this study there is 1 (one) dependent variable, namely the satisfaction of the people, and five (5) independent variables, namely intentions, goals, plans or proposals, programs and accuracy of fee collection policies cleanliness. Under these conditions, the methods of analysis used is multiple regression, the equation can be expressed as follows:

$$Y = \alpha + \beta_1 X_1 + \beta_2 X_2 + \beta_3 X_3 + \beta_4 X_4 + \beta_5 X_5 + E$$

Specification

Y= Satisfaction community
X_1 = intention
X_2 = Destination
X_3 = plan or proposal
X_4 = Program
X_5 = Accuracy policies levy charged cleanliness
α = Konstantan Regression
$\beta_{1,}\beta_{2,}\beta_{3,}\beta_{4,}\beta_{5,}$ Regression coefficient =
e = *Random Error*

Later in the regression analysis coefficient of determination. The coefficient of determination ($R^{2)}$essentially measures how far the model's ability to explain variations in the dependent variable. Coefficient of determination is between zero and one.R value^2 smallmeans the ability of independent variables in explaining the variation is very limited dependent variables. A value close to the mean of independent variables provide almost all the information needed to predict the variation of the dependent variable (Ghozali, 2006: 83).

2.4 Coefficient Determination

The coefficient of determination ($R^{2)}$essentially measures how far the model's ability to explain variations in the dependent variable. Coefficient of determination is between zero and one.R value^2 smallmeans the ability of independent variables in explaining the variation is very limited dependent variables. A value close to the mean of independent variables provide almost all the information needed to predict the variation of the dependent variable (Ghozali, 20106: 83).

3. Findings and Discussion

This research builds on a theoretical framework regarding the cleanliness retribution policy evaluation in an effort to increase the PAD in Pekanvaru City based on the groove of literature review. The theories used in this research are the Theory of Public Administration of Garcia and Khator (2010: 34) as Grand Theory, Theory of Public Policy of Dunn actualized by Fischer (2014: 229) as Middle Range Theory and Theory of Public Policy Implementation of Dunn actualized by Fischer (2014: 229) as Applied Theory.

This study resulted in two (2) models i.e. model of policy evaluation and model of public satisfaction towards the cleanliness policy of Pekanbaru City which can be seen in the following table:

Tabel 1: Model of Policy Evaluation and Model of Public Satisfaction

Model of Policy Evaluation	Model of Public Satisfaction
Y = [illegible]	Y = [illegible]
Noted:	Noted:
Y = Evaluation of Cleanliness Policy of Pekanbaru City	Y = Public satisfaction
X_1 = Effectiveness	X_1 = [illegible]
X_2 = Efficiency	X_2 = [illegible]
X_3 = Adequacy	X_3 = [illegible]
X_4 = Equity	X_4 = [illegible]
X_5 = Responsiveness	X_5 = Appropriateness Policy of Cleanliness [illegible]
X_6 = Appropriateness	

From those above mentioned model, then it can be discussed the evaluation of cleanliness policy of Pekanbaru City influenced or associated with effectiveness, efficiency, adequacy, equity, responsivenness and appropriateness. Appropriateness criteria interrelated with other criteria started from the effectiveness and efficiency, adequacy and equity, and responsiveness. All of which will result cleanliness for cleanliness policy implementation of Pekanbaru City. A policy becomes useless if it can not achieve the expected goals and necessary to find out what the errors of its implementation. For this reason policy evaluation is needed. From that model, appropriateness and responsiveness are two (2) criteria with the largest regression coefficient. This shows that people in Pekanbaru City more expected the appropriateness of the policies and responsively of the implementer officer of cleanliness. The appropriateness is more directed at appropriateness of goal, objectives and targets. Besides also the appropriateness of the fit between Pekanbaru people's habits who throw garbage in the daytime with working schedule garbage collector officers who work at night. This is what must be considered by the Cleanliness and Gardening Office of Pekanbaru City.

IOP Conf. Series: Earth and Environmental Science **156** (2018) 012011 doi:10.1088/1755-1315/156/1/012011

Overall, the influence of criteria/variables of effectiveness, efficiency, adequacy, equity, responsivenness and appropriateness on cleanliness policy evaluation of Pekanbaru City amounted to 69.6%, while the remaining of 30.4%, influenced by other criteria/variables outside the model, namely *Model of Public Satisfaction.* This model of public satisfaction, the evaluation of cleanliness policy of Pekanbaru City, the public satisfaction influenced or associated with intentions, goals, planning or proposal and appropriateness policy of cleanliness retribution collection. Similarly on the model of policy evaluation, in this model of public satisfaction, the appropriateness is also more directed at appropriateness of goal, objectives and targets. Besides also the appropriateness of the fit between Pekanbaru people's habits who throw garbage in the daytime with working schedule garbage collector officers who work at night. This is what must be considered by the Cleanliness and Gardening Office of Pekanbaru City. Overall, the influence of criteria/variables of intentions, goals, planning or proposal and appropriateness policy of cleanliness retribution collection on public satisfaction of cleanliness policy evaluation of Pekanbaru City amounted to 52,7%, while the remaining of 47,3% influenced by other criteria/variables outside the model.

From the above models of policy evaluation and the model of satisfaction of the cleanliness policy of Pekanbaru City, there are influenced by other criteria/variables outside the model. In this case the researchers consider that the criteria/ variables outside the model is public participation. The reason the obligation to manage the garbage is not just the responsibility of the government only. Of course the public concern to participate in carrying out garbage management is very needed because of the presence of garbage in the Pekanbaru City which accumulates and is not managed properly also very disturbing the public convenience.

4. Conclusions

Based on the formulation of the problem, research objectives and the research that has been described above, it can be concluded matters as follows:

1) Evaluation of the effectiveness, efficiency, responsiveness, adequacy, equity and accuracy in order to improve the satisfaction of the people of Pekanbaru is made to the accuracy of precision directed towards policy objectives and the needs of society. Cleanliness or solid waste policy should be able to meet the needs of society. Gone are the days of society should be submissive and obedient to the bureaucracy. Precisely bureaucracy should listen to the aspirations of the people. If it is the habit of throwing garbage Pekanbaru City community during the day, it should have followed people's habits is by picking up trash in the daytime. Changing habits of the people who have been going on for years or even centuries, it is difficult to do. So it should waste collection patterns are changed, which is certainly easier than changing people's habits, the most important policy objective of cleanliness which makes clean Pekanbaru city, especially during the day can be achieved.
2) Factors that affect the cleanliness of the policy implementation by Regulation No. 10 Year 2012 on Service Fees Waste / Hygiene in Pekanbaru City implemented through communication factor, support resources, attitude and disposition of forces, bureaucracy and community support structures. Implementation is done in a coordinated manner between the Department of Hygiene and Pekanbaru City and Regional Revenue Office Pekanbaru. Withdrawal / levy charged under the authority of the Regional Revenue Office Pekanbaru, while cleaning services is the responsibility of the Department of Sanitation and the city of Pekanbaru. However the quality of sanitation services has not been satisfactory, people felt that what was paid at the discretion of cleanliness retribution, not in accordance with hygiene services provided by Pekanbaru City Government. The regulatory process is not optimal yet resulted in the levy charged hygiene is not optimal. For collection by officers not in accordance with established rules of which are still there sanitation fee collection exceeds the established provisions and regulations.

3) Hygiene policy implementation model that can improve people's satisfaction with the city of Pekanbaru is revising the existing policy, improve the effectiveness and efficiency to achieve policy objectives. Revision of Pekanbaru City Regulation No. 10 In 2012, the above is more focused on compliance / accuracy of garbage collection performed by the Department of Hygiene and Pekanbaru city with people's habits in the trash in order to achieve the policy objectives that city of Pekanbaru clean of trash, especially in the afternoon.

5. References

[1] Creswell, John W. 2010. *Research Design Pendekatan Kualitatif, Kuantitatif, dan. Mixed.*Yogyakarta : Pustaka Pelajar

[2] Fischer, F. 2014. *Reframing public policy: Discursive Politics and Deliberative Practices.* Oxford: Oxford University Press.

[3] Garcia, Jern Claude-Zomar and Rew Khator. 2010. *Public Administration in the Global Village.* USA: An Imprint-Greenwood Publishing Group, Inc.

[4] Indrawan, Rully dan Yaniawati, R. Poppy. 2014. *Metodologi Penelitian: Kuantitatif, Kualitatif dan Campuran untuk Manajemen, Pembangunan dan Pendidikan.* Bandung: Refika Aditama

[5] Maman, Abdrurrahman. 2009. *Analisis, Korealasi, Regresi dan Jalur dalam Penelitian.* Bandung: CV Pustaka Setia.

[6] Moleong, Lexy. 2010. *Metode Penelitian Kualitatif.* Bandung: PT. Remaja Rosdakarya.

[7] Muller, Michael. 2013. "Municipal waste management in Berlin". *Berlin Senate Department for Urban Development and the Environment Communication,* Vol. 6.

[8] Raffel, Anthonius F., Arief Rachmansyah, Soemarno, dan Veronica A. Kumurur. 2015. "Priority Scale of Waste Management Policies that Support Environmental Hygiene in City of Ternate". *International Journal of Ecosystem*, Vol. 5, No. 1, pp. 9-21.

[9] Rochdianingrum, Widhi Ariestianti. 2011. "Potensi Retribusi Kebersihan Sampah Rumah Tangga dalam Pendapatan Asli Daerah Studi Kasus di Kota Semarang" *Disertasi* Fakultas Ekonomi Universitas Diponegoro Semarang.

[10] Supardi. 2010. "Efektivitas Peningkatan Restribusi Sampah terhadap Pencapaian Pendapatan Asli Daerah pada Dinas Kebersihan dan Pemakaman Kota Palembang" *Ilmiah,* Vol. I1, No.2, hlm. 1-6

[11] Sugiyono. 2012. *Struktural Equation Modeling: Teori, Konsep dan Aplikasi.* Semarang: Badan Penerbit Universitas Diponegoro

[12] Yandra, A. 2017. Penyelamatan Hak Pilih Warga Perbatasan Jelang Pilkada. *POLITIK, 13*(1).

IOP Conf. Series: Earth and Environmental Science **156** (2018) 012012 doi:10.1088/1755-1315/156/1/012012

Utilization of Cocoa Pod Husk Waste Composting by *Tremella Sp a*nd *Pleurotus Sp* as A Medium to Growth of Cocoa Seedling

Iradhatullah Rahim[1], A Nasruddin[2], T Kuswinanti[3], L Asrul[4], B Rasyid[5]
[1]Faculty of Agriculture, Animal Husbandry, and Fisheries, Universitas Muhammadiyah Parepare, Jendral Ahmad Yani Street 90235 Parepare, South Sulawesi, Indonesia.
Email:iradhat76@gmail.com
[2]Department of Plant Pests and Diseases, Faculty of Agriculture,Hasanuddin University, km 10 Tamalanrea,Makassar 90245, Indonesia.
Email:andinasruddin@yahoo.com
[3]The Institutions of Research and Community Service, Hasanuddin University, km 10 Tamalanrea Makassar 90245, South Sulawesi Indonesia.
Email:koeswinanti@yahoo.com
[4]Department of Soil Science, Faculty of Agriculture, Hasanuddin University km 10, Makassar 90245 South Sulawesi Indonesia.
Email:otheasrul1963@gmail.com
[5]Department of Soil Science, Faculty of Agriculture, Hasanuddin University km 10, Makassar 90245 South Sulawesi Indonesia
Email:burrasyid@unhas.ac.id

Abstract. Cocoa pod husk waste is a problem in the cocoa field, but it potentially as a source of organic matter to improve soil fertility.The paper discuss about the ability of *Tremella sp* and *Pleurotus sp* on producing phytohormone and on degrading cocoa pod husks waste. The research start with isolation, screening, and propagation of rot fungi were collected from decayed cocoa plants. The measurement of IAA is according to the method of Glickman and Dessaux (1995), by addition of L-Tryptophan 0.1 g l^{-1}, whereas the Gibberellic Acid content was measured by using the method of Borrow *et al.,* (1955). Composting process of cocoa pod husks waste was revealed during 40 days. This research showed that the IAA and GA_3 content in compost fermented with *Tremella* sp was higher than treatment with *Pleurotus sp*. Similarly, the result was also observed in the ability of hemicellulose degradation. However, *Pleurotus sp* was capable to produce compost with higher nutrient levels. Compost fermented by rot fungi gave significant effect to the growth of cocoa seedlings. Nevertheless the difference in varieties of cocoa had no effect on growth of cocoa seedlings. Cocoa pod husk waste composted by *Tremella sp* and *Pleurotus sp* gave the significant effect on Leaf Area Index (LAI), Net Assimilation Rate (NAR), Crop Growth Rate (CGR), Root-shoot ratio, and root dry weight of Cocoa seedling.

Keywords: Cocoa, *Tremella sp, Pleurotus sp*

1. Introduction

Cocoa is one of the leading commodities in Indonesia. The development of wide and cocoa production has increased during the last 5 years. In 2007 the land area of cocoa 1,379,280 had increased to 1,677,254 hectares by 2011. However the production tends to decrease 2013 [1], in 2010, 2011, 2012, and 2013 it was successively in tons of 837,900, 712,200, 777,540, and 740,510 [2]. The decline in production is closely related to the level of soil fertility due to low soil organic matter. Organic materials in land area of Indonesia are very low that less 2%. Research by Munandar et.al [3] in cocoa showed that the higher organic matter, the more efficient plant to use of water. Soil organic matter increased to 6.09% causing growth and cocoa production continued to increase in linear. Cocoa plants require minimum levels of organic matter of 3.5% (approx. 2% Carbon) on 0-15 cm depth [3]. Source of organic matter are easily found around us. The cocoa pod husk waste has a composition and nutrient compounds that very potential as a medium growing plants, however it has not been processed optimally.

Water content of cocoa is about 86%, and organic material levels of around 34.6% [4]. According to Didiek and Yufnal (2004), cocoa pod compost has a pH of 5.4, N total of 1.30%, C-organic 33.71%, P_2O_5 0.186%, K_2O 5.5%, 0.23% CaO, and 0.59% MgO. However, the experiment results indicated that cacao pod compost had not significant effect yet to crop growth rate, the high of plants, stem diameter, dry weight, root dry weight, stem + leaf dry weight, and total dry weight of cocoa seed cultivars Upper Amazone Hybrid [5]. It is can be caused by nutrient on the cocoa pod husk has not absorbed by plants. Source of the nutrient in the cocoa pod husk waste can be utilized through the decomposition process. The cocoa pod husk waste, that has decomposed when it was returned to cocoa fields, will be a source of nutrients which play a role in the growth process and the production of cocoa as well as the sustainability of the land. The relationship between organic matter and plant growth can be directly or indirectly. Organic material of natural substrates for saprophytic microorganisms and indirectly provide nutrients for the plants through the activity of the microorganisms [6]. During the decomposition process of organic material, it results organic acids such as fulvic and humic, also phytohormones like Indole Acetic Acid (IAA), Gibberellin (GA), and cytokine. Hormones and organic acids are useful for cocoa plant growth. Saprophytic microorganisms which play a major role in the decomposition process of organic materials are rot fungi. The purpose of this research was to examine the capabilities of producing phytohormone, degradation ability of cocoa pod husk waste by *Tremella sp* and *Pleurotus sp,* and the growth of cocoa seedlings on the cocoa pod composting by *Tremella* sp and *Pleurotus* sp.

2. Method

2.1. Isolation, Screening, and Propagation of Fungi

Isolates of *Tremella sp* and *Pleurotus sp* were obtained from decayed stems of cacao in cocoa plantation in the Bila Village of Pitu Riase, Sidrap district, South Sulawesi, Indonesia. The fruiting body of fungi were stored in the paper bag until it isolated in the laboratory. Pieces of fungal fruit bodies (1 cm x 1 cm in size) were surface sterilized with 70% alcohol, rinsed 2 times with sterile water and placed on to sterile filter paper. Each piece (ø 7 mm) cultivated aseptically on PDA medium and incubated at room temperature. After sub culturing and purification, isolates are then coded according to the kind of fungi [7].

2.2. Measurement of IAA and GA_3 Content

A Ø 5 mm disc of seven days old of rot fungi isolates were incubated on Potato Dextrose Broth (PDB) added L-Tryptophan 0,1 g l^{-1} shaken out with speed 150 rpm/min during 7 days at temperature room. Suspension of rot fungi centrifuged with speed of 5000 rpm for 25 minutes. Five ml supernatant was added 1ml Salkowski reagent (12 gl^{-1} $FeCl_3$ in 429 ml l^{-1} H_2SO_4), following this, it was stored for 24 hours at temperature room in the dark conditions. The quantity test was done by measured absorbance supernatant with spectrophotometer wavelength of λ 535 nm [8]. Supernatant colour change into pink indicates the presence of IAA. IAA concentration measured using a standard curve with the regression equation Y = x + 0.064 0.09, where $R^2 = 0.995$. Equations obtained at one of the dilution of stock solution of IAA range 0 – 5.0 mgl^{-1}. Concentration in the culture filtrate was compared to the standard curve.

The production of GA_3 used isolates from PDA's media taken 5 pieces with cork borer, grown on a medium of PDB, and incubated at room temperature for 7 days. Then, the culture was centrifuged at 8000 rpm for 10 minutes, and it was transferred 15 ml into test tubes and added 2 ml of aqueous solution of zinc acetate. After 2 minutes, added 2 ml of a solution of Potassium Ferro cyanide and centrifuged at 8000 rpm for 10 minutes. Five ml supernatant added 5 ml of hydrochloric acid 30% and incubated at room temperature for 75 minutes. Blanco prepared with hydrochloric acid 5%. Absorbance is measured at a wavelength of λ 254 nm using the spectrophotometer. GA concentration compared to the standard curve of GA_3 (Sigma-Aldrich) in the range of 0.25-2.25 ppm.

IOP Conf. Series: Earth and Environmental Science **156** (2018) 012012 doi:10.1088/1755-1315/156/1/012012

2.3. Composting Cocoa Pod Husk Waste by Fungi

Isolates of *Tremella sp* and *Pleurotus sp* were sub cultured on the dried corn soaked media and sterilized by autoclave temperature 250°C for 1 hour. Isolates on reproduction media incubated at room temperature for 7 days. Cocoa pod husk waste were chopped then mixed with sawdust, and chaff. Compost medium were moisten with water until 30% of RH, spread on plastic sheets up to a thickness of 60 cm pile and covered with plastic tarpaulin. The temperature was measured daily, when it reached 50°C the mix was reversed. Compost were fermented for 40 days.

2.4. Measurement of Nutrient and Lignocellulose Content

Mature cocoa pod husks waste compost were mixed and sieved (Ø 2 mm), separated and weighed then analyzed for it total N, pH, organic C, the ratio C/N, P, Mg, and S according to Soil Research Hall (2009). Measurement of lignin, cellulose, and hemicellulose content was also revealed 40 days after fermentation using the method [11].

2.5. Experiment Design

The study used Split-Split Plot design, consist of the cocoa varieties as main plot:Sulawesi 1 (v1) and Sulawesi 2 (v2). Kind of fungi treatment compost as sub plot: without rot fungi (j0), *Tremella sp* (j1), and *Pleurotus sp* (j2). The use of P fertilizer as sub sub-plot: without P fertilizers (p0), and with P fertiliser (p1). There were 12 treatment combinations repeated in 3 groups, so there were 36units of treatment. Each unit consists of 4 poly-bags. The data were analyzed by using the analysis of kovarian (ANOVA), if the treatment was significant, the advanced test using Tukey Honest Significant Difference (HSD) test for means comparison.

2.6. Data Collection

Observation on the growth of seedlings of cocoa on cocoa pod compost inoculated *Tremella sp* and *Pleurotus sp* measured in 90 days after planted, namely:

a. Leaf Area Index (LAI)[12]:
   LAI = Leaf area total/ plant area
b. Net Assimilation Rate (NAR), by Gardner, 1985:
   NAR = (W2-W1) / (T2-T1) x (ln ld2-ln ld1)/(ld2-ld1).
c. Crop Growth Rate (CGR) (Gardner, 1985) :

$$\overline{NLD} = \frac{1}{T2 - T1}\int_{T1}^{T2} Id/w = \frac{Id2 - Id1}{W2 - W1} \times \frac{\ln W2 - \ln W1}{Id2 - Id1} \text{ cm2/mg}$$

Where: ld2 = leaf area in the end of experiment (63 days)
ld1 = leaf area in first measurement (42 days)
W2 = Total dry weight in the end of experiment
W1 = Total dry weight in the first measurement.
T1-T2 = Period of growth (21 days).

3. Results and Discussion

3.1. Morphological characterization of Tremella sp and Pleurotus sp

The morphological characterization of two kinds of rot fungi isolated from decayed cacao stem showed on Figure 1.

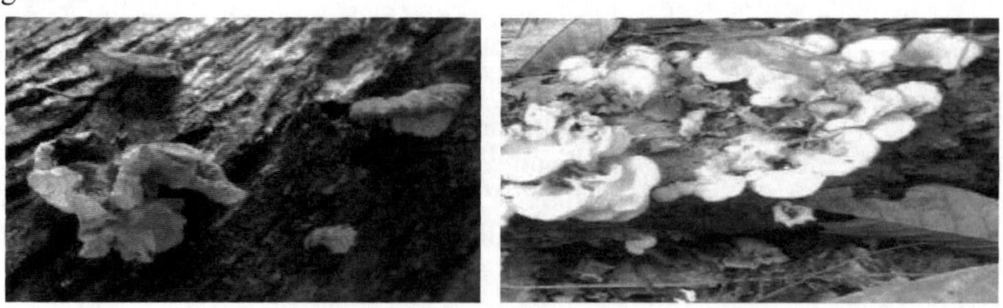

Figure 1. Fruiting bodies of fungi *Tremella sp* (left) and *Pleurotus sp* (right) collected from decayed cocoa stem.

Tremella sp have orange-coloured fruit body, thin, and dry. The edge of the cap rolled and attached on one side on a wooden rod. The upper surface of the hood is darker than the surface of the bottom. The rot fungi is included in the division of Basidiomycota, sub divisions of Agaricomycotina, Tramellomycetes class, order Tramellales [13]. The fruiting body is yellow, orange, and brown. Variations in the intensity of the colour of fruit body could be caused by variations in light intensity by carotenoid [14]. *Pleurotus sp* is also known as oyster mushrooms. Have the stalk and a thick covering of flesh and soft. The surface is smooth and feels feathery soft. The spores are white to creamy. The diameter of basidiospores were in 8.7 – 11.2 × 2.3 µm, cylindrical, thin walls, and soft [15]. This fungi has the ability to kill the nematode known toxin called trans-2-decenedioic acid [16] and linoleic acid [14].

3.2. Phytohormone Production

Tremella sp and Pleurotus sp has the ability to produce secondary metabolites in the form of the hormone grows IAA and GA_3. Measurement results showed concentrations of the phytohormones generated *Tremella sp*was higher than *Pleurotus sp*. The Figure 2 clearly showed thatconcentrations of IAA and GA_3 produced by *Tremella sp* was 2,444 μgL^{-1} and 4,111 μgL^{-1}, while the IAA and GA_3 produced by *Pleurotus sp* respectively are 1,794 μgL^{-1} and 2,551 μgL^{-1}. Hormone GA, such as IAA, is very useful for the process of growth and physiology of plants. This process includes the seed germination, seedling emergence, growth of stems and leaves, the flower induction and growth of flowers and fruit [17],[18],[19]. Challenged and was also involved in root growth, abundance of hair roots, the inhibition of differentiation of flower buds on the island, setting up the generative and vegetative bud dormant, and delaying senescence in many organs in various plant species [19].

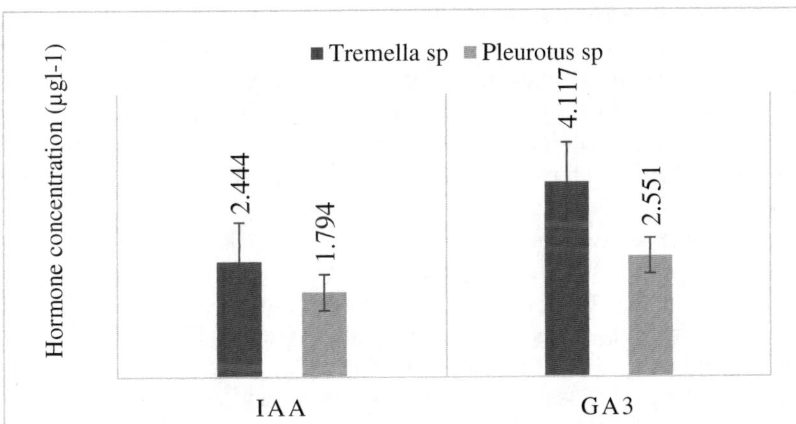

Figure 2. Production of IAA and GA_3 by fungal rot *Tremella sp*and *Pleurotus sp.*

3.3. Cocoa Pod Husk Waste degradation by Tremella sp and Pleurotus sp

Cocoa pod husk waste fermented for 40 days with isolates *Tremella sp* and *Pleurotus sp* changing colours and textures become more dark and smooth.

Figure 3. Degradation rate of Lignocellulotic of cocoa pod husk composted by *Tremella sp* and *Pleurotus sp*.

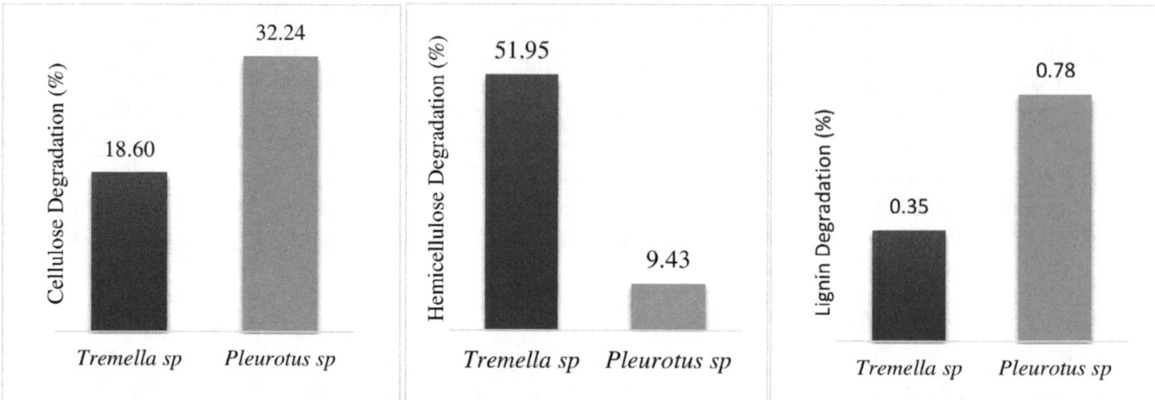

This is due to the activity of the microorganisms. The compost is organic matter that has decomposed [6], referred to as the dark topsoil and quite stable, and is defined as a complex of ligno-protein or a complex of amino acid-lignin. Polyphenols compounds is the result of the decomposition of compound lignocellulose and Quinones became the next reaction with amino acid-forming compounds, or humicfulvic dark (black) [20].

The cocoa pod decomposed by rot fungi became the major constituent components of plant cell component, namely cellulose, hemicellulose, and lignin. The components obtained from results of photosynthesis, i.e. glucose. Cellulose and hemicellulose were formed of glucose units. Whereas, lignin is composed of fenilpropan units. The percentage degradation of cellulose and lignin were the highest on the inoculated composting Pleurotus sp., i.e. each 32.24% and 0.78%. However, the degradation of hemicellulose was the highest on compost *Tremella sp* i.e. 51.95% (Figure 2). According to Brady (1984) [21], dried biomass contains 60% carbohydrates, 1-5% sugar and starch, 10-30% of hemicellulose, 20-50% cellulose, 10-30% lignin, 10% protein, and 1-8% fat, wax, and tannins. Lignocelluloses is the main component of woody plants, consisting of 40% cellulose, hemicellulose 20-30%, and the 20-30% lignin [22].

This research showed that hemicellulose has the lowest levels compared to cellulose and lignin. Hemicellulose is easier than degraded other lignocellulosic components in biomass [23]. Rot fungi which is inoculated on the skin of the fruit of the cocoa has different capabilities degrade hemicellulose. Hemicellulose degradation by highest isolates *Tremella sp* and low on *Pleurotus sp* isolates. Hemicellulose is a polysaccharide that is an important part of cell walls and absorbs water with strong [24]. Hemicellulose enzymes like xylanase and beta-D-xylosidase is a polymer of sugar pentose. Enzyme lignocellulosic of boletus is bio-an important commercial product used in a variety of industries and applications including environmental chemicals, fuel, food, wine, brewing and animal feed, textiles and laundry, pulp and paper, agriculture and bioremediation [25]. This application is an indication of *Tremella sp* on the pulp and paper industry waste.

Cellulose is a component in the cell wall of the plant as micro fibril. Biological treatment causes changes in microstructures of cellulose. Rice straw on the crystalline index decreased from 44% to 15% after application of P. Crysosprorium [23]. In this study, *Pleurotus sp* degrades cellulose cocoa pod husk waste to 32.24%. Lignin is the hardest component to degrade. This can be seen from the low percentage of lignin degradation cocoa fruit skin than other components. Many factors affect the ability of rot fungi degrade lignocellulosic components. These factors, among others, the difference in strain of rot fungi and substrate. The research shows *Pleurotus sp* has the ability to degrade lignin from straw, but not able to degrade lignin in birch hardwood or soft wood pine [26]. Likewise, in this study, *Pleurotus sp* has the ability to degrade hemicellulose lowest cocoa pod. But the highest on the degradation of cellulose compared *Tremella sp*.

The decomposition of cellulose and hemicellulose by white rot fungi with the same speed, whereas lignin break down more quickly. Cowling (1961) describes a hypha of rot fungi first attack the fingers or vessels or directly penetrate the walls of the cells. The enzyme helps the penetration. The growing hypha in the cavity of the cell degrades the secondary wall from within and beyond the walls of the tertiary outwards. Materials produced from decomposition are complex and can be directly absorbed hypha [24].

3.4. Nutrient content of the Cocoa pod Compost Decomposes Tremella sp and Pleurotus sp

Nutrient levels of cocoa pod compost inoculated with rot fungi shows the content of the nutrient was able to exempt by microbes. Nutrient content of the cocoa pod is quite high, and reported that 61% of total cocoa fruit nutrients stored in the peel[27]. Table 1 shows the nutrient content in the compost inoculated *Pleurotus sp.* higher than compost *Tremella sp* on 40 days of fermentation, except Mg and S. Nutrient content of compost, however without the rot fungi much lower than both. This shows that there is the ability of *Tremella sp* and *Pleurotus sp* enhance the availability of nutrients in composted cocoa pod husk waste. Nutrient content of compost shows categories are up to very high. Nutrient analysis of compost on this experiment shows moderate to total N content is high, very high P_2O_5, K_2O similarly to enter categories is very high. These categories are retrieved from the level of nutrient elements[28]. The C-organic content of compost on the study also fall into the category of very high.

Nevertheless, the results are higher, namely 33.76% C-organic and N 0.22% after cocoa pod of *Pleurotus pulmonarius* inoculated and incubated for 30 days [33]. Cocoa pod compost pH ranges between 6.89 and 7.69. The reaction of organic material effect on nutrient availability required of microbes. General microbes thriving and active at neutral pH to alkalis (6.5 – 8.5), whereas the process of mineralized optimum at pH around 7.0 [24]. Magnesium and Sulphur is a nutrient that is released in the process of microbe mineralized in addition to N, P, K, and Ca. Mg is an important element as a constituent of chlorophyll and function as cofactors in enzymes. While, S was the instrumental in the formation of disulphide bonds between protein chains [29].

Nutrient content of cocoa pod compost inoculated *Tremella sp* and *Pleurotus sp* has meet the standards laid down by the compost [30], which according to the Indonesian National Standard 19-7030-2004, compost quality standards contain minimum N 0.2%, while the cocoa pod compost generated has rates of N of 0.65-0.60%. Similarly, the levels of P_2O_5 at least 0.1, whereas compost is generated contain P_2O_5 of 4.65% – 6.70%. Standard quality compost for K_2O and C-organic minimum each 0.2% and 9.8% respectively. Nutrient of K_2O and C-organic cocoa pod inoculated composting *Tremella sp* and *Pleurotus sp* respectively range 0.71 - 0.59% and 16.91 – 17.58%. Similarly, the requirement of INS compost Mg levels maximum is 0.6%, while cocoa pod compost has the highest levels of Mg 0.32%. This shows the resulting compost on research is quite standards compliant nutrient quality compost which determined the National Standardization Agency.

Table 1. Nutrient content of Cocoa Pod decomposed by rot fungi after 40 days of fermentation.

Nutrient Content	Without Fungi	*Tremella sp*	*Pleurotus sp*
pH	8.02^{a}	8.21^{a}	8.12^{a}
C-organics (%)	10.36^{a}	16.91^{b}	17.58^{b}
N-total (%)	0.50^{a}	0.60^{a}	0.65^{b}
C/N	24.00^{a}	25.00^{a}	26.00^{a}
P_2O_5 (%)	4.65^{a}	5.84^{ab}	6.70^{b}
K_2O (%)	0.33^{a}	0.59^{ab}	0.71^{b}
Mg (%)	0.10^{a}	0.32^{ab}	0.28^{b}
S (%)	0.09^{a}	0.22^{b}	0.19^{b}

Note: Number followed by the same letter in the same column are not significantly different according to Tukey HSD at 95% confidence level.

3.5. Growth of Cocoa Seedling

Application of compost is fermented by rot fungi as the medium to growth of two varieties cocoa seedling show in Table 2.

Table 2. Growth analysis two varieties of cocoa seedling in the medium of cocoa pod husk waste inoculated by rot fungi

Growth Analysis	Variety		Cocoa pod husk waste: Without Fungi	Tremella sp	Pleurotus sp
LAI	Sulawesi 1	Without P	1.21^{cA}	0.83$^{a\,B}$	0.88$^{a\,A}$
		With P	1.09^{bcB}	1.15$^{d\,A}$	0.80$^{a\,A}$
	Sulawesi 2	Without P	0.83^{aA}	0.86$^{a\,A}$	0.92$^{ab\,A}$
		With P	0.93$^{ab\,A}$	0.79$^{a\,A}$	0.86$^{a\,A}$
Root-shoot ratio (g)	Sulawesi 1	Without P	2.90^{bA}	3.96$^{f\,B}$	3.06$^{c\,A}$
		With P	3.69^{dA}	2.22$^{a\,A}$	4.02$^{f\,A}$
	Sulawesi 2	Without P	3.91$^{e\,B}$	3.75$^{de\,A}$	3.04$^{c\,A}$
		With P	2.54^{bA}	3.59^{dA}	4.65$^{g\,B}$
Root dry weight (g)	Sulawesi 1	Without P	0.35^{bA}	0.31$^{ab\,B}$	0.42$^{c\,A}$
		With P	0.33$^{ab\,A}$	0.46$^{c\,A}$	0.29$^{a\,B}$
	Sulawesi 2	Without P	0.33$^{ab\,B}$	0.32$^{ab\,A}$	0.33$^{b\,A}$
		With P	0.47^{cA}	0.32$^{ab\,A}$	0.35$^{a\,A}$
NAR ($g.m^{-2}\,day^{-1}$)	Sulawesi 1	Without P	0.51^{aA}	0.52$^{a\,A}$	0.52$^{a\,A}$
		With P	0.52^{aA}	0.51$^{a\,B}$	0.52$^{a\,A}$
	Sulawesi 2	Without P	0.52$^{a\,A}$	0.52$^{a\,A}$	0.52$^{a\,A}$
		With P	0.52^{aA}	0.51$^{a\,B}$	0.51$^{a\,A}$
CGR ($g.m^{-2}\,day^{-1}$)	Sulawesi 1	Without P	0.66^{aB}	0.82$^{a\,B}$	0.87$^{b\,A}$
		With P	0.71$^{ab\,B}$	0.70$^{ab\,B}$	0.72$^{ab\,A}$
	Sulawesi 2	Without P	0.74$^{ab\,B}$	0.59$^{a\,B}$	0.74$^{ab\,A}$
		With P	0.76$^{ab\,B}$	0.66$^{a\,B}$	0.63$^{a\,A}$

Notes: Number followed by the same letter in the same column (capital letters)and the same row (lower case letters)are not significantly different according to Tukey HSD at 95% confidence level.

Table 2 shows the highest LAI of cocoa seedling varieties Sulawesi 1 on compost *Tremella sp.* The production of IAA and GA_3 hormone by *Tremella sp* was higher in comparison with those from *Pleurotus sp.* (Figure 1). The hormone has essential role in the growth and development of roots and shoots, even in stress of conditions. Research showed sorghum seeds germination rate and seed of wheat reached 75% with the application of IAA produced by *F. oxysporum,* and stress on salinity[35]. *Tremella sp* compost application on cocoa seedling varieties Sulawesi 1 also showed the highest root dry weights. Compost derived from *Tremella sp* induced the production of GA_3 higher than compost of *Pleurotus sp.* (Figure 1). Acid forms lateral roots reviving [12]. The existence of the rooting ability of plants absorb pushing nutrient around the plant. This is closely related to the ability of seedlings of cocoa obtained the nutrient for growth. Mobilization of N, P, and K of compost to the system increases with the presence of rooting materials topsoil. This led to the high rate of NAR on the compost by*Tremella sp* on varieties Sulawesi 1 (Table 2). Figure 2 clearly depicted that the type of type of compost has a good capability in degrades cellulose and hemicellulose. The ability of these elements will liberate degrades the C, H, and N were needed as a source of energy for the microorganisms.

Generally, the different varieties of cocoa seedlings do not cause physiological response differences. It can be seen from the difference between the values of their CGR which is not much different. CGR correlation with age of the plant is dry weight increase of plants per cm^2 of area per day with a period of 3 weeks. Value added is measured at the time of growth (formation of vegetative parts) cocoa plants on the compost and giving of various media P.The highest CGR in cocoa seedling varieties Sulawesi 1 on the medium compost *Pleurotus sp.* Nutrient analysis of *Pleurotus sp* compost indicated a higher nutrient content than compost *Tremella sp.* It can be seen at Table 1. The content of N, P, and K were higher on compost *Pleurotus sp.* which needed by crops. N is an important element of the constituent amino acids, nucleotides, and nucleoprotein, as well as essential for the cleavage and the enlargement of cells for growth [12]. P is one of the essential macro nutrient required in carbohydrate metabolism. Preparation of sucrose and hexoses requires high energy phosphate required leaf cells [32]. K plays an important role in photosynthesis because it directly increases the growth and yield of photosynthesis from the leaves translocation [12].

However, the nutrient contained in the organic material can be used only when in the form provided. The decomposition of organic materials in microbiology is an important step to release the nutrients in remaining organic material in a form that can be utilized plant [6]. Growth of shoots hypha or mycelium to cause physical pressure coupled with spending on an enzyme that dissolves cell walls of the wood chain [32]. Some enzymes degradation organic materials produced by *Pleurotus eryngii, P. ostreatus*, and *Bjekanderaadusta* [33]. Nutrient content of compost by*Pleurotus sp.* showed the highest of shoot-root ratio on cocoa seedlings. Shoot-root ratio showed how big the result of photosynthesis that accumulated on the parts of plant organ. This shows the ideal plant growth that reflects the process of absorption of nutrient elements. Dry weight through the process of photosynthesis more translocation to the section of heading from to plant roots [12]. The researched that given 100 g cocoa pod husk per poly-bag increased the shoot-root ratio of cocoa seedlings [34].

The difference varieties do not give the significant affect to growth of cocoa seedling. Sulawesi 1 and Sulawesi 2 actually has long been known by cocoa farmers in Sulawesi Island, about 30 years ago. Both these varieties released in 2008 as the superior varieties. Sulawesi 1 which was also known as PBC 123, was optimal production in the fifth year after planting with the potential production of 1.8-2.5 tonnes/ha. It was quite tolerant of pests of Cocoa pod borer and diseases of the Vascular Streak Dieback (VSD). Morphology of the grooves are less assertive, rounded fruit shape, fruit tip obtuse, base fruit blunt without bottlenecks, seasonal harvest, harvest time is long, the young leaves colour is red maroon, a young fruit of brownish red colour, the colour of a ripe fruit is orange, branches that form leading up to the top.

Sulawesi 2 variety also known as BR 25, potential production of clones around 1.8-2.7 tonnes/ha in the fifth year, it is quite tolerant of VSD's attack. The grooves of the fruit is clearly, pointed fruits, bear fruit almost all year long, rough skin surfaces, colour flus was red yellow, harvest time is short, branching a lot more lead laterally. The colour of the fruit ripe is orange, fruit resembling a blunt and the base of the neck of the bottle [35].

4. Conclusions

Tremella sp had the higher levels of IAA and GA_3, higher than *Pleurotus sp*. The cconcentrations of IAA and GA_3 produced by *Tremella sp* were 2,444 μgL^{-1} and 4,111 μgL^{-1}, while the IAA and GA_3 produced by *Pleurotus sp* respectively are 1,794 μgL^{-1} and 2,551 μgL^{-1}. It was similarly with the ability to degrade hemicellulose on the cocoa pod. However, *Pleurotus sp* capable of producing compost with higher nutrient levels and had the ability of degradation of cellulose and hemicellulose higher than *Tremella sp.* The use of compost fermented with rot fungi medium of cocoa seedling has the significant effect to the LAI, NAR, CGR, root-shoot ratio, and root dry weight of cocoa seedling, compared the compost without fungi. Nevertheless the difference in varieties of cocoa had no significant effect on growth of cocoa seedlings.

5. References

[1] Asrul, L., 2013. Agribusiness of Cacao. Media Bangsa Publisher, Jakarta. ISBN 978-602-77570-7-3 pp.3-8.

[2] Central Berau of Statistic, 2014. Plantation Production According To Provincial And Plant Type. http://www.bps.go.id/linkTabelStatis/view/id/1672.

[3] Munandar, D.E, Abdoellah, S., Mulyanto, D., Soekodarmodjo, S., dan Maas, A., 1995. The Effect of Organics Matter and Water Potential to Growth of tCacao. Pelita Perkebunan. Vol.11 No.3Desember 1995:168-180.

[4] Soedarsono, Abdoellah, S., Sulistyowati, E., 1997. Spreading of Cocoa Husk as A Source of Soil Organis Matter and its Effect on Cocoa Production. Pelita Perkebunan. Vol.13 No.2 August 1997.

[5] Sudirja, R., Solihin M.A., Rosniawaty, S., 2005. Research Report Influence Of Cocoa Skin And Kascing Compost Against Repair Some Chemical Properties FluventicEutrudepts. UniversitasPadjadjaran, Bandung.

[6] Rao, S., 1986. Soil Mikroorganism and Plant Growth. Oxford & IBM Publishing CO.

[7] Choi, Y.W., Hyde, K.D. and Ho, W.H. *Single spore isolation of fungi*. Fungal Diversity 1999; 3: 29-38.

[8] Glickman E and Dessaux, Y, 1995. A Critical Examination Of The Specificaty Of The Salkowski Reagent For IndolicCompoumds Produced By Phytopatogenic Bacteria. ApplEnvironMicrobial 1995; 61:793-796 (Accessed on October 2015). (Achttps://www.ncbi.nlm.nih.gov/pmc/articles/PMC1388360/)

[9] Borrow, P.W.B., V. E. Chester, P. J. Curtis, H. G. Hemming,, Catherine Henehan, E.G.J., P. B. Lloyd, I. S. Nixon, G. L. F. Norris And, Radley, M., 1955. Gibberellic Acid, A Metabolic Product Of The Fungus Gibberella Fujikuroi : Some Observations On Its Production And Isolation. J. Sci. Food Agric 6. DOI 10.1002_jsfa.2740060609.

[10] Soil Research Hall, 2005. Technical Chemical Analysis Of Soil, Plant, Water, And Fertilizer. Agricultural Research And Development Agency, The Department of Agriculture, Jakarta.

[11] Van Soest, P.J., 1976. New Chemical Methods for Analysis of Forages for The Purpose of Predicting Nutritive Value. Pref IX International Grassland Cong.

[12] Gardner, F.P., R.B. Pearce, R.L.Mitchel, 1985. Physiology of Crop Plants. The Iowa State University Press.

[13] Chen C.J,1998. Morphological And Molecular Studies In The Genus Tremella. Berlin: J. Cramer:. 225.

[14] Webster J., dan Weber., R.W.S., 2007. Introduction to Fungi. Third Edition. Cambridge University Press.

IOP Conf. Series: Earth and Environmental Science **156** (2018) 012012 doi:10.1088/1755-1315/156/1/012012

[15] Junior, N.M., Asai. T., Capelari, M., Meirelles, L.D.P., 2010. Morphological and Molecular Identification Of Four Brazilian Commercial Isolates Of *Pleurotus* Spp. And Cultivation On Corncob. Braz. arch. biol. technol. vol.53 no.2 Curitiba Mar.
[16] Kwok, O. C. H., Plattner, R., Weisleder, D. & Wicklow, D. T.,1992. A Nematicidal Toxin From*Pleurotus Ostreatus* NRRL 3526. Journal of Chemical Ecology, 18: 127-136.
[17] King RW, Evans LT (2003) Gibberellins And Flowering Of Grasses And Cereals: Prising Open The Lid Of The "Florigen" Black Box. Annu Rev Plant Physiol Plant MolBiol 54:307–328. 10.1146/annurev.arplant.54.031902.135029.

[18] Pharis RP, King RW (1985) Gibberellins and reproductive development in seed plants. Annu Rev Plant Physiol 36:517–568. DOI: 10.1146/*annurev*.pp.36.060185.002505.
[19] Bottini, R., F.Cassan, Piccoli, P., 2004. Gibberellin production by bacteria and its involvement in plant growth promotion and yield increase. Applied microbiology and biotechnology 65: 497-503.DOI 10.1007/s00253-004-1696-1.
[20] Stevenson, F.J. 1994. Humus Chemistry, Genesis, Composition, Reaction. John Wiley & Sons. New York.
[21] Hanafiah, K.A., 2005. Fundamental of Soil Science. Publisher by PT RajaGrafindo Perkasa, Jakarta.
[22] Hasan, H.A.H, 2002. Gibberelin and Auxin Production by Plant Rot-Fungi and Their Bio Synthesis Under Salinity-Calicium. Rostlinna Vyroba, 48. 2002 (3): 101 – 106.
[23] Isroi, 2007. Composting the Waste of Cocoa. Paper presented at the training of Trainers the Cultivation of coffee and cocoa. Research Center for Coffee and Cocoa, Jember.
[24] Musa., Y, danNasaruddin, 2013. Plant Physiology. Masagena Press. Makassar.
[25] Tuomela, M., M.Vikman, A.Hatakka, M.Itavaara., 2000. Biodegradation of lignin in a compost environment: a review. Bioresource Technology 72:169-183.
[26] Hatakka, A.I., 1993. Pretreatment of Wheat Straw By White Rot For Enzymatic Saccharification Of Celluse. Eur J. Appl.Microbiol.Biotechnol 18: 350-357.
[27] Directorate General of plantation, 2009. Technical Manual Cultivation of Cocoa (*Theobroma cacao L.*). Publisher by Department of Agriculture Republic Indonesia, Jakarta.
[28] Roskarman, A.,Yuwono, N.W., 2002. Science of Soil Fertility. Publsher by Kanisius, Jakarta. ISBN 978-979-21-0468-4.
[29] Adenikum, C.O, and Dada, O.J. (2013). Biodegradation of Three Agricultural Wastes by a white-Rot Fungus, Pleurotus pulmonarius (Fries) Quelet. Journal of Nature and Science, 11, 19-25.
[30] Nasional Standardization Agency, 2004. Specifications and Domestic Organic Waste Composting. SNI 19-7030-2004.
[31] Isroi, R. Millati, S.Syamsiah, C. Niklasson, M.N. Cahyanto,K.Lundquist, M.J. Taherzadeh, 2011. Biological pre-treatment of lignocelluloses with white-rot fungi and its applications: a review. BioResources 6(4), Nov: 1 – 23.
[32] Saraswati R., Santoso, E., Yuniarti, E., 2006. A degraded Organism of Organic Materials in Biological Fertilizer and Organic Fertilizer . (eds) R.D.M. Simanungkalit, Suriadikarta D.A., Saraswati R, Setyorinis D, Hartatik W. BadanBesarLitbangSumberdayaLahanPertanian. BadanPenelitiandanPengembanganPertanian. Jakarta.
[33] Lankinen, P. 2004. Ligninolytic Enzymes of TheBasidiomycetous Fungi *AgaricusBisporus* And *PhlebiaRadiata* On Lignocellulose-Containing Media. Academic Dissertation in Microbiology.http://www.u.arizona.edu/~leam/lankinen.pdf. (Accessed on August 17, 2015).
[34] Yoseva,S., Ardian, C.Mariana., 2013. Compost Cocoa Pod Husk Utilization On The Growth Of Hybrid Cocoa Seedlings (*Theobroma cacao* L). J. Agrotek. Trop. 2 (1): 23-27.
[35] Hasan, H.A.H, 2002. Gibberelin and Auxin Production by Plant Rot-Fungi and Their Bio Synthesis Under Salinity-Calicium. Rostlinna Vyroba, 48. 2002 (3): 101 – 106.

Controlling DC-DC Buck Converter Using Fuzzy-PID with DC motor load

Jumiyatun Jumiyatun[1] and Mustofa Mustofa[2]

[1]Electrical Engineering Department, Tadulako University, Palu, Indonesia
Email: jum@untad.ac.id
[2]Mechanical Engineering Department, Tadulako University, Palu, Indonesia
Email : mustofa@untad.ac.id

Abstract: Circuit of DC-DC converter is a device converting from remains DC input voltage to variable output voltage. The components used for switching converter can be MOSFET, IGBT, Thiristor, GTO and others. This paper discusses DC-DC buck converter to reduce the voltage by using MOSFET switching that is used to control the output voltage of buck converter remain constant with DC motor load. To control the output voltage the buck converter remains constant and stable, it is required proper control to regulate the duty cyclenya. The methods are combination of conventional control methods namely Proportional Integral Derivatives (PID) and intelligent control of Fuzzy logic. It was simulated by using MATLAB software. The simulation results show the combination of PID control and Fuzzy Logic can improve the performance. Small error is almost zero and the output voltage of the buck converter remains constant even though the motor load is changed.

Keywords: Buck Converter, fuzzy PID kontroller, Motor DC

1. Introduction

To get an output voltage that has a high quality and efficiency, it usually uses a switching system rather than linear systems that have power losses. This paper discusses the control of output voltage DC-DC buck converter which serves to decrease the voltage till remain constant, when DC motor load is altered using MOSFET switching. Combination of PID control and Fuzzy logic are used that regulates their cycle duty. On switching converter,MOSFET, IGBT, Thiristor, GTO and others components are applied. PID method is a conventional control that has been widely used both for research and applications in the industry. While *Fuzzy like PD* is a fuzzy control intelligent whose behavior resembles PD.

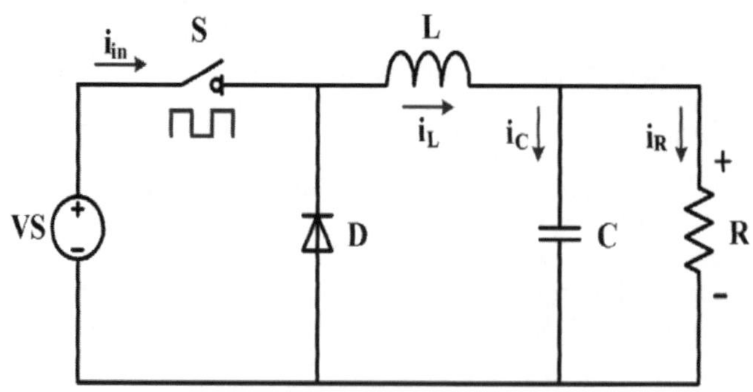

Figure 1. Buck converter circuit

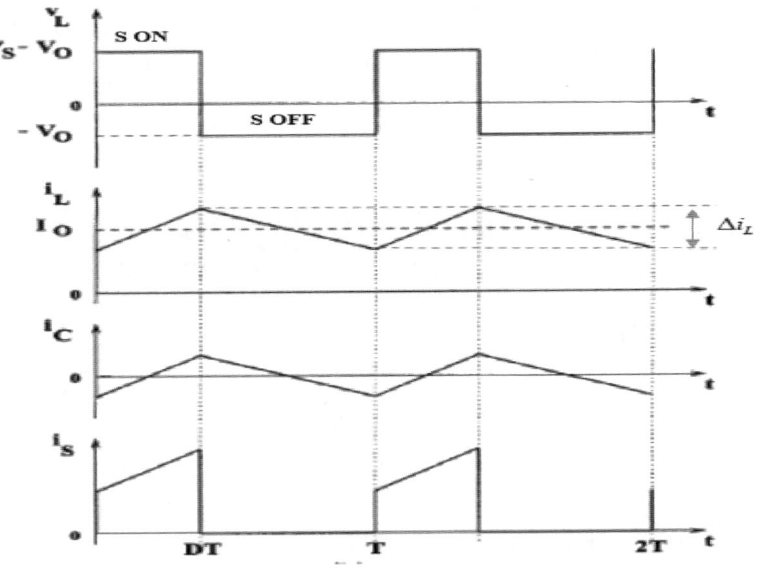

Figure 2. Wave response of buck converter

Some studies that have been done related to buck converter control.Amit, et al. [1] discusses speed control driven by FOPID DC motor to see performance and robustness orvarying loads. Bhagyashri et al. [2] figure out load changes and input voltage reducing efficiency and power of the converter. The function of fuzzy control is to maintain the output voltage constantly even though the input and load voltages are changed in fast.Sutedjo et al. design construct DC-DC converter module with PI controller. It can control the output voltage of buck and boost converter circuit to be stabler even with the condition of input voltage is not balanced, when the circuit works [3].Jumiyatundkk., compare two methods; conventional PI method and Fuzzy-PI Logic to control the output voltage of buck converter remaining constant. Controlling using Fuzzy-PI controller is better performance and faster response than conventional method [4].

2. Method.

In this study, DC-DC converter is designed by determining the correct parameters value of the main components of buck converter, namely inductor (L), Capacitor (C), Resistor (R), duty cycle, and DC motor instead of R load. During simulation, combination of Fuzzy logic control and PID performs better output steady state response.

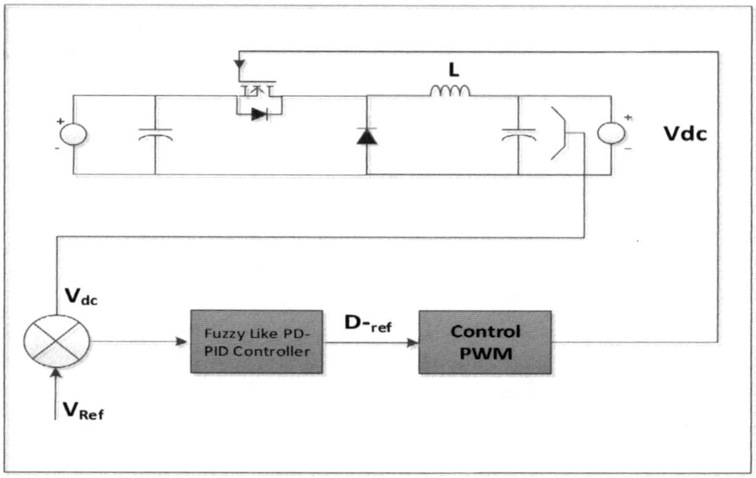

Figure 3. System diagram

As can be seen in Fig3, the system diagram above the buck converter receives the input and current voltage from the DC source, then converts the output voltage lower than the input voltage and the result is fed to the load. Thus, the voltage that flows into the load can be valued according to the set point that has been determined and then output from buck converter in feedback to control.

2.1. Buck converter [5]

In designing the system, the parameter values of each component of the buck converter as seen in Table must be set so that the whole system runs well.

Table 1. Parameters of buck Conventer

Parameter	**Value**
Input voltage (Vin)	500 Volt
Output voltage (Vo)	200 Volt
Ripple (ΔV)	5%
Arus Output (Io)	12 A
Ripple Arus (ΔI)	10%
Frekuensi Switching	10 KHz

- Determine of duty cycle values (1)

$$Son = Soff$$

$$(\text{Vin} - \text{Vo})\text{ton} = \text{Vo.toff}$$

$$\text{Vin.ton} - \text{Vo.ton} = \text{Vo.T} - \text{Von.ton}$$

$$\text{Vin.ton} = \text{Vo.T}$$

$$\frac{\text{ton}}{\text{T}} = \frac{\text{Vo}}{\text{Vin}}$$

$$D = \frac{Vo}{Vin} = \frac{200V}{500V} = 0{,}4$$

- Determine resistor value (2)

$$R = \frac{Vo}{Io} = \frac{200V}{12A} = 16{,}667\Omega$$

- Determine inductor value (3)

$$\Delta IL = 10\%x12A$$
$$\Delta IL = 1{,}2A$$

$$L = \frac{Vin(1-D)D}{\Delta ILxfs}$$

$$L = \frac{500V(1-0{,}4).0{,}4}{1{,}2Ax10Khz}$$
$$L = 0{,}01H$$

- Determine capacitor value (4)

$$\Delta V = 5\%x200V \quad R = \frac{Vo}{Io} = \frac{200V}{12A} = 16{,}667\Omega$$
$$\Delta V = 10V$$
$$C = (D.Vin + \Delta V)\frac{(1-D)}{fs.R.\Delta V}$$
$$C = (0{,}4.500V + 10)\frac{(1-0{,}4)}{10000x16{,}667x10}$$
$$C = 0{,}0000756\ \mu F$$

Table 2. Parameters of Motor DC

Parameter	**Unit(SI)**
Mechanical power($P_{mekanik}$)	5 HP
Anchor voltage (E_s)	240 V
Nominal speed Nominal ($\omega_{nominal}$)	1220 Rpm
Anchor resistance (R_a)	0,6 Ω
Anchor current (I_a)	16,2 A
Anchor inductance (L_a)	0,012 H
Inertia (J)	I $Kg.m^2$

IOP Conf. Series: Earth and Environmental Science **156** (2018) 012013 doi:10.1088/1755-1315/156/1/012013

Designing controller method

To get the output voltage constant buck converter used to supply DC motor load using combination of Fuzzy like PD and PID control, the steps are as follows:

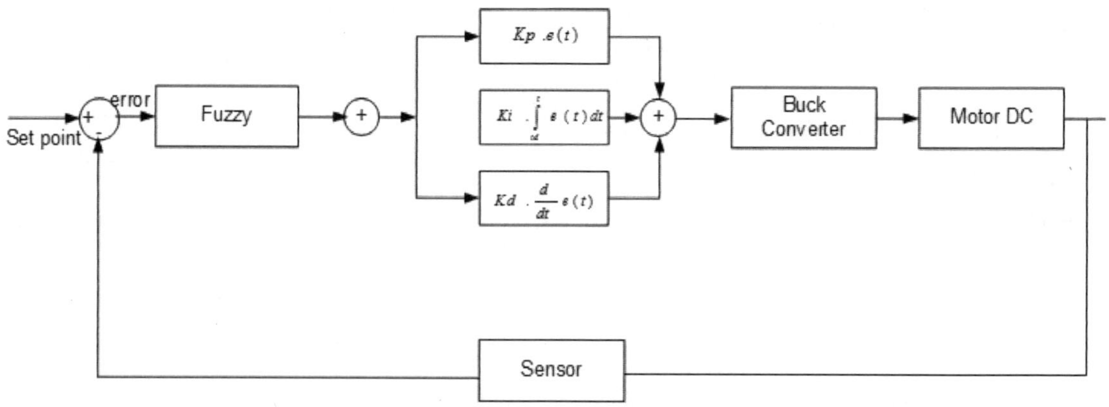

Figure 4. Block diagram of combination Fuzzy & PID Logic Control at Buck Converter with DC Motor Load

A. PID (Proportional Integral Derivative)[6]

Proportional Integral Derivative (PID) is a controller to determine the precision of an instrumentation system with the characteristics of feedback on the system. The PID control component consists of 3 types: Proportional, Integral and Derivative. All three can be used simultaneously or independently depending on the response we want to a plant. In the PID control, the error signal e (t) is the control input, while the control output is the control signal u (t). The relationship between the control input e (t) and the control output u (t) is:

$$u(t) = K_p \left\{ e(t) + \frac{1}{\tau_i} \int_0^t e(t)dt + \tau_d \frac{de(t)}{dt} \right\}$$

Or in form of Laplace Transformation :

$$U(s) = K_p \left(1 + \frac{1}{\tau_i s} + \tau_d s \right) E(s)$$

Thus, transfer function of PID controller is :

$$\frac{U(s)}{E(s)} = K_p \left(1 + \frac{1}{\tau_i s} + \tau_d s \right)$$

Block diagram of PID controller, namely :

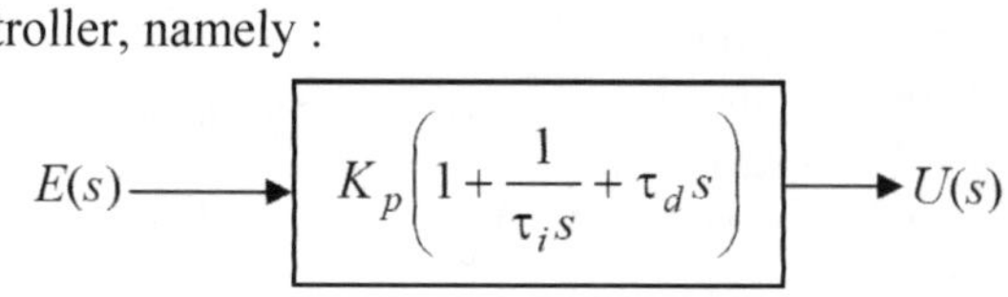

which :

K_p = Proportional reinforcement

τ_i = Integral time

τ_d = Differential time

IOP Conf. Series: Earth and Environmental Science **156** (2018) 012013 doi:10.1088/1755-1315/156/1/012013

The function of the PID component is as follows:

1. Proportional control functions to accelerate the rise time.
2. Integral control functions to eliminate steady state errors.
3. Derivative control functions to improve system stability and reduce overshoot.

B. Fuzzy like PD controller [7].

Fuzzy like PD controller is a fuzzy controller that has a behavior resembling a PD controller. Design consists of the process of quantization, fuzzifikasi, inference, and defuzzifikasi. In the Fuzzy control type used in this design, the formation of Ke and Kde is used to normalize the boundary E and ΔE into the limit of -3 to +3. I used to denormalize the control signal limits from 0 s / d 3 to 0 s / d U_{max} or -3 s / d 3 to U_{min}s / d U_{max}. The design of the fuzzy control system is shown in Figure 5.

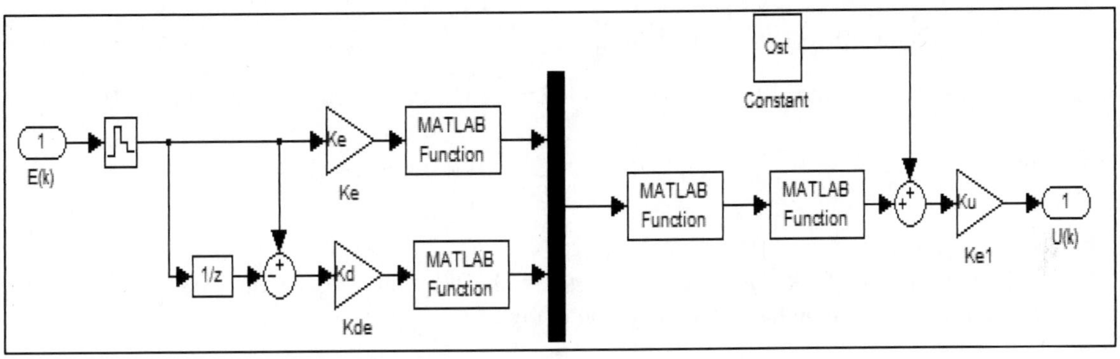

Figure 5. Block diagram of logic control on buck converter

Figure 6. A linear representation, membership level mapping that defines membership functions for E, ΔE and control signals in a normalized form on triangular shapes with 7 support sets.

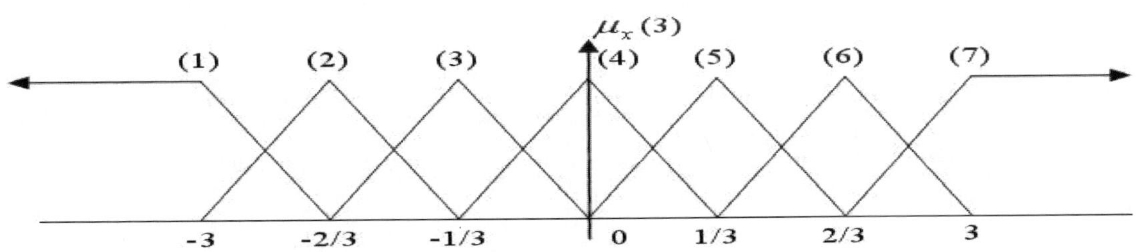

Figure 6. Triangle function

Create a basic fuzzy rule that is used to determine the control signal with the help of inference. The preparation of fuzzy control rules with the membership set for each of the seven input entries is shown in Table 3.

Table 3. Preparation of Fuzzy Rules

E/dE	1	2	3	4	5	6	7
1	1	1	2	2	3	3	4
2	1	2	2	3	3	4	5
3	2	2	3	3	4	5	5
4	2	3	3	4	5	5	6
5	3	3	4	5	5	6	6
6	3	4	5	5	6	6	7
7	4	5	5	6	6	7	7

Inference here isMamdanimethods in which the equation is:

$$\mu_y(k) = \max\left[\min\left\{\mu_u(k), \mu_R(E(i), \Delta E(j))\right\}\right]$$

To get values of action control (u) , it needs defuzzification process that is calledCOA (*Center Of Area*).

3. Results and Discussion

The simulation test of DC-DC buck converter with DC motor load with load variation using Fuzzy-PID control is shown on the response graph below:

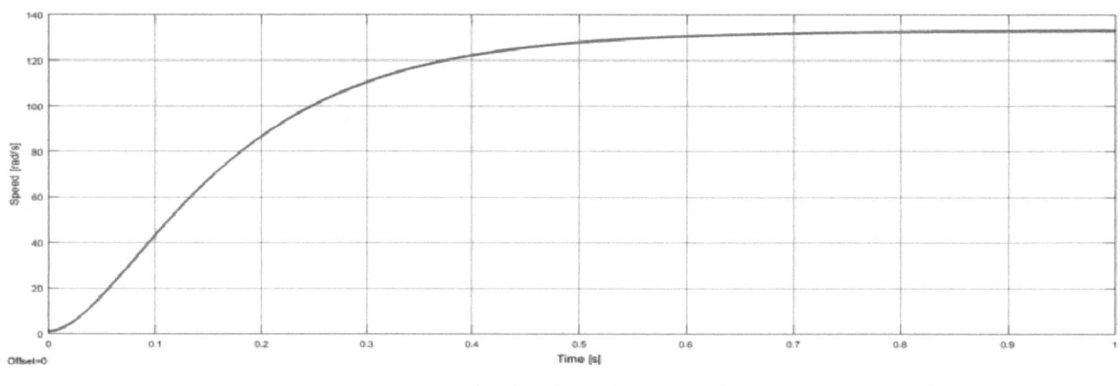

Figure 7. Response on zero load and nominal speed (ωnom) = 133.2 rad / s

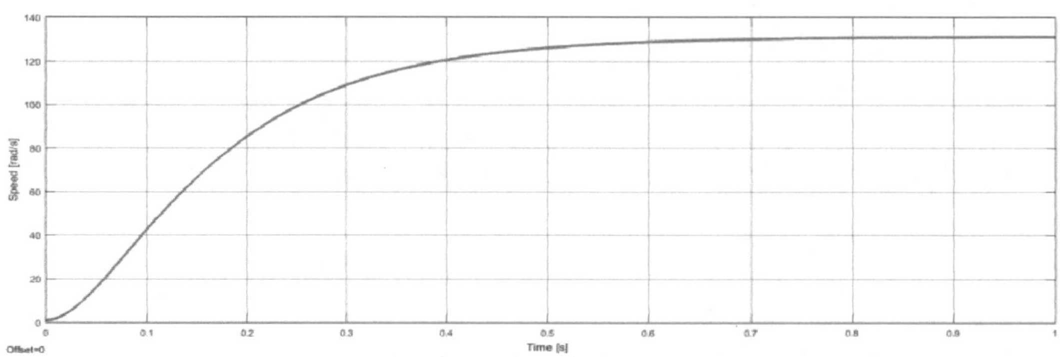

Figure 8. Response on Load 10 and nominal speed (ωnom) = 131.3 rad / s

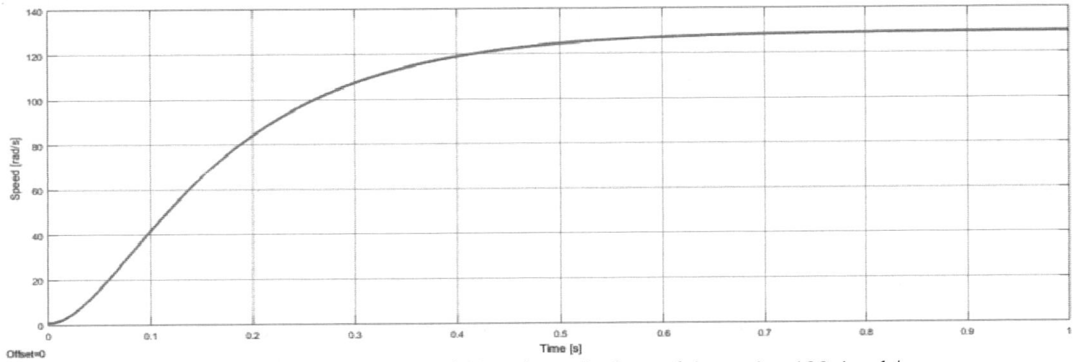

Figure 9. Response on Load 20 and nominal speed (ωnom) = 129,4 rad / s

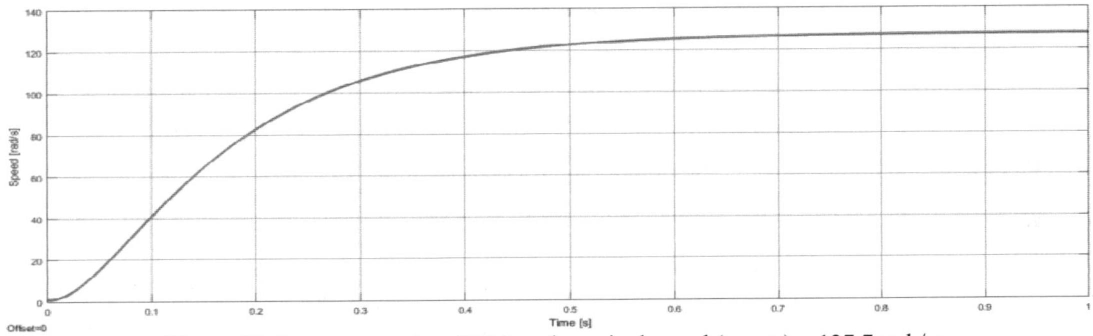

Figure 10. Response on Load 29,2 and nominal speed (ωnom) = 127,7 rad / s

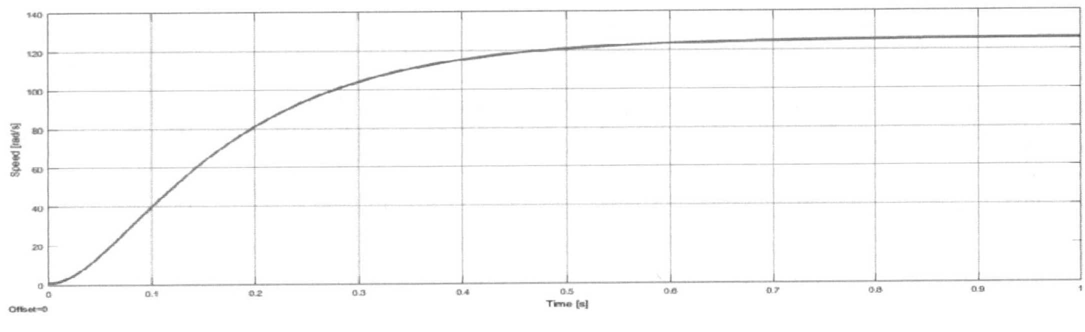

Figure 11. Response on Load 40 and nominal speed (ωnom) = 125,7 rad / s

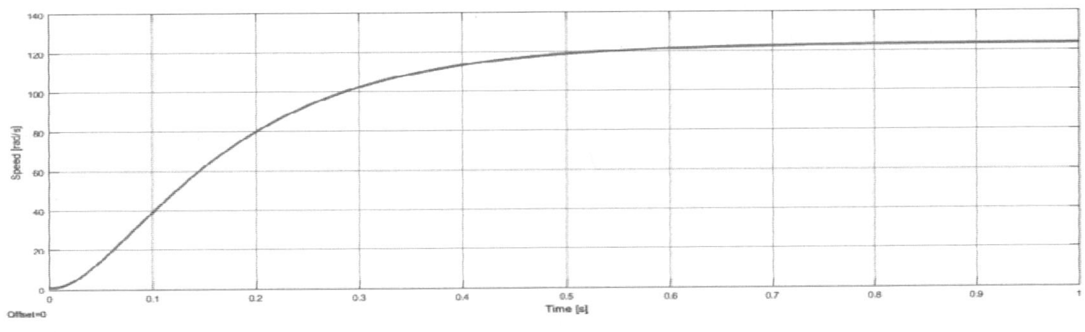

Figure 12. Response on Load 50 and nominal speed (ωnom) = 123,9 rad / s

Table 4. Testing on Open Loop

V.Input (v)	V.Output (v)	Torse(Load)	Nominal Speed (rad/s)
500	240,2	0	133,1
500	239,9	10	131,3
500	239,9	20	129,4
500	239,8	29,2	127,7
500	239,8	40	125,7
500	239,8	50	123,8

4. Conclusion

The simulation result of controlling buck converter with DC motor load using Fuzzy-PID control with load variation shows better performance than when open loop. The output voltage of the buck converter remains constant at 240 volts although the load is variable, in which the nominal speed of the motor follows the given load..

5. References

[1] Amit S. Chopade, Swapnil W. Khulbalkar, A.S. Junghare and M.V.Aware," Fractional Order Speed Controller For Buck – Converter Fed DC Motor,"International Conference on Control, Measurement and Instrumentation, IEEE. 2016

[2] Bhagyashri U, Patil, SatyawanR.jagtap. International Conference on Computation of Power Energy, Information and Communication. 2015

[3] Sutedjo, Zaenal Efendi, Dina Mursyida. PoliteknikElektronikaNegeri Surabaya-ITS

[4] Jumiyatun, Andi Fatmawati, "Komparasi Metode Pi Dan Fuzzy-Pi Pada Pengendalian Tegangan Output DC-DC Buck Converter", Prosiding SNTE Vol 2. 2017

[5] Rashid, Muhammad. Power Electronic Handbook .Akademik Press.Canada.2001

[6] Gamayanti, Nurlita.Kontrol PID. Institut Teknologi Sepuluh Nopember. Surabaya.2007

[7] Ana Ratna Wati, Dwi. Sistem Kendali Cerdas. Cetakan Pertama.Graha Ilmu.2011

[8] Ogata, Katsuhiko. Teknik Kontrol Automatik. Jakarta, Erlangga. 1996

[9] Rashid, Muhammad H.Terjemahan Elektronika Daya.PT prehallindo.Yogyakarta.1993

[10] Mohan, Ned. Power Electronic Handbook : Converters, Aplication, and design, Second Edition. 1995

Design of the Node Coordinator Based on WSN Network as Fisherman Vessel Monitoring System

Sarono Widodo[1], Rizal Budi Cahya[2], Yusron Nasrullah[3], Eko Supriyanto[4], Gatut Rubiono[5]

[1] Electro Departement, Semarang State Polytechnic, Center of Java , 50275 Indonesia
Email: sarono.widodo@polines.ac.id
[2] Electro Departement, Semarang State Polytechnic, I Center of Java , 50275 Indonesia
Email: rizal.b.cahya@gmail.com
[3] Electro Departement, Semarang State Polytechnic, Center of Java , 50275 Indonesia
Email: yusronn@gmail.com
[4] Electro Departement, Semarang State Polytechnic, Center of Java , 50275 Indonesia
Email: ekosupri2000@gmail.com
[5] Mechanical Engineering, PGRI University, Banyuwangi, East Java, 68418 Indonesia
Email: g.rubiono@fdi.or.id

Abstract- Fishermen in Indonesia in general are traditional fishermen who use boats weighing less than 30 GT and are not equipped with adequate navigation equipment. This becomes a technical obstacle when important information could not be directly obtained from the fishermen. An alternative that can be used to identify the position of traditional fishing boats is to a vessel monitoring system (VMS). A component in a fishing boat monitoring system is a node coordinator using Wireless Sensor Network (WSN) technology. By installing the node coordinator, the position information of the fishing boat sent by the boat node in the GPS data can be accessed within the scope of the coordinator. The coordinator node has two LoRa devices as the receiver and sender of data and an Arduino as a data processor. Data received from the boat node will be processed and added to the location of the coordinator crossed by the boat in the boat data format and then sent to the central in real time. The coordinator node has a coverage area of 0.45 miles in radius (coordinator A) and 0.35 miles (coordinator B) with 96% accuracy rate of data reception and 100% data processing.

Keywords: *WSN, VMS, real time, node coordinator*

1. Introduction

Traditional fishermen using fishing vessels weighing less than 30 GT with the sea reaching less than 4 miles from the coastline are estimated to be 80% of the population [1]. There are still many fishing boats that have not been equipped with navigation tools so that when the fishermen are fishing, their position could not be monitored in real time. Indonesian government regulatedvessel monitoring system in *Permen Nomer 42/PERMEN-KP/2015 tentang Sistem Pemantau Kapal Perikanan. SistemPemantauKapalPerikanan* (Fishery Vessel Monitoring System). The Fishery Vessel Monitoring System is not set up for fishing vessels weighing under 30 GT [2].

However to provide information about the existence of traditional fishing boats is necessary because in general there are many boats that are not equipped with navigation equipment.Investment in improving the Indonesian Vessel Monitoring System(VMS) towards real time monitoring and improved system interface has very large net benefits and is a very viable investment for theIndonesian government [3].The application of navigation systems and communication activitiesfishery at Bitung Fishery Port is still lackingoptimal where the system runs partially which cause overlapping information [4].

Vessel Monitoring System has been used as a fishing vessel monitor in various countries. The research related to the fishing boat monitoring system is conducted by Duc-Tuyen (2013) that examines the improvement of ad hoc wireless networks based on GPS capable of providing emergency location and services for small fishing boats by fixing the weakness of radio links from sea to land from ship to center station on land [5]. The research review of Manoufali (2013) examines wireless communications technologies to build a reliable maritime mesh network, challenges and requirements to ensure high quality maritime mesh communications [6]. There are also other system based on GPS as intelligent system for tracking and border alert for fishermen [7,8,9,10] and secured maritime alert system based on RSSI localization scheme [11].

Limitations for traditional fishermen to provide monitoring tools are a matter to be solved. An alternative fishing boat monitoring system should have at least a function that can monitor the movement of fishing boats. The system used is based on WSN (wireless system network). Node coordinator is a tool of a fishing boat monitoring system that can receive position information of a fishing boat in the form of GPS data within the scope of the coordinator and forward it to the central node for data processing.

2. Method

This research was conducted in three stages which are system design, model making and testing.The global design of fishing boat monitoring system is shown in Figure 1. There are four parts: sensor nodes (fishing boats), coordinator nodes, central and monitoring servers.

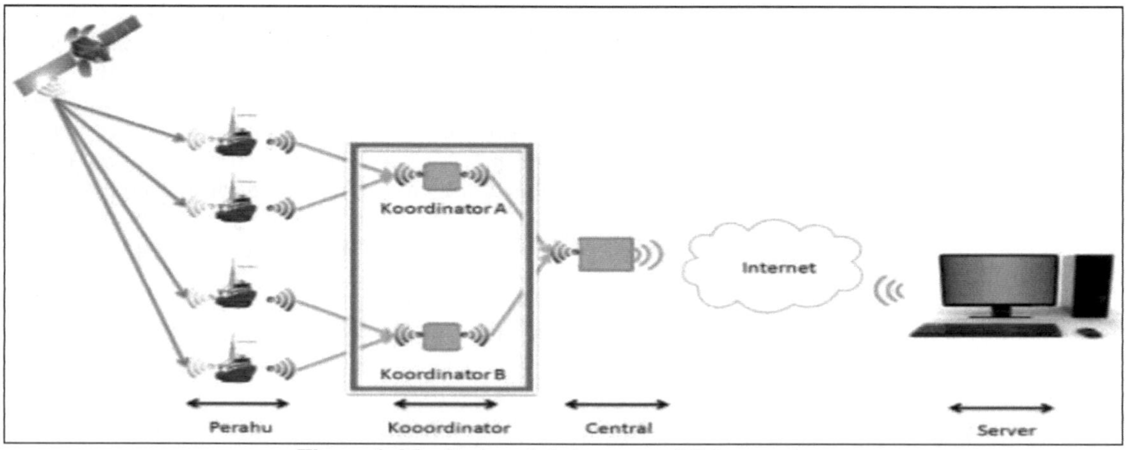

Figure 1. Monitoring global system of fishermen boat

The nodes coordinator as part of the research are designed and created using Wireless Sensor Network (WSN) technology to communicate between the sensor nodes with the node sink as the data collection center of the sensor node [12]. Sink node as well as base station is the interface between user and network. Node sensors that have movement can be connected to GPS devices and location information so that location information can be monitored by the base station [13]. The WSN method is chosen because it is possible to get boats movement data that are difficult to obtain by conventional method. The sensor nodes usually consist of four main components: the sensor itself, the processor, the radio, and the energy supply [14]. The coordinator node is designed using the LoRA shield component as a transceiver and Arduino as a data processor equipped with an energy source for the device.

The node coordinator that acts as a transceiver uses two Arduino mounted in series form using I2C multipoint serial communication and two LoRa shields are used as a receiver and a transmitter. Figure 2 shows the block diagram of the coordinator node design.

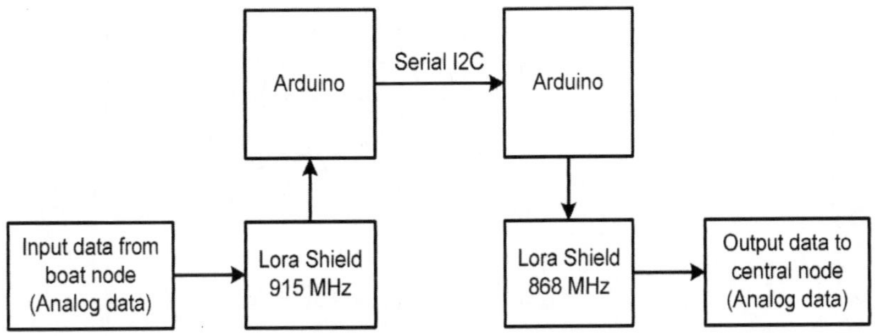

Figure 2. Node coordinator block diagram

The data input and the data form output of the nodes coordinator are shown in Figures 3 and 4.

(ID ship)|(Latitude),(Longitude)|(Year-Month-Date)|(Hour:Minute:Second)

Figure 3. Data input form

(ID ship)|(Latitude),(Longitude)|(Year-Month-Date)|(Hour:Minute:Second)|(Coordinator Code)

Figure 4. Data output form

The Arduino program includes input data received from the sensor node (boat) as in Figure 3. Arduino via LoRa shield checks whether there is input data received by Lora working at 915 Mhzin frequency. The received data will be stored in the buffer and then sent to Arduino via I2C serial interface. The Arduino connected to LoRa. The Sender will process for the addition of the coordinator code. Flowchart of data receiving program to coordinator and data transmission between arduino and sending data by Lora sender are shown in Figure 5 and Figure 6.

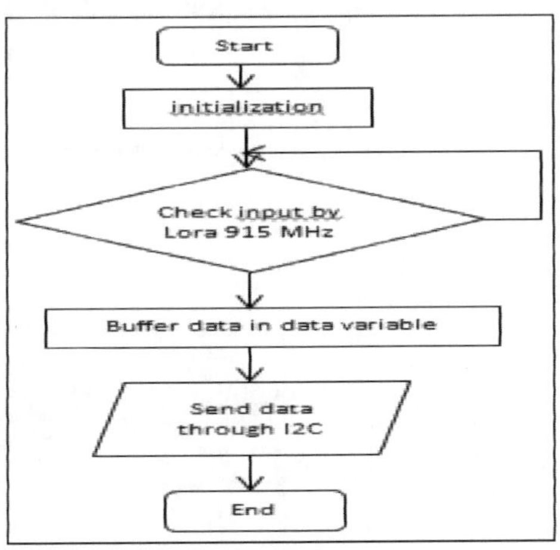

Figure 5. Input flowchart

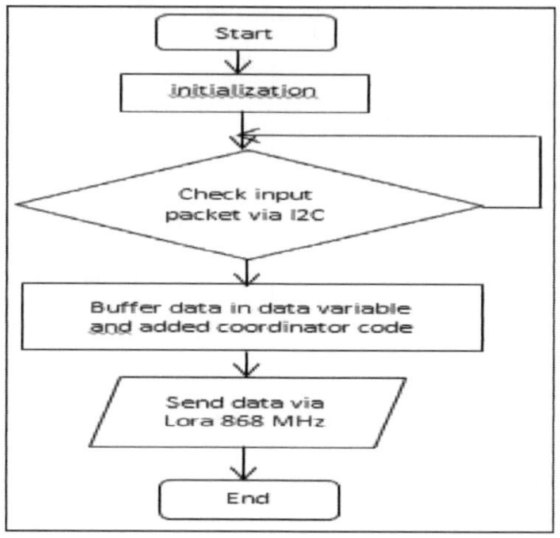

Figure 6. Output flowchart

3. Result and Discussion

Testing the system is aims to know that the device node coordinator has been running in accordance with the provisions and to know the device running according to the program created.

3.1 Data Receiving Test

The data receiving test is aims to know the data sent by multipoint-to-point node boats can be received by the coordinator node. The description of the data reception test can be seen in Figure 7.

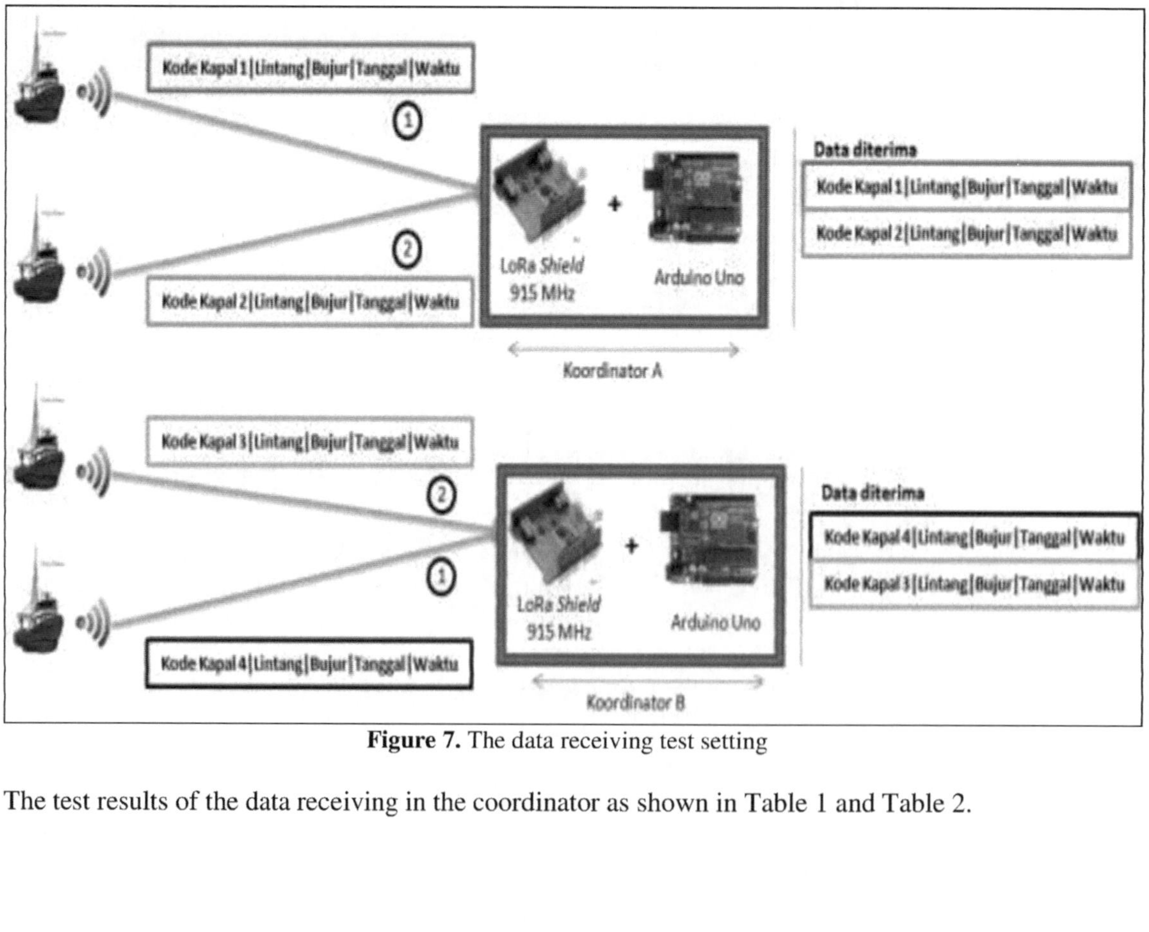

Figure 7. The data receiving test setting

The test results of the data receiving in the coordinator as shown in Table 1 and Table 2.

Table 1. Data Receiving in The Coordinator A

No	KODE KAPAL	LINTANG	BUJUR	TANGGAL	WAKTU
1	KN004	-6.812568	110.526237	2017-08-21	15:39:26
2	KN001	-6.812970	110.527008	2017-08-21	15:39:34
3	KN004	-6.812383	110.525505	2017-08-21	15:39:56
4	KN001	-6.812722	110.527046	2017-08-21	15:40:04
5	KN001	-6.812927	110.527465	2017-08-21	15:40:34
6	KN002	-6.809547	110.528686	2017-08-21	15:49:36
7	KN002	-6.809526	110.528640	2017-08-21	15:50:06
8	KN004	-6.811773	110.523635	2017-08-21	15:50:29
9	KN002	-6.809512	110.528579	2017-08-21	15:50:37
10	kI004	-6.811773	110.5‡1¶15	0000-00-00	0:00:00
11	KN002	-6.809506	110.528518	2017-08-21	15:51:07
12	KN002	-6.809514	110.528450	2017-08-21	15:51:37
13	KN002	-6.809516	110.528396	2017-08-21	15:52:07
14	KN002	-6.809522	110.528289	2017-08-21	15:53:07
15	KN002	-6.809545	110.528236	2017-08-21	15:53:38
16	KN002	-6.809569	110.528190	2017-08-21	15:54:08
17	KN002	-6.809584	110.528144	2017-08-21	15:54:38
18	KN002	-6.809587	110.528030	2017-08-21	15:55:38
19	KN001	-6.809767	110.529121	2017-08-21	15:55:58
20	KN002	-6.809587	110.527969	2017-08-21	15:56:08

Table 2. Data Receiving in The Coordinator B

No	KODE KAPAL	LINTANG	BUJUR	TANGGAL	WAKTU
1	KN004	-6.812088	110.524795	2017-08-21	15:40:27
2	KN004	-6.811777	110.523628	2017-08-21	15:49:29
3	KN004	-6.811773	110.523635	2017-08-21	15:49:59
4	KN003	-6.811683	110.523605	2017-08-21	15:50:56
5	KN003	-6.811683	110.523605	2017-08-21	15:51:26
6	KN004	-6.811773	110.523635	2017-08-21	15:51:30
7	KN004	-6.811773	110.523635	2017-08-21	15:52:00
8	KN003	-6.811695	110.523567	2017-08-21	15:52:03
9	KN004	-6.811773	110.523635	2017-08-21	15:52:30
10	KN002	-6.809514	110.528343	2017-08-21	15:52:37
11	KN004	-6.811777	110.523635	2017-08-21	15:53:00
12	KN003	-6.811695	110.523567	2017-08-21	15:53:11
13	KN004	-6.811778	110.523635	2017-08-21	15:53:30
14	KN004	-6.811780	110.523628	2017-08-21	15:54:00
15	KN003	-6.811698	110.523567	2017-08-21	15:54:12
16	KN004	-6.811782	110.523628	2017-08-21	15:54:30
17	KN004	-6.811785	110.523628	2017-08-21	15:55:31
18	KN004	-6.811785	110.523628	2017-08-21	15:56:01
19	KN003	-6.811717	110.523574	2017-08-21	15:56:13
20	KN004	-6.811787	110.523628	2017-08-21	15:56:31

Note: The sample data is taken the first 20 data from 126 data received. The red data is defect data.

3.2 Data Processing Test

Data processing test is aims to know the position information data sent node boat can be done addition of data format coordinate code or not to be forwarded through I2C serial communications to be sent to the central node. The result is shown in Table 3 and Table 4

Table 3. Data Processing in The Coordinator A

No	KODE KAPAL	LINTANG	BUJUR	TANGGAL	WAKTU	KODE KOORDINATOR
1	KN004	-6.812568	110.526237	2017-08-21	15:39:26	A
2	KN001	-6.812970	110.527008	2017-08-21	15:39:34	A
3	KN004	-6.812383	110.525505	2017-08-21	15:39:56	A
4	KN001	-6.812722	110.527046	2017-08-21	15:40:04	A
5	KN001	-6.812927	110.527465	2017-08-21	15:40:34	A
6	KN002	-6.809547	110.528686	2017-08-21	15:49:36	A
7	KN002	-6.809526	110.528640	2017-08-21	15:50:06	A
8	KN004	-6.811773	110.523635	2017-08-21	15:50:29	A
9	KN002	-6.809512	110.528579	2017-08-21	15:50:37	A
10	kI004	-6.811773	110.5I1¶15	0000-00-00	0:00:00	A
11	KN002	-6.809506	110.528518	2017-08-21	15:51:07	A
12	KN002	-6.809514	110.528450	2017-08-21	15:51:37	A
13	KN002	-6.809516	110.528396	2017-08-21	15:52:07	A
14	KN002	-6.809522	110.528289	2017-08-21	15:53:07	A
15	KN002	-6.809545	110.528236	2017-08-21	15:53:38	A
16	KN002	-6.809569	110.528190	2017-08-21	15:54:08	A
17	KN002	-6.809584	110.528144	2017-08-21	15:54:38	A
18	KN002	-6.809587	110.528030	2017-08-21	15:55:38	A
19	KN001	-6.809767	110.529121	2017-08-21	15:55:58	A
20	KN002	-6.809587	110.527969	2017-08-21	15:56:08	A

Table 4. Data Processing in The Coordinator B

No	KODE KAPAL	LINTANG	BUJUR	TANGGAL	WAKTU	KODE KOORDINATOR
1	KN004	-6.812088	110.524795	2017-08-21	15:40:27	B
2	KN004	-6.811777	110.523628	2017-08-21	15:49:29	B
3	KN004	-6.811773	110.523635	2017-08-21	15:49:59	B
4	KN003	-6.811683	110.523605	2017-08-21	15:50:56	B
5	KN003	-6.811683	110.523605	2017-08-21	15:51:26	B
6	KN004	-6.811773	110.523635	2017-08-21	15:51:30	B
7	KN004	-6.811773	110.523635	2017-08-21	15:52:00	B
8	KN003	-6.811695	110.523567	2017-08-21	15:52:03	B
9	KN004	-6.811773	110.523635	2017-08-21	15:52:30	B
10	KN002	-6.809514	110.528343	2017-08-21	15:52:37	B
11	KN004	-6.811777	110.523635	2017-08-21	15:53:00	B
12	KN003	-6.811695	110.523567	2017-08-21	15:53:11	B
13	KN004	-6.811778	110.523635	2017-08-21	15:53:30	B
14	KN004	-6.811780	110.523628	2017-08-21	15:54:00	B
15	KN003	-6.811698	110.523567	2017-08-21	15:54:12	B
16	KN004	-6.811782	110.523628	2017-08-21	15:54:30	B
17	KN004	-6.811785	110.523628	2017-08-21	15:55:31	B
18	KN004	-6.811785	110.523628	2017-08-21	15:56:01	B
19	KN003	-6.811717	110.523574	2017-08-21	15:56:13	B
20	KN004	-6.811787	110.523628	2017-08-21	15:56:31	B

3.3 Area Coverage Test

Area coverage test is aims to find out how far the coverage area node coordinator. The results of coverage area testing and node coordinator placement at sea can be seen in Figure 8.The tests result are shown in Figure 9.

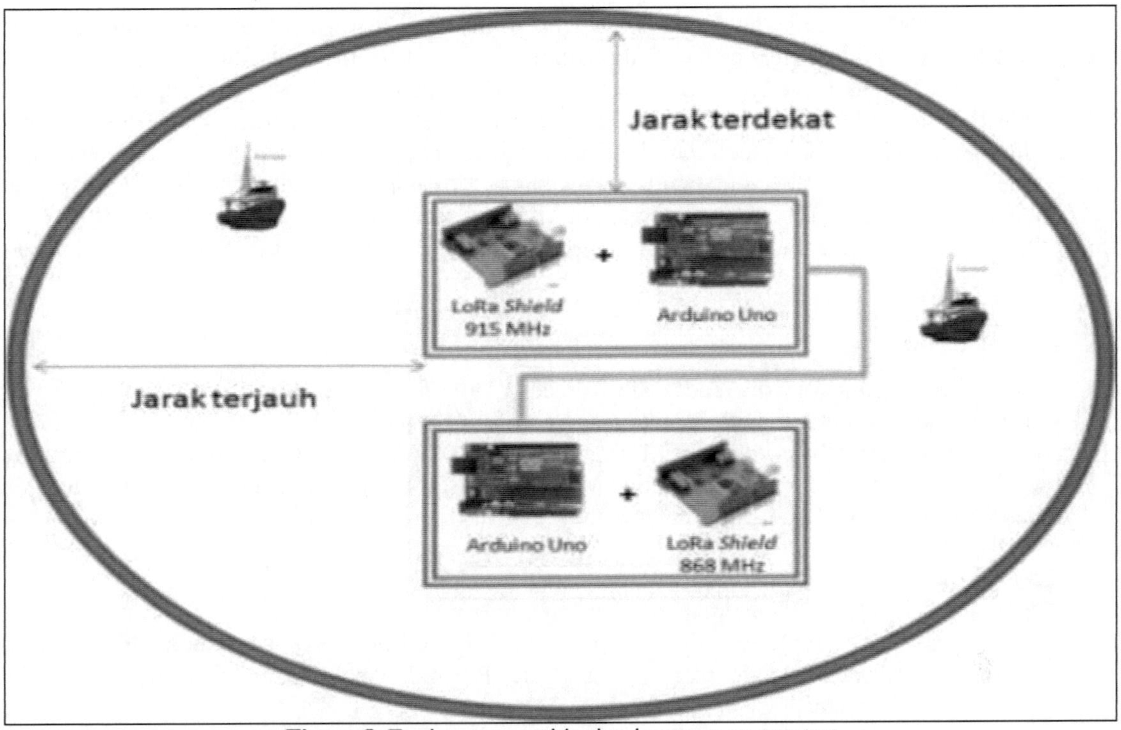

Figure 8. Equipments positioning in area coverage test

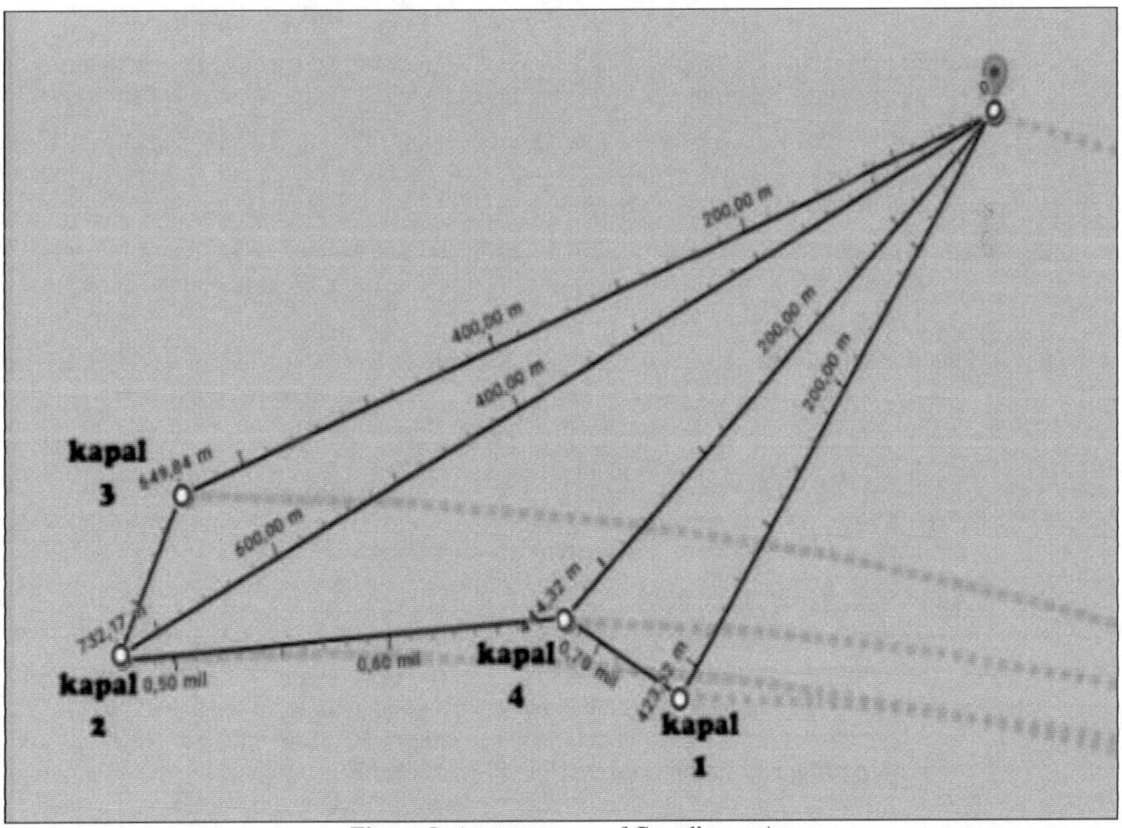

Figure 9. Area coverage of Coordinator A

IOP Conf. Series: Earth and Environmental Science **156** (2018) 012014 doi:10.1088/1755-1315/156/1/012014

The data receiving test result shows that at the coordinator A of 20 data sample test, there is 1 data defect when the process of receiving data from the node of the fourth boat, whereas at coordinator B of 20 test sample data, data can be accepted with no defective data.The data processing test result shows that from 20 sample data from the test results of each coordinator obtained that the process of adding coordinator code to the input data can run well. The defect data received also gets the addition of the coordinator code (automatically in the same format and system).The area coverage testresults obtained that each coordinator has a range of different areas of distance range of coordinator A is 0.45 miles (732.12 m) and the farthest range of coordinator B is at a radius of 0.35 miles (566.11 m) in radius.

4. Conclusion

The coordinator node successfully receives the boat or ship position information within the scope of the coordinator with a success rate of 96%. The node coordinator successfully added coordinator code on boat position information with 100% success rate. Adding node coordinator is useful to mark a part of territorial waters. Maximum distance of area coverage of coordinator A is 0.45 milesand coordinator B is 0.35 miles in radius.

5. References

[1]. Rauzatul N. (2016). Prototipe Sistem Pelacak Jejak Kapal Nelayan Pancing Tradisional Secara Otomatis Menggunakan Visualisasi Berbasis Web. http://repository. ipb.ac.id/ jspui /bitstream/ 123456789/ 80139/1/2016ma.pdf

[2]. Peraturan Menteri Kelautan dan Perikanan Republik Indonesia (2015). Sistem Pemantau Kapal Perikanan (Berita Negara Nomor 2025). http://djpsdkp.kkp.go.id/rule/file/26/permen-kp-nomor-42-tahun-2015-tentang-sistem-pemantauan-kapal-perikanan-.pdf/

[3]. Suhendar, M. (2013). Comparison of Vessel Monitoring System (VMS) between Iceland and Indonesia [final project].http://www.unuftp.is/static/fellows/document/suhendar12prf.pdf

[4]. Wahab R.A, (2014). Penggunaan Alat dan Perangkat Telekomunikasi dalam Sistem Navigasi dan Komunikasi Aktivitas Perikanan di Pelabuhan Perikanan Bitung, Buletin Pos dan Telekomunikasi, 12(4): 279-290

[5]. Ta DucT.,Tran DucT.,Do DucD. (2013). Efficient and Reliable GPS-Based Wireless Ad Hoc for Marine Search Rescue System.Lecture Notes in Electrical Engineering May 2013. DOI: 10.1007/978-94-007-6738-6_112.

[6]. Mohamed M., Hamada A., Peng-Yong K.,Shihab J. (2013). Technologies and Networks Supporting Maritime Wirelss Mesh Communications. IFIP WMNC'2013.

[7]. Arunvijay D, Yuvaraj E, (2014). Design of Border Alert System for Fishermen Using GPS, International Journal of Students Research in Technology & Management 2(02): 67-70

[8]. Yogesh Kumar A, A. BharaniVasan, S.P. ValanArasu (2015). GPS Based Tracking of Maritime Line of Control Monitoring System, INT J CURR SCI 2015, 14: E 12-18

[9]. Arull K, Asha J, Mohamed Nizar S, Malathi M, (2015). A Review on GPS Tracking and Border Alert System for Fishermen, International Journal of Science Technology & Engineering 2(5): 1-3

[10]. R.Raja Nandhini, S.Malarvizhi, A.Praveen, C.Mohanraj, R.Srinivasan, (2016). Intelligent Navigation System for Fishing Boats Using GPS, International Research Journal of Engineering and Technology(IRJET) 3(1): 1169-1172

[11]. M.Vanitha, S.P. Muthulekshmi, R.Srilakshman, D.Kiran Kumar, (2016). Secured Maritime Alert System Based on RSSI Localization Scheme, International Journal of Engineering Trends and Technology (IJETT) 34(1): 28-31

[12]. Annisa Y. (2017). Pengaruh Mobilitas Sink Node pada Wireless Sensor Network (WSN) untuk Pemantauan Aktifitas Pergerakan Gajah Dalam Area Penangkaran. Bandar Lampung: Universitas Lampung.

[13]. M.A.Matin.(2012). Chapter 1: Overview of Wireless Sensor Network.http: //cdn.inte chopen.com /pdfs /38793/InTechOverview_of_wireless_sensor_network.pdf

[14].M.A.Perillo (2007). Role Assignment in Wireless Sensor Networks: Energy-Efficient Strategies and Algorithms. Department of Electrical and Computer Engineering University of Rochester, Rochester, NY, USA

IOP Conf. Series: Earth and Environmental Science **156** (2018) 012015 doi:10.1088/1755-1315/156/1/012015

Stimulation of Changes, Collective Commitment and The Patterns of Group Formation in Community Development in South Sulawesi

Syafiuddin Saleh
Faculty of Agriculture University of Muhammadiyah Makassar 90221,Indonesia
Email :syafiuddinsaleh@gmail.com

Abstract: This study aims to examine the pattern of group formation, related to the stimulation of change through the empowerment of farmers and poor fishermen The pattern of group formation is the basis for sustainable development. The research method used is qualitative descriptive method and relevant research type such as case study and triangulasi. The results of the study showed that (1) stimulation of changes made through development programs or community empowerment in the areas studied both among farm households and poor fishermen households for some programs received positive response from farmers and fishermen. However, the collective commitment to the breeding is relatively weak, since the group formed in each program is not done through good planning and concepts. (2) there are two patterns of group formation that are natural and formed formations initiated by outsiders. Groups that are naturally formed are more institutionalized and have characteristics such as intense and relatively routine interaction, strong mutual trust, and have a common form or mechanism shared for common purposes. The group can form the basis for sustainable development in improving the welfare of the poor.

Keywords: *change, collective commitment, empowerment, society, poor*

1. Introduction

The implementation of modernization as the paradigm of development in all fields, including in Indonesia for 32 years, on the one hand has brought progress.On the other hand, however, it raises various concerns such as creating inter-nations, inter-regions, and inter-communities dependency.Exploitation of natural resources without considering the aspects of continuity and sustainability has caused damage to the environment; economic growth without equal distribution of income causes gaps, in which the rich enjoys a better ratio of development while the poor get poorer and drowned in a sense of helplessness.Nationally, the focus of the development emphasizes on the development of supporting facilities and infrastructure for urban areas, with the expectation to instill the enthusiasmof market players. The implementation of trickle-down effect policy to improve development resources since the 60'scauses the activities in the local area not to develop well, and particularly weakens economic activity in the local market. Another drawback lies on the ability of the local administration system to provide services for the community. Both drawbacks continuously weaken the local community, including local poor households [1].

The spirit of regional autonomy in Indonesia is reflected in the Law No. 32 Year 2004, stressing that regions have the authority to organize, manage and develop themselves. According to the Law, parts of the responsibility of a region are to improve the quality of life of its community, to actualize justice and equity, and to improve basic services. The spirit is embodied in various programs and efforts to address poverty that havebeen conducted every year. The process of community development or empowerment has been carried out in various ways; however, the efforts made have not fulfilled what is expected.

However, the approach of community empowerment in development should mean thatthe community is positioned as the actor and not merely as the beneficiary of the development process; the community searches for solutions and achieves better results; the community is thus able to improve the quality of independencein order to overcome the problems encountered. Community empowerment should be able to improve the quality of human resources, especially in shaping and changing people's behavior in order to achieve a higher quality of living standards. The behavioral changes in both sectoral dimension and social dimension should reach all strata of society, the poor in particular [2].

The research conducted by [3] shows that the effort to reduce poverty needs to involve the community itself in a more participatory form in order to enable learning process in the community as well as the process of behavioral change for a more dignified life to occur. The finding [4] describes that in a poor family happiness has a connection with what is thought and felt within the family related to several dimensions such as: education, health, wealth, and others. Therefore, the success of an empowerment program is the result of the interaction between the elements of development and empowerment strategies applied. Empowerment effort and strategy is a pendulum between evolutionary paradigm and revolutionary paradigm that complements each other in a proportion that is in line with the condition and institution of the farmers [5]. If empowerment is a strategic model to eradicate poverty, then it should be a multidimensional and multifaceted process, capable to mobilize various aspects or elements of resources as well as the capacity and potential of the community concerned; it needs mutual cooperation between all existing elements in order to encourage the local micro-economy [6].

The empowerment effort should be able to increase the ability of the community to organize itself, in a sense it is able to organize and manage problems and existing potential in order to adapt and cope with the occurring changes [7]. Lately, there is an indication of the strengthening of problem or the weakening of poverty reduction in Indonesia. According to the World Bank in [8] the indication is shown by the weakening of poverty indicator, not only of income but also on the circumstances such as the level of education, health, infant mortality, etc. and the growing gap in the performance and utilization of development's result. Poor population in rural areas increases from 19.5 percent to 21.29 percent, while poor population in urban areas increases from 11.4 percent to 13.4 percent. The condition is also exacerbated by the state of growth and equity disparities between regions, groups and individuals.

The main problem in this research is the tendency of the development at the local level in rural areas that has not been capable to increase institutional capacity and formation for rural community, especially among the poor or underprivileged households. Various programs have been carried out,yet there are still issues that need to be improved. Furthermore, the weak institutional capacity and formation originates from a sense of powerlessness, achievement orientation, weak collective action, the lack of an alternative vision and low stimulation changes. This study examines specifically how stimulus change and collective commitment can be an important element for strengthening the institutional poor rural household and how the pattern of group formation that can be the basis of institutional strengthening within the framework of sustainable development.

2. Method

This study was conducted in Jeneponto Regency as an area with dry-land farming community. In this area, the research location was more specific in rural areas with the highest population (households) concentration, and had been given empowerment programs undertaken by the government as well as by other stakeholders, such as Non-Governmental Organizations (NGOs) and the private sectors. This study used a qualitative descriptive approach by applying case study research combined with triangulation technique. In the pre stage of this study, it enumerated the poor communities who were subjected to this study. The research procedures included: (1) identifying the location of poor households through secondary data, (2) identifying the development or empowerment programs which have been conducted,

(3) collecting baseline data on the conditions and characteristics of poor households, and mapping them based on poverty indicators such as; insufficiency of household elements such as: into insufficient production assets, consumption assets and managerial assets and (4) indepth interviews to get a picture of how stimulus of change and collective commitment can be an important element for strengthening the institutions of rural poor households, what kind of group patterns can be the basis for institutional strengthening within the framework of sustainable development. The main analysis method used was qualitative data analysis.

Qualitative data analysis was defined as an attempt of analysis based on words arranged into an expanded text form.

3. Results and Discussion

3.1 Overview of Research Sites

This research was conducted in two regions in South Sulawesi namely Jeneponto Regency and PangkajeneKepulauan Regency (Pangkep). Jeneponto Regency is one of the regencies in South Sulawesi Province located in the southern part of Makassar City. This area has an area of 749.79 sq km, and administratively covers 11 districts. The management of this regional government is centered in Bontosunggu City in Binamu Sub-district. Based on the administrative division, the number of villages and sub-districts spread throughout the sub district are 83 villages and 30 urban villages. The data shows the population in this area about 361,000 people, which spread in each district. This area is dominated by dryland farming areas. Of the existing rice paddies manage rice crops and corn. For corn farming, it is generally only harvest twice, whereas rice is harvested only once a year. This situation causes the area to be categorized as dry areas with a large population of poor people. The poor household classification in Jeneponto District is made up of very poor, poor, near poor and vulnerable poor households. The data shows the number of poor households in this area of 54 072 households. [9] (Jeneponto Regency BPS, 2014). Bangkala sub-district includes the largest number of households, following Binamu Sub-district, and Tamalate Sub-district. So the choice of study focus is set in the sub-district of Bangkala, i.e., in Kapita Village.

While *PangkajeneKepulauan* Regency (Pangkep) is a regency in South Sulawesi Province which is located north of Makassar City. This area as its name sebahagian consists of small islands located in the sea of Flores and Makassar Strait. Overall the total area of this area 1,112.29 sq km. Among the area is largely marine territory. Pangkep regency has 11 districts. with central government in Pangkajene. Based on data on the number of poor or pre-prosperous homes in this area around 84948 households. Several sub-districts in *Pangkep* District can be divided into two divisions, namely the dominant sub-districts in the Mainland and the dominant subdistricts in the island archipelago. [10] (BPS Pangkep Regency, 2014).

Based on the consideration of the easiness and coverage of the area observed, it was initially chosen or focused on poor fisherman households in Tupabiring sub-district i.e.,*MatiroDeceng* i.e., in *Badi* Island and *Pajenekang* Island. This island can be reached by sea using motorized boat with 2 hours travel time from Paotere Port of Makassar City or from *Pangkajene* City. However, *MatiroDeceng Village* is very far from the central government of Pangkep regency, so rarely gets a touch of empowerment program and for the adequacy of data related to local development or empowerment of poor households, the observation area is moved to North *Tupabiring*Subdistrict in *MatiroBombang* Village which consists of *Salemo* Island, *Sabangko* Island, *Sagara* Island and *Sikuala* Island.

3.2 Characteristics of Poor Households

In general it can be described that the condition and characteristics of poor rural households both farmers and fishermen from the point of view of participatory social development, in the two districts of *Jeneponto and Pangkep*District are not much different. The only difference is in the use of assets related to the physical production activities used by both groups in the type of livelihood. However, the state of the resource or asset in each condition is limited according to the state of the region and the community. While other resource elements such as human resources, finance for production activities are basically the same ie the number is limited by productive workers, working capital that most borrowed from other parties or investors whose terms of return positioned poor households in a fairly difficult situation.

Similarly, the assets or resources associated with consumption activities, kedaaanya not different like kitchen equipment (plates, glasses), jirigen water, furniture potluck, walled house and bamboo flooring, electricity from neighbors and other equipment of modest consumption. In terms of human consumption is shown in the number of dependents that many average 3-5 people even up to 7 people and generally less productive. In terms of fiansial prominence that is the lack of savings, and the tendency to have debt and unable to send children to a higher level. This situation causes poor households unable to grow and develop. So from the aspect of managerial activities for various purposes, especially those that are economically profitable can not be done.

In this connection, it can be illustrated that the conditions and characteristics of poor farmer households from the perspective of participatory local social development can be illustrated in the following household matrix Table 1:

Table 1. Matrix of Nine Elements of Poor Farmer Households in Jeneponto Regency

Resource / Activity	Physical	Human	Financial
Production	**Farmer:** Land (limited) Crescent, hoe, machete and other tools that are also limited **Fisherment:** Fishing line, single device wire, fuel, boat (loan), crab fishing trap <200, strap span for seaweed <150	1 productive laborer The rest do not work	Funds, limited capital Most of the capital is in loans form
Consumption	Kitchen equipments (plates, glasses), water jerry cans, makeshift furniture, walled house and bamboo flooring, electricity from neighbors, electric hired from neighbors	3–4 burden people	No savings, Education fund is limited
Managerial	Existing physical assets are not maintained for various purposes	Other laborers are not attempted to assist the household	The investment effort for various purposes is not possible because of limited funding

Further on the institutional side, both on farmers and on fishermen have a similar relationship. Both are only differentiated by the types of commodities that are managed. The form of natural groups like gotongroyong and arisan still exist in each community. Initially this activity is more to strengthen social relations, but along with the development and empowerment, the situation has begun to shift to economic goals as happened in KUBE and other empowerment programs. However, in reality the collective group or formation is largely incapable of continuing due to various causes, and when empowerment and assistance are still dependent on government funding. In a long-term perspective, farmer groups experiencing the process of empowerment (community development) should continue to shift and position themselves close to the market system and establish themselves as autonomous or independent groups [1] and show their performance as a sustainable institutionalized groups [11]. In long-term perspective it is expected to change closer to the market system, and become an independent group or institution. These positions bring many consequences from an institutional standpoint and then become part of the observations in the next study.

In general it can be argued that (1) poor peasant households in *Jeneponto* generally manage dry land according to the surrounding natural bounty, by cultivating corn and rice. Corn is grown 2 times a year whereas paddy is only one time a year and farmers generally are cultivators and some own a very narrow land. In fishermen community in *Pangkep* most is fishing fisherman, crab and seaweed. The main resources are fishing rods and boats. *Bubu* and rope and span, boats (dugboats) (2) The business capital of each poor household is generally owned by the other party, obtained by borrowing and returning in the form of deposits of production and subsequent profit sharing entirely governed by the owner of capital. Among farmers and fishermen, cultural savings have not existed or have not developed. This is due to limited income, also due to the urgent need for family life. (3) Institutions in poor farming communities and poor fishermen form a natural relationship between patron-client institutional and profit sharing arranged by patrons. Empowerment towards institutional formation that benefits farmers has not been adequately pursued. [4]

3.3 Stimulation of Changes

Every year, the local government through the Local Government Work Unit (i.e., known in Indonesian acronym as SKPD) was tried to empower the rural poor communities. In the area of this study, i.e., Jeneponto Regency area, the planning plot of empowerment program in principle was started by the proposals from the village. These proposals included the number of target groups and the number of members in each group that would be as the implementers of the poverty alleviation program. Those proposals from each village were proposed to the government through the relevant SKPD. In addition, there were programs arising from SKPD initiatives or related ministries through certain SKPD and consulted with the community through the village apparatus.

Among the programs in the village observed was Kapita Village as an agricultural village, in the last 5-7 years, there had been conducted several programs related to the poor communities. These programs included, namely cultivation of food crops, i.e., there was a water pump and tractor assistance program, in the field of livestock, i.e., there was a collective business group (KUBE) coaching program that specializes in goat farming and then in the field of forestry that was in the form of community forestry program through the groups system. Besides, there was also coaching group of joint ventures, such as furniture groups and workshop groups.In the fishermen community in MatiroBombang Village, Pangkep District observed, there are several programs in several groups such as: the program of making fish abon that organize themselves in groups of women or housewives; coaching program of seaweed farming group; a floating net cage program and a crew fishing fisherman group building program.

The empowerment approaches and methods for each program were different based on the implementers of SKPD and the program types being conducted. But, generally each program used a group approach. In that area, several SKPDs involved in empowerment and poverty alleviation programs, namely SKPD from the Department of Social Affairs, the Department of Community Empowerment and Regional Development (i.e., known as Indonesian acronym as BPM-PD), the Department of Agriculture, and the Department of Forestry,but in KabupatenPangkep more emphasized on the Department of Fisheries and Marine. and Social Service Department

The total of funds and resources in each activity or in each group, based on interview results with the group leader, each response was different. There were in the form of money and there was also in the form of aid goods, seeds, training and other facilities relevant to the needs of the groups. For example, in coaching of KUBE goat breeding in Kapita Village, each group received IDR 20 million and subsequently each group which was under the supervision of a companion, bought their own goats to be kept together in groups.

The delivery mechanism of development funds that was conducted in each empowerment program can be divided into two, namely (1) the SKPD submits the resources or funds directly to the groups which are guided and forwards to the group membership, and (2) SKPD submits the development fund through Village Head for subsequent resources/coaching funds received by the group to be managed during program implementation. Both mechanisms systems based on the recognition that group members were not entirely accepted in accordance with the agreements depending on the escort of related SKPD. But, overall the involvement of the Village Head as an administrator representing the government at the village level is a supervisor charged with determining whether the grants and resources in each program are properly used and accepted in accordance with the applicable rules.

The involvement of other parties, such as government, private sector, universities, international agencies and NGOs during the study conducted for some of the mentioned programs did not occur collaboratively because the whole program was implemented only by SKPD, except in some activities where the program beneficiaries were recruited from community leaders and also from NGOs. The funds receiving mechanism in the group was usually accepted by the group leader. If the resources or assistance were in the form of funds or down payment system, the group leader would submit to the group treasurer. But if the aid program was a tool or seeds, then it forwarded to each group member.

In each program, there was always involvement of poor household communities. In each group for each empowerment program, there was a pattern of group membership that showed the proportion of members involved. Although program schemes which were implemented were aimed to alleviate poverty for their target groups, but in groups, they were not entirely composed of poor farmers' households, but it was generally combined with more prosperous households, because poor farmers were generally poorly educated. There were conditions where members' involvement can be shared by 50 percent of poor farmers and 50 percent of the more prosperous members of the community, but there were also some 70 percent of poor households but the rest were either prosperous households or vice versa. In the early stage of program implementation, the groups' condition was always intact with the agreement and the collective rules under the assistance of the officers who were entrusted by SKPD but along with the program progress, the interaction of group members decreased following the interests of the individual entrusted manager.

The poor fishermen households' response to stimulus changes by the government and outsiders is short-term. The efforts of fishermen (cultivators) in managing the assistance provided are not long-term meaning. For example, in a group of seaweed farmers 4 years ago received the help of strap ropes and seedlings. Each member at that time gained 200 stretch. Among the 200 members of the group, only about 10 people or 5% are able to develop themselves well and respond positively to change, by means of each year trying to increase the number of strings that he runs. Among the cultivators who initially received only the same aid of 200 stretches, it has now grown to 800 expanses. One of them is *Asr* (42) by setting aside little by little of his income to increase investment. According to *Asr's* acknowledgment, each stretch can produce 5 kg of dry at a price of Rp 8000 per kg then in 2 months it can generate 32,000,000 rupiah. This amount does not include the acquisition of catching crabs and catching fish. Harvest if the production increases, it tries to set aside its income to increase investment and can finally succeed until now.

Farmers' household responses towards the changes stimulus which was conducted by the government and other parties were in term of short-term or long-term. In some programs, such as water pump assistance, tractor assistance and goat farming, initially they still managed the resources collectively but after the project was ended, the activities were dominated by certain farmers, and returned to individual activities, so that the group's institutions were not sustainable. Generally, the relationship pattern between them was in an imbalanced relationship between patrons and clients, where group leaders oversee several members and place their members in the position of hired labor, not as partners in an activity in order to improve common welfare.However, based on interviews with farmers, there were positive things that can be found in the empowerment as occurred to farmers around the forest area in *Kapita* Village. They considered that the government's efforts to stimulate each empowerment program involving them, is as a long-term opportunity to improve the economic condition of the society. For example, several years ago the government (SKPD) gave the help of *gemelina* tree seedlings to be planted around the field, now it is expanded with other plants, such as mahogany and cashew nuts. The program was successfully developed. There were some farmers who were not as the group members after seeing their neighbors in the field succeeded, now they are consciously trying to plant these plants on their land by imitating their neighbors' system, although there are still many farmers who are not concerned with the sustainability of the business.

In some poor fishermen's households whose livelihoods were looking for crabs in the past few years there was a help from the iron-braided distribution program, each with 150 vats per iron-made fisherman but the durability of the fishing gear did not last long. For the fishermen the help of the program helps them, especially the poor but can not grow and only lasts 1 year. Only interesting, in making bubu develop new innovation which originally *bubu* of iron changed with material from bamboo material. In addition to more durable is also lighter, so the tool is more quickly adopted by fishermen. It's just the cost of making for one *bubu* need cost Rp 25.000 / unit in installed condition. This means that for 200 *bubu* required investment of Rp 50.000.000.

This is felt heavy by poor crab fishermen. Especially when the investment is borrowed at the retainer who will consequently be paid by depositing the production. Return on investment is often difficult to pay in the short term, even the debt can last all time. However, in terms of getting capital both fishermen and poor farmers experienced the same thing. In the village to get capital, the only easy alternative is through the punggawa (village financiers). This tendency defeats the option of obtaining capital, for example from the Bank. When farmer or fishermen are asked about the two options they suggest reasons for the reason such as the retainer is close to the residence, easy to do because the courtier persists, the rules are not complicated, payment is not bound by time no need collateral (collateral)

3.4 Collective Commitment

Almost all programs which have been conducted by SKPD used a group approach in distributing support. However, the group's approach to each program in its practice is more varied. The fact that, each program is completed by a group that originally proceeded with the rules agreed with SKPD, and it is not sustainable as expected, except in the group formed on the initiative of the community itself. Based on the observation in that region, the communities' responses, especially the poor household communities for each empowerment program in stimulating change, are relatively not disappointing, especially among the younger people, however the efforts made in collective self-development in fact are still very low. Function of groups as a place where societies gather and as a place to experience-based learning is finally only as a tool to get aid. After the aid is received, each individual is dissolved with his/her activity, while waiting the assistance proposal or appeal for next year. The weakness of collective commitment to any empowerment program can be due to several factors, such as:

a) Some programs were conducted hastily without further reviewing to the targeted group bases in those programs. Group formation was often carried out where SKPD simply accepted the proposal of society or the district or village. The group that became the direct program implementer was accommodated without considering the proximity of the working area of the farmers or fishermen to be nurtured. There were group formations conducted by recruiting based on residence area, while the working area was still separated apart.
b) Learning from the experience of more functional guided groups and now they are still sustainable early establishment based on natural mechanisms, such as mutual suport or resources pool (i.e., known in Indonesian term as *arisan*) mechanism, for example, in those groups, there are elements of mutual support that formed among them by giving priority to togetherness and trying to avoid the mutual domination that caused conflict between them. Those groups still exist and are developed, but they are only given less opportunities for coaching.
c) There was a group that initially started at the initiative of richer farmers and subsequently recruited poorer households. Recruitment was not a working partner to jointly achieve common goals but those poorer households were employed. The relationship system between one and another was not equivalent so that the tool aids for example in the program was ultimately dominated by richer individuals. Group conditions were formed like this in each program which tends to be *Toukenism* as termed by [1] where each member is recruited only as wage laborer. Therefore, the degree of member participation of those groups did not improve.

d) Generally, collective commitment was not well established because in the beginning, the groups were formed without a strong interest and concern among its members to accomplish the goals jointly. Besides, the knowledge of society about how to manage the group was still very poor.

The problem solving jointly and group decision making in the groups were not appeared because the groups which got program assistance were generally no longer shown in the activity after each individual received the first and second year assistance funds. This commitment was not developed, due to the low sensitivity of members to the benefits and usefulness of the groups in the long term and less growing attitudes toward others, the low attitude of sharing and not growing trust to other members. Based on the interview results and all previous descriptions, there were two patterns of group formation occurred at both sites observed which are presented in Table 2.

Table 2. General Characteristics of Sample Groups on Two Patterns of Poor Household Groups, in Both Locations (Jeneponto and Pangkep Regency) in 2016

No	Patterns of Group Formation	General Characteristics	Sample Groups
1	Form groups are formed by natural mechanisms	• Groups are formed because of the urgency of common needs. • Managing resources jointly for mutual consciousness. • There is a natural mechanism as a group rule. • They are managed democratically and participatory. • Member interaction is relative routine. • There is *arorosseng*principle. • There is a shared desire to progress jointly. • Relations with outsider parties are less. • The ability to access resources from outside is still poor.	• Farmer Group (KUBE) "Jaya Bersama" Tombololoe, Kapita Village • The *Arisan*group of workers around the forestry area.
2	Form groups are based on external parties' initiatives.	• It has long been known by external parties (SKPD) or NGOs • It is relatively able to get additional resources from external areas. • It is managed by applying *tokenism* (poor households are generally as a complement or labor). • It only utilizes groups to get assistance. • After getting assistance, the group members tend to leave groups and the resources are managed individually. • The results of group effort are not long-term. • Generally, the groups disperse after program or project finish. • It tends to be dominated by individuals.	• Farmer groups who receive tractor aid, • Beneficiary group of water pump • Group of furniture makers • KUBE group of goat breeding

Table 2 describes that initiatives on the initial formation of each group illustrate different characteristics of the group's sustainability and performance in terms of its institutional. The naturally patterned groups have a more solid group performance and practice shared values, share values and members' solidarity based on individual honesty, whereas in the pattern of group formation on external initiatives applies on the contrary.

These characteristics show that the farmer communities still maintain the local wisdom which is as the guides for social life among them. Based on the interview results with some members of farmer group "Jaya Bersama", they stated that the growing collective awareness among them is influenced by a long-lived principle in society, namely *a bulosibattangacerasitongka-tongka* (working together like a bamboo, and feeling the same fate). In addition, there is also the work principle of mutual support, which is known as *arroroseng (makassar), Asedisedi (buginess).* This principle is better known as mutual cooperation or*gotongroyong*[12]

In the long run, there should be a development (empowerment) effort that focuses on the dynamics of the community itself. [5] describes that the success of an empowerment program is a resultant interaction of development elements with the empowerment strategy applied. Efforts and empowerment strategies are contingencies between the evolution paradigm and the complementary paradigm of revolution in proportion to the farmers' conditions and institutions. In addition, the important thing is that in a success program also dominantly determined by the relationship between the funder in this case the government (SKPD) with poor households as the recipient. As [1] [13] points out however, at the project level, the relationship between donor and recipient is fundamental to the success of any intervention.

4. Conclusion

Based on all results descriptions, it can be concluded that:

1. Stimulation of changes made through development programs or community empowerment in the areas studied both among farm households and poor fishermen households for some programs received positive response from farmers and fishermen. However, the collective commitment to the breeding is relatively weak, since the group formed in each program is not done through good planning and concepts.
2. There are two patterns of group formation that are natural and formed formations initiated by outsiders. Groups that are naturally formed are more institutionalized and have characteristics such as intense and relatively routine interaction, strong mutual trust, and have a common form or mechanism shared for common purposes. The group can form the basis for sustainable development in improving the welfare of the poor.

5. References

[1] Ohama, Y. (2001). The Participatory Local Social Development (PLSD) Concept and Frame Work. Nagoya: Nihon Fukushi University-The Government of Japan.

[2] Karsidi, R. (2002). PemberdayaanMasyarakatPetani Dan Nelayan Kecil. Research Paper presented at SemilokaPemberdayaanMasyarakat. Semarang: BadanPemberdayaanMasyarakatJawa Tengah. (Online) retrieved from: eprints.uns.ac.id.

[3] Murjanayasa, I.G.W. (2008) PenanggulanganKemiskinanBerbasisMasyarakat, JurnalEkonomidanSosial 1(2) :86-91.

[4] Syafiuddin, &Mahsyar, A. (2016). Condition and Characteristics of Poor Farmer Households in the Perspective of Participatory Local Social Development in Jeneponto Regency, South Sulawesi, Indonesia. European Journal of Social Science. 51(1) : 133-140.

[5] Suradisastra, K. (2016). StrategiPemberdayaanKelembagaanPetani. In Forum Penelitian Agro Ekonomi 26(2) : 82-91.

[6] Prasetyo, P. E., &Maisaroh, S. (2014). Model StrategiPemberdayaanEkonomi Rakyat SebagaiUpayaPengentasanKemiskinan. Trikonomika Journal, 8(2) : 103-116.

[7] Sharma, P.N, &Ohama, Y. (2007). Participatory Local Development, an Emerging Discipline. Delhi: Bharat Book Centre.

[8] Rusastra, I. W., &Napitupulu, T. A. (2007). KarakeristikNelayandanKeluargaMiskin di Pedesaan. Bogor: PusatAnalisisSosialEkonomidanKebijakan.

[9] Jeneponto regency Statistic. 2015. https://jenepontokab.bps.go.id/. Accsessed 20th September 2017

[10] Pangkep Regency Statistic. 2015. https://pangkepkab.bps.go.id/. Accsessed 20th September 2017

[11] Uphoff, N. T. (1992). Local institutions and participation for sustainable development. London: Sustainable Agriculture Programme of the International Institute for Environment and Development.

[12] Mubyarto. (1989). PengantarEkonomiPertanian. Jakarta: LP3ES.

[13] Jacobs, C. (1998). Institutional Strengthening And Technical Co-Operation: Developing A Best Practice Model. Journal of International Development, 10(3), 397-406.

Identification of Knowledge Maritime Based Songs of Children in Banyuwangi District

Sri Surahmi[1], Burhani Buhani[2]
[1]Conseling Departement, of University of PGRI Banyuwangi, JL. Ikan Tongkol NO.22. Banyuwangi, 41482, Indonesia
Email: g.rubiono@fdi.or.id
[2]Mathematic Education Department, University of PGRI Banyuwangi, JL. Ikan Tongkol No.22. Banyuwangi, 41482, Indonesia
Email: bburhani@gmail.com

Abstract –Music and songs have effect due to development of the children. Maritime based songs can be used to encouraged children about the spirit of maritime. It is an important thing to build maritime awareness to children regarding Indonesia as a maritime nation. The research is done by interviewing 40 pupils from elementary schools located in inland and coastal area. The result shows that pupils know 1-2 maritime based songs (1,6 in average in inland area and 1,45 in coastal area). This result can be used as recommendation to enrich children maritime based songs.

Keywords:*maritime, song, children, Banyuwangi*

1. Introduction

Music has a vast influence over the nations and peoples on this planet. It has been used in every culture, and is often connected with anxiolytic and analgesic properties [1].Music can be found in every culture all around the world. Music has become such a big part of our lives, that researchers can't help but want to study how music affects people, especially children [2]. Music plays an important role in the socialization of children and adolescents. Listening to popular music is considered by society to be a part ofgrowing up [3].

Music can represent a developmental resource and psychological research on music opens up a scientific window to the psychological, social, and cultural needs [4]. Exposure to music and music education can have a positive influence on child functioning [5].There is a great deal of research supporting all the positive effects of early musical experiences on young children's lives. Within the first few years of a child's life musical experiences can help to develop a child's literacy, language and math skills, social interactions, and emotional well-being [6].

From the time they enter preschool, all children have musical knowledge – they have preferences for certain kinds of music, some songs that they can sing and a basic understanding of rhythm and melody. Through musical activities, young learners are encouraged to focus their attention on various aspects of the sound structure of language which helps them to improve their listening skills and to develop metacognitive and metalinguistic skills [7].

Research about music and children have already done for the exposure to music and cognitive performance [8], placing music at the centreof literacy instruction [7],exploration of the impact of music [5, 6, 9], potential role of music [10], the influence of music on the development of children [2] and determining the cultural identity of a child through folk literaturesuch as folk songs [11].

In agricultural sector, folk media such as folk songs can be involved in rural development [12]. Folk media has role in nation building. Folk media also has role in national integration and conservation of cultural heritage [13]. Cultural identity is recognized and determined through the interplay of different identifying elements such as location, language, ethnicity, myths, religious culture, nations, history, art and traditions (such as songs, traditional costumes, dietary habits) that make up the national cultural heritage [11].

Indonesia has a large ocean territory which is a potential resource for the future and for the next generations. It is important to encourage the children about maritime spirit. This effort can be done by several ways. One of this effort is through music, especially songs based on maritime theme. By learning and knowing maritime based songs, children insight of maritime can be develop.

Folk culture is an important factor for the outline of the concept of cultural identity, which seals and keeps alive the soul of a people [11]. Folk songs have played a major role in conveying developmental and socially relevant messages most frequently and most effectively [14]. Young children have a remarkably selective sensitivity to shared cultural knowledge. Shared knowledge may be a powerful determinant of children's social preferences, both because it underpins effective communication and because it is conveyed by others through social interactions and therefore can serve as a marker of social group identity [9].This research is aimed to identify the knowledge of maritime based songs in children.

2. Method

The research is done by direct interview to 40 pupils of elementary school in Banyuwangi area with 10-11 years of age. The pupils are chosen by random order. Schools are selected based on the location of the inland and coastal areas. Pupils are ask how many maritime based song he/she know. The kind of song is either children song, national song and folk song. They also have to tell the title of the song and sing part of it. The answers are categorized by the similarity.

3. Result and Discussion

The results show that most of pupils know about 1 or 2 songs. In inland area schools, pupils know 1,60 in average. In coastal schools, pupils know 1,45 in average. The total average is 1,525.

Table 1. Interview Results

Number of songs	Number of pupils				Average
	Inland area	%	Coastal area	%	
1	10	50	12	60	55.0
2	8	40	7	35	37.5
3	2	10	1	5	7.5

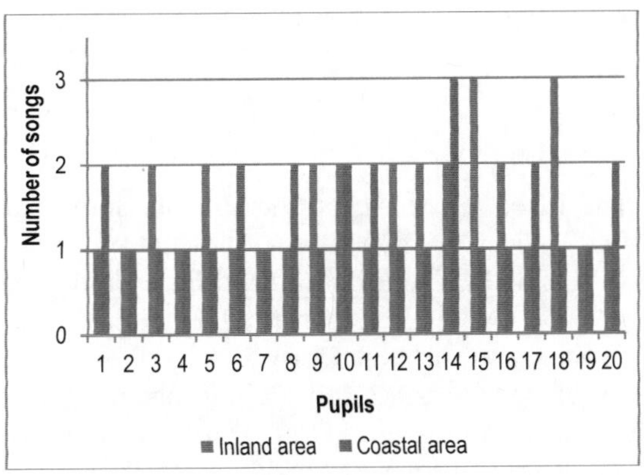

Figure 1. Graph of respondens answers

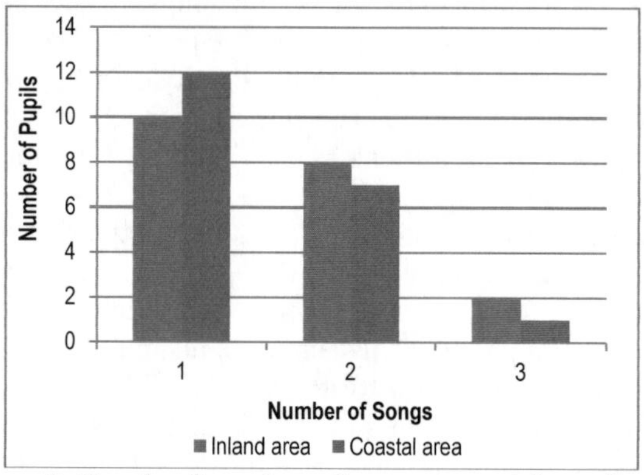

Figure 2. Graph of number of songs vs number of pupils

The result also shows that 55% of the pupils respondents only know one song. The songs that mostly know by the pupils is song by the title of "*NenekMoyangku*" which is children song about the ancestors who became brave sailor. The second song is song by the title "*KapalApi*" which is children song about steamship. Pupils learn these songs in kindergarten. The third song is "RayuanPulauKelapa" and "*NyiurMelambai*" which are national songs. Pupils also mention some songs that included as pop or adult songs.

The number of maritime based songs that known by the pupils are relatively poor. The average of pupils who know more than 3 songs is only 7,5%. This condition is should be improved regarding Indonesia as maritime nation. Folk songs such as "*TondukMajeng*", a Madura ethnic maritime based song should be more introduced to the children. This Madura song has moral and educational values [15].

The condition is also can caused by the number of existing maritime based songs. There are only a few maritime based songs for the children. Song writers must encourage to write maritime based songs, especially for the children. Reproducing more maritime based songs will be enrich the reference for the childrenEducation plays a preponderant role in the development of children. With the consolidation of pedagogical practices that mediate the relationships between children and culture, it is possible to organize activities focusing on different dimensions of human development and personality [13]. It is important to educate children about maritime based songs in school in order to encourage maritime spirit.

4. Conclusion

The number of maritime based songs that known by the pupils are relatively poor. It is important to enrich maritime based song for the children and encourage the maritime spirit through education.

5. Reference

[1] Kent D, (2006). The Effect of Music on the Human Body and Mind, Thesis, Honors Program, Liberty University

[2] Chau C, Riforgiate T, (2010). The Influence of Music on the Development of Children, A Senior Project, Psychology and Child Development Department, College of Liberal Arts, California Polytechnic State University, San Luis Obispo

[3] American Academy of Pediatrics, (2009). Policy Statement-Impact of Music, Music Lyrics, and Music Videos on Children and Youth, Organizational Principles to Guide and Define the Child, Health Care System and/or Improve the Health of all Children

[4] Miranda D, (2013).The Role of Music in Adolescent Development: Much More Than the Same Old Song, International Journal of Adolescence and Youth 18(1): 5-22

[5] Hogenes M, van Oersb B, Diekstra RVW, (2014). The Impact of Music on Child Functioning, The European Journal of Social &Behavioural Sciences, 135: 1507-1526
[6] Mead, RS, (2008). Children and Music: An Exploration of the Impact of Music on Children's Lives, Thesis, The Degree Master of Arts in Education, Marietta College.
[7] Bolduc J, Fleuret C,(2009).Placing Music at the Centre of Literacy Instruction, Research Monograph #19, a partnership between The Literacy and Numeracy Secretariat and the Ontario Association of Deans of Education.
[8] Schellenberg EG, Nakata T, Hunter PG, Tamoto S, (2007).Exposure to Music and Cognitive Performance: Tests of Children and Adults, Society for Education, Music and Psychology Research 35(1): 5–19
[9] Soley G, Spelke ES, (2016).Shared Cultural Knowledge: Effects of Music on Young Children's Social Preferences, Cognition 148: 106–116
[10] Zeromskaite I, (2014)The Potential Role of Music in Second Language Learning: A Review Article. Journal of European Psychology Students, 5(3) : 78-88
[11] Stavrou EP, (2015).Determining the Cultural Identity of a Child through Folk Literature, American Journal of Educational Research 3(4): 527-534
[12] Daudu S, Anyanwu AC, (2009). Folk Media Preferences and Prospects for Agricultural Extension Service Delivery in Benue State, Nigeria, Tropical Agricultural Research & Extension 12(1): 47-50
[13] Kumar S, (2012).Role of Folk Media in Nation Building, Voice of Research 1(2): 1-6
[14] Bissoli M deF, (2014).Development of Children's Personality: The Role of Early Childhood Education, Psicologia em Estudo, Maringá19(4): 587-597.
[15] Wakil A, (2015). An Analysis of Moral and Educational Values on Madura Folk Songs, JurnalIlmiahBahasadanSastra, 2(1): 47-57

Implementation of Markerless Augmented Reality Technology Based on Android to Introduction Lontara in Marine Society

Mila Jumarlis[1], Mirfan Mirfan[2]

[1]STMIK Bina Adinata, Jl. Serikaya No. 8 Kab. Bulukumba
Email: milajumarlis@almitra.ac.id
[2]STMIK Handayani, Jl. Adyaksa Baru No. 1 Kota Makassar
Email: mirfan@handayani.ac.id

Abstract. Local language learning had been leaving by people especially young people had affected technology advances so that involved lack of interest to learn culture especially local language. So required interactive and interest learning media for introduction Lontara. This research aims to design and implement augmented reality on introduction Lontara on mobile device especially android. Application of introduction Lontara based on Android was designed by Vuforia and Unity. Data collection method were observation, interview, and literature review. That data was analysed for being information. The system was designed by Unified Modeling Language (UML). The method used is a marker. The test result found that application of Augmented Reality on introduction Lontara based on Android could improve public interest for introducing local language particularly young people in learning about Lontara because of using technology. Application of introduction of Lontara based on Android used augmented reality occurred sound and how to write Lontara with animation. This application could be running without an internet connection, so that its used more efficient and could maximize from user.

Keyword: Augmented Reality, Markerless, Android and Lontara

1. Introduction

Recent years, Augmented Reality (AR) was an important field of research. AR was a merger of objects which exist in virtual to reality within two as well as three dimensions that can be touched, seen and heard. AR gave great opportunities in science and technology because this disciplines emphasized directly on training [1]. Through AR technology, someone could get a sensation of exploration and learn with different and unique ways because he directly involved in Augmented Reality.

Local language learning had left by people. Only definite people were still looking up and defending local culture value. Currently, people particularly young people had influenced of technology advances, so that involved lack of interest to learn culture especially local language. With the advent of smartphone based on Android, so that involved lack of interest to learn by the book. Therefore required application based on Android which could be learning and companion media [2].

Augmented Reality (AR) was a term for the environment which combines reality and virtual with made by computer so that the boundary between reality and virtual became very thin. AR had one of the advantages was that could implement extensively in various media. As the technology develops, used of technology AR on learning process of student in primary school was expected to stimulate student in order not to quickly bored, more understand material and improve student's comprehension with supporting program of local learning especially introduction of Lontara by educational institutions, primary school, in order to make class be interest, simple, creative and interactive. So required learning media which be interactive and interest in introduction Lontara [3].

2. Method

2.1. Markerless Augmented Reality

Augmented Reality had two methods of target recognition were used for marker and markerless. The marker was an image with unique pattern which could be taken by the camera and recognized by AR application. The marker could be the photo of a real object or artificial image with the unique pattern. This marker used a marker recognize technique or fiducial marker. Markerless was a tracking method where AR used an object in reality as a marker. AR with the markerless method used tracking technique with nature feature. This technique used edge, corner detection and texture from image or object. The marker had sustained several times evolution, among others :

- Barcode.
- QR Code.
- Printed AR Marker.
- Natural Printed AR Marker.
- Real Life Marker as Human Face [4].

The marker which natural printed AR and human face was a markerless category. On Vuforia, Marker called image target.

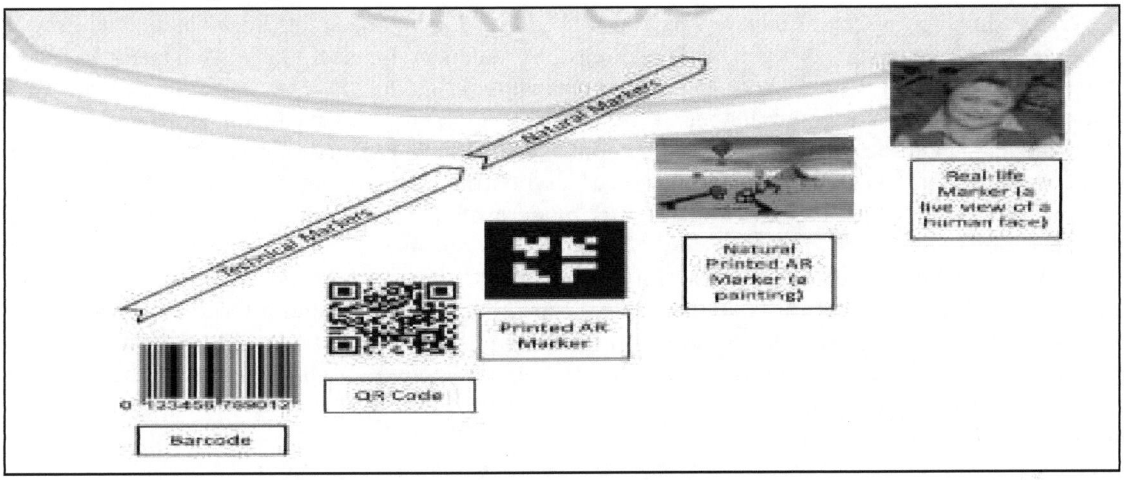

Picture 1: Evolution of Marker

Image target can be tracking and ideal image target by AR system comparises: Features with a complicated pattern, for example, outlook image, group of people image, collage, etc. The contrast was good. There isn't repetition pattern, for example, grass field, plaid. The color graph was 8 or 24 bit with PNG or JPG format, its size lack of 2 MB, and must RGB or grayscale. One of many Augmented Reality methods currently involved was Markerless Augmented Reality method. Through this method, the user didn't need to use a marker which is boxy and black to show digital elements. Like recently involving by greatest Augmented Reality company in the world, Total Immersion. They had made various of markerless tracking technique as their pledge technology like Face Tracking, 3D Object Tracking and Motion Tracking [5]. *Face Tracking:* With using algorithms which they developed, the computer could recognize human face generally with recognizing eye, nose, mouth, human position then would ignore other objects around like home, tree, and others object. *3D Object Tracking:* 3D object tracking could recognize all of the surrounding shape's object, like the car, table, television, etc. *Motion Tracking:* On this computer, technique could capture motion. Motion tracking had used extensively for producing movies which stimulate motion. For example on Avatar movies where James Cameron used this technique for making that movie and using in real time. *GPS Based Tracking: It* recently began popular and considerable to develop on smartphone application (iPhone and Android). By using GPS features and compass which exist on the smartphone, this application would take data from GPS dan Compass then show in shape direction which we want in real time, in fact, there is some application show it in 3D [6].

2.2. Lontara

Lontara was a script for peoples of South Sulawesi. Lontara took from Java and Malay language was lontar or palm tall (Borassus Flabellifer). Therefore Lontara was a script which wrote in leaf tall. The tradition also had done by people of Sunda, Java, Bali in writing script by their lontar. There was also an opinion that epistemology said Lontara composed from two words by *raung* (leaf) and *talak* (lontar). Word of *raung talak* sustained evolution became Lontara. There was a unique lontar, like audio or tapes. Its text was written one line on a narrow leaf lontar which roll up, can be only written when the roll was turned back. Articles on roll moved in front reader up, from left to right. One of *Lontara gulung* was La Galigo, an original epic by people of Bugis, estimated which written on the 14th century before Islam. This opus had numbered 6.000 pages, with five syllables. Background of location of La Galigo's stories was in luwu which kingdom considered as location where people of Bugis came up. Lontara (someone called *Lontaraq* or *Lontarak*) was genuine Lontara of people of Bugis, Makassar, and Mandar on South Sulawesi. Actually, still there was an old letter of Makassar, its age older than Lontara. But then the sustainability was Lontara. There was an opinion that this Lontara different with another script in Indonesia like Bali, Jawa, Lampung, Sunda. According to philology Lontara had a relationship with *pallawa's script* from India. But there was an opinion said this script came from *pallawa*. Besides their script, people of Bugis used their dialect which called "*Bahasa ugi*". Meanwhile, another people in South Sulawesi was *Saqdan Toraja*, not having a tradition of writing, only having a tradition of oral.

According to history, Lontara was introduced by *Sabbannarak* or *Syahbandar* of Gowa' Kingdom, Daeng Pammate. When Kingdom of Gowa ordered by King of Gowa IX, Daeng Matanre Karaeng Manngutungi who called Karaeng Tumapakrisik Kallonna, Daeng Pammate got two positions in this kingdom were Sabannarak besides Tumailalang (minister of home affairs). At the time, Karaeng Tumapakrisi Kallonna gave a command to Daeng Pammate for creating script which using for writing. In 1538, Daeng Pammate succeeded writing Lontara consist of 18 letters and also old Makassar letters. Finally, this Lontara was modernized so that its word became 19, due to influence Arabic Language. Lontara was traditional script people of Bugis. The form of Lontara according to humanist, Prof Mattulada (Alm) came from "*sulapa eppa wala suji*". *Wala Suji* came from word of *Wala* which mean separator, fence, keeper, and *Suji* which aim princess. *Wala Suji* was a bamboo fence in ritual events which form part of *Ketupat*. *Sulapa eppa* was mystical trust of classic Bugis which symbolized an order of the universe, fire-water-wind-sail. This Lontara in general used to write a rule of procedure of government and society. One of the scripts which the most famous on Bugis was La Galigo.

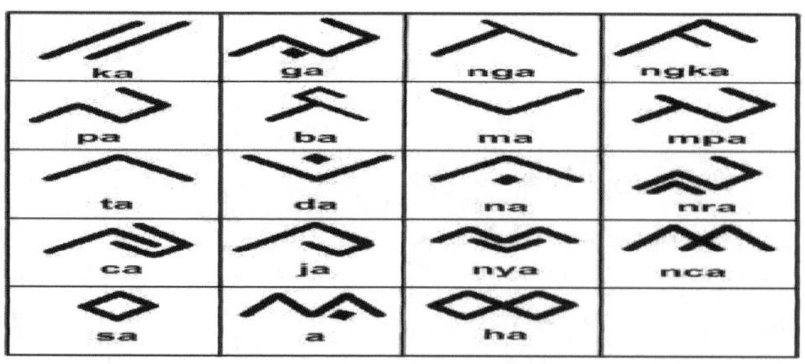

Picture 2: Lontara

Design of digital application of Lontara was an application which could show object virtually with available size on 36 adjusted with a marker which had been inputted to library vuforia SDK.

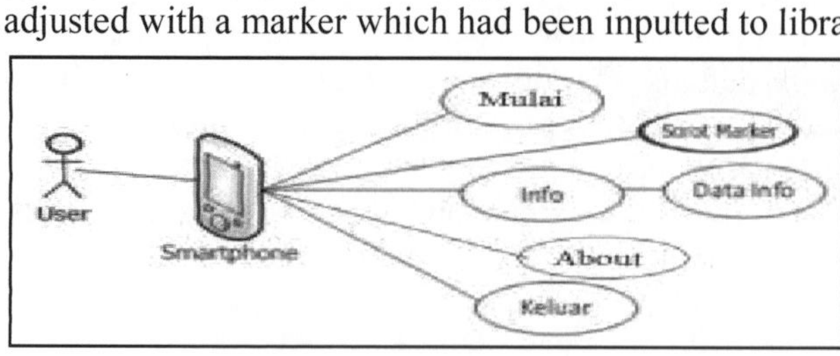

Picture 3: Architecture Design of System

On this picture described that android phone camera would highlight object where eventually if a marker was recognized that showing image object 2D/3D for trying virtually. As for design marker which used can show in picture 4.

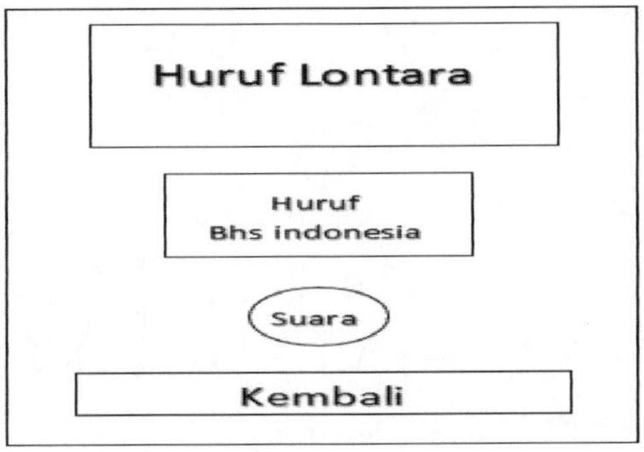

Picture 4: Design of view of Lontara

Lontara came out if we were clicking one of a button on the form of Lontara's menu and there was also Indonesian language and sound button. This Augmented Reality marker didn't use many forms, when the program was running, the first page showed logo, after that showing main interaction page which showing view of behind camera, when camera detected marker, then coming up a virtual object on a screen.

2.2.1. *Icon view on the main page*

The following was icon view of implementation application of Augmented Reality which recognized Lontara on a smartphone screen.

Picture 5: Icon View of AR Lontara

Picture 6: The First View when running

After camera active, highlight that camera to marker, if user click button start that would come out three menus of word choice, the way of writing Lontara and button out. And if user click button of a word that would come out word choice of Lontara.

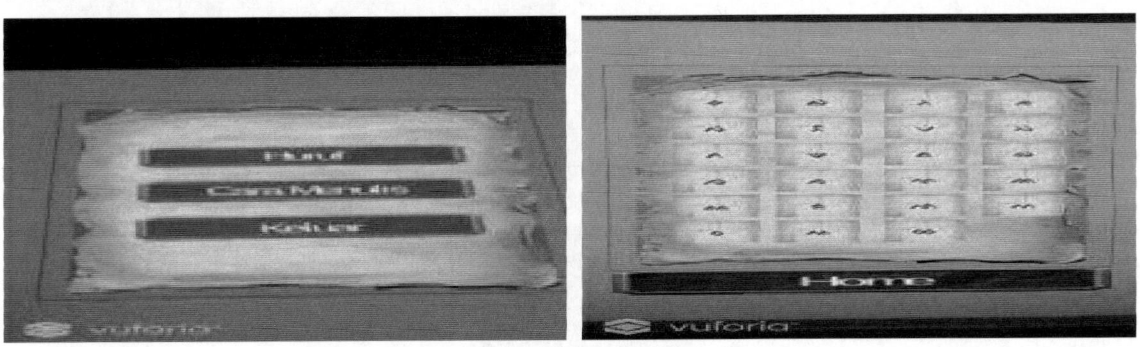

Picture 7: Output Choice Menu of Lontara

IOP Conf. Series: Earth and Environmental Science **156** (2018) 012017 doi:10.1088/1755-1315/156/1/012017

If user clicks one of Lontara that came out 3D Lontara, the button of audio and back. Further, if user click one of choice menu way of writing a word of eating would come out in this animation view wrote a word of "*KA*".

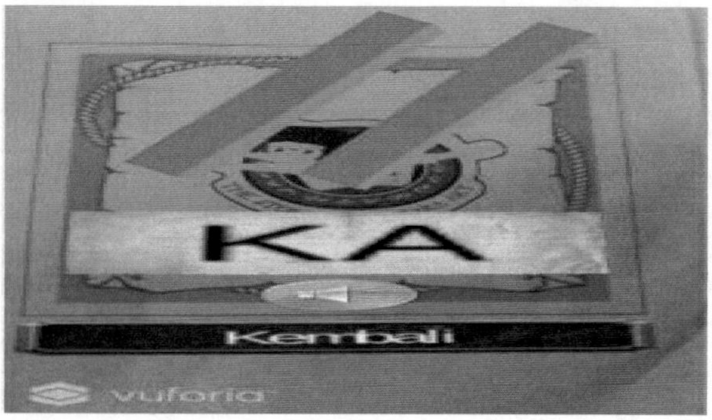

Picture 8: View of Lontara

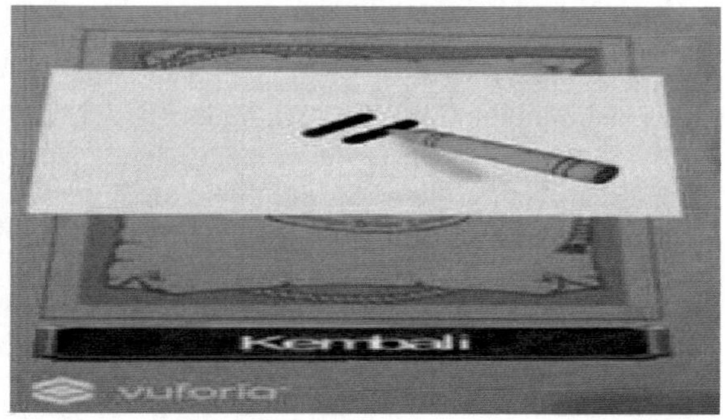

Picture 9: Animation wrote "*KA*"

The application which used to :. Blender and Unity 3D. Blender was used to describe the 3D object and Unity 3D was used to make virtual reality application.

3. Result and Discussion

This research used to experiment with software which builds with using Black Box method.

Table 1: Experiment Results of Range of Marker

No	Range of Marker	Achieved Result	Experiment Result
1.	< 30 Cm		[x] Accepted [ ] Unaccepted
2.	< 30 Cm		[ ] Accepted [x] Unaccepted

Table 2: Experiment Result of Radiance of Marker

No	Radiance of Marker	Achieved Result	Experiment Result
1.	With Radiance		[x] Accepted [ ] Unaccepted
2.	Without Radiance	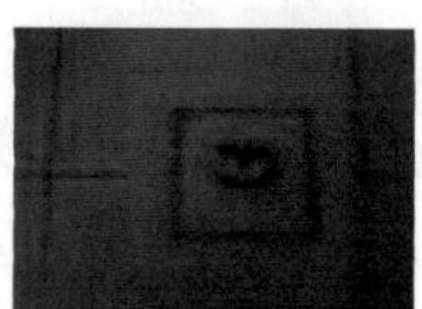	[ ] Accepted [x] Unaccepted

Based on this experiment table, Black Box overall could be taken the conclusion that marker could be recognized if range between marker and phone was 30 cm – 69 cm, meet the functional requirements however on recognizing the process of an object which determined still there was deficiency so that needed repairing on an object before highlighting.

4. Conclusion

From this experiment, result could be concluded that there were companion media like practice application of Augmented Reality on Lontara based on Android application which could be improving interest to introduce local language especially young people in learning Lontara because of using technology besides the printed book and the application could give benefit in the learning process about Lontara. Application of introduction of Lontara used Augmented Reality could be media to introduce for knowing kinds of Lontara, there were voice and way to write Lontara with animation with introduction media can be used without an internet connection, so that its used more efficient and could be maximized by the user.

5. References

[1] Auliawati Buchari, Steven R. Sentinuwo, dkk. Implementasi *Augmented Reality* warisan Budaya Berwujud di Museum Propinsi Sulawesi Utara . Teknik Informatika Universitas Sam Ratulangi, Manado (2015).

[2] Dedynggego, dkk. Perancangan media pembelajaran interaktif *3d* tata Surya menggunakan teknologi *augmented reality* untuk Siswa kelas 6 sekolah dasar sangria.Stmik Bina Mulia Palu [online]. Tersedia : http://jesik.web.id/index.php/jesik/article/viewFile/13/12.[1 Januari 2017] (2015).

[3] Efendi, dkk Penerapan teknologi AR (*augmented reality*) pada Pembelajaran energi angin kelas IV sd di rumah Pintar al-barokah. (2016).

[4] Helmi Irhamie Apristio,. Aplikasi multimedia Pembelajaran Soal Ujian nasional Tingkat Smp . Program studi sistem informasi Fakultas teknik dan ilmu komputer Universitas komputer indonesia Bandung. 2015.

[5] Shena Febrian,. Aplikasi Simulasi Teknik Sepak Bola Berbasis Android. Program studi sistem informasi Fakultas teknik dan ilmu computer Universitas komputer Indonesia Bandung 2013.

[6] Tahta Alfina Lutfiyati,. Aplikasi Augmented Reality Pembelajaran Pengenalan Hardware Komputer Untuk Sekolah Menengah Pertama Dengan Metode Transformasi Geometri. Tugas Akhir Jurusan Teknik Informatika Fakultas Sains Dan Teknologi Universitas Islam Negeri Maulana Malik Ibrahim Malang (2016).

[7] Indra Wihanry, Phie Chyan . Perancangan aplikasi pembelajaran aksara Lontara dengan metode game based learning . Program Studi Teknik Informatika, Fakultas Teknologi Informasi, Universitas Atma Jaya Makassar (2015).

IOP Conf. Series: Earth and Environmental Science **156** (2018) 012018 doi:10.1088/1755-1315/156/1/012018

Exotic Heritage from Coastal East Java of Batik Bayuwangi

Syaiful Hadi[1], Ikhwanul Qiram[2], Gatut Rubiono[2]

[1]Pancasila and Citizenship Education Departement, University of PGRI Banyuwangi, Jl. Ikan Tongkol 22 Banyuwangi68416, Indonesia
Email:shadi@gmail.com
[2]Mechanical Engineering Departement, University of PGRI Banyuwangi, Jl. Ikan Tongkol 22 Banyuwangi68416, Indonesia
Email:_ikhwanulqiram@gmail.com
[2]Mechanical Engineering Departement, University of PGRI Banyuwangi, Jl. Ikan Tongkol 22 Banyuwangi68416, Indonesia
Email:grubiono@gmail.com

Abstract-Banyuwangi batik is included in type of Pesisiran (coastal area) batik. This batik has not been thoroughly studied, although has tourism attractiveness and the manufacturer production population indicated a great potency. This research is aimed to describe the characteristic of Banyuwangi batik. The research is done by literature and field study. The result shows that the batik has philosophical meaning which represent about the life of the people and natural environment. The motif objects are things that can be found in Banyuwangi area such butterfly, marine vegetation, coconut, etc. The motif pattern is in straight or diagonal alignment. Banyuwangi batik also have contrast color such a combination between red, yellow, etc.

Keywords: Batik, Banyuwangi, pesisiran, characteristic

1. Introduction

Batik is an Indonesian cultural richness which has become the daily life of society [1]. Batik as a one of nation's artworks still existing, practice and develop [2]. Batik has become world fashion market attractiveness since UNESCO crown batik as "Humanitarian Heritage for Oral and Non Object Cultures" [1,3,4]. In addition to its function as fashion, batik also use as household accessories such as tablecloths, pillowcases, bed linen and house decoration [2].

Batik in Indonesia is very diverse in every region. Batik has a variety of types, patterns, motif and style in accordance with the areas origin [1]. Batik has a characteristic which influent by age, environment and geographical aspect [5]. Batik motif has culture values that represent social culture life of the community [4]. Batik motif is a representation of visual imaging [6]. Batik also has spiritual beauty that shown by decoration and pattern formation which full with philosophical meaning [7].

Research in batik has done for the aspect of history, motif, technical production, marketing and product development. The research also done for imaging media such as visual communication, interior product, architecture design and other field [8]. Batik motif research has done for Surakarta batik [9], Druju batik of Malang city [10], Pesisiran (shore area) batik [2] and Semarang batik [11].

The research also has been done for symbolical meaning of batik such as Kliwonan batik from Sragen city [12], Situbondo batik [1] and Kudus batik [5]. Development effort has done for Pasirsari batik of Pekalongan [13] and East Kalimantan batik [7]. The research also done for motif identification method [4], Gedog batik of Tuban city as tourism attractiveness [3] and Tuban batik illustration design book [4].

Banyuwangi as a town in east end of Java island also has batik culture. Banyuwangi batik is included in type of *Pesisiran* (shore area) batik [2,3] which has characteristic in material and motif [2]. Pesisiran batik is different from *Keraton* (palace) batik [1]. Pesisiran batik has potency to develop business opportunity and working field [15].

Research in East Java batik has been done for the area of Tuban, Sidoarjo, Madura and Surabaya [16]. Banyuwangi batik has not been thoroughly studied, although has tourism attractiveness [17]. Research has only been done for Gajah Oling motif [18], one of several types. This research is aimed to describe the characteristic of Banyuwangi batik.

2. Method

The research is done by literature and field study. Departemen Perindustrian, Perdagangan dan Pertambangan kabupaten Banyuwangi (Industry, Trading and Mining Departement of Banyuwangi City) indentified more than 82 small medium enterprises (SMEs). This population shows the potency of Banyuwangi batik.

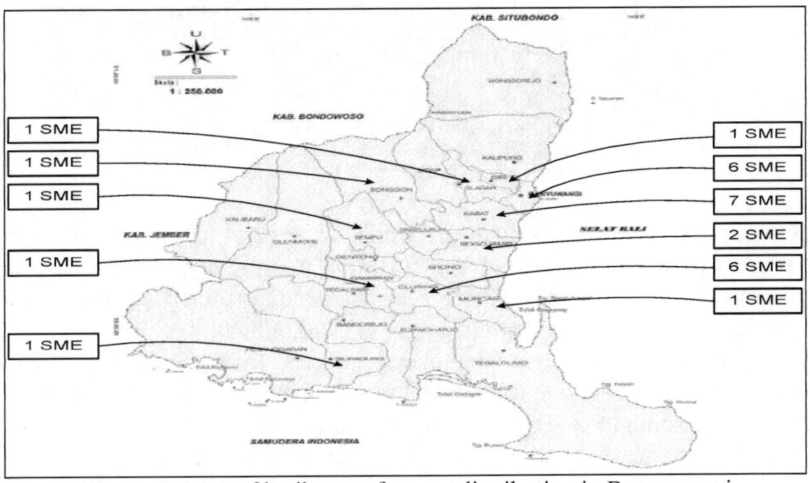

Figure 1. Map of batik manufacturer distribution in Banyuwangi

3. Result and Discussion

Banyuwangi batik motif is a reflection of natural wealth that causes different character from others [19]. There are more than 20 origin motifs which are Gajah Oling, Kangkung Setingkes, Gedekan, Sembrok Cacing, Kawung, Manukan, Galaran, Kluwungan, Semanggian, Moto Pitek, Jenon Kacangan, Blarakan, Sekar Jagad, Totoghan, Kopi Pecah, Sosi, Sisek, Paras Gempal, Sisek Papak, Latar Puteh, Jenon, Alas Kobong, Maspon, dan Ungkel. Gajah Oling is the most famous motif.

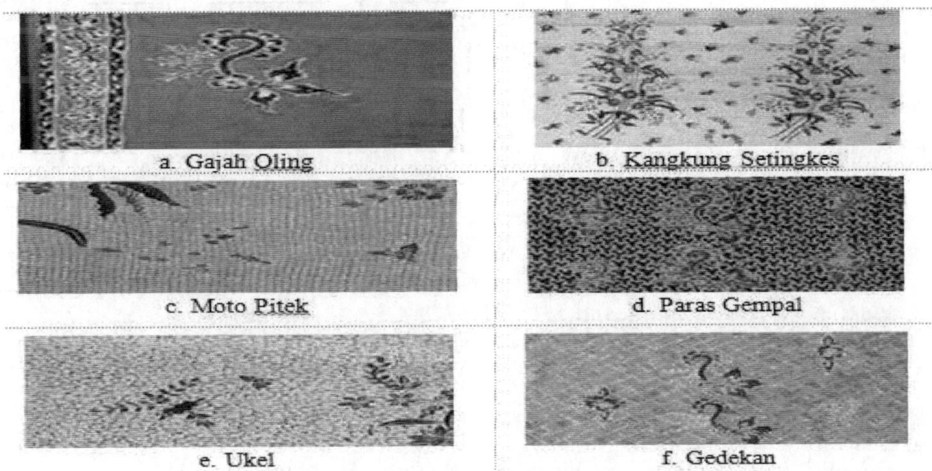

Figure 2. Some motif of Banyuwangi batik

One of the characteristic of Banyuwangi batik can be found in Gajah Oling motif. Gajah Oling is believed as the oldest and the original motif that symbolizes the form of power that grows in the identity of society. This motif is similar with question mark which philosophically represented the shape of elephant trunk (Gajah means elephant) and also the shape of a giant eel (Oling means giant eel or moa). Gajah means something big and Oling comes from word "*eling*" that means remember. Gajah Oling defined as spiritual aspect that people have to always remember to God.

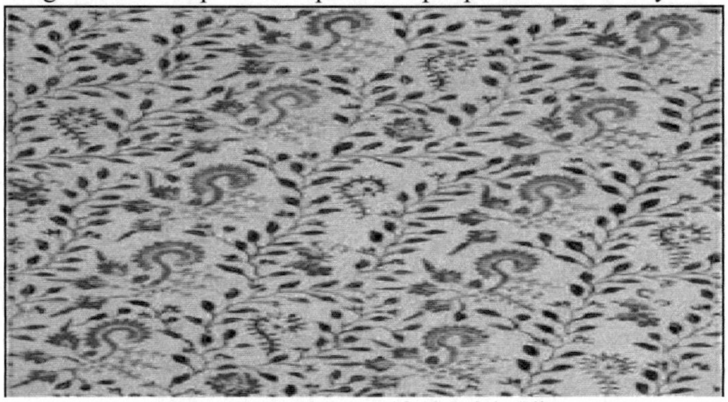

Figure 3. Another motif of Gajah Oling

Gajah Oling figure character is also surrounded by several nature objects as shown in figure 3 that can be found in Banyuwangi which are butterfly, marine vegetation, coconut flower etc. The design of this motif is symmetrical pattern which balancing the space around the question mark shape. The pattern can be found in straight or diagonal alignment.

Banyuwangi batik also has color characteristic. The batik tends to have contrasting color. The color combination tends to describe the diversity of life and natural environment. The characteristic is very different comparing to other Pesisiran batik from coastal area of East Java. Situbondo batik, west area of Banyuwangi, has characteristic that describe the livelihood of most of the population which are fishermen and farmers [1]. Tuban batik, more further west area, has characteristic that influent by Javanese, Islamic and Chinese cultures [4].

4. Conclusion

Banyuwangi batik has philosophical meaning which represent about the life of the people and natural environment. The motif objects are things that can be found in Banyuwangi area. The motif pattern is in straight or diagonal alignment with has contrasting color.

5. Aknowledgment

This article is a part of a research conducted under Minister of Research and Technology of Higher Education (Menristek Dikti) funding in Penelitian Produk Terapan scheme year of 2017.

6. References

[1]. Kifrizyah R, Sudarmawan A, Witari NYS, 2013, *Batik Situbondo di Desa Selowogo Kecamatan Bungatan Kabupaten Situbondo*, Jurusan Pendidikan Seni Rupa, Universitas Pendidikan Ganesha Singaraja, Bali

[2]. Poerwanto, Sukirno ZL, 2012, *Inovasi Produk dan Motif Seni Batik Pesisiran Sebagai Basis Pengembangan Industri Kreatif dan Kampung Wisata Minat Khusus*, Jurnal Al-Azhar Indonesia Seri Pranata Sosial 1(4): 217-229

[3]. Handini YD, Sisbintari I, 2013, *Batik Gedog Tuban: Mempertahankan Warisan Budaya Melalui Penciptaan Pengetahuan dan Pengembangan Desa Wisata*, Jurnal Ilmiah Pariwisata18(2): 74-89

[4]. Nugroho CSA, Fianto AYA, Hidayat W, 2014, *Perancangan Buku Ilustrasi Batik Tuban Sebagai Upaya Pelestarian Nilai-Nilai Budaya*, Art Nouveau 3(1)

[5]. Maryanto SA, 2013, *Bentuk dan Makna Nama-Nama Batik Kudus*, Skripsi, Jurusan Bahasa dan Sastra Jawa, Fakultas Bahasa dan Seni, Universitas Negeri Semarang

[6]. Mulaab, 2010, *Ekstraksi Fitur Motif Batik Berbasis Metode Statistik Tingkat Tinggi*, Makalah Seminar Nasional Informatika 2010 (semnasIF 2010) UPN "Veteran" Yogyakarta: A69-A75

[7]. Noviana M, Hastanto S, 2014, *Penerapan Metode Quality Function Deployment (QFD) untuk Pengembangan Desain Motif Batik Khas Kalimantan Timur*, J@ti Undip IX(2): 87-92

[8]. Tresnadi C, Irfansyah, Prihatmanto AS, 2008, *Perancangan Game Batik "Nitiki" Berteknologi Multi-Touch Screen,* ITB J. Vis. Art & Des. 2(3): 221-236

[9]. Ediwati M, 2007, *Motif Batik Tulis Kreasi Baru Produksi Batik Merak Manis di Surakarta* (*Sebuah Tinjauan Estetika*), Skripsi Jurusan Kriya Seni/Tekstil, Fakultas Sastra dan Seni Rupa, Universitas Sebelas Maret, Surakarta

[10]. Mulyanti P, 2012, *Studi Tentang Motif Batik Druju Dusun Wonorejo Kabupaten Malang*, Artikel Hasil Penelitian, Program Studi Pendidikan Seni Rupa, Fakultas Sastra, Universitas Negeri Malang

[11]. Santoso R, 2011, *Proses Inovasi Motif Batik Semarangan di Batik Semarang 16*, Skripsi, Jurusan Manajemen, Fakultas Ekonomi, Universitas Katolik Soegijapranata, Semarang

[12]. Puryanti, 2010, *Batik Kliwonan di Kabupaten Sragen (Studi Nilai-Nilai Filsafati Jawa dalam Batik Kliwonan)*, Skripsi, Program Pendidikan Sejarah, Jurusan Pendidikan Ilmu Pengetahuan Sosial, Fakultas Keguruan dan Ilmu Pendidikan Universitas Sebelas Maret, Surakarta

[13]. Soekesi AEM, 2013, *Upaya Peningkatan Kualitas Produk Batik untuk Meningkatkan Daya Saing di Pasar Global pada Sentra Batik Pasirsari Pekalongan*, Prosiding Seminar Nasional 2013, Menuju Masyarakat Madani dan Lestari, ISBN: 978-979-98438-8-3

[14]. Poerwanto, Sisbintari I, Handini YD, 2010, *Inovasi Produk dan Motif Seni Batik Pesisiran Sebagai Basis Pengembangan Industri Kreatif dan Kampoeng Wisata Minat Khusus*, Abstrak Laporan Hasil Penelitian Stranas

[15]. Wardani LK, Sitindjak RHI, 2013, *Ragam Hias Batik Jawa Timur dan Implementasinya dalam Elemen Desain Interior Modern (Kajian Estetika dan Perancangan Desain*), Laporan Akhir Penelitian Hibah Bersaing, Tahun ke-1, Universitas Kristen Petra, Surabaya

[16]. Furyana SA, Wahyudi E, Handini YD, 2013, *Inovasi Produk Batik Pesisiran pada Perusahaan Batik Virdes di Banyuwangi*, Artikel Ilmiah Hasil Penelitian Mahasiswa, Ilmu Administrasi Bisnis, Fakultas Ilmu Sosial dan Ilmu Politik, Universitas Jember

[17]. Zehan M, 2012, *Studi Desain dan Motif Hias Batik Gajah Oling Produksi Sanggar Batik Sayu Wiwit Banyuwangi*, Artikel Ilmiah, Jurusan Seni dan Desain, Fakultas Sastra, Universitas Negeri Malang

[18]. Anonim, 2012, *Sejarah Batik*, http://www.tinashue.blogspot.com, accessed in April 30th, 2017

Children Nature Education About Names of Ocean Fish in Banyuwangi

Triana Kartika Santi
Biological Education Department, University of 17 Agustus 1945 Banyuwangi East Java, 68416, Indonesia
E-mail: g.rubiono@fdi.or.id

Abstract - Children can learn about biology from the nature. Children in fishing area of Muncar sub distric in Banyuwangi learn about name of fish from community in daily life. This research is aimed to identify children learning about the names of ocean fish in Muncar Banyuwangi. The research is conducted by interviewing the children, the fishermen and the local society. The result shows that there are a lot of local names of fish found in learning process. The children tend to be familier with these names rather than biological names. It is also found that names of fishes could be categorized in three different terms.

Keywords*: Children, learning, fish, Banyuwangi*

1. Introduction

Children education is important matter in society. The education of the children can be done in formal institution such as school or education in family and in the community. The education provide knowledge fo the children to help them in the future. Schooling can affect strategies used to recall information, by changing the brain organization of cognition in a myriad of ways [1].

Children also learning things from the society and the natural environment. Recent research also suggests that even views of nature can affect children's cognitive capacities, in particular their ability to concentrate [2]. Children are inherently connected to nature and fascinated by living things; there is a significant body of research discussing the benefits of children interacting with nature on a regular basis [3]. Early childhood appears to be a critical point in maximising the chances of such formative experiences by facilitating access to natural areas. Even among older children, though, there are profound benefits to encouraging time spent in natural habitats [4]. a feeling of connection to nature explained a portion of the effect that time spent outdoors has on environmental stewardship behaviors [5].

Banyuwangi is a district with coastal area. Muncar is a sub district with relatively large fishermen community. Muncar is a fishing base that provide 90% ocean fishing product in Banyuwangi. This sub district has 13 km of coastal line with 5,5 km fishing harbour. Geographically Muncar District is located at position $08^0.10$ '- $08^0.50$ LS or $114^0.15$' - $115^0.15$ 'east and has a bay called Pangpang bay by surrounding community. The population is 130,280 and until 2014 there are 13,203 people living as fishermen [6]. Muncar also the biggest fishing production in East Java province [7].

Biologists have long been concerned about the quality of undergraduate biology education [8]. Espesially in marine biology, maritime english teachers must improve the skill with maritime knowledge to optimize the learning process [9]. A research shows that fishing expeditions are fundamental channels of children's education and of women's empowerment that is essential for maintaining the social cohesion [10].

IOP Conf. Series: Earth and Environmental Science **156** (2018) 012019 doi:10.1088/1755-1315/156/1/012019

Research in biology education in Indonesia espesially in fish subject has done for inventory types of fish [11, 12, 13], props for fish organs [14], and android application [15]. In the learning process, the research has done for learning process in maritime education [16] and language difficulties faced by students [17].

Maritime biology is an important education subject in Indonesia. A large coastal area in Indonesia is a potensial resources for the future generation. The learning process is important to studied in order to give the children about maritime knowledge. This knowledge becomes provision for the marine resources management for the next generations.

Children in Muncar are familier with fishing activities among the fishermen. Biology learning espesially in names of fish is hapen in daily activities. This research is aimed to identify children learning about the names of ocean fish in Muncar Banyuwangi.

2. Method

The resarch is conduct by direct interviewing method with children and fishermen in fishing area. The interview is also done local community citizen in surrounding area. The interview is aimed to identify the learning process and the names of fish that learn by the children.

3. Result and Discussion

The interview result shows that children learn about names of fishes from daily life activities. As a fishing area, fish is a common thing in Muncar. Children learn from community such as family, fishermen, from each other and from their playground in the nature as children are play in the coastal area.

There are fish names that categorized as local name such as *lemuru*, *layang*, *mrenying*, *banyar*, *putihan*, *belanak*, *tongkol*, *layur*, *mungsing*, *lujung*, etc. The children tend to be familier with these names rather than biological names. Local language is used in the learning process which are Javanesse and Maduranesse languages which are ethnical residence in Muncar.

The names of the fish can be categorized in three different types which are:

a. The name is the same term in local, Indonesian and Latin name such as *lemuru* (local), *lemuru* (Indonesian) and *Sardinella lemuru* (Latin).
b. The name is the same for local name and Indonesian name such as *layang*, *tongkol*, *cumi* and *belanak*. These names can also found in other fishing area in Indonesia.
c. The name is different in local, Indonesian and Latin such as *mungsing* (local), *pari* (Indonesian) and *Mystacoleucus marginatus* (Latin).

Muncar nature surrounding environment as fishing area has provide knowledge of fish name for the children. Children learn directly from the fish that caught by fishermen or the fish in the fish market. Fishermen's children learn from their parents when welcoming their parent from sea. Caught fish become the direct learning media. Children can directly learn about the name of the fish and the diferent among the fish.

Non fishermen's children learn from the family and the community. They learn about the name of the fish from daily food consumption. The children also learn it from formal school. With the different types of the names, the school teachers must complement their knowledge with the local names of the fish and povide the right media. This is necessary because the children mostly prior learn the name of the fish from the nature than from formal education in school.

4. Conclusion

Chidren in Muncar subdistrict Banyuwangi learn the name of fish from daily life activities. The knowlegde they get from each other, family and community. The children tend to be familier with local names rather than biological names. The name of the fish can be categorized as three different terms.

5. References

[1]. Reyes-García V, Pyhälä A, Díaz-Reviriego I, Duda R, Fernández-Llamazares Á, Gallois S, Gueze M, Napitupulu L, 2016, Schooling, Local Knowledge and Working Memory: A Study Among Three Contemporary Hunter-Gatherer Societies, 11(1): 1-18

[2]. Strife S, Downey L, 2009, Childhood Development and Access to Nature: A New Direction for Environmental Inequality Research, Organ Environ. 2009 March, 22(1): 99–122

[3]. Chawla L, 2012, The Importance of Access to Nature for Young Children, Article, Bernard van Leer Foundation: 1-4

[4]. Russell MH, 2014, Connecting Children to Nature in a Montessori Primary Environment, Thesis, Master of Science in Education – Montessori

[5]. Andrejewski R, Mowen AJ, Kerstetter DL, 2011, An Examination of Children's Outdoor Time, Nature Connection, and Environmental Stewardship, Proceedings of the Northeastern Recreation Research Symposium

[6]. Aliyubi FK, Boesono H, Setiyanto I, 2015, Analisis Perbedaan Hasil Tangkapan Berdasarkan Warna Lampu pada Alat Tangkap Bagan Apung dan Bagan Tancap di Perairan Muncar, Kabupaten Banyuwangi, Journal of Fisheries Resources Utilization Management and Technology 4(2): 93-101

[7]. Pratama MAD, Hapsari TD, Triarso I, 2016, Faktor-Faktor Yang Mempengaruhi Hasil Produksi Unit Penangkapan Purse Seine (Gardan) di *Fishing Base* PPP Muncar, Banyuwangi, Jawa Timur, Jurnal Saintek Perikanan 11(2): 120-128

[8]. Singer SR, Nielsen NR, Schweingruber HA, 2013, Biology Education Research: Lessons and Future Directions, CBE—Life Sciences Education 12: 129–132

[9]. Zi-hua D, 2015, The Existing Situation and Training about Maritime English Teachers in China, Journal of Shipping and Ocean Engineering 5 (2015): 266-270

[10].Gallois S, Duda R, 2016, Beyond productivity: The Socio-Cultural Role of Fishing Among the Baka of Southeastern Cameroon, *Revue d'ethnoécologie* [Online], 10: 1-28

[11].Diniyah A, 2012, Inventarisasi Jenis Ikan Air Tawar di Bendungan Sampean Baru Kecamatan Tapen Bondowoso Sebagai Media Pembelajaran Biologi di SMA, Skripsi, Program Studi Pendidikan Biologi, Jurusan Pendidikan MIPA, Fakultas Keguruan dan Ilmu Pendidikan, Universitas Jember

[12].Janurianda FV, 2013, Inventarisasi Ikan Hasiltangkapan Nelayan di Danau Bekat dan Implementasinya Pembuatan Buklet Keanekaragaman Jenis, Artikel Penelitian, Program Studi Pendidikan Biologi, Jurusan Pendidikan MIPA, Fakultas Keguruan dan Ilmu Pendidikan, Universitas Tanjungpura, Pontianak

[13].Maidika D, Siburian J, Hamidah A, 2016, Studi Jenis Ikan di Pasar Parit 1 Kuala Tungkal sebagai Bahan Buku Ilmiah Populer Biologi SMA, Bioedukasi 9(2): 27-33

[14].Budiyanto A, 2015, Pengembangan Alat Peraga Sederhana Struktur dan Organ Dalam Ikan Untuk Mempermudah Pembelajaran pada Praktikum Ikhtiologi Perikanan, Jurnal Kelautan 8(2): 1-6

[15].Moedjahedy J, Bokang A, Raranta A, 2017, Aplikasi Pengenalan Ikan Hias Predator Air Tawar Menggunakan Teknologi Augmented Reality Berbasis Android, Cogito Smart Journal 3(1): 91-99

[16].Dirgayasa IW, 2014, Survey of English Teaching and Learning Process in Maritme Education and Training in Indonesia: A Case Study in Private MET in Indonesia, English Language Teaching 7(7): 111-119

[17].Akedeiwei ME, Friday MD, Kingsley KA, Odunze CK, Siefere BO, Caiga BT, 2015, Language Difficulties Faced by Nigerian Maritime Students in one Asian University, Asia Pasific Journal of Maritime Education 1(2): 37-41

Cost Efficiency of Sea Freight and Lowering Cost of Consumption Goods

Muh Rum
Academy of Public Manajemen Makassar, Jl. Maccini Tengah No.48, South Sulawesi 90144, Indonesia
Email: r12um@yahoo.co.id

Abstract. The subject of this research are administrative processes related to loading and unloading costin term of a ship's arrival and departure to seaports,typically attributed to elevated levels of cost of labor, handling cost, dwelling time, and port fee which required to complete the related administrative tasks. Research design by comparative method in administrative way, which compare with many way implemented of previous practitioner export and import. In the previous phases of the research, an average expected cost of the administrative labor cost in traditional seaport clusters in eastern Indonesia was identified and quantified on an hourly basis. This research continues in its aim by using the results of the previous research as a starting point, and Find that enterprise resource planning and improvement, transactional due diligence and merger integration, positively affect and not associated with cost reduction. Moreover variable of reducing inventory cost affect and associated with reducing cost. The main hypothesis is that the usage of such a new model will result in a measurable decrease of the required freight cost in sea port, which indirectly reduce the consumption goods.

Keywords: sea freight, cost efficiency, unit cost handling, consumption goods

1. Introduction

Indonesia is the largest archipelagic country in the world, has an area of 7.7 million km2, with an area of two-thirds of Indonesia's sea, and the fourth longest coastline in the world along 95,181 km, and has 17,480 islands having economic potential in sea transport services is very large, because it is inevitable that sea transportation (ships) is a major means of transportation to reach and connect the islands in the archipelago, creating inter-island connectivity in Indonesia. Due to the vital importance of transportation for the economy, sea transport must be developed properly and properly to support economic growth. If sea transportation is disrupted, then the national economy will also be disrupted. One of the strategies of acceleration and expansion of national economic development is to promote strengthening connectivity between islands, especially outer islands. This connectivity can only be realized if the marine transportation in the archipelago country continues to be played significantly. Since the days of the Hindu-Buddhist kingdom, the Port has become one of the means connecting the trade and defense resources of the Kingdom in Indonesia. As the times progressed, the role of ports was needed to support various fields in a country. Definition and role of Harbor according to Law no. 21 of 1992 concerning shipping is a place consisting of the land and the surrounding waters with certain limits as a place of government activity and economic activity which is used as a place to lean anchor ship, up and down passengers and loading and unloading of goods equipped with safety facilities and activities of shipping Supporting ports as well as intra and intra-modal transport.

The function of sea transportation is essentially to transport passengers or goods from one place to another separated by the territorial waters. With the existence a sea transportation it can help the creation of national distribution patterns. However, in order to achieve this it is necessary to have an effective, efficient and safe marine transportation system. The movement or movement of passengers and goods is the basis of trade. Through the means of sea transportation, raw materials and goods produced from one region can be marketed to other areas. Indonesia, as the largest archipelago country in the world, desperately needs sea transport that can reach all its territory. Indonesia is an archipelagic country with a vast sea area, two-thirds of which is waters and located in a strategic location because it is located in the stop of world trade route. As an archipelago, the role of the port is vital in the Indonesian economy. The presence of an adequate port plays a major role in supporting the mobility of goods and people in this country. Ports become the most important means to connect between islands and between countries. The port is one of the most important trading chains of all trade processes, be it inter island or international trade. As the intersection of land and sea transportation, the role of the port becomes vital in promoting economic growth, especially its hinterland areas where the movement of goods and humans in large quantities. As part of the transportation system, ports play an important role in the economy.

The port in Indonesia is very important in the flow of the economy in Indonesia, even in the world. This is where the process of export and import goods. Because the activities at the harbor are vital to the economy, then all processes must be done perfectly. If the port service occurs dysfunctional then it will affect the price of goods to be received by consumers. In the practice of port services, known as the process of loading and unloading of goods is also called the dwelling time or the residence time of goods within the port area. The length of the dwelling time process depends on the infrastructure, the maintenance of the license, and from the company itself. Ports can play a role in stimulating the growth of economic, commercial, and industrial activities from their spheres of influence. But the port does not create such activities, but only serves to grow and develop these activities. Such activities enhance the role of the port from merely as a vessel to the center of economic activity. In principle, the relationship of development activities by humans in the sea cannot be separated with on the beach even on land entirely. Ports become a means of the rise of inter-island trade and even trade between countries, ports in an area will be more stimulating wheel rotation of the economy, various types of business

There are many reasons to improve transit evaluation. Current transportation evaluation practices tend to overlook and undervalue many transit benefit categories, such as parking cost savings, increased safety from reduced vehicle travel, and reduced chauffeuring burdens on drivers [1] . More comprehensive analysis includes more impacts and so is more accurate. This is not to suggest that every transit project is cost effective or that transit is always the best solution to every transport problems. However, transit improvements tend to provide significantly more value to society than conventional models indicate. There are four general categories of transit improvements to consider:

a. Increased service (more transit vehicle-miles)
b. Improved service (more comfortable, convenient, reliable, etc.).
c. Transit use incentives (lower fares, commuter financial incentives, marketing, etc.).
d. Transit oriented development (land use patterns designed to support transit, including more compact, walkable, mixed development around transit stations and corridors).

The port volume is called to be one of the main indicators of port performance. It is easiest to perform performance comparisons between one port and the other. The port of a connectivity, its comparability can be seen from the increase in volume, if the port is said to be good but there is no increase in volume that is nonsense, said Minister of Transportation Budi Karya Sumadi. Efforts to increase the volume, should be encouraged by improving the various port services, such as increased security and service to stakeholders. Budi also emphasized the importance of improving services to shipping line and producers in order to increase volume. The need for ports to make alternative services for foreign shipping line more interested to come.

The opening of transportation by sea will open the economic path. The principle of trade follow the ship is one effort to stimulate the Indonesian economy. Increased volume at the port is one indicator of the running of the economy in an area, moreover the sea toll is working to increase the inflows and exports of goods, especially from eastern Indonesia. And the sea toll with logistics warehousing can increase the volume of trade in eastern Indonesia, the increase in port volume can boost economic growth in Indonesia, hoping the economic increase could be 5.7 percent or 5.8 percent[2].

South Sulawesi is the driving force of the economy in eastern Indonesia. This is because the export potential of South Sulawesi is very diverse. Among others in the fields of agriculture, fisheries (shrimp, fish and seaweed), tourism, agriculture (cocoa, coffee beans, corn, and cloves), mining (nickel) and forestry. One of South Sulawesi's mainstay is cocoa potential whose production reaches an average of 330,000 tons per year and become the largest contributor to export. Even President of Republic Indonesia Jokowi is pleased to hear that export products have expanded to 27 export commodities. Ranging from Frozen Shrimp, Tuna Frozen Crab Meat, Octopus Frozen, Fresh Fish, Cocoa Beans, Cocoa residues, Cocoa Powder, Coffee Arabica, Fruit Passion Fruit, Corn, Marble, Wood Processing, Seaweed, rubber, Live Fish, Fish Fillets , Oil Nut, Cashew Peel, Skin Ari Mete, Wheat Flour, Wheat Bran, Incense, Flying Fish Eggs, Reptile Skin, Cement to Nickel. Export destinations also varied to 24 destination countries. Namely the United States, Japan, South Korea, Italy, Puorterico, Germany, Malaysia, Singapore, Hong Kong, Philippines, Denmark, Britain, China, Poland, Dubai UAE, Kuwait, Saudi Arabia, Ukraine, Spain, Israel, the Netherlands, Vietnam, Australia , And Timor Leste. Diversification of export destination countries from traditional markets to emerging markets is urgently needed in the future [3][6]

Feel looks lazy when spoken about the prospect of his business in eastern Indonesia, such as the clothing trader with the huge cost of delivering goods from Jakarta to areas in Eastern Indonesia. In addition to shipping costs are quite expensive, moreover the goods that are sent often do not get in time. This is caused by the lack of modes of transportation. The same complaint is also of a resident from Merauke, Papua about the procurement of instant noodle product shopping, sugar, wheat, in the area is quite expensive. Just because the goods are from the factory on the island of Java that we must buy expensive. Complaints both become opaque portraits for other business world, so less tasteful in aiming for investment to the east of Indonesia. Indeed, inadequate infrastructure such as electricity, clean water, road and port conditions, became a ghost inhibiting investment. Commonly the supply of goods to the eastern regions due to the condition of port infrastructure and roads that are less conducive, resulting in swelling logistics costs. Port development and road infrastructure improvement, we hope to support the business growth.

The gap of development of the western and eastern regions, as a result of the implementation of efficient logistics management as an archipelagic country. The sea transportation system from Jakarta and Surabaya to some areas in Eastern Indonesia has not been well integrated. Ideally to build eastern Indonesia, it must start from the revamping of port infrastructure. As the Port of Makassar, dubbed the East Indonesia entrance, needs to be pushed into a major port in Eastern Indonesia, thus overcoming the unbalance constraints of the loading and unloading of goods that generally occur in ports in Eastern Indonesia.

The importance is about connectivity between ports in Indonesia, as the strategies of acceleration and expansion of national economic development is to promote the strengthening of inter-island connectivity, especially the outer islands and maintain the balance of the western and eastern regions of Indonesia.

2. Method

The research design is based on descriptive and associative method how to reduce freightcost in sea port, which can affect to the cost of consumption goods in central and eastern Indonesia. This survey consists of the factors that previous studies stated which might affect sea port cost. This survey was distributed over a random sample of practitioner import in Makassar Port, through the number of vessel loading during the year of 2016. Sample data collected about 30 of stevedoring activity in Makassar Port, applied technical research by multiple regression models.

Research design by doing comparative method among the several exporters and importers from many countries with the importer and exporter in the Port of Makassar. This comparison In terms of loading and unloading time and loading and unloading costs So it can be known whether the mechanization of unloading equipment loading and warehouse procurement able to improve performance loading and unloading method like what is optimal and profitable to the port management. The comparative implemented method showed in table 1, which the author can combine the whole method to reduce freight cost in Makassar port even in eastern Indonesia area of port. These factors should be considered when evaluating public transit benefits and costs:

IOP Conf. Series: Earth and Environmental Science **156** (2018) 012020 doi:10.1088/1755-1315/156/1/012020

a. Public transit can provide various types of impacts. Comprehensive evaluation should consider all significant benefits and costs.

b. Many transit services (those that operate at times and places with low demand) exist mainly to provide basic mobility for non-drivers. Although relatively costly per trip, they are often cheaper than alternatives such as taxis and chauffeuring (drivers making special trip to carry non-drivers, which often requires empty return trip), or inadequate mobility for non-drivers.

c. High quality (relatively fast, convenient, comfortable and integrated) transit can attract discretionary travelers who would otherwise drive, which reduces traffic problems including congestion, parking costs, accidents and pollution emissions. Transit that attracts discretionary travelers provides consumer welfare (surplus) benefits, since they would not change mode if they did not consider themselves better off overall.

d. High quality transit can stimulate transit-oriented development, compact, multi-modal neighborhoods where residents tend to own fewer vehicles, drive less and rely more on alternative modes than in more automobile-oriented communities. This can leverage additional travel reductions and benefits (besides just the travel shifted to transit).

e. Traffic congestion tends to maintain equilibrium: it increases until delays discourage additional peak-period vehicle trips. High quality, grade-separated transit can reduce traffic congestion costs by reducing the point of equilibrium, offering travelers an alternative to driving, and by supporting compact development which reduces travel distances.

f. Highway expansion tends to induce additional vehicle travel which increases external costs such as downstream congestion, parking demand, traffic risk, barrier effects, and pollution emissions, costs that are avoided if travelers instead shift to public transit. These impacts should be considered when comparing roadway expansions with transit improvements.

g. Transit travel time unit costs (dollars per hour or cents per minute) vary significantly depending on travel conditions and user preferences. Many travelers prefer high quality transit even if it takes longer than driving because they can work or rest.

h. These impacts and benefits tend to increase if transit improvements are implemented with support strategies such as walking and cycling improvements, more compact development, transportation demand management programs, and efficient road and parking pricing.

i. Since active transport (walking and cycling) and public transit are complements, transit travel tends to increase public fitness and health.

j. Public transit services have three features that justify public support and underpricing: they help achieve social equity objectives, they experience scale economies, and they can reduce various external costs including traffic congestion, accident risk and pollution emissions.

k. Current demographic and economic trends (aging population, rising fuel prices, urbanization, changing consumer preferences, increasing health and environmental concerns) are increasing demand for transit and transit-oriented development, and therefore their benefits.

Table 1. Comparative Method of Practitioner in Seaport

Name	Background	Implemented Method	Variable Output
David Martinelli	Independent Principal with Blossom Growth and specializes in: Manufacturing and operational efficiency, Supply chain effectiveness, ERP and software implementation and improvement, Transactional due diligence and merger integration.	ERP (Enterprise Resource Planning) and software implementation and improvement, Transactional due diligence and merger integration.	Enterprise Resource Planning and improvement, Transactional due diligence and merger integration.
Bryan Mattimore	The Cofounder and Chief Idea Guy at Growth Engine, Innovation Agency. He's the author of "Idea Stormers, How to Lead and Inspire Creative Breakthroughs	Create a consortium of buyers (our client and several of their suppliers) to buy needed logistics supplies (i.e. transportation fuel) at a reduced cost that comes with buying in greater quantities	Create a consortium of buyers
Brian Stutzman	Investor in franchise ownership with BlueGrace Logistics. With an operation location in Indianapolis, IN but with a worldwide reach, Brian and his team are focused on becoming a part of a new standard in the third party logistics industry.	Implement more stream-line processes or reporting that saves your transportation and accounting teams time	Implement more stream-line processes
Scott Stone	Director of Marketing for Cisco-Eagle, a provider of integrated material handling systems for industrial operations. Scott has over 23 years experience in industrial operations and marketing.	Truly understand its costs—per pallet, per sku, per order shipped, etc, and Every project should be seen through the prism of labor costs which can be addressed in multiple ways.	Unit Cost Handling
Brian Sutter	Director of marketing at Wasp Barcode Technologies. He succeed to minimized the Inventory cost through holding goods in stock	Reduce your inventory cost, chances are you're stocking too much inventory. Know Your Up-to-Date Inventory Levels: Keeping track of your inventory levels is the most straightforward way to prevent overstocking inventory and, as a result, reduce inventory cost.	Reducing your inventory cost
Richard McGirr	Marketing manager at floship, an e-commerce logistics company based in Hong Kong	Have one distribution center in a major logistics hub, such as Hong Kong, that can service a very large area or possibly even the entire world from one location.	Build a distribution center in a major logistics
Bob Shirilla	The Owner / Manager – Simply Bags – Leading US distributor of personalized and custom tote bags.	Our marketing starts with an analysis of supplier locations and potential clients in the same region. For example, we are doing a marketing campaign targeting Beach Wedding Planners and have developed a business relationship with two manufactures in Florida. This gives us a competitive price advantage over other suppliers that have higher shipping cost.	Analysis of supplier locations and potential clients in the same region

Chuck Intrieri	Management consultant at Charles M. Intrieri Consulting. A highly experienced supply chain professional.	Collaborate and partner with Suppliers to help reduce costs. By create a consortium of buyers (a client and several of their suppliers) to buy needed logistics supplies (i.e. transportation fuel) at a reduced cost that comes with buying in greater quantities.	Create a consortium of buyers
Martin Murray	A senior supply chain management consultant and author of several books on supply chain using enterprise resource planning (ERP) software, Martin Murray has been the About.com expert for logistics and supply chain since 2008	Use a single source for a range of products that a single vendor can provide. The same can be achieved for transportation. By offering all transportation out to bid, via a RFQ (Request for Quotation), a company can provide carriers with a detailed explanation of what they require, which may fall outside of what is normally provided by a common carrier.	Use a single source for a range of products
Rob O'Byrne	Owner and group managing director of specialist management consulting firm Logistics Bureau, based in Australia and southeast Asia	Cost reduction starts with fully understanding all costs and separating them into variable and fixed costs. Variable costs fluctuate with volume (truck fuel, direct labour, packaging, etc.) while fixed costs don't (rent). The major cost components we will review in this class include: labor (direct, indirect and temporary staff vs. permanent staff) stock loss distribution packaging the cost of low productivity the cost of poor quality utilities (electricity, heat, water)	Fully understanding all costs and separating them into variable and fixed costs.
Alex Stark	Modern Marketing Expert and Supply Chain Collaboration Evangelist for Kane Is Able, Inc.	Allocating available inventory against current orders creates problems	Integration of sales and inventory data

Source: Angela Stringfello [10]

3. Research Hypotheses

After reviewing the literature that covered the topic of how to reduce the freight cost in sea port, the researchers developed the following hypotheses that were set out to achieve the study objectives:

H 1: Enterprise Resource Planning and improvement, Transactional due diligence and merger integration affects the sea freight cost efficiency.

H 2: Create a consortium of buyers affects the sea freight cost efficiency.

H 3: Implement more stream-line processes affects the sea freight cost efficiency.

H 4: Unit Cost Handling affects the sea freight cost efficiency.

H 5: Reducing your inventory cost affects the sea freight cost efficiency.

H 6: Build a distribution center in a major logistics affects the sea freight cost efficiency.

H 7: Analysis of supplier locations and potential clients in the same region affects the sea freight cost efficiency.

H 8: Use a single source for a range of products affects the sea freight cost efficiency.

H 9: Fully understanding all costs and separating them into variable and fixed costs affects the sea freight cost efficiency.

H 10: Integration of sales and inventory data affects the sea freight cost efficiency.

4. Data Analysis Procedures

This research implements a number of statistical techniques and procedures that help to examine research hypotheses. These techniques include reliability and validity test, frequency analysis, independent sample t-test, descriptive statistics, correlation matrix, linear regression, and simple regression. All statistical procedures were estimated using path analysis with Smart PLS implemented properly.

5. Models

The research model is estimated is direct to reduction cost measure in statistical model $\eta_1 = \Phi_1\xi_1 + \Phi_2\xi_2 + \zeta_1$ or noted as Red Cost= Φ_1ERP + Φ_2Cons +$\Phi_3$line Proc + Φ_4 Unit Cost+ Φ_5 Red Invent+ Φ_6Dist Cent+ Φ_7Supp Loc+ Φ_8Sing Source+ Φ_9Sept Cost+ Φ_{10}Integ Sel Invent $+\zeta_1$. The partial Least Square (PLS) Estimation Parameters Estimation parameters of structural equation modeling with partial least square approach was obtained through a three-stage process of iteration and at every stage of producing estimates. The first phase resulted in estimated weight w_{jh} Weight estimation of w_{jh} weights obtained through two ways, namely mode A and mode B. Mode A is designed to obtain the estimated weight of the types of indicators reflexive, whereas the B mode is designed to obtain the estimated weight of the types of indicators formative.

In modeAweightswjhisthe regression coefficientofZjinsimple regressionmodelsXjhoninnerestimationZj, $X_{jh} = w_{jh}Z_j + e_{jh}$. Estimatesfor the model1is obtainedthroughOLSin a wayto minimize thesum of squaredejh, asfollows:$e_{jh} = X_{jh} - w_{jh}Z_j$, $\sum_{h=1}^{J} e_{jh}^2 = \sum_{h=1}^{J}(X_{jh} - w_{jh}Z_j)^2$. Then thesum of the squares ejhloweredtothe face wjh in order to obtainweightsformodeA: $\widehat{w}_{jh} = \frac{Cov(X_{jh}, Z_j)}{Var(X_j^2)}$. Mode2of theweightingvectorswj of wjh is theregressioncoefficientvectorofZj atthe center ofthe manifestvariables ($X_{jh} - \overline{X}jh$) areconnected toeach otherlatent variablesξj:, $Z_j = w_jX_j + \varepsilon_j$and $\varepsilon_j = Z_j - w_jX_j$. Generating of $\varepsilon_j^T\varepsilon_j$: $\varepsilon_j^T\varepsilon_j = (Z_j - w_jX_j)^T(Z_j - w_jX_j) = Z_j^TZ_j - 2w_jX_j^TZ_j + w_j^Tw_jX_j^TX_j$. Then $\varepsilon_j^T\varepsilon_j$ lowered to w_{jh} in order to obtain weights for mode B: $\widehat{w}_j = (X_j^TX_j)^{-1}X_j^TZ_j$, Where$X_j$is a matrixwith columnsdefined bythe manifestvariables ($X_{jh} - \overline{X}jh$) linking $\xi_j$latentvariablej. Innerweight vectormodel is a$w_{jh} = (Var(X_j))^{-1}Cov(X_{jh}, Z_j)$ with $Var(X_j)$ is covariance matrix Of X_j and $Cov(X_{jh}, Z_j)$ Is the column vector between variabels of X_{jh} and Z_j . The second stage produces estimates obtained path through the model inner and outer estimationmodels.

- Inner Model Estimation

  By following thePLSalgorithmofWold(1985) andwhichhas been improvedbyLohmoller's (1989), the estimated inner$Z_j$modelsofstandarizedlatent variables$(\xi_j - m_j)$is defined by $Z_j \propto \sum_{i;\xi_i \text{ connected to } \xi_j} e_{ji}Y_i$. Whereinthe weightinnere_{ji}modelscan beselected viathreeschemes, namely:

- Path scheme

  Latent variablesconnected to$\xi_j$are dividedintotwogroups, namely: the latentvariablesthat explain$\xi_j$andis followedbythe variablesdescribed byξ_j. If$\xi_i$described by$\xi_j$then$e_{ji}$is multiple regressioncoefficient between Y_i and ofY_j. Latentvariablesconnectedto$\xi_j$aredividedintotwogroupsitems, namely: the latentvariablesthatexplain $\xi_j$and isFollowedby thevariablesDescribedbyξ_j.

$$e_{ji} = \begin{cases} \text{multiple regressioncoefficient of } Y_i \text{ from } Y_j \text{ ,if } \xi_j \text{described by} \xi_i \\ Cor(Y_iY_j)\text{ , } \xi_j \text{described by} \xi_i \end{cases}$$

- Centroid schema
  Inner model weight of e_{ji} as the sign correlation between Y_i and Y_j , can be writtenasfollows:$e_{ji} = sign[Cor(Y_i Y_j)]$
  Factor schema Inner mode weight of l e_{ji} as the sign correlation between Y_i and Y_j, can be writtenasfollows: $e_{ji} = Cor(Y_i Y_j)$
- Mean estimation of m_j
  Mean obtained as qutaions as $\xi_j = Y_j + m_j + e_j$ and $\xi_j - m_j = Y_j + e_j$ with $Y_j = \sum_{h=1}^{J} \widetilde{w}_{jh}(X_{jh} - \bar{X}_{jh})$ then $\xi_j - m_j = \sum_{h=1}^{J} \widetilde{w}_{jh}(X_{jh} - \bar{X}_{jh})$. Equations analog as : $\hat{\xi}_j = \sum_{h=1}^{J} \widetilde{w}_{jh} X_{jh} = Y_j + \widehat{m}_j$So that: $\widehat{m}_j = \sum_{h=1}^{J} \widetilde{w}_{jh} \bar{X}_{jh}$. Where$\widetilde{w}_{jh}$is definedasthe weightof theoutermodels,withall themanifestvariablesthatobservations onthe samescale of measurement. According toFornell(1982), says that if the estimation of latent variables on the original scale $\xi_j^* = \frac{\sum_{h=1}^{J} \widetilde{w}_{jh} X_{jh}}{\sum_{h=1}^{J} \widetilde{w}_{jh}}$. The above equationis possible when all the weighting of outerpositive model. Ofteninreal applications, the estimation of latent variables require scale0-100scalein order to have a benchmark to compare with an individual score. Soin the case of the I series observation, the easily obtained through the transformation as follows: $\xi_j^{0-100} = 100\, x \frac{\xi_j^* - x_{min}}{x_{max} - x_{min}}$. Where$x_{min}andx_{max}$are the minimum and maximum values of common measurement scale for all variables manifest.
- Parameter Constant
  In general,the path coefficient $b_{ji}$is multiple regressioncoefficientof$Y_j$endogenous latent variables were standardized inthe explanatory latent variables (exogenous) Y_i , $Y_j = \sum_{i=1}^{J} b_{ji} Y_i + e_j$. At the time of converging latent variables (non-centered) $\hat{\xi}_j$ is equal to $Y_j + \widehat{m}_j$. the regression equation when the latent variable $\hat{\xi}_j$ not converge is: $\hat{\xi}_j = b_{j0} + \sum_{i=1}^{J} b_{ji} \hat{\xi}_i + e_j . e_j^2 = \left(\hat{\xi}_j - (b_{j0} + \sum_{i=1}^{J} b_{ji} \hat{\xi}_i)\right)^2 = \hat{\xi}_j^2 - 2\hat{\xi}_j b_{j0} - 2\hat{\xi}_j \sum_{i=1}^{J} b_{ji} \hat{\xi}_i + \left(b_{j0}{}^2 + 2b_{j0} \sum_{i=1}^{J} b_{ji} \hat{\xi}_i + \sum_{i=1}^{J} b_{ji}^2 \hat{\xi}_j^2\right)$. $\frac{\partial e_j^2}{\partial b_{j0}} = \hat{b}_{j0} = \hat{\xi}_j - \sum_{i=1}^{J} b_{ji} \hat{\xi}_i$ with $b_{j0} = \widehat{m}_j - \sum_i b_{ji} \widehat{m}_i$, So the location parameter is a constant b_{j0} for endogenous latent variables and the average $\widehat{m}_j$for exogenous latent variables.

6. Finding and Discussion

Recorded the realization of ship traffic during the first quarter from January to March 2016 in Pelindo IV Makassar Branch of 1,408 calls or equivalent to 7,343,565 Gross Tonnage (GT) with absorption rate of 26.88 percent, the realization of ship visit which was managed by Pelindo IV Makassar Branch in the first Quarter increased compared to the same period in 2015. In the first quarter of 2015 the number of ship visits in PT.Pelindo Makassar reached 1,289 calls or equivalent to 6,555,894 GT up to 1,408 call do this first semester. The number of vessel traffic is still dominated by domestic ships (Indonesian flag), which is 1,355 calls or equivalent to 6,587,527 GT. While the number of foreign ship visits (foreign-flagged) only recorded at 53 calls or equivalent to 756,038. Viewed from ship types, the increase also occurred in the flow of freight vessels during January to March 2016 through the port of Makassar as many as 755 calls or equivalent to 1917.571 GT an increase that previously only reached 672 calls or 1,627,611 GT in the same period in 2015. The increase also occurred in the type of container vessel, i.e during January until March 2016 amounted to 377 calls or equivalent to 1917,571 GT, compared to the same period in 2015 which only reached 334 calls or equivalent 2,355,269 GT[11]. Based on the following table we can make the following observations:

IOP Conf. Series: Earth and Environmental Science **156** (2018) 012020 doi:10.1088/1755-1315/156/1/012020

a. The number of observations for each variable was 30 persons involve in port handling, which reflect that the respondents have answered all questions concerning study variables.
b. The average value for Enterprise Resource Planning and improvement, Transactional due diligence and merger integration was 86.4%, indicating that the respondents believe that Enterprise Resource Planning and improvement, Transactional due diligence and merger integration in the port handling is relatively high.
c. The average value of the independent variables ranged from 72% to 82%, which means that the respondents assume that Enterprise Resource Planning and improvement, Transactional due diligence and merger integration is highly affected by those variables.
d. Standard deviation for all variables was relatively low, which indicate that the respondents" answers are consistent and close to each other [12]

The following sections shed some lights on the concept of each statistical procedure used in the research. Results of testing that the whole independent variables showing positively associated with the likelihood of reducing cost. The value of loading (λ) to be valid if it has a latent variables to be measured ≥ 1,96, that way in Table 3, the whole indicators apparently valid to be continue as good indicator to support latent variable. Enterprise resource planning and improvement, transactional due diligence and merger integration more influence the reduction cost in sea freight cost, as well as reducing inventory cost

Determinants model (R Square)

The estimated coefficients for reduction cost is 60.6 percent. We attribute the prominent of significance in statistical model.

Table .2 Model Summary

R	R^2	Adje R^2	DW	F	Sig.
0,779	0,606	0,399	1,709	2,925	0,021

Source: Output of Test of Bootstraping SPSS

In Table 2.Results of the autocorrelation test shows Durbin-Watson value of 1.709 which means that data is processed free from autocorrelation. In Table 2, to detect whether there is multicoloniearity between independent variables can be seen VIF value in colinearity statistic as illustrated in table 5.13 above which shows VIF value exceeds tolerance value, meaning no multicolinearity.

Table .3 Analysis of Varience

Variables		λ	C.R	B	SE	Beta	t	Sig.	Tolerance	VIF
C				0,793	1,180					
ERP	X_1	.829	7.514	0,829	0,248	0,876	3,348	0,003	0,303	3,300
Cons	X_2	-.639	-6.481	-0,639	0,313	-0,755	-2,043	0,055	0,152	6,596
Line Proc	X_3	.020	.281	0,020	0,141	0,033	0,141	0,890	0,381	2,625
Unit Cost	X_4	-.204	-1.973	-0,204	0,456	-0,230	-0,447	0,660	0,078	12,747
Red Invent	X_5	.794	7.739	0,794	0,437	0,902	1,818	0,085	0,084	11,877
Dist Cent	X_6	.121	1.449	0,121	0,207	0,169	0,585	0,565	0,249	4,019
Supp Loc	X_7	-.064	-.846	-0,064	0,118	-0,099	-0,544	0,593	0,630	1,587
Sing Source	X_8	.293	3.326	0,293	0,187	0,388	1,561	0,135	0,336	2,974
Sept Cost	X_9	.001	.006	0,001	0,345	0,001	0,002	0,998	0,200	4,996
Integ SelInvent	X_{10}	-.308	-2.468	-0,308	0,374	-0,288	-0,825	0,420	0,170	5,865

Source: Output of Test of Bootstraping Smart PLS

7. Hypotheses Tests

Based on Table 3, results of testing hypothesis, that enterprise resource planning and improvement, transactional due diligence and merger integration, positively affect and not associated with cost reduction with estimated path coefficient 0,829, and C.R value is 7.514. Variable of reducing inventory cost affect and associated with reducing cost with estimated path coefficient 0.794, and C.R value is 7.739.

8. Findings

Our finding is that the wider the scope of the reduction cost will be an effort to develop the enterprise resource planning and improvement, transactional due diligence and merger integration, positively affect and not associated with cost reduction with estimated path by arranging online system before ship bank, embark and disembark. Moreover, the using virtual money more effectively in port field, avoid physical money in transaction, or avoid extra payment, regarding to illegal payment. Variable of reducing inventory cost affect and associated with reducing cost of storage, cost of maintenance, and cost of order through just in time applied system.Refer to Yoshitsugu Kanemoto [7] that, transportation relate to distribution and consumption of society. That way the government should prioritize the importance of transportation as one of the drivers of the economy, and lowering cost of consumption goods

9. Conclusions

If one of your goals is to reduce freight spend, and hope that applying competitive pressure to carriers will encourage them to reduce port rates, and you may want to think again and looking for another determinant factors, such as information system, system and procedure, management system, and inventory management. Every year is a good idea as you will be able to align your needs with the carriers that find it most attractive at that point in time management. However, this presumes some carrier flexibility. Start by challenging organizational beliefs about different carriers with data and facts about performance. After all, it may be necessary to change carriers to save money.

Regional carriers typically offer a lower price point, with transit times often equal to or faster than national carriers. Using a national carrier for regional shipments typically leaves money on the table.You should have taking advantage of the fact that carriers tend to have some geographical pockets that are more competitive than others.

Many shippers, however, do not have such technology, so use a routing guide, which tells them the carrier to use based on shipment origin and destination. In order to generate that routing guide, a comparison between carriers needs to be conducted that takes into account about pricing factors of tariff discounts, minimum charges, fuel surcharges, and transit times.

A robust comparison requires a deep dive into historical data of amount of practice importer, about their experience in port and inventory handling to reduce freight spends. Many shippers' routing guides only account for tariff discounts, and do not take into account the fact that often the best-priced carrier does not have the best discount.Judging the true low-cost option is only possible when you have a solid handle on your shipping patterns.

10. References

[1] United States Department of Transportation (USDOT).2016. Passenger Travel Facts and Figures 2016. Bureau of Transportation Statistics

[2] Mansyur Faqih.2017.Pelabuhan Didorong Tingkatkan Volume dan Layanan Alternatif (http://www.republika.co.id/berita/nasional/umum-pelabuhan-didorong-tingkatkan-volume-dan-layanan-alternatif). Accessed 16th May 2017.

[3] Ministry of State Secretary of the Republic of Indonesia. Kementerian Sekretaris Negara Republik Indonesia.2015.Bangga Sulsel Ekspor 27 Produk Komoditas ke-24 Negara, Presiden Jokowi: Pemda Perlu Buat Terobosan. http://www.setneg.go.id/Accessed 03rd August 2017.

[4] Lubis, et.al.2005. Multimodal transport inIndonesia: recent profile and strategy development, Proceedings of the Eastern Asia Society for Transportation Studies, 5: 46 – 64.

[5] Kreutzberger, Konings, R, and Aronson, L, D. 2006. Evaluation of the cost performance of pre- and post-haulage in intermodal freight networks. In Jourquin, B., Rietveld, P., Westin, K. 2006. Towards Better Performing Transport Networks, Routledge, Oxon.

[6] Noor Mahmudah et.al.2011.Pengembangan Metodologi Perencanaan Transportasi Barang Regional. Jurnal Transportasi Vol. 11 No. 3 Desember 2011 : 173-182

[7] Yoshitsugu Kanemoto.2006.Urban Transport Economic Theory.Companion to Urban Economics.Edited by Richard J. Arnott, Daniel P. McMillen. Publishing Ltd.

[8] Ostrovsky, Michael.2008. Stability in Supply Chain Networks. American Economic Review 2008, 98(3) : 897–923

[9] Parkinson, M, 1981, The Effect of Road Investment on Economic Development in the UK, Government Economic Service working paper no.43, Department of Transport.

[10] Angela Stringfello.2017.How to Reduce Logistics Costs: 19 Experts Reveal Ways Organizations Can Cut Their Logistics Transportation and Carry Costs. https://www.camcode.com/asset-tags/how-to-reduce-logistics-costs/. © Copyright 2017 Horizons, Inc. - All Rights Reserved.

[11] Rizman Rumi.2017.Info Loker dan Bisnis .https://siletmks.blogspot.co.id/2016/04/ kunjungan-kapal-di-pelabuhan-makassar.html

[12] Diamond, D and Spence, NA.1989.*Infrastructure and Industrial Costs in British Industry*, Research conducted for the Department of Trade and Industry.

IOP Conf. Series: Earth and Environmental Science **156** (2018) 012021 doi:10.1088/1755-1315/156/1/012021

Habitus Disposition within Culture of Land and Marine Sasi in Moluccas

Faidah Azuz[1], Farida Mony[2]

[1]Bosowa University, Jenderal Urip Sumoharjo KM.4, Sinrijala, Makassar, Sulawesi Selatan, 90231, Indonesia.
Email : faidah.azuz@universitasbosowa.ac.id
[2]Darussalam University, Raya Tulehu, Km 24, Tulehu, Salahutu, Maluku, 97582, Indonesia
Email : fmony@gmail.com

Abstract. The Moluccan embraces a local wisdom namely sasi. It functions is a tool to manage their natural resources. The land and marine sasi belongs to the customs territory, yet along with the development of the community, this culture experiences various transformations. Therefore, it is necessary to explore, identify, and compare the transformations of sasi on the land and the marine. The research was conducted at Morella village and Haruku, both are within the administrative territory of the District of Central Moluccas. Observations and in-depth interviews were held to obtain data. The result of this research shows that there is a shifting in the implementation of sasi which was held by the customary leader (habitus disposition). Nevertheless, the shifting from custom to ordinary tradition is only found on plant sasi. Meanwhile, the marine sasi remains within the customs territory. A unique finding has shown at land sasi which has a connection with religion. Meanwhile, the marine sasi does not have many links to religion. Finally, this research provides recommendations to conduct further identification and analysis of sasi shifting on wider area coverage such as in Southeast Moluccas. It also recommends conducting further research on the implementation of sasi on more diverse natural resources.

Keywords: *Habitus disposition, Land Sasi and Marine Sasi.*

1. Introduction

The study aims to identify and compare the changes occur on the implementation of Sasi on land and sea. As a custom base local culture, the sasi experiences improvement to adjust itself along with the development of its society. Many studies which observe sasi's improvement as an institution has been conducted. However, a question on how a change process happens within the territory of land and marine sasi has not been yet conducted. Therefore, this study focuses on the effort to identify the figures whom involve in the sasi implementation as well as the changes within. The habitus disposition analysis uses post-structuralist approach under a reference of main theory by Bourdieu on habitus. While the theoretical reference to identify land and marine sasi on this research employs the common-pool resources approach from Ostrom. *Sasi* has been widely understood as the local wisdom of Mollucan to manage the natural resources. The mechanism to manage the natural resources is based on three main basic procedure, they are, open sasi (permission to take the harvest), close sasi (prohibition to take the harvest in certain period of time, unless the permission from customary leaders are granted) and sanction for any violations [1,2,3]. The prohibition and permission to take harvest are under the control of customary leaders. It is a signal that the power or authority in the village is dominated only by a certain people. The control figures are not free from vested interest including political and economic interest [4]. Custom romance on harmonization of nature that involved the community in the village at the end will be confronted with the power relations owned by a few people who act as the defining actors for the sustainability of *sasi.* Various degree of interests between the community and the determinant actors will continuously take place and against to each other. The habitus disposition of *sasi* in the Moluccas has occurred in this context.

In regard to custom, several studies show the diversity of custom terminology. Custom is understood as a form of traditional cultural form and perceived as aftermost and directly linked to one certain tribe, restricted on territory, or on owned cultural homogeneity [5,6]. People of Moluccas have their own rules to identify their place as custom village. The identification is due to the typical clan in the area, local language, the local leader is called a *bapa raja* (King), and has a *pela* (brotherhood/sisterhood) with the other custom villages [7]. On the other study, custom is a tool of struggle to recapture people's rights who have been taken by the Government in the name of people's interests. The same thing happened when the custom lands in Bali island experiencing price spikes due to development of tourism sector. Custom was again used to suppress the speed of land selling in Bali. Besides land, Bali traditional dance experienced commodification that is transformation from sacred to commercial dance to promote the tourism [5,8]

Sasi as the Moluccas cultural product experiences powerful tug on religion interest and then evolves to be economical interest. Indubitably on the tug of the interests there are figures whose play central role. They are the controller of the sasi cultural activity continuous progress. What happen to the actor of sasi who always experiences transformation is the most important part in this study. The answer of the question of what happen to the actor of sasi is the discussion of habitus disposition in the Bourdieu's references.

The explanation about habitus disposition in analyzing the transformation of sasi culture in Moluccas is believed to be contributed to the efforts of revitalization local culture over wider platform. Revitalization in turn provides positive influence to the natural resource management on local basis of land as well as sea for the interest of the people. At this certain platform, the importance of a study of habitus disposition of land and marine sasi is offered.

2. Theories

Habitus is and individual or group behaviour which continuously occur and involving individual and community experiences over a social reality. The continuous behaviour is called as disposition terminology. Thus habitus disposition is a transformation process of individual or group behaviour to achieve certain goals [9,10]. At last, the habitus disposition creates classes within society due to the habitus character's restriction. A research by Purwanto [11] uses habitus concept and capital utilization toward the people of Kasongan in Yogyakarta proves how certain design and type of raw materials become a characteristic which cannot be replicated by other groups. Each group forms their own habitus at last. The fight between habitus occurs when they scramble ceramics market especially the marketing lane of exports.

Field of sasi in this study is not at the marine or land zoning. Field or place of sasi's fight is at where each figure or leaders are expanding their territory to obtain domination in sasi. Every single means is used by individual or group to win the fight that is gaining the most powerful authority in managing the sasi. The effort needs habitus' intelligence in using various capitals. Bourdieu divided capitals into four groups, they are capital of culture, social, economy and symbolic [12].

The marine and land sasi's positions are identified through the concept of natural resources utilization systems those are *commons-pool resources* and *common property right.* The natural resources management on community basis is what [13] called as *community-based resource management* (CBRM) *system.* In the context of natural resources utilization there are four good classifications they are *public goods, common-pool resources, tool goods,* and *private goods.* Marine sasi is applied for several types of marine animals among other sea cucumber and certain fish and every person is able to make use of that natural resource without any ownership bond. Therefore the marine sasi is categorized in *common-pool resources*. Unlikely on land sasi whereas all types of managed plants are own by individuals (private goods). To organize the maximum utilization of common-pool resources, people must obey the common rules called *common property right.* Although the concept of common property rights by Ostrom applies only to common pool resources, however in reference to sasi's law, the implementation of land sasi (plant) can also be bound by *common property right* [14]. Therefore, the rules of sasi is the implementation of what known as *common property right*.

3. Method

This article was composed based on the data which was collected through qualitative approach. The data collection process was commenced by holding identification over the area which conducting marine sasi, it is the Haruku village and land sasi at the Morella village. Geographically, Haruku and Morella are located on two different islands. Haruku is on Haruku island, whilst Morella is located on Ambon island. Nevertheless, both villages are administratively located at the Central Moluccas district. The people of Haruku villages implement the marine sasi on lompa fish whilst the nutmeg plant sasi is operated in Morella. In dept interview is a way to record information related to the focus of this study. Research was conducted in January 2016.

4. Result and Discussion

This part is to review two main issues they are elements identification inside the sasi implementation and secondly, habitus disposition analysis found in the implementation of *lompa* fish sasi in Haruku and nutmeg plant sasi in Morella. This part will be ended by providing recommendation for the government, community at large and academia.

4.1. The Elements of Land and Marine Sasi

In addition to the basic regulations of sasi implementation (open-close and sanction), this study found several other elements which even show the diversity of sasi culture implementations in Moluccas. The elements are type of commodity, local community involvement, leaders or the determinants of sasi implementation, form of sasi and time of implementation. The major differentiation between marine and land sasi is at the zoning whereby the sasi is implemented (marine and land). This study shows that the zoning differentiation is in fact the beginning of series of quite influencing differentiation affected to other elements.

Certainly there is no individual ownership on commodity nor area (territory) in marine sasi (*common pool resources*), meanwhile the ownership in land and nutmeg plant sasis is private (*private goods*). Lompa fish sasi is a custom sasi, therefore a consequence of custom sanction is applied to the offenders. In depth interview manages to reveal that custom sanction form is psychological not physical sanction. The offender will be considered as people who do not respect the customs. The punishment makes the offender be ashamed and isolated from local community. This kind of sanction looks effective at the local community who upholds the custom. The timing of lompa fish sasi is conducted once a year and its period is two months. The stipulation of open sasi is marked by a sacred custom ceremony. Normally, lompa fish sasi is conducted in October. Because the lompa fish sasi is a custom sasi, therefore the authorized leader for sasi ceremony is only the custom leader who assisted by seven custom stakeholders.

The people of Morella constructs nutmeg sasi as ordinary sasi (not custom). The controller of nutmeg sasi is not appointed by custom but auction mechanism. The auction winner normally comes from the economical player. Due to the agronomical character considerations, a sasi authority holder has the right for three harvesting times within a year. Through in depth interview reveals information that there is almost no case of harvest failure found because of the pest. Harvesting failure due to thefts is the biggest threat, and the climate change threat comes after. Consequently, nutmeg plant safety from thefts is the main issue in the activity of nutmeg plant production. The people pay supervisory service to the authorized sasi holder at a ratio of 1:10. It means from ten nutmeg yields (usually on kilogram measure), the sasi manager of whom represented by *kewang* (plant supervisor) reserves the right one portion as the rewards for the service of supervising the nutmeg plant from theft (Azuz, 2017). Although there is now obligation to become sasi participant, all the nutmeg plant is participating in nutmeg sasi when the research is conducted in Morella.

Table 1. Comparison of Land and Marine Sasi based on the Sasi Element

Sasi Element	Lompa Fish Sasi	Nutmeg Plant Sasi	Remarks
Basic Regulation	Open-Close-Sanction	Open-Close-Sanction	Same
Zoning	Sea	Land	Different
Commodity	Lompa Fish	Nutmeg Plant	Different
Type of sanction	Moral or Custom	Physical and Fine	Different
Form of sanction	Custom	New tradition	Different
Controller Figure	Custom Leader	Auction winner	Different
Community involvement	All member of the community	Individual	Different
Time	Once a Year	Three times a Year	Different
Reward	None	1:10	Different

Source: Primary data

4.2. Habitus Disposition of Land and Marine sasi in Moluccas

On Table 1 shown there are nine elements works on *lompa* fish and nutmeg plant sasi. The discussion over habitus disposition is available at the element of sasi controller figure and the involvement of community. The history of *lompa* fish has been inherited through generations and formed an exclusive class for the custom leader as the controller of sasi. The custom leader's position whom held by key informant was not shifted to other people as long as this figure is alive. In social reality of Moluccas traditional community, the descendant plays an important role as marked by clan identity. There is a clan whom from long time ago has been in charge to manage lompa fish sasi. The person in charge needs to make various efforts to maintain his power. In case he does not make the efforts there is a possibility that the position shifts to other person within the same clan.

This study reveals that one way to strengthen the position is by inserting environmental issue inside the *lompa* fish sasi culture. Whenever *lompa* fish sasi is linked to the environmental issue, there is a slot to establish cooperation with various parties. The local custom leader joins a national indigenous organization. This involvement makes *lompa* fish sasi sounded within the rhythm of environmental issue gains a lot of appreciation from national and international environmentalists.
This success in one hand meant as the effort to lift up local wisdom values for the environmental sustainability. Yet at the same time, in the sociology point of view, the effort will strengthen the bargaining position of custom leader to the community. The stronger the position, the more solid the habitus class formed and the more difficult to penetrate by other parties. This position strengthening process is known as habitus disposition.

The habitus disposition as seen on nutmeg plant sasi is unlikely as seen on *lompa* fish sasi. If *lompa* fish sasi has the effort to keep its position through establishing cooperation and voicing the environmental issues, at the other side the nutmeg sasi the main issue sounded is the effort to keep nutmeg plant safe from thefts. The geographical condition causes nutmeg plant is vulnerable from stealing threats. In such situation those who are able to guarantee the community's nutmeg plants are safe from thefts, would deserve to hold the power of nutmeg sasi. The power to hold nutmeg sasi in Morella formally gained through offering during the auction of nutmeg sasi.

Nutmeg sasi which take place for more than ten years has yielded changes of regulations from to time. The sasi power holder along with *kewang* ((plant supervisor) always renew regulations aiming to ensure the yield certainty. The certainty of yield in turn provides the guarantee of financial income for the nutmeg plants owner community. The change of the regulations actually strengthens the sasi power holder position as a party who is able to provide protections. The continuous strengthening of position is in fact the habitus disposition to forming new social class whose walls are hard to penetration.

Through the study of lompa fish and nutmeg plant reveals that there is a difference main issue used to ensure the sustainability of sasi. The main issue results continuous changes which more strengthening to the position of the sasi controller figures position for the continuity of the habitus. Through the study of *lompa* fish and nutmeg plant have shown the main different issue to ensure the *sasi's* sustainability. The main issue has resulted in the continuous changes that strengthen the position of the control figures for the continuity of *habitus*. The discussion of land and marine *sasi* at the Moluccas has shown that the study on *sasi* is still paramount important. The study should broaden its scope to include another commodity. It is vital to obtain the full picture of the changing process on local culture from time to time. At the end, it may provide the picture of culture reproduction within the society.

5. Conclussion

Marine and land sasi have similarity in its basic regulation. There are quite a lot of differences on the element of sasi.

Habitus disposition occurs on two elements of sasi they are on the figure of controller. The main issue exhaled on the lompa fish sasi is environment. Meanwhile the assurance of to yield the nutmeg plant is the issue on land sasi. The sasi controllers use the main issue to strengthen their position in the community which means the process of habitus strengthening is taking place.

6. Acknowledgement

I would like than to Ismail Suardi Wekke who support me writing scientific paper. Bapak Eliza Kisya dan Morella society as my informan for this research.

7. References

[1] Azuz, Faidah and Aylee Christine AS, 2017. The Role of Economic Agent and Traditional Institution on the Sustainability of Sasi Culture in Maluku, Indonesia. International Journal of Science and Research (IJSR). 6 (7) :25-32

[2] Adhuri. Dedi S., 2013. Selling the Sea: A study of conflict over marine tenure in Kei Islands, Eastern Indonesia. ANU Press. Canberra.

[3] Benda-Beckmann, Franz and Keebet Von Benda-beckmann, 2007. Social Security Between Past and Future: Ambonese Networks of Care and Support. Lit Verlag Dr. W. Hopf. Berlin.

[4] Zerner, Charles, 1994. "Through a Green Lens: The Construction of Customary Environmental Law and Community in Indonesia's Maluku Islands". Law & Society Review, Symposium: Community and Identity in Sociolegal Studies. 28 (5) : 1079-1122.

[5] D'Andrea, Claudia, 2013. Kopi, Adat, dan Modal; Teritorialisasi dan Identitas Adat di Taman Nasional Lore Sindu Sulawesi Tengah. Tanah Air Beta, Sayogyo Institute, Yayasan Tanah Merdeka. Indonesia.

[6] Suartika, Gusti Made Ayu, 2007. Territoriality and the Market System_Adat Land vs. Sate Regulations on Land Matters in Bali. Journal Habitat International 31. Pp. 167-176. Elsevier Ltd.

[7] Latinis, D. Kyle, (2004). Protohistoric archaeology and settlement in Central Maluku, Eastern Indonesia. Thesis for the Degree of Doctor of Philosophy Southeast Asian Studies Programme. National University of Singapore. Singapore.

[8] Avonius Lena, 2003. Reforming Adat, The Asia Pasific Journal of Anthropology, 4(2) : 123-142.

[9] Bourdieu, Pierre, 1984. Distinction; A Social Critique of the Judgment of Taste. Translated by Richard Nice, Roultledge & Kegan Paul Ltd. UK.

[10] Crossley, N, 2003. From Reproduction to Transformation Social Movement Fields and the Radical Habitus. Theory, Culture & Society. 20(6) : 43–68. Sage Publication.

[11] Purwanto, Antonius, 2013. Peranan Jaringan Sosial dalam Klaster Industri. *Jurnal Ilmu Admintsrasi (JIA)*. 9(3) : 1-11

[12] Vincent Carol and Stephen J.Ball, 2007. 'Making Up' the Middle-Class Child: Families, Activities and Class Dispositions. *Journal Sociology. Volume 41(6): 1061–1077.* Sage Publication.

[13] Novakzek, 2001. "An Institutional Analysis of Sasi Laut in Maluku" Journal *Indonesia.*

[14] Ostrom, E., Roy Gardener, and Jimmy Walker, 2006. *Rules, Games, and Comon Pool Resources.* The University of Michigan Press. Ann Arbor.

IOP Conf. Series: Earth and Environmental Science **156** (2018) 012022 doi:10.1088/1755-1315/156/1/012022

Islamic Education Values in Tinasuka's Customary on Wawonii Tribe of Konawe Islands

La Hadisi[1] Muhammad Alpin[2]

[1]Institut Agama Islam Negeri Kendari, Jl. Sultan Qaimuddin No 17 Baruga, Kendari, Sulawesi Tenggara, 93563, Indonesia
Email: lhadisi@gmail.com
[2]Institut Agama Islam Negeri Kendari, Jl. Sultan Qaimuddin No 17 Baruga, Kendari, Sulawesi Tenggara, 93563, Indonesia
Email: muh.alpin321@gmail.com

Abstract: This study aims to know and understand the values of Islamic religious education in *Tinasuka* custom of Wawonii community in Konawe Islands District. This research uses qualitative descriptive research. Data was obtained in the form of opinions, views or another expression of thoughts through interviews, then analyzed by compiling data. To determine the validity of data obtained, this study used triangulation of sources and data. The results show that: "*Tinasuka* in Wawonii Tribe of Konawe Islands Regency has a close relationship with the values of Islamic education". *Tinasuka* comes from wawonii language. It means dowry which way of fulfilment and its kind is different from another area, the dowry type is coconut tree, and the fulfilment depends on social strata of the woman. Historically, according to the Wawonii community, the philosophy of *Tinasuka* customary originated from the meaning of wawonii island and the facts about the source of the wawonii tribe's life dominated by coconut trees compared to other types of agriculture. So the value of Islamic education in *Tinasuka* custom is to prioritize tolerance and humanism in the fulfilment of dowries, are required to work hard especially in planting coconut trees, and cooperation. After the end of the dominant system, the *Tinasuka* custom continues to grow and undergo some adjustments coupled with the reduced land to grow coconut trees, so that the *Tinasuka* in the form of coconut trees can be converted in money.

1. Introduction

The primary function of culture is to make the community still have unity in equally and have its identity [1]. Thus, it can be said that members of society to be an active component in building integration, must have cultural awareness and historical awareness. History is a past that serves as a guide to know the culture or traditions that exist in a each region, especially Konawe Island Regency or often called Wawonii. Wawonii society's culture is a reflection of life [2] concerning the history in this *Tinasuka*/ dowry in the form of coconut trees.

In the history of the kingdom described by the late Wahinda (grandson of king mbeoga) that the customs in the Wawonii society have been born since the reign of King Mbeoga based in the village of Tangkombuno around the 16th century. The custom is still valid until now for example Kalapaeya. This custom is only intended for the mokole/king of Wawonii or the royal family with varying degrees in its use. But in its development, Kalapaeya has become a tradition for all including ordinary people. Tinasuka, this custom is also common place called the dowry given the groom to the bride in the form of a coconut tree. One of the joints of life associated with it is *Tinasuka* custom or marriage custom of wawonii tribe. Wawonii tribe is located in Konawe Regency of Southeast Sulawesi Province. Wawonii itself is an acronym of "WAWO" which means above and "NII" means coconut [3]. Hence, *Tinasuka* became one of the requirements in marriage because the wawonii tribe is based on coconut plants. *Tinasuka* has a sum of thirty coconut trees which are generally benefited while forty coconut trees are benefited to the king, and ordinary people/ benefited fifteen or ten coconut trees depending on the social strata of the bride. It means that the higher the degree of a woman, the more coconut tree is mentioned as a marriage dowry. After the end of the royal system, some people decided to balance the value of the marriage dowry on the grounds of humanity. Nevertheless, most people, especially the custom leaders, still adhere to the predetermined dowry values. This is due to the philosophy of the birth of *Tinasuka* custom which is derived from the meaning of Wawonii island and the facts about the source of Wawonii's life dominated by coconut trees compared with other types of agriculture.

According to the Batumea village community, *Tinasuka* comes from the wawonii language which means dowry. *Tinasuka* word cannot be separated. If it is separated into two words; Tina and Suka then will make a different meaning. That is Tina (female) while Suka (like) so that word *Tinasuka* cannot be separated into two words. *Tinasuka* is also only in the allotment of women who come from wawonii tribe because this is a tradition of the local population (wawonii). As for those outside the wawonii tribe who are domiciled in wawonii then they will adapt to the *Tinasuka* custom itself, it is pointed out that the native wawonii people and people outside the wawonii tribe uphold the value of ethics and mutual respect.

Tinasuka was born as an agreement in the past initiated by the king of Mbeoga in determining the dowry of a groom against the bride. According to Bunggulawa in his article on the history of the discovery of Wawonii island, the philosophy of the birth of *Tinasuka* custom which uses coconut trees is at that time the largest plantation commodity on Wawonii island is coconut tree, so this is the possible alternative to be made as dowry. The customs of *Tinasuka* contain many elements of education based on the teachings of Islam, which still require further study to create a complete understanding that the customary marriage of the Wawonii tribe contains elements of Islamic education both studied from historical aspects and other approaches.

2. Method

This research uses qualitative research. The data was obtained in the form of opinions, views or another expression of thoughts through interviews [4], then analyzed by arranging data. To determine the validity of data obtained, this study using triangulation of sources and data [5]. To analyze data, the writers utilized content analysis.

3. Finding and Discussion

3.1 Tinasuka (dowry) and its function

Based on the observation and interview, the position of *Tinasuka* is very sacred in the marriage because it is not in a marriage proceed without the dowry. *Tinasuka* has also been determined by tradition where *Tinasuka* or dowry, in general, is thirty coconut trees. There are several divisions of Dowry according to Wawonii tribe culture. (1) Raja is 40 coconut trees, (2) Descendants of the king is 30, (3) Ata or ordinary people is 10-15 coconut trees. Asnung *Tinasuka* is a language taken from the tribe Wawonii which means dowry, where setiapa men are obliged to pay the dowry on the side of women to be able to establish a sakinah mawaddah and wa Rahma relationship appropriate atuaran belaku. Ihwan says that *Tinasuka* is one of the most urgent parts of the Wawonii tribal marriage. *Tinasuka* is also translated as giving a dowry to marriage, especially adat Wawonii. *Tinasuka* comes from wawonii which means dowry. Dowry is not in the form of money but the coconut tree. So, it is not a surprised news when the dowry in wawonii tribe marriage is in the form of coconut trees, because the birth of *Tinasuka* originated from the name of Wawonii island. After the end of the royal system, some people decided to balance the value of the marriage dowry on the grounds of humanity.

IOP Conf. Series: Earth and Environmental Science **156** (2018) 012022 doi:10.1088/1755-1315/156/1/012022

Nevertheless, most people, especially custom leaders, still adhere to the predetermined the dowry. Because they think culture or customs are very closely related to society. Everything contained in society is determined by the culture owned by the community itself. Life and civilization at the modern level are not able to eliminate habits in society. Only those bound in the process of advancement of the customary age can adapt to the circumstances of the times.

Cultural change in a society occurs due to the mixing of different cultural elements [6] that are united by the assimilation process of the culture brought by each of different regions [7]. The change is influenced by its importance regarding the positive benefits to be changed. *Tinasuka* in indigenous marriage of Wawonii in Wawonii district, Batumea village is conducted after the whole process of marriage to the implementation phase of *ijab qabul* in which mentioned the value and form of its Tinasuka. Based on the current price of a coconut, a tree is equal to one hundred thousand, and in 1982 the price of one coconut tree is thirty thousand to fifty thousand. If it is included with the land, it depends on the agreement. But if Tinasuka/Dowry around thirty coconut trees are included with the land, so the price of Tamaruka/dowry for the bride is replaced with the money equals to three Million, but that's all not including the giving of money on the cost of marriage. For example Tinasuka/Dowry is thirty coconut trees, it is patented. That is the cost in the aisle.

According to H. Kalamuddin, Tinasuka/dowry's function is to legalize marital relationship, because the marriage is not executed without any dowry. *Tinasuka* also shows the responsibility of the husband in the marriage life by providing for a living. Therefore, men are the top leaders of women in their household life, and to get the right, it is natural that the husband should remove his property so he should be more responsible and not arbitrary against his wife.

3.2 Processing Process of TinasukaTradition in Batumea Village, Middle Wawonii District

The process of execution of *Tinasuka* tradition in Batumea village community in subdistrict of middle Wawonii become habit from generation to generation because *Tinasuka* tradition has been existed since ancient time. *Tinasuka* in Batumea village is basically coconut tree but can also be converted with real money. This is addressing attitude of tolerance, because this tradition was born with the agreement between the bride and groom's representative. The implementation of *Tinasuka* tradition in Batumea village community has a nuance in the acculturation system through local culture. The name of identity is maintained so that the substance of value still pays attention to the similarity with other areas in southeast Sulawesi. *Tinasuka*/ Dowry in the Wawonii tradition of Batumea village is a series of wedding processions with the commencement of the stage of Melamasi or apply up to the pompepanga/proposal event.

(a) Melamasi is the language of Wawonii which means applying, which means that before carrying out the marriage proces, a representative of the groom went to the household to ask women metukana the willingness of the women. The metukana/ask process is as follows. (1) In Metukana/ask, male parent or representative is called pu'utobu/ the speaker goes to the women's place to ask the child's willingness, the example of question in melamasi. (2) Male parent: dadi ai kaku leu dahamiu metukana kida anatalu miu kida mokea us umowu. It means (so this I came to ask if there is a girl if we want to marry her). The woman's parent: ai mana dah dahomo miano mia umowuo dadi rigid tukanao pirou anangku balipo kidamo answeranno anangku dakusi kumuwai komiu. This means (this may already have a person who has so I will ask my daughter later if there is an answer from my daughter later in the news). With the completion of the process of applying which begins metukana process, then the men return and wait for the answer because in the process of metukana, if there is positive answer from the women after there is news that the men accepted then they will conduct the process of *peminangan.*

(b). Application. The proposal is a continuation series after the completion of the event procession. The appointment is also the process of determining Tamaruka/dowry, besides the course of the Proposal is very important because at this time, it is used to be negotiated and decided everything related to the wedding ceremony, such as Mopatantu kinta kawia or Determination of marriage. In the process of offering, the men bring my kolungku/custom contents.Tthis kolungku is round and wrapped in putty cloth, and inside contains lime, areca nut, leaf series, and cigarettes. The contents of kolungku/adat contents have different meanings. (1) *chalk / ngapi meaningful:* mobulano ngapi / lime mobulaopo laro mami tumarima komiu, meaning seputi-pitih kapur / ngapi more puti contents in our heart to receive you. (2) *Pinang / wua meaningful:* bobotono wua / pinang bobotopo pelaro mami mokea mepoko asade dahamiu, meaning round areca / wua more rounded content in our heart to unite with you. (3) Leaf Siri / lewe seua meaningful: omboleno lewe seua / leaf Siri pepoko ombolento petutuwai, meaning width-width leaves Siri / lewe seua marihlah we widen this family. (4) Cigarette / Tabako meaning: pomone ahuno dadi kundo dadi ki'ti dadi usa, meaning the rise of cigarette smoke so cloud, kalu down so the rain so that feeling to be peaceful. The condition of my collung or adat content is only a symbol of life or describes the variations of life that must be taken by the bride in their life.

In the process of *Tinasuka* proposal, they already mentioned how the *Tinasuka* will be and what kind of type it will be, whether coconut or money. And *Tinasuka* will be given during the wedding process. In the marriage process, *Tinasuka*or dowry is mentioned at the time of marriage ceremony, then given to the family or who represent the family in the presence of custom leader, religion figure and society who attend the place. And the process of marriage kolungku or custom contents still brought because in the process of marriage, kolungku play an important role because kolungku / contents of customs are an important point in the delivery *Tinasuka*/ dowry. The delivery kolungku or custom contents should be together with *Tinasuka*/ dowry.

3.3 The Process of Giving a Marriage

The giving of *Tinasuka* tradition in Batumea society has nuance in acculturation system through local culture, but its characteristic and its identity is maintained as an example of the process of giving *Tinasuka* in *Ijab Qabul* Wawonii. (prince: Mr rahmin, I hereby to marry you with Evi Yanti binti Majid, with thirty coconut trees as dowry are paid in cash because of Allah. Rahmin Replied: I received the marriage of Evi Yanti binti Majid with thirty coconut trees as dowry paid in cash because of Allah). The coconut tree will be awarded by the marriage contract, from the male family and women's families and traditional figure, Religion figure, village heads and other communities will be witnesses when they go to the country to appeal to the mentioned of Tinasuka/dowries. But otherwise, if real money becomes the Tinasuka, then it will be given when the process of marriage. In the administration of *Tinasuka* in the marriage process, there are some indivisible customs that have a value of equality with *Tinasuka* as disclosed by Gasari that, in the process of granting dowry, there are some customs that cannot be left behind because this is one of the important conditions in giving dowry, such as; (1) 1). Tinasuka/thirty coconut trees, if replaced with money then the extension of the granting of *Tinasuka*still based on the reference of *ijab qabul*. (2) Pu'uu / Gamis, this robe adjusts to the giving of Tinasuka, if *Tinasuka*is mentioned at the time of qabul ijab equals to thirty coconut trees then the three pieces of robe/gamis which are given to the bride, all depends on *Tinasuka*of how it is has been mentioned in Qabul process. (3) Rane-rane pa'aa / one sheet Sarong, the sarong given to the bride, does not depend on the value of *Tinasuka*/ dowry because the sarong will be given to the bride is only one sheet. (4) Pompepanga / Ring, This Ring as a bond in a marriage because building a family begins with a holy intention for Allah SWT, to continue the generation and as a means of worship him; then the ring is a proof in marriage when the process of ijab qabul.

According to Arif, giving *Tinasuka*in Batumea Village there should be clarity in pronunciation of *qabul*, because the pronunciation of ijab qabul which he witnessed is wrong, in the pronunciation of Kabul. He witnessed that it is not included with coconut as dowry but money .. Example "I marry you with her with the dowry of thirty coconut trees is paid in cash in the name of God ", Meanwhile, it should like this 'I marry you with her with dowry thirty coconut trees with the value of money three million rupiahs paid cash in the name of God. If the price of coconut one hundred thousand for one tree. And according to traditional figure, community, Religion's figure and village head's deal, the price for one coconut tree is one hundred thousand. And he argues that in the pronunciation of kabul, it is wrong. Coconut value should be mentioned in the value of the money. With the review of Islamic law, the provision of *Tinasuka*custom is still wrong and must be added to the sentence by traditional granting that *Tinasuka*is correct. Because in the law of marriage, the value of the dowry must be mentioned at the time of qabul ijab, if it is unclear then according to Mohammad Arif is not a valid marriage.

3.4 The Values of Islamic Education in Tinasuka or Dowry Customary

The Values of Islamic Education in *Tinasuka* or Dowry Customary are tolerance, hard work, responsibility, the honor to the women, and cooperation. The following is the the explanation of the values.

3.4.1 Tolerance

Tolerance According to the Indonesian dictionary, tolerance comes from the word "tolerant" itself which means to be tolerant (respect, let, allow), opinions (opinions, views of customary beliefs and so on) that are different or contrary to their stance [8]. Tolerance in Arabic (tasamuh) means is "equally good, gentle and forgiving each other" [9]. In the sense of the general term, tasamuh is a morally appropriate attitude in the association, where there is mutual respect between fellow human beings within the boundaries outlined by the teachings of Islam. So that Islamic education has special characteristics that distinguish the formal education system in schools and nonformal in the community. Although both are elements integrated into the national education system. Therefore, the study of theoretical conception in this research is more oriented to the education of Islam in *Tinasuka* custom. In the view of Islam, *Tinasuka* is a sincere tribute to women and it is obligatory for men to pay dowry/Appeals to women as a sign of their altitude and as a prelude to a loving base that gives the bridegroom to the beloved bride by their respective family agreements. The giving Dowry is justifying the marital relationship. So the commitments agreed upon since the beginning that there is no force in determining *Tinasuka* and tolerance upheld by Batumea villagers.

The meaning of tolerance is also to let other people think differently, do things that do not agree with us, without us disturbing or intimidating them. In terms in a social, cultural and religious context, it means attitudes and actions that prohibit discrimination against different groups. In determining Tinasuka, tolerance takes precedence because it does not rule out that in giving *Tinasuka* there is an element of coercion or has been determined by the bride even though customary law of *Tinasuka* has been determined to be thirty coconut trees and can be converted into money with a value of three million. But the problem is the cost of weddings. In the granting of dowries, of course, consensus is held so in having a party, men do not feel burdened in determining the dowry and the cost of marriage.

3.4.2 Hard Work

Hard work is to work with teeth and earnestly. The success of bias obtained by someone with sincerity while failure in general due to laziness. Working in the view of Islam is free, any work is allowed that is important does not violate Islamic law and have halal results. In Islam, a good worker is a hard worker. It is said that the mass employment of all the people of wawonii is demanded hard to plant coconut trees as a source of livelihood because the income of wawonii tribe is in the coconut tree. In the old days of Batumea's village view of the lifestyle of society, there are important values to emulate such as help, mutual respect and mutual help, and that is still taught by our parents, what else when the event marriage of parents is working together in the event. The aims are solely aimed so that they will know how to live the best life. Because of every parental guidance can be realized that this is an education or provision that must be applied in life and has positive value. Thus hard work is to do something seriously, and in every way, especially to seek fortune.

In hard-working based on Islamic teachings, it is obliged by all his people, because Islam encourages his people to achieve the desired goals. Hard work has also been performed by the Prophet Muhammad and his companions, Rasulullah Saw work hard by trade to help the economy of Abu Talib, the wawonii community planting coconut trees as a source of livelihood and as well as dowry.

3.4.3 Responsibility

Human life whether personal, family, or society and nation must have some needs that must be fulfilled, whether the needs of clothing, food, boards, means of living up to self-development. And these needs can be fulfilled when the availability of sufficient funds. Therefore, each person should be able to meet the needs of his economy and better yet, until it can help meet the economic needs of others. Because meeting the needs of the economy is a must for every person and in the context of the family, the primary responsibility lies with the husband or wife so that in the giving of the responsibility of a groom is fully responsible for being able to support the bride. Behind the simplicity of the symbol of thirty coconut trees converted into money, it turns out its meaning gives ease to the requirements of paying off the dowry. But there is a big responsibility behind it. That is the form of a husband who is ready to give a living to the wife and guide his wife in terms of religion. In addition, ensuring that the family is always on a straight path.

3.4.4 The Honor of Women

In the process of marriage, there are one thing obligations that must be submitted by the men to the prospective wife. In the language of Wawonii known as *Tinasuka* (Dowry), this is done with a sign of love as well as *Tinasuka* is also one of the symbols of respect and the glorification of women required by Allah as a gift of men to the bride proposed when the groom want her to become life companion as well as the respect and his honor. The above description shows that the dowry is one of shari'a in Islam. However, in its development (one of them is because the majority of the population of Indonesia is Muslim), this *shari'at* eventually became customary in marriage in almost all regions in Indonesia [10]. This dowry becomes the wife's right completely, so the form and value of this dowry are determined by the will of the wife. Islamic Shari'ah does not specify minimum or maximum limits for dowry weddings. And it could be a dowry in the form of money, goods or services.

3.4.5 Cooperation

A joint effort is between individuals or groups to achieve common goals. Cooperation is the most important interaction because man cannot live alone without others so that he always needs others. Cooperation can take place when the individuals concerned have the same interests and have the awareness to work together to achieve their interests. To determine a family, in a state of harmony and prosperity, the cooperation between family members is one of the main factors that must be implemented. Because it will affect the personal formation of each member of the family in living their daily lives. So in determining Tinasuka, cooperation in the family is very needed and according to the results of the observation that the Batumea village according to the family system. When there is a wedding party, of course from the bridegroom's family come to contribute to thinking, giving solutions and financial assistance to provide *Tinasuka* to the bride as a wedding dowry. In terms of the value of Islamic education, cooperation in specifying the work specify the dowry, of course, the cooperation between the family, the group is needed because Allah Almighty in the Quran says *"And please-help you in doing good and piety, and do not help-in helping to sin and transgression."* (Al-Maidah 5:2) There is no doubt that in every society, there are the poor and the needy, the unemployed and insufficient incomes to meet their needs. Noting that according to the Islamic viewpoint, all human beings are Allah's creatures and all wealth is His, we must meet the needs of these individuals within acceptable and acceptable limits. This issue proves how important it is to work together with each other in the family.

In custom, there are human relationships that always built harmoniously and complete each other to create a happy life. As Mohammad said wisely that humans want the happiness of the world and afterlife in a marriage, Man living on earth is not alone, but with other creatures, namely plants and animals. Other living beings are not just living friends who live together neutrally or passively towards humans, but human life is closely related to them. Without them, man cannot live, as this is the picture of life in marriage. Based on the above explanation it is clear that human life cannot be separated from

IOP Conf. Series: Earth and Environmental Science **156** (2018) 012022 doi:10.1088/1755-1315/156/1/012022

each other, it means to complete each other. This has been done by villagers of Batumea Village of Middle Wawonii in Konawe islands, in a marriage

4. Conclusion

Based on the above description, researchers can take the following conclusions. (1) *Tinasuka* comes from the language of Wawonii which means dowry. The dowry is not an object but a plant that is thirty coconut trees, but the granting of this dowry can be converted into money which equals to three million because the coconut tree planted in Batumea village worth one hundred thousand requests according to the custom's figure, religion, community and village head. Granting Tinasuka/dowry is at the time of a qabul ijab process. When the bride's dowry is thirty coconut trees then finished the marriage process by the traditional's figure, Religion, and the people who witnessed will plant the garden to show its Tinasuka/dowry. But if the *Tinasuka* is replaced with money, it will be given at the time of the consent process. (2) The value of Islamic education contained in *Tinasuka* is tolerance, hard work, responsibility, the honor to the women, and cooperation.

5. References

[1] Bruner JS 1996 *The culture of education*. Harvard University Press.

[2] Parker L, Creese H 2016 The stigmatisation of widows and divorcees in Indonesian society.

[3] Noorduyn J 1991 The languages of Sulawesi. *Pacific Linguistics. Series A. Occasional Papers*. **81**, 137-150.

[4] Jick T D 1979 Mixing qualitative and quantitative methods: Triangulation in action. *Administrative science quarterly*. **24**(4) 602-611.

[5] Patton MQ 2005 *Qualitative research*. John Wiley & Sons, Ltd.

[6] Inglehart R, Baker WE 2000 Modernization, cultural change, and the persistence of traditional values. *American sociological review*. 19-51.

[7] Sam DL, Berry JW 2010 Acculturation: When individuals and groups of different cultural backgrounds meet. *Perspectives on Psychological Science*. **5**(4) 472-481.

[8] Moore HK, Walker CA 2011 Tolerance: a concept analysis. *Journal of Theory Construction & Testing*. **15**(2) 48.

[9] Dragon M S 2015 *The Spirit of Tolerance In Islam Religion*. Sakura Dragon SPC.

[10] Idriss MM, Abbas T (Eds.) 2010 *Honour, violence, women and Islam*. Routledge.

IOP Conf. Series: Earth and Environmental Science **156** (2018) 012023 doi:10.1088/1755-1315/156/1/012023

Religious Study of *Katoba* Tradition and Its Function in Character Building of Muna Society

Hardin Hardin,[1] Sitti Hermina[2]
[1]Universitas Halu Oleo Kendari, Kampus Hijau Bumi Tridharma, Anduonou, Kambu, Kendari, Kambu, Kota Kendari, Sulawesi Tenggara 93132, Indonesia
Email: hhardin@yahoo.co.id
[2]Universitas Halu Oleo Kendari, Kampus Hijau Bumi Tridharma, Anduonou, Kambu, Kendari, Kambu, Kota Kendari, Sulawesi Tenggara 93132, Indonesia
Email: hermina_sitti@yahoo.co.id

Abstract: This research aims at studying the function of *Katoba* tradition to the character-building of the society of Muna ethnics. The research was conducted in Kecamatan Lawa, West Muna District. This research includes descriptive-qualitative research. Data collection techniques used are observation, interview, record, and documentation. The object of this study is the speech or advice in the *katoba* tradition related to character education. From the research results found that the ethnic Muna community has a tradition of *katoba* in shaping the character of society. The value of the character in question is the formation of respectful values of honour, the value of character and ethical values in the person of the child who has been tried. It expected that this article will be able to inspire and guide people to have a good personality, to behave and to behave wisely to their family and social environment.

1. Introduction

The society of Muna ethnics has the tradition of *Katoba* that is regarded to be sacred by the community as its supporters [1]. The tradition of *Katoba* is one of the oral traditions of the society of Muna ethnics that is conducted at the time of Islamization process of a child at the age of 6-10 years [2]. This culture is considered to be sacred by the society of Muna ethnics because it is made as the medium of Islam values installation towards the child who has been grown up to actuate his life well in the future [3]. Hadirman states that *katoba* is one of the oral traditions of Muna tribe in Muna Regency and West Muna Regency in Southeast Sulawesi Province [4]. The implementation of this ritual has adat and cultural expressions spoken verbally by a speaker (village faith) to the child being staged (di-*katoba*). The series of ceremonies and the delivery of moral and ethical information to the Catholic child are the things that must be created in this tradition.

The tradition of *katoba* has very useful and beneficial lessons for the young generation specifically the children who will do *katoba.* The lessons that are found in tradition of *katoba* are: (1) the lessons about the things that may be done and may not be done; (2) the lessons about the rules of making ourselves holy; (3) the lessons of actuating Islamic law; (4) the lessons around the customs; (5) the lesson of appreciating parents, the elder persons, elder brothers/sisters, and of loving younger brothers and sisters and children

[5]. According to La Niampe, *katoba* tradition is part of the Islamic procession for young (adult and adult) children (7-10 years old) [6]. Circumcision stage precedes the *Katoba* Tradition. After going through this procession then declared legitimately undergo the process of studying the Islamic religion, especially learning to read the holy book of the Qur'an and learn to perform obligatory prayers and customary learning, especially beginning with listening to the advice or teachings of both parents.

There are four main elements in the process of *katoba*, namely; Religious leaders, adat leaders or *katoba* advisers), children who tried (the object of speech), the closest relatives who held the child at the time in *katoba*, and the closest families who acted as witnesses to the implementation of the *katoba* procession. Speech delivered by the adat stakeholders to the children in *katoba* is a media in the character formation of the child to become a person of noble moral character, independent, and responsible. For example, not to steal, not to gamble, not to drink alcohol, not to commit adultery, and so forth. Another thing that is taught to children is to always care about the environment, for example, should not damage the plants/trees, get rid of stones in the road, and so forth. Then emphasised also that if it has been in *katoba* (Islamised), then the child can no longer show his nakedness.

2. Method

The research was conducted in Lawa, West Muna District. This research includes descriptive-qualitative research. Data collection techniques used are observation, interview and documentation. The observations made by researchers down directly to observe the location of the study. Interviews were used to obtain valid data on the process of *katoba* traditions conducted and experienced by informants. Document study was conducted to find information relating to the object of research through books, research results, journals, and online media (internet). The technique of data analysis of this study is done by descriptive-qualitative.

3. Findings and Discussion

3.1. The value of Religiosity in the *Katoba* Tradition

The implementation of the *katoba* tradition is a good tool for introducing moral and religious teachings to children from an early age. In the *katoba* tradition, a child must perform a process of repentance by reciting istighfar three times. This istighfar reading is done to ask forgiveness of Allah SWT, parents, and fellow human beings for the sins that have been done so that the child undergoing the *katoba* is expected to be clean or holy from the sin that has been committed during his life. Hardin says that the *katoba* tradition is the starting point for a child to learn the teachings of Islam in general [7]. *Katoba* tradition has religious values associated with Islamic teachings as beliefs held by the Muna ethnic community. Spiritual value states the existence of a close relationship between humans as being with Allah SWT as the khalik. It can be seen in the following verbal discourses.

1 *Dosahadha be Debasa Dhoa*
"Uttering two sentences of shahadah and reciting a prayer".

Saying two sentences shahadat in ceremony *katoba* done in Arabic lafadz and Muna language. Lafadz in Arabic as follows. *Asyhadu an lailaha ilallah wa asyhxadu anna Muhammad rasulullah 3x*

2 *Aini dasumakusiimo. Miina bhe ompu nisomba sokati Allah Taala. Maka dasumakusiimo tora. Anabi Muhammadhi itu katuduno Allah Taala.*
"I testify that no god is worshipped besides Allah Taala. And I testify also that the Prophet Muhammad is the messenger of Allah Taala".

IOP Conf. Series: Earth and Environmental Science **156** (2018) 012023 doi:10.1088/1755-1315/156/1/012023

Saying two sentences creed means that we as ordinary people should worship Allah SWT, not associated with the creatures, must carry out all his orders and stay away from all the restrictions. We must also admit that the Prophet Muhammad is the messenger of Allah who is the last prophet.

The oral speech of *katoba* traditions conveyed by the most dominant speakers is instruction or command. The command is addressed to children who will try to do according to the norms that have been shown by both their parents. Speakers or religious leaders firmly inform the children who will try that in social life should always pay attention to the norms of religion and customs norms. Through the *katoba*, children are advised to know what actions are liked and disliked by Allah SWT, Prophet Muhammad, parents, brother, fellow age, and under our age. Good deeds must be exercised. Evil deeds must be abandoned, the elderly respected, the fellow appreciated, and the younger ones cherished and nurtured.

The religious function conveyed through the *katoba* is to guide and simultaneously alert the children who are being tried, to always be honest with oneself. They are reminded not to take the goods of others who are not hers. They should always try to restore the goods to their owners or their families. Children who are being tested are always reminded to keep oral or speech from all dirty words or words that are not useful, both to the teacher, both parents, older siblings and to his sister. They are reminded that all actions in the world after will be accounted for the creator.

3. 2 Function of *Katoba* Tradition in Character-Building

Wisudariani said that character education becomes a system of installation of character value for the citizens of school (students) including societal citizens in general that comprises the components of knowledge, awareness and willingness, and action for carrying out the values mentioned above [8]. Character education has the same essence and meaning as moral and behavioural education, that aim at forming children's personality, becoming good humans, good societal citizen, and good country citizen that are influenced a lot the culture of its nation and country. The society of Muna ethnics also has a tradition of *katoba* that contains a critical value in building society's character to be a good human having noble character. This tradition owns an essential function in building human's civilisation to be an independent human having good character. The purpose of character that is found in the tradition of *katoba* is morals as seen in the oral text below.

3 *Amando dotehie felo kabolosino Allah Taala*
"The real father feared, regarded as a substitute for Allah SWT"

The oral speech (3) contains the moral values that are expected to be executed by the child in the life of the world, especially in the family environment, can know his biological father well. Knowing the birth father in the sense of fear of all that is forbidden and obedient to everything he commanded, that is the picture of a child who is afraid of Allah's prohibition and obedient and obedient to all the commandments of Allah SWT with all his greatness.

4 *Sano fomasighoo itu kamokula moghane nofomasighoomo dua Allah Taala. Sano fomoraku itu kamokula moghane nofomorakumo dua Allah Taala.*
"When loved by our male parents have been also cherished by Allah SWTTo be hated by a male parent means that Go has hated us."

The oral speech (4) implies that God's wrath will be inflicted upon the child if the child does anger or rebellion against the male parent. Therefore, through the process of *katoba* tradition, the child is expected to understand every content of *katoba* delivered by Imam or religious leaders. Because the consequences of God's wrath on the child depend on obedience, obedience in carrying out every command of both his parents and away from any restrictions.

5 *Inando dotehie felo kabolosino* Prophet Muhammad
"The real mother feared, regarded as the successor of Prophet Muhammad"

The oral speech (5) implies that how much the mother's service to the child. Since the start of the fetus he has taken care of us, then when will be born into the world how suffering he endures pain. He risked lives between life and death for our salvation. After birth, he cares for us with great affection. Through the *katoba* a child is reminded to always obey and obey his mother's command because God's wrath depends on muriya mother's mother. How sad we as a child after adulthood show disobedience to the birth mother.

6 *Isando dotehie felo kabolosino* Malaekati
'Dreaded sister, think substitute Angel'

The oral speech (6) implies that as a child who is obedient to God must always appreciate his brother. As a good sister must always carry out every command that leads to the right things and away from all that bannedWhich leads to our mental or moral damage. In the *katoba* tradition, it is emphasised that an elder brother is always required to provide examples that can be modelled by his siblings, as well as the contents of the *katoba* that consider the older brother to be the successor of the angel who always teaches goodness and prevents the man from bad deeds.

7 *Aindo doasiane, dopiarae felo kabholosino Muumini*
'Younger siblings are cherished, nurtured, assume the substitute of the believers'

The text (7) implies a sister expected to always love, nourish her younger siblings. Through this *katoba* also expected between an older brother and a younger brother can love each other, mutual respect, mutual obedience, and mutual respect. To appreciate the sister is to have affection and nourish the self-esteem of fellow believers.

4. Conclusion

Based on the result of research, it is found that tradition of *katoba* is a system of religion that functions as the process of Islamisation admittance towards the children in the society of Muna ethnics. This tradition has been inherited from generation to generation and still survives up to now. The tradition of *katoba* possesses a valuable function in it. The content of value that is found in culture of *katoba* functions as the builder of character for the society of Muna ethnics since the period of our ancestor until now. It is expected to the religious figure, customary figure, and the society of Muna ethnics, on the whole to keep maintaining the survival of *katoba* tradition to be made as a medium of the continual character-building for the next young generations.

References

[1] Bauto LM 2016 Socio-cultural values as community local wisdom katoba muna in the development of learning materials social studies and history. *Historia: Jurnal Pendidikan Sejarah dan Sejarah.* **14**(2) 195-218.

[2] Hadi H 2016 Tradisi Katoba Sebagai Media Komunikasi Tradisional Dalam Masyarakat Muna (Perspektif Komunikasi Ritual). *Jurnal Penelitian Komunikasi dan Opini Publik.* **20**(1).

[3] La Fua J 2015 Etnopedagogi Katoba Sebagai Bentuk Pendidikan Karakter Dan Pendidikan Lingkungan Hidup Pada Etnis Muna. *Zawiyah: Jurnal Pemikiran Islam.* **1**(1) 97-119.

[4] Amin H 2017 Model Pendidikan Karakter Berbasis Tradisi Katoba dalam Budaya Masyarakat Muna di Sulawesi Tenggara. *Etnoreflika: Jurnal Sosial dan Budaya.* **2**(1) 65-76.

[5] Hadirman 2016 Tradisi *Katoba* Sebagai Media Komunikasi Tradisional dalam Masyarakat Muna. *In the Journal of Communication Research and Public Opinion*, **20**(1).

[6] Niampe La 2008 Tuturan Tentang *Katoba* dalam Tradisi Lisan Muna. *Paper presented at International Seminar in Wakatobi*. Southeast Sulawesi. 1-3 December.

[7] Hardin 2008 Ritual *Katoba* Sebagai Produk dan Praktek Budaya Islam yang Mencerminkan Seperangkat Norma dan Nilai Sosial - Budaya Bagi Masyarakat Etnik Muna. *Paper presented in the National Student Writing Competition in Makassar.* 29-30 March.

[8] Karmini et al (*ed*) 2013 Mengurai Tradisi Lisan Merajut Pendidikan Karakter. *Cakra Press.* Denpasar.

IOP Conf. Series: Earth and Environmental Science **156** (2018) 012024 doi:10.1088/1755-1315/156/1/012024

In Vitro Selection of a Drought Tolerant Callus of Dwarf Napier Grass *(Pennisetum purpureum Cv. Mott)*

Marhamah Nadir[1], Rinaldi Syahrir[2] and Syamsia[3]

[1]Faculty of Animal Science, University of Hasanuddin
Email: marhamahnadir@unhas.ac.id
[2] Applied Plant BIO-SCIENCE Laboratory, Teaching Industri University of Hasanuddin
Email: rinaldi.sjahrir@gmail.com
[3]Faculty of Agriculture, University of Muhammadiyah Makassar
Email : :syamsiatayibe@unismuh.ac.id

Abstract : Drought stress causes 50% yield losses of plant production in a tropical area like Indonesia. In vitro selection was a method to the identified somaclonal variation of dwarf Napier grass by using callus induction. The non-ionic water-soluble polymer polyethylene glycol (PEG) of molecular weight 6000 was used as osmoticum to simulate water stress for in vitro selection. This study aimed to use somaclonal variation to select drought-tolerant plants of dwarf Napier Grass. Multiple shoot clump (MSC) of sterilized explant were cultured on Murashige and Skoog (MS) medium containing 2 mgL-1 2,4-dichlorophenoxyacetic acids (2,4-D), 50 ml/l coconut milk (callus growth medium; CGM) with 30 g/l sucrose and 3 g/l phytagel for callus induction. The calli were grown on CGM with five concentration of PEG for selected medium (0, 40 μM, 80 μM, 120 μM and 160 μM) for 6 weeks. There were significant differences between fresh and dry weights of callus in different concentrations of PEG. Callus regeneration of dwarf Napier Grass occurred at the concentration of 40, 80, 120 and 160 μMof PEG and all regenerated MSC well rooted on all concentration PEG. In order, regenerated shoot occurred at the concentration more than 80 μM PEG solutions.

Keywords: *Callus, Drought Tolerant, In Vitro, PEG, Napier grass*

1. Introduction

Drought is a crucial problem in Indonesia, 144.470.000 ha dry land, it is covering 67,2% of marginal land and 16,57% dry land in Sulawesi island. The negative impact of drought stress has been recognized to pose a growing threat to sustainable agriculture, particularly under the global climate change. Water stress is a major factor of environmental stresses such as drought, salinity, and low temperature. Whereas, 40–60% of the agricultural land around the world suffers from drought [1, 2]. To anticipate widely of the dry land regarding of climate change and reducing agriculture land for residential and manufacture sector, construct new cultivar crop adapted to climate change is one of the important strategies for adapting agriculture to climate change. Breeding program by using biotechnology mostly needed for environmental engineering. Increase in drought periods may require the development of drought-tolerant crops [3, 4]. .

Breeding for water stress tolerance by traditional methods is a time consuming and inefficient procedure. Tissue culture as a possible technology for obtaining the desired characteristics of variants can lead the variation to the expected outcomes while the probability accomplish an in vitro selection depends on the accessibility of an effective regeneration system associated with an efficient selective agent [5, 6, 7, 8 and 9]. The most of popular forages in Indonesia is dwarf Napier grass, its grown widely environment and climate. Dwarf Napier grass contains several positive characters which are beneficial for animal feeding, i.e., they can be grown for years, having high yield, highly nutritious, selectively chosen as feed by animals, and can be grown in a wide variety of soil types. However, there are several negative properties which prevent this grass to be developed as planned [10].

Polyethylene glycol (PEG) is one of the most popular approaches is to use high molecular weight osmotic substances. PEG used for selection callus and plantlet were in vitro drought stress induction. PEG6000 is a non-penetrable and nontoxic osmotic substance which is used to lower the water potential of the culture medium and it has been used to simulate drought stress in cultured plant tissues [11]. Successful in vitro selection for drought tolerance using polyethylene glycol (PEG), sorbitol, mannitol and agar as selection agents has been also applied to several crops such as (rice, sugarcane, sun flower, tomato, and *Tagetes minuta* [12, 13, 14, 15, and 16].

Selection drought tolerance by using tissue culture technique has been widely used for the breeding purpose [17]. The lack research on drought tolerance of Napier grass by using PEG, in order, there are several genera of grass were reported such as sugarcane, sorghum, and wheat [18, 19, 20].

2. Method

Multiple shoot clump (MSC) from the clone of *Pennisetum purpureum* cv Mott were used to induce embryogenic callus. Explant was sterilized in 96% ethanol solution for 2-3 minutes and burn on Bunsen for 2-3 times [21]. To initiate embryogenic callus, MSC cultured on solid MS medium (Murashige and Skoog) supplemented with 1 mg/l NAA (α-naphthaleneacetic acid), 0.5 mg/l 2.4 D (2,4-dichlorophenoxyacetic acid), 1 mg/l BAP (6-benzilaminopurine), 5% v/v coconut milk, 30g sucrose and 0.8% w/v gelatin. The inhibitory effects of polyethylene glycol PEG 6000 at 0, 40 μM, 80 μM, 120 μM and 160 μM, concentrations in MS medium on initiation callus were evaluated in a completely randomized design experiment. Each experimental unit consisted of three explants cultured in one culture vial (250 ml) containing 25 ml of MS- medium. Five replicates were prepared for each of PEG concentration. This study used liquid PEG-supplemented MS-medium to initiate callus as agar would not solidify in the presence of a high concentration of PEG 6000. Sterilization of the media used standard heat sterilization methods at 120 °C, 1,5 kg/cm2 using an autoclave. The EC were transferred to fresh medium every 2 weeks.

3. Results and Discussion

Callus induction was observed following 6 weeks of culture. The percentage of callus compact and viable showed the calli was growth. Napier grass explants exhibited a significant interaction with PEG stress levels. In control (0 PEG), almost all explants initiated callus. With the increasing level of PEG, the mean percentage of callus induction reduced significantly. The increasing level of PEG from 0 to 40 μM, 80 μM, 120 μM and 160 have the lowest effect of differentiation of callus, rooting and shooting. These results indicating their relative tolerance for drought stress whereas the highest effect of PEG on medium selection. In vitro response of Napier grass callus to PEG stress was shown in Table 1. There are three responses of explant from MSC after transferred to treatments medium by using PEG.

Table 1. Percentage Organogenesis of Dwarf Napier grass on Medium Selective Polyethylene glycol (%)

Treatments (μM)	Induction callus	Shoots	Rooting
Kontrol	85.7	50	85,7
PEG K1	71.4	40	71,4
PEG K2	83.3	25	83,3
PEG K3	66.6	0	66.6
PEG K4	66.6	0	66.6

Description: K0 (Without PEG/Control); K1 (40 μM PEG); K2 (80 μM PEG); K3 (120 μM PEG); K4 (160 μM PEG). The observation was started since explant planted on callus initiation medium until organogenesis on regeneration medium age 45 days. Table 1 Showed induction percentage on callus induction medium containing 0 to 40 μM, 80 μM, 120 μM and 160 μM PEG was 85.7, 71.4, 83.3, 66.6 and 66.6% respectively. Generally, the average of callus induction decreased under higher PEG treatments, except in treatments 80 μM, the percentage callus induction higher than 40 μM PEG. The decrease in callus induction is a typical response of explants of crop genotypes when subjected to PEG stress (Matheka et al. 2008; Biswas et al. 2002; Abdel-Ghany et al. 2004). To maintain osmotic balance to assist initiation of callus cells under severe stress conditions or might be due to either water shortage which led to profuse mutation in cellular metabolism including protein functioning and alteration in the number of proteins

The results of induction callus showed that somaclonal variation from embryonic callus might tolerance or resistance to environmental stress were modified using PEG. Callus is the initial step in the cellular plant to initiate somatic embryogenic. If embryonic callus, not resistance to drought condition and water stress the callus will necrosis on PEG medium at least 1 to 3 day after callus induction. In order, the percentage of callus for shoot formation more decrease than the percentage of callus induction. Table 1 showed, on medium treatments for 120 μM and 160 μM PEG, not formation shoot, but on medium 0, 40, 80 μM PEG was 50, 40, 25% respectively. Generally, the average of shoot formation is under 50% on treatments medium.

Next step after callus induction for in vitro selection was callus rooting. Early and rapid elongation of roots is an important indication of drought tolerance. Continued elongation of root under the situation of drought conditions was the remarkable character of resistant cultivars. Until 160 μM of PEG, equivalent to 40 g/l concentration of PEG, embryogenic explants ability to rooting for stress conditions. This may be due to their initial ability to fight with water stress condition and stay green. Initial rooting appears in 2 weeks after callus induction, Table 1. Showed that all treatments of PEG concentrations rooting, despite highest concentration of PEG decreased the number of roots. Highest root percentage of treatments were recorded by drought-resistant whereas 85.7, 71.4, 83.3, 66.6 and 66.6, when tested at 0, 40, 80, 120 and 160 μM PEG solution, respectively. The lowest percentage of root was similarly on medium with concentrations 120 and 160 μM of PEG.

Studied showed that PEG solution, not the inhibitor for rooting formation of Napier grass callus, in order on shooting formation, increasing concentration of PEG more than 80 μM could not shoot formation. Studies indicated that root is more affected by drought condition than shoot. Balanced growth was observed in drought-resistant of callus Napier grass on same treatments of PEG solution. Drastic reduction in shoot growth was observed with increasing PEG concentration, which was considerably lower in the percentage of the shoot.

From these studied we resulted that without gamma irradiation, induction of callus on PEG solution for assessing drought tolerant successfully in Napier grass under in vitro selection. Tissue culture increased somaclonal variation, therefore, new clonal will appear under stress condition for genetic variation [22]. Selection on cell phase more effective for one's factors selection. So the targeting of agent selection on in vitro selection such as drought tolerance, salinity, and disease under a controlled environment. Therefore somaclonal variation generated by using in vitro selection allow mutation on cellular phase, thus new variants that are genetically tolerant to the abiotic stress like water stress are generated in the initial phase of plant or cell formation [23]

4. Conclusion

The average percentages of callus formation were 100% on the selective medium of PEG solution, in order concentration more than 80 μM PEG decreasing percentage of rooting and shooting of callus dwarf Napier grass (*Pennisetum purpureum cv.Mott*).

IOP Conf. Series: Earth and Environmental Science **156** (2018) 012024 doi:10.1088/1755-1315/156/1/012024

5. Acknowledgment

This research was supported by Ministry of Higher Education and Research Technology, as supporting grant of National Strategic Research (Hibah Penelitian STRATEGIS NASIONAL 2016-2017). We would also like to show our gratitude to the researcher for sharing their pearls of wisdom during the course of this research. We are also immensely grateful to Forum Dosen Indonesia (FDI) for their comments on an earlier version of the manuscript, although any errors are our own and should not tarnish the reputations of these

6. References

[1] E.A. Bray, Plant response to water deficit, Trends Plant Sci. 2 (1997) 48–54.

[2] M.W. Nabors, Environmental stress resistance procedure and applications, in: J.D. Philip (Ed.), Plant Cell Line Selection, VCH, Weinheim, 1990, pp. 167–185.

[3] Altmant, A. 2003. From plant tissue culture to biotechnology. Scientific revolutions, abiotic stress tolerance and forestry. In vitro cell dev. Biol-Plant, 39: 75-84 p.

[4] Ahmed, M., M. Asif, M. Sajad, J. Z. K. Khattak, W. Ijaz, W. A. Fayyaz-ul-Hassan and J. A. Chun. 2013. Could agricultural system be adapted to climate change? A review. Aust. J. Crop Sci. 7: 1642- 1653.

[5] Jain, M., 2001. Tissue culture-derived variation in crop improvement. Euphytica 118, 153–166.

[6] Predieri S (2001) Mutation induction and tissue culture in improving fruits. Plant Cell Tiss Org Cult 64: 185 – 210

[7] Dita, M.A., Rispai, N., Prats, E., Rubiales, D., Singh, K.B., 2006. Biotechnology approachesto overcome biotic and abiotic stress constraints in legumes. Euphytica 147, 1–24.

[8] Predieri S, Di virgilio N (2007) In vitro mutagenesis and mutant multiplication. In: Jain SM,Haggman H (eds) Protocols for micropropagation of woody trees and fruits. Springer, 323 – 333

[9] Rai MK, Kalia RK, Singh R, Gangola MP, Dhawan AK (2011) Developing stress tolerant plants through in vitro selection – An overview of the recent progress. Environ Exp Bot 71(1): 89 – 98

[10] Pongtongkam P, Peyachoknagul S, Arananant J, Thongpan A, Tudsri S.(2006). Production of salt tolerance dwarf napier grass (*Pennisetum purpureum* cv *Mott*) using tissue culture and gamma irradiation. Kasetsart J. Nat.Sci. 40: 625-633

[11] Muhammad, H., Khan, S.A., Shinwari, Z.K., Khan, A.L., Ahmad, N., and Lee, J. (2010). Effect of polyethylene glycol induced drought stress onphysio-hormonal attributes of soybean. Pakistan Journal Botany . 42 (2): 977-986.

[12] Biswas, J., Chowdhury, B., Bhattacharya, A., and Mandal, A.B. (2002). In vitro screening for increased drought tolerance in rice. In Vitro Cell Developmental Biolology . 38 :525–530.

[13] Patade, V.Y., P. Suprasanna and V.A. Bapat, 2005. Selection for abiotic (salinity and drought) stress tolerance and molecular characterization of tolerant lines in sugarcane. Proceeding on The National Conference on Biotechnological Aspects Towards Cultivation, Utilization and Disease Management of Plants, Dharmabad, Nanded during 24-25 December 2005

[14] Turhan, H and I, Baser, 2004. In vitro AND In vivo water stress in sunflower (*Helianthus annuus* L.). HELIA, 27, Nr. 40, p.p. 227-236,

[15] Kulkarni M and U.D. Deshpandei,Iin vitro screening of tomato genotypes for drought resistance using polyethylene glycol (PEG), Veg. Sci. 32(1) : 11-14 (2005)

[16] Mohamed M.A.-,. P.J.C. Harris a, J. Henderson, 2000. In vitro selection and characterisation of a drought tolerant clone of Tagetes minuta. Plant Science 159 :213-222

[17] Encheva J., Tsvetkova F. and Ivanov P. (2004) Heritable tissue culture induced genetic variation in sunflower (Helianthus annuus L.) as a tool for crop improvement. Helia 27: 163-172

[18] Yadav PV, Suprasanna P, Gopalrao KU, Anant BV (2006) Molecular profiling using RAPD technique of salt and drought tolerant regenerants of sugarcane. Sugar Tech 8: 63– 68

[19] Tsago, Y, M. Andargie2 and A. Takele. (2013) In Vitro Screening for Drought Tolerance in Different Sorghum (Sorghum bicolor (L.) Moench) Varieties. Journal of Stress Physiology & Biochemistry Vol. 9 No. 3

[20] K. Dorffling, H. Dorffling, G. Lesselich, In 6itro selectionand regeneration of hydroxyproline-resistant lines of winter wheat with increased proline content and increased frost tolerance, Plant Physiol. 142 (1993) 222–225.

[21] Nadir, M., Sjahrir, R., Budiman. (2017). Perakitan kultivar rumput gajah unggul (*Pennisetum purpureum cv. Mott*) tahan kekeringan dalam rangka perbaikan genetik dengan teknik kultur jaringan. Laporan Hasil Penelitian Strategis Nasional Ristek Dikti.

[22] Shah, M.M., Q. Khalid, U.W. Khan, S.A.H. Shah, S.H. Shah, A. Hassan and A. Perves, 2009. Variation in genotypic responses and biochemical analysis of callus induction in cultivated wheat. Genet. Mol. Res: 8(3): 783-793

[23] Wenzel, G,B. Foroughi-Wehr. 1993. In Vitro selection (pp. 45-59) in Hayward, MD, NO. Bosemark and Romagosa, editor. Plant Breeding. Principle and Prospects. Chapman & Hall, London

IOP Conf. Series: Earth and Environmental Science **156** (2018) 012025 doi:10.1088/1755-1315/156/1/012025

Analysis of Culture Shock Experienced Towards Indonesian Sailors in Overseas

Ana Rosida

FEIIS, English Literature Study Program,Fajar University.
Correspondence Address: Jl. Azalea B/C3 Maros, South Sulawesi.
Telephone: 081342915122
E-mail: ann.rosyed@gmail.com

Abstract: This research aimed to find out the culture shock experienced towards Indonesian sailors who sailed to overseas. This research also focused on Indonesian sailors who experienced the cultural shock. There were 60 students as population and 10 students as responded to the questioners who studied Seafarers Training Grade IV Nautical and Machine Majors in BP2IP (Balai Pendidikan dan Pelatihan Ilmu Pelayaran) shipping education and training center Barombong Makassar Indonesia. The method in this research was descriptive qualitative. According to the culture research, it found four kinds of cultural shock (U-curve). They are Honeymoon, Culture Shock (Crisis/Frustration), Adjustment, Mastery. The implication of this research is to give understanding about the phase of culture shock who would like to sail to overseas.

Keywords: culture shock, u curve, indonesian sailors.

1. Introduction

A sailor is someone who works on passenger ships, freighters, and tanker ships, navigating sea-going vessels and assisting with maintenance, operation, and service of these vessels. Being a seaman is not the easy of career choices. Sailors are required to be at sea for extended periods of time. Most sailors are away from their families for several months before they come back to shore. While this is not a career for everyone, many people continue to find that the positives far outweigh the negatives. It is important to understand that being a sailor is sometimes dangerous work.

The workplace environment is not the most comfortable, as the living quarters on most vessels are often small and cramped with little to no privacy. Working with different people in countries also absolutely creates different perspectives and feeling rather than work in their own country. Therefore, culture shock is becoming the most issue among sailors who sailed in overseas.

Culture shock becomes an interesting topic to be discussed. Every country has a different culture. Culture cannot be separated from humans' life. Culture affects people in their society because of their ideas, values, attitudes, and normative or expected patterns of behaviour.

According to Taylor, culture or civilization is whole complex which includes knowledge, belief, art, morals, law, custom, and any other capabilities and habits acquired by man a member of society [1]. According to Margaret , it means the whole complex of traditional behaviour which has been developed by the human race and is successively learned by each generation [2]. Therefore, it conclude that culture is attitudes, human behaviour, a way of life of people and everything that people have, think, and do as members of a society.

Culture shock is the anxiety and emotional disturbance experienced by people when two sets of realities and conceptualizations meet [3]. Poejiastuti stated that culture shock is relatively short-term feeling of disorientation of comfort due to unfamiliar of surroundings.The differences between cultures can make it very difficult to adjust to the new surroundings. People may encounter unfamiliar clothes, weather, and food as well as different people, schools, values, struggling to do things in new surroundings that were easy back home. Dealing with the differences can be very unsettling; those feelings are part adjusting to a new culture [4].

IOP Conf. Series: Earth and Environmental Science **156** (2018) 012025 doi:10.1088/1755-1315/156/1/012025

Culture shock is not a clinical term or medical condition. It is simply a common way to describe the confusing and nervous a person who may have after leaving a familiar culture to live in a new and different culture. Adjusting to a new environment takes a long time. People who cross cultures usually are uncomfortable at first and talk of feeling confused, discouraged, lonely and anxious. As they become more comfortable, however, many people find that learning in a new culture is stimulating and broadening. Based on the background above it is formulated research question "how are the phases of culture shock toward Indonesian sailor in overseas?". From that question, it gives two objectives in this research, namely: to describe the culture shock and to reveal the phase of culture shock towards Indonesian sailors in overseas. Meanwhile the significance of this research is to give contributions to the readers and scientist to analyze culture shock in other perspectives. To avoid broaden analysis, this research limits the topic of the culture shock and the theory that used to analyze is culture theory by Lysgaard psychology of literature approach by Sigmun freud [5].

2. Literature Review

Brisbin saw cultural sensitivity as the ability to perform accordance with both collectivism and individualism. In his study, cultural sensitivity is defined as the ability to identify whether individual are behaving with a collective or individualistic mindset, to be open-minded to other cultures, and to be flexible enough to behave in an appropriate manner in other culture [6].

Stefanie Baier in *International Students: Culture Shock and Adaption to the U.S Culture,* Teacher Education Department Eastern Michigan University. This study analyzes to what degree cultural background, gender differences, language proficiency, self-confidence/self-efficacy, and social support networks impact the adjustment process of international students to the US culture [7].

The two researchers were analyzing the culture shock in different context. However, this research was focus on the phase of culture shock that experienced by Indonesian sailors who sailed to overseas.

2.1 Culture Theory

The theory of culture refers to the process and time it takes a person to assimilate to a new culture. It is not always an easy transition. Living in a new culture can be jarring and very difficult to process, especially if it is drastically different from your own. A person that is being introduced into a new culture can feel a variety of different sensations. For example, one may feel frustrated because she or he cannot communicate or become angry because certain cultural norms do not make sense. In some cultures, spitting in public is considered bad manners. It may make you feel uncomfortable when the people of that culture think nothing of it. Cultural adaptation theory offers an explanation for these emotions. As a process of cultural adjustment, hypothesis of *U-Curve* by Lysgaard (1955 : 51), 1996) following figure of *U-curve* shows the four main stages of adjusting to a foreign culture. This is used as an original model of cultural adjustment.

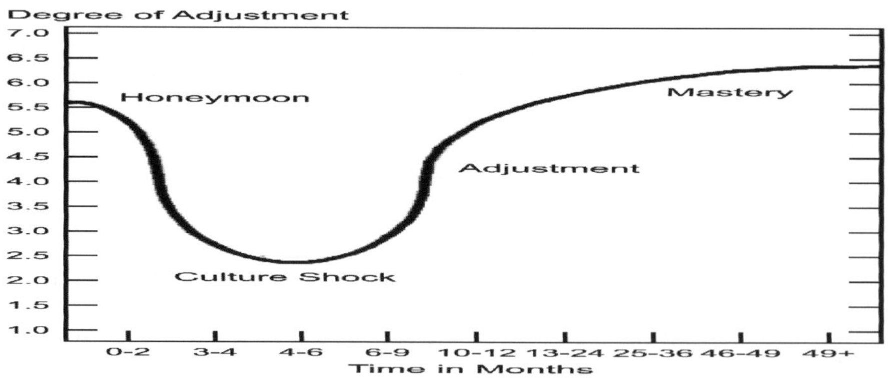

Figure 1. figure of *U-curve* shows the four main stages of adjusting to a foreign culture

IOP Conf. Series: Earth and Environmental Science **156** (2018) 012025 doi:10.1088/1755-1315/156/1/012025

2.2 Honeymoon

In this phase, people usually see minor problems and look forward to getting new experiences. During the time, cultural differences are seen as attractive and charming and they tend to see only the positive aspects of the difference of the new culture. This phase can last from a few days to about six weeks. People could be interested, open-minded and ready to accept a new culture in this first stage.

2.3 Culture shock (Crisis / Frustration)

This stage starts when people begin to see the situation they are in less idealistic and more realistic terms, so they see the more negative aspects of the host culture. The feelings people may experience are frustration, irritability and confusion. Especially the difference in language is a big issue as well as values and beliefs of the country. People find themselves in a situation of disappointment and refection of the host culture due to the inconsistency between expectations and reality. From this time, people need to start to adjust themselves to a new culture.

2.4 Adjustment (Recovery)

This stage is a turning point. During this stage, people return home or they get the understanding and adjust to the host culture. The adjustment involves acquiring greater knowledge of language or culture of host country, and understanding the local situation. Besides, people have the great sense to expect things and to well control themselves.

2.5 Mastery

In the final stage, people have a sense of dual cultural identity and can get on with both host and home cultures, which is called biculturalism. They accept the new culture, and even more, start to think differences or similarities are valuable things that can be found only among the people who have experienced culture shock.

3. Method

An open questionnaire was developed for the purpose of the study and was administered as students Seafarers Training Grade IV Nautical Majors in BP2IP (Balai Pendidikan dan Pelatihan Ilmu Pelayaran) shipping education and training center Barombong Makassar Indonesia. In the year 2017 for 27 grades, it has two classes with 30 students each. One class is for nautical majority and machine the other one.

The sample was only for sailor students who experienced sailing to overseas. However, it found 5 students in nautical majority and 5 students either in machine majority. All ten students had different places to sail, namely: Singapore(1 student), India (2 students), Netherland (1 student), Qatar (2 students), United Arab Emirates (2 student), Spain (1 student), Malaysia (1 student).

Some points were revealed and asked in the questionnaire:

- How long did you enjoy new experience in new place of your work?
- When did you feel homesick in your work place?
- When did you adjust to the host culture in your work place?
- When did you get on with both host and home culture in your work place?

Each subject completed questionnaire individually at his own convenience. The questionnaires used Bahasa Indonesia to ease the respondents to answer freely and sincere. However, the writer transferred into English to bridge source language into the target language.

The obtained data was analyzed descriptive qualitative as well in the form of narrative presentations. Everything was presented as it was. The data and responses were neither subtracted nor added to maintain the originality of the story. The next step was to find out the respondents' common answers then to create classification. If the respondents' answer had a great variety and proved inconsistent one another then the data was presented as it was.

4. Result

4.1 Honeymoon

The question was how long they enjoyed new experience in new work place. Many respondents answered between 2 weeks until 4 weeks. These are 3 examples respondents' answers;

When I joined the company and they sent me to Qatar, it was such as a joy experienced. I found new co-worker, new habits and new environment which it was so different when I worked in Indonesia. It just happened for 3 weeks.

I sailed to India and my position was an abbey for one year. Firstly, it felt really nice and joyful because I found new place and new friends but it just happened two weeks. I got difficulty to adjust with Indians' co-worker. They seemed un-tidy and dirty. That eased me to lost nice experience.

That was my first time to sail to Singapore. My contract was one year and Alhamdulillah I could finish and still joined that company even though I continued and upgrade my study here. It was nice and excellent experiences. It felt 4 weeks or one month.

As above comments, the honeymoon phase as the first step in culture shock was joyful and attractive. However, it found different period to enjoy the new experience in new work place of work. Other respondents were mostly answer between second weeks into the fourth weeks.

4.2 Culture shock (Crisis / Frustration)

The next question was when they felt homesick in a new work place. This question had short and long answers, it was short because the respondents just answered by writing the numbers, meanwhile it was long description because they shared their feeling. However, the writer just wrote the numbers in the tabulated data below;

Table 1. Respondents responded toward culture shock

Initial Name	Country	Feeling homesick
MHR	India	It was in the fourth weeks on board
MA	India	It was in the third weeks on board
IR	Qatar	It was fifth weeks on board
MB	Singapore	It was in Second month
AB	Netherland	It was in Second month
AK	Spain	It was fifth weeks on board
MMW	Malaysia	It was in Second month
MIM	United Arab Emirates	It was in second month on board
AMR	Qatar	It was in sixth weeks
AR	United Arab Emirates	It was in Second month

From the data above, it concluded that it was between third weeks until eighth weeks the respondents felt homesick or culture shock in a new work place. Some of them mentioned feeling homesick to families and food as the basic reason.

4.3 Adjustment (recovery)

The third question was asking about the length of the respondents to adjust their new environment towards their life during their running contract. We could see the data below;

Table 2. Adjustment Phase

Initial Name	Country	Adjustment
MHR	India	The third month
MA	India	The third month
IR	Qatar	It was in the third month
MB	Singapore	It was in the third month
AB	Netherland	It was in the third month
AK	Spain	It was eighth weeks on board
MMW	Malaysia	The third month
MIM	United Arab Emirates	The third month
AMR	Qatar	It was in tenth weeks
AR	United Arab Emirates	The third month

From the respondents' answer, most of them were in the third of months to adjust in their new work place. However, there were also in eighth and tenth weeks on board.

4.4 Mastery

In the final stage, people had already accepted the new culture, and even more, start to think differences or similarities are valuable things that can be found only among the people who have experienced culture shock. The question was when they got on with both host and home culture in a new work place. The data was tabulated below:

Table 3. Mastery Phase

Initial Name	Country	Adjustment
MHR	India	In the fourth month
MA	India	In the fifth month
IR	Qatar	In the fifth month
MB	Singapore	In the fourth month
AB	Netherland	In the fifth month
AK	Spain	In the fifth month
MMW	Malaysia	In the fourth month
MIM	United Arab Emirates	In the fifth month
AMR	Qatar	In the fifth month
AR	United Arab Emirates	In the fourth and half month

The table above shows the respondents' answers to the fourth question. Based on the table above, in mastery phase of culture shock was between in fourth and fifth month.

5. Discussion

As it has been stated before that living in the different places in the long period away from family and natural habits from the mother land cannot be avoided for the Indonesian sailors to work in overseas to earn money as their profession and responsibility to their family. People from one context moved to the new context because of several reasons. And, of course mobility might change the nature of our society, and it also affects the individuals involved. This present findings had revealed the phase of respondents' culture shock experienced. The phase used U-curve used to analyze culture theory by Lysgaard and psychology of literature approach by Sigmund freud.

The first phase is honeymoon. It is the phase where people usually see minor problems and look forward to getting new experiences. During the time, cultural differences are seen as attractive and charming and they tend to see only the positive aspects of the difference of the new culture. From respondents' answer, it found between two until four weeks to gain attractive and joyful to work in overseas. In addition, they also shared several reasons, namely: getting new environment, new co-workers from different countries, food, culture, and also the condition on board was different from Indonesia.

The second phase is culture shock (crisis/ frustration).The data found that they were in the critical situation where the most respondents began to see the situation more negative aspects of the host culture. Actually, in the beginning of their arrival, they did not pay any attention to the language, where English language became their communication on board. Meanwhile, in this second phase they did in frustration and getting disappointed towards language, missing their families, home food, and the culture due to the inconsistency between expectations and reality. The data shows, it was between third weeks until eighth weeks the respondents felt stress, homesick and culture shock in a new work place.

The third phase is adjustment (recovery). Recovery means people have great sense to control themselves to see and learn about the language and culture of host culture. Their adjustment involves greater knowledge to understand their surroundings. The questionnaires described the respondents' answers that most of them were in the third of months to adjust in their new work place. However, there were also in eighth and tenth weeks on board. Some of respondent just wrote the numbers, meanwhile there were also mentioned their reason about getting adjustment in work place, namely; professionalism, career, experience and money. They stated that overseas has good money rather than working in Indonesia.

The last phase is mastery. In this stage, respondents had a sense of dual cultural identity and language that could get on with both host and home cultures and language. They had already accepted the new culture, and even more, start to think differences or similarities are valuable things that can be found only among the people who have experienced culture shock. The respondents' answer was getting in mastery phase is between in fourth and fifth month which they had enjoyed their work and life on board. In addition, several respondents noted also on their questionnaires that they are even still in contract towards that company and now they are in off position, after finishing their grade, they would back to sail to overseas.

6. Conclusion

The result of this study indicates that Indonesian sailors sailed to overseas also experienced culture shock. The phase of having culture shock was described in the U-curve, namely; honeymoon, culture shock (crisis/ frustration), adjustment (recovery), and mastery.

It had been revealed that sailors in honeymoon phase were between two until four weeks to gain attractive and joyful to work in overseas. In cultures shock phase, the data shows, it was between the third week until eighth week the respondents felt stress, homesick and culture shock in a new work place. However, the third phase is adjustment (recovery) where the respondents began to adjust and learn about the language and culture of the host culture. It was in the third of month to adjust in their new work place. However, there were also in eighth and tenth week on board. The last phase is mastery, the respondents answered that it is between fourth and fifth month to enjoy their work and master the condition of their surroundings.

It is strongly recommended for the sailors who want to sail to overseas for the first time to browse and learn about the host culture. And also, it is recommended to practice and learns about English language as daily conversation on board.

IOP Conf. Series: Earth and Environmental Science **156** (2018) 012025 doi:10.1088/1755-1315/156/1/012025

7. References

[1] Tylor. (1870). *Primitive Culture: Researches Into the Development of Mythology, Philosophy, Religion, Art, and Custom*. Newyork: Universitas Oxford.

[2] Mead, Margaret. (1937). *The Changing Culture of an Indian Tribe*. New York: Columbia University.

[3] Oberg. (1960). *Cultural Shock: Adjustment to New Cultural Environments*. Pratical Anthropology 7.

[4] Poedjiiastuttie. (2009). *Culture shock experienced by foreign students studying at Indonesian University*. Jurnal Humanity Volume IV nomor 2.

[5] Lysgaard. (1955). *Adjustment in a foreign society:* Norwegian Fulbright grantees visiting the United States. International Social Science Bulletin.

[6] Brisbin, R. (1993). *Understanding Culture's Influence on Behaviour.* Fort Worth IX. Harcourt Brace Jovanovich.

[7] Baier, Stefanie. (2005). *International Students: Culture Shock and Adaption to the U.S Culture,* Teacher Education Department Eastern Michigan University.

Need Analysis of Coastal Fisherman empowerment Based on Economics education and Potential Coastal in Minahasa Regency of North Sulawesi

Edwin Wantah[1], Ery Tri Djatmika[2], Mit Witjaksono[2] and Hari Wahyono[2]
[1]Post-Graduate Program, State University of Malang, East Java Indonesia & Faculty of Economics State University of Manado, Indonesia
Email: edwinwantah@yahoo.co.id
[2]Post-Graduate Program of State University of Malang, East Java, Indonesia

Abstract: This research article aims to describe the need analysis process of coastal fisherman empowerment in North Minahasa Regency of North Sulawesi Province through internalization of economics education and technical training based on potential coastal.This research used descriptive qualitative research design by using survey, field observational interview and Focus Group Discussion (FGD) method. The research subjects were 40 coastal fishermen in North Minahasa Regency spread in two sub-districts, namely Wori Sub-district and Kema Sub-district in North Minahasa Regency, who have capture capacity of 10 gross ton sand below. The results of needs analysis were identified in observation, in-depth interview, and Focus Group Discussion (FGD), which was confirmed by the survey results indicating that 87.5% of the coastal fishermen need knowledge and understanding of characteristics, attitudes and principles to become successful entrepreneurs, which can be implemented in productive business activities on coastal area, while 92.5% of the coastal fishermen require an understanding of creativity and innovation and its implementation, 90% of the coastal fishermen require knowledge of business diversification based on coastal area and marine potentials, 90% of the fishermen stated that they need knowledge and understanding of the process of processed fish products because of the abundant raw materials, 80% of the coastal fishermen said that they need an understanding of the way to establish business partnerships and business networks with other business groups, 90% of the coastal fishermen stated that they need an understanding of the way to form fishermen joint business groups and the way to develop the business group. 92.5% of the coastal fishermen need an understanding of business capital and the way to access business capital, and 85% of the coastal fishermen said that they need to understand money management, the way to plan proper financial allocations, and saving procedures

Keywords: Need Analysis, Fisherman Empowerment, Economics Education, Potential Coastal

1. Introduction

So far, coastal community is considered to be a part of the poorest and marginalized group [1]. Most of them work in fishery sector or work as fishermen. Fisheries development program that focuses on economic growth through capture fishery industry does not always have a positive impact on the economic income of poor fishermen [2]. Coastal community empowerment program is a

IOP Conf. Series: Earth and Environmental Science **156** (2018) 012026 doi:10.1088/1755-1315/156/1/012026

necessity for the development of coastal resources comprehensively [1]. One of the ideas of the empowerment process is through the development of human resources, which in turn are able to manage the coastal resources they have been involved in so far. Basically, coastal community has diverse characteristics, but generally they work as fishermen with different levels of fishery technology used. The dominant profession of coastal community is in the sector of marine-based resource utilization, namely fishermen and fish farmers, both with aquaculture and mariculture method [3]. Another characteristic is that most of the coastal fishermen are traditional fishermen who generally have the same characteristics, namely low level of education because they think that there is no need to get high education to look for fish at sea and they are more concerned or rely on their energy and experience [1]. With low level of education, fishermen find difficulties to switch into other jobs besides fishermen.

Poverty and economic limitations that become the main problem of coastal fishermen, especially traditional fishermen, are caused by many factors. In addition to low education level, other natural factors such as weather and fishing seasons, very simple fishing technology, competition with modern fishermen or fishery corporations, the great cost of fishing due to expensive production facilities and infrastructure, expensive fuel for fishing, dependence of traditional fishermen on moneylenders or middlemen, the absence of alternative livelihoods other than traditional fishing, the economic income of traditional fishermen that is uncertain and far from decent living [4].

The poverty of fishermen is also caused by external and internal factors. Kusnadi states that the poverty of fishermen is due to internal factors, namely limited qualified human resources among fishermen, limited business capital, limited fishing technology, unharmonious relationship between fishing boat owners and fishermen as well as fishing boat workers, dependence on fishing season, and lifestyle or consumptive behavior of fishermen [5]. For external factors, Kusnadi stated that the poverty of many fishermen is caused by the policies of fisheries development that are still not in favor of fishermen, the marketing system of fishery that is only beneficial to middlemen, damage of marine ecosystem through sea water pollution, coral reef destruction, the use of fishing gear which is not environmentally friendly, limited technology for post-harvest processing, and limited employment opportunity in non-fisheries sector (off fishing) in the fishing villages [6]. The livelihood of fishermen still depends on sea, which is increasingly difficult as a means of the fishermen to improve the quality of life. On the other side, the catch which is the main source is not sold to consumers directly but to middlemen or to other fishermen who have better economy (fish traders), who has two functions: as fish traders and moneylenders [4]. The absence of alternative livelihoods other than fishing leads fishermen to remain trapped in poverty.

Government programs related to fisherman's life such as coastal community empowerment program (PEMP) and small-scale fishing business development program (PUPTSK) have not been maximally answered and not equally felt by all traditional coastal fishermen in North Minahasa Regency. Assistance programs of production means cannot address the needs of fishermen due to the lack of identification and need analysis of fisherman in advance so, that empowerment programs experience many failures. The need analysis of coastal fishermen in Minahasa Regency is very required in identifying the problems and the main needs of fishermen, so that the approach of coastal fishermen empowerment program in North Minahasa Regency can be done maximally.

Socio-economically, coastal and marine areas have important significance in national development since most of the territory of the Unitary State of the Republic of Indonesia is maritime territory. About 75 percent of Indonesia is waters, with a total of 17.508 islands. Indonesia is the largest archipelagic country in the world with the second longest coastline after Canada, namely 81.000 kilometers. Being flanked by two continents, Asia and Australia, and being flanked by two oceans, pacific and hindia, make Indonesia a very strategic country and have abundant marine natural resources, ranging from various types of fish, minerals, diverse coral reefs, petroleum, and many other marine [7]. Abundant potential of maritime resource is not accompanied by the increase of coastal and marine community welfare due to the management of marine resources that have not been maximized and not in favor of coastal communities. Abundant potential of coastal resources is also owned by

North Minahasa Regency of North Sulawesi Province, but it has not been managed maximally so that it does not give much impact to the income of coastal fisherman.

2. Objectives of Problem Identification and Need Analysis of Coastal Fisherman Empowerment Program Based on Economics Education and *Off Fishing* Concept

Problem identification and need analysis of coastal fishermen empowerment program in North Minahasa Regency of North Sulawesi Province aims to identify problems and needs of fishermen in order to answer every problem identified in coastal fishermen in North Minahasa Regency. Need analysis is done as the basis for the material preparation of empowerment program of coastal fishermen in North Minahasa Regency.

3. Literature Review

3.1 Concept of Out-of-School Education with Society Empowerment.

Several actualization programs of Out-of-School Education (PLS) are literacy, equity, homeschooling, training, course, life skills, Early Childhood Education (PAUD), street children and society empowerment. One of the actual programs of out-of-school education is society empowerment program. Society empowerment program is a program that will always be faced with various social problems because this program aims to make a change in the targeted community of learners. Society empowerment program is really in need of transformative process [8]. The need for out off-school education through community empowerment is very essential in order to improve the standard of living and literacy of a community. Out-of-School Education (PLS) is needed by coastal fishing communities including fisherman families in order to increase the literacy of coastal and marine natural resources utilization, so that coastal fishermen and their families live in prosperous life.

3.2 Concept of Non-formal Education in Coastal Fisherman Empowerment Program

Non-formal education is any type of educational activity which is organized outside formal education system, either implemented separately or as a part of larger activity, which serves particular learners with specific learning objectives. According to Evans, non-formal education is an organized educational activity outside formal education system. Evans places non-formal education as a part of overall integrated concept of education system. In that concept, Evans emphasizes the following features: very broad distribution, participatory, involving the work of community organizations, private associations, more concerned about action at local level [9]. For those who do not have the opportunity to study at school with various reasons, non-formal education is required as a substitute for equal education in earning a living [10], while for those who have an opportunity tostudy at school, non-formal education serves as complement and gains additional knowledge and particular skill because at schoolthey only obtain a little knowledge and skill or even not at all. All learning needs required by these groups are meant to improve their life skills, so that they can overcome difficulties they find or in other words, they can overcome the environmental challenges, both physical and social environment.

4. Method

This research is a series of beginning or early stage of 6 stages of research development of coastal fishermen empowerment model based on economics education in North Minahasa Regency of North Sulawesi Province. The initial stage of this research is a part of design based research (DBR) stage of DBR model adopted from Reeves model [11]. Design based research contains 6 stages of developmental research. This article will illustrate the results of problem identification and need analysis of coastal fishermen empowerment model based on economics education and off fishing concept.

4.1 Type of Research

This research used descriptive qualitative research design by using survey, observation, and focused discussion method to identify and analyze the problems and needs of coastal fisherman on empowerment model based on economics education and utilization of coastal resources with off fishing concept or based on livelihoods other that fishing.

4.2 Research Subjects

In this research, the research subjects were coastal fishermen in two sub-districts, namely Kema Sub-district and Wori Sub-district in North Minahasa Utara with 40 research samples of traditional fishermen with capture capacity of 10 gross tons and below.

4.3 Data Collection Technique

In collecting data and research information, the techniques used were direct observation, survey, in-depth interviews and Focus Group Discussion (FGD). Researchers involve themselves directly in the process of collecting data and research information. Direct observation was done by researchers and experts of community empowerment to see the activities of coastal fisherman directly. In the survey, the questionnaires were filled in by coastal fishermen. As for Focus Group Discussion method, researchers involved fishermen, empowerment facilitators, community leaders, and representatives from Department of Marine and Fisheries. In Focus Group Discussion, the information about the problems and needs of coastal fishermen for the model or program of fishermen empowerment based on economics education and utilization of coastal resources based on off fishing concept is explored.

4.4 Data Analysis Technique

The data was analyzed by using qualitative descriptive data analysis technique in the form of simple percentage analysis. The data which was analyzed through percentage analysis through survey on the research subjects was supported by the analysis results of in-depth interview and the responses and inputs on Focus Group Discussion (FGD).

5. Results and Discussions

In the early stage of the study, the researchers conducted problem identification and need analysis. Problems can be identified based on inputs from coastal fishermen through observation and in-depth interview. From the identification of coastal fishermen's problems, the needs of coastal fishermen can be analyzed and identified. The results of observation and in-depth interview were then discussed again in Focus Group Discussion to confirm the prioritized basic needs. Focus Group Discussion (FGD) was conducted with groups of fishermen, government, fishermen empowerment facilitators, related department, and academics monitoring fisherman empowerment program. After conducting Focus Group Discussion, the questionnaires were arranged and distributed to the research samples to find out the priority of coastal fishermen's needs for the empowerment model, which would be identified in-depth interview and Focus Group Discussion.

5.1 Education Level of Coastal Fishermen in North Minahasa Regency Who BecameResearch Samples

In this study, the researchers explored information about the education level of 40 coastal fishermen in two sub-districts, namely Kema Sub-district and Wori Sub-district, who became research samples. Data of formal education level and non-formal education of the coastal fisherman was based on survey through questionnaires distributed to 40 fishermen who were members of fishermen groups in those two sub-districts. The education level of 40 fishermen who became the subjects of this research is presented in the following table.

Table 1. Education Background of Coastal Fishermen Being Research Subjects

Formal Education (%)						Non-Formal Education (%)		
Never study at school	Did not finish primary school	Graduated from elementary school	Graduated from junior high school	Graduated from senior high school	Bachelor	Have followed training	Never followed training	Follow othernon-formal education
0	15%	22.5%	45%	17.5 %	0	30 %	65 %	5 %

Source: Data processed byresearchers, 2017.

Based on data in the table above, it can be concluded that all the fishermen who became the subjects of this study have formal education background, although there are 15% who did not complete primary school. Mostly, the coastal fishermen (45%) are graduates of junior high school, 22.5% of them are graduates of elementary school, and 17.5% of them are graduates of senior high school. In general, all fishermen who became the subject of research have been able to read and write. The most of the fishermen (65%) have never followed non-formal education in the form of education and training outside formal education activities, 30% of them have followed technical training, and 5 percent of them have followed other non-formal education. One of the obstacles in the coastal fishermen empowerment program in North Minahasa Regency so far is that the technical training and non-formal education conducted have not concerned on the priority of fishermen's needs because there was no need analysis first, so it does not answer their problems and needs.

5.2 Knowledge, Literacy, and Entrepreneurship Skill

In exploring the need for knowledge, literacy, and entrepreneurship skill for fishermen, data was collected through questionnaires, FGD, and in-depth interview with existing fishermen community. From survey data, it can be seen that 65% of fishermen have never got education and training at all, 30% of them have got education and training, but the education and training process is not concerned on building knowledge, literacy and entrepreneurship skill, and 5% of the respondents have got other non-formal education and training which is not related to entrepreneurship skill. Therefore, it can be concluded that some coastal fishermen in North Minahasa Regency, particularly who became the objects of this study, do not have the knowledge, literacy and skill about entrepreneurship. From the information obtained in Focus Group Discussion, most of the fishermen want to have alternative livelihoods other than fishing. They expect technical training of entrepreneurship and the assistance in entrepreneurship.

5.3 Knowledge and Skill in Processing Fish

From the results of observations and in-depth interview, researchers found that most fishermen do not have skill in processing fish into high-value products. From the results of Focus Group Discussion which was conducted with groups of fishermen from Kema Sub-district and Wori Sub-district, it was obtained information that the catch of fish is often abundant but the price of fish becomes cheap and fishermen are having trouble marketing fish because fishermen and their families do not have skill and expertise to process fish into processed food products which are durable and have high economic value. From the results of focus group discussion and in-depth interview, it can be concluded that there is very high demand for processed fish products and abundant catch of fish, but fishermen and their families do not have expertise and skill in processing fish into products which have high economic value.

5.4 Knowledge and Understanding of Fisherman Family on Financial Management

From the observation conducted by the researchers about economic behavior of fisherman family, most of fisherman families still behave consumptively when they get income or fishing results. The results of survey to the research samples show that 35% of fishermen who had participated in technical training did not get training on financial management of fishermen family. Only 5 percent of fishermen who had got other non-formal education and training had attended a financial management course, but they did not really understand the financial management of family, while 65 percent of the samples never attended any education and training including financial management education and training for fisherman families.

The income of coastal fishermen in North Minahasa Regency is allocated mostly for household consumption.Most of them even to get capital for fishing by borrowing money to other people or usurers. From Focus Group Discussion which involved representatives of fishermen groups in Kema Sub-district, Department of Marine and Fisheries of North Minahasa Regency, and fisherman empowerment facilitators, there was information that some coastal fishermen in Kema Sub-district did not know how to manage their family finance, had never planned to expenditure allocate well. Most of them have no savings and other investments. The conclusion of Focus Group Discussion with need analysis session of coastal fishermen in Kema Sub-district of North Minahasa Regency indicates that fishermen need family financial management training in order to increase their knowledge about how to manage money, including how to save and invest money well. Focus Group Discussion obtained data about the desire of fishermen not to get stuck on consumptive behavior and, usurers, and *pengijon* who provide capital loan for fishing with high rates of interest, which must be paid by selling their catch of fish with very low price.

5.5 Knowledge on Business Capital and How to Get It

Knowledge of fishermen on access to capital in financial institutions or cooperatives that offer low interest is minimal. The result of in-depth interview with a fisherman of Lilang Village in North Minahasa Regency, stated that so far the capital for fishing is only obtained from loan to usurers, to the owner of pajeko boat (big fishing boat), or to fish trader. Fishermen also find difficulties in returning it due to high interest rates of loan; therefore, they really need capital loans for fishing and other coastal businesses from banks with easy schemes of returning money, more flexible period of time, and small interest rate. Fishermen need an understanding of business capital and how to obtain business capital credit assistance from banks.

5.6 Knowledge on Business Diversification Based on Coastal Resources

The results of observation and in-depth interview show that most coastal fishermen in North Minahasa Regency have no alternative livelihood other than fishing, so they are highly dependent on fish and other marine resources. This becomes a problem if they do not go to sea or face fish famine. Information about this problem was found by the researchers from the results of observation and in-depth interview with fishermen. This becomes more difficult because most of them do not have ability in processing coastal resources, which are raw materials available on the coast and sea, into something with economic value. Some businesses that can be developed by fishermen and their families because the raw material is available is processing fish *surimi* (fish gel), fish balls, fish nuggets, fish crackers, fish chili sauce and shredded fish. Business diversification of processed products made from raw fish can be used as a reference of business or other livelihood for fishermen and their families. Results of FGD and in-depth interview conducted by researchers show that most of fishermen and their families do not have skill to process fish.

5.7 Knowledge on Establishment and Legitimization of Joint Business Group of Fishermen.

One of findings from observation and in-depth interview of researchers in the study site shows that most coastal fishermen in North Minahasa Regency are not actively involved in the organization or group of fishermen. Most of them work individually in fishing activities. Although they are not involved in group, they often help each other and have strong social interaction in their social life, both among fishermen and with other groups such as farmers, traders and other communities. The results of survey in research samples show that 80% of them do not understand the procedure of establishing a group or economic institution such as joint business group of fishermen or fishermen cooperatives. Only 20% of them have the ability to organize group formation and group legitimization.

The results of Focus group discussion show that fishermen actually have a desire to interact within groups or institutions of fishermen in order to improve their skill and ability together and to help each other. Findings in observation show that indirectly, coastal fishermen implemented *mapalus* cultural value or mutual assistance, which is the social capital of society in North Minahasa Regency. Mutual assistance and trusting each other are the main cultural values of *mapalus*, which are naturally implemented in their life. Results of FGD also show that they want to establish fishermen's business institutions in the form of joint business group of fishermen because by grouping or forming formal group of fishermen, they are entitled to the assistance of production facilities from the government. The obstacles they face are the lack of knowledge to form a group, group legitimization, group administration, group programming, including knowledge in adopting cultural values of *mapalus* in developing groups or fisheries business institutions.

5.8 Knowledge on Establishing Business Partnership

One of the findings obtained by the researchers from observations and interview with fishermen is that some of them do not understand how to establish partnerships with fishermen business groups or larger business partners. The results of Focus Group Discussion show that fishermen have no knowledge on identifying business partners as well as the forms of cooperation that they can do with business partners which are beneficial t them. From the results of observation, interview and FGD which have been conducted, it can be concluded that fishermen need knowledge about building business partnerships and business networks, maintaining business relationship with business partners, and knowledge of all forms of business cooperation that can be done by coastal and marine business planned by fisherman group.

Table 2. Summary of the Results of Problem Identification and Need Analysis of Coastal Fishermen

Problem Identification and Need Analysis	Results of Problem Identification
Observation, interview through discussion with fisherman group, Department of Marine and Fisheries in North Minahasa Regency, and survey through questionnaires of fishermen's problems	• Low education level • PEMP program conducted by Department of Marine and Fisheries is not maximal. • No alternative job in fish famine and in not capture season • Inappropriate assistance of boat operational cost because fishing cost is increasing • Low family financial management literacy • Lack of skill in processing coastal and marine resources • Loan to usurers and *pengijon* • High consumptive behavior • Low motivation in entrepreneurship • Cannot establish business partnership well • Do not know financial institutions well • Minimal economic institution of fishermen due to minimal skill in establishing and legitimizing economic institution.
Focus group Discussion (FGD) with fisherman group, Department of Marine an d Fisheries, Experts ofFisherman Empowerment	**Results of Need Analysis** • Entrepreneurial literacy based on coast and marine. • Literacy of marine resource process • Literacy of skill in livelihood based on off fishing concept

	• Literacy of Business Diversification of processed fish products • Literacy of family financial management • Literacy of establishing joint business group (KUB) or economic institution

Data Source: Processed by Researchers.

Results of survey of need analysis of fisherman empowerment program based on economics education and off fishing concept can be summarized in the following table:

Table 3. Results of Need Analysis of Fishermen in Empowerment Model through Learning and Training for Coastal Fishermen in North Minahasa Regency

No	Identified Component	Fishermen's Attitude	
		Agree	Disagree
1	Fishermen need an understanding on the attitude, characteristics, and principles of successful entrepreneurs trough learning	87.5 %	12.5 %
2	Fishermen need an understanding on creativity and innovation as well as their examples in coastal and marine business through learning and training	92.5%	75 %
3	Fishermen need an understanding on diversification (extension) of coastal and marine business as alternative job though training	90 %	10 %
4	Fishermen need an understanding on how to make processed fish products utilizing the whole parts of fish through training	90 %	10 %
5	Fishermen need an understanding on establishing business partnership and network though learning	80 %	20 %
6	Fishermen need an understanding on the procedures of establishment and development of fisherman business group through learning	90 %	10 %
7	Fishermen need an understanding on business capital and procedures to get business capital	92.5 %	7.5 %
8	Fishermen need an understanding on family financial management, expenditure allocation planning, and saving through learning and training	85%	15%

Data Sources: Processed by the researchers,2017.

6. Conclusions

The results of identification and need analysis coastal fishermen empowerment model in North Minahasa Regency can be illustrated as follows:

1. Focus Group Discussion obtained a lot of information about most of coastal fishermen who did not understand entrepreneurial knowledge and skill, so that they found difficulties and they were not motivated at all to do business diversification based on coast and marine. The survey results of need analysis of fishermen, there are 87.5 percent of research samples expect entrepreneurial education and training, while 92.5 percent of fishermen want training on creativity and innovation in running coastal and marine business. The researchers suggest that there should be education and technical training for fishermen in order to train their creativity and innovation in managing marine products. One of the technical training that is required is processing fish into several derivative products such as fish *surimi*, fish balls, fish nuggets, and fish chili sauce. Technical

training is also directed to how the technique of packaging and quality improvement of processed fish products into high-value economic products.

2. The results of observation and in-depth interview with coastal fishermen, researchers found that most fishermen do not have the expertise and skill in managing fish into high-value products. From the results of Focus Group Discussion which was conducted with groups of fishermen from Kema Sub-district and Wori Sub-district, it was obtained information that the catch of fish is often abundant but the price of fish becomes cheap and fishermen are having trouble marketing fish because fishermen and their families do not have skill and expertise to process fish into processed food products which are durable and have high economic value.. From result of need analysis revealed in FGD and survey of coastal fishermen's responses about coastal fisherman empowerment need, it can be concluded that fishermen really need skill to manage fish to some products with high economic value like fish ball, fish *suruimi*, fish nugget and other processed products. This information was supported by the results of survey which was conducted after FGD that 90% of coastal fishermen expect training of fish processing into high-value products. Therefore, fishermen need fishermen empowerment program through education and non-formal training on processing fish into products based on zero waste production (no waste).
3. The research findings show that most of fisherman families still behave consumptively when they get income or fishing results. The results of survey to the research samples show that 65% of the research samples never attended any education and training including financial management education and training for fisherman families, 35% of fishermen who had participated in technical training did not get training on financial management of fishermen family. Only 5 percent of fishermen who had got other non-formal education. From the results of the survey on the research samples, 95% or most of the fishermen who became the research samples really need coastal fisherman empowerment model based on economics education and off fishing concept through education and training of financial management for fisherman family. Fishermen are suggested to participate in family financial management training and live economical life. Another suggestion is that fishermen should have saving for the family's future, long-term plan, and plan to invest through savings, invest in production factors such as boats, fishing gear, coastal farmland, and avoid excessive consumption.
4. The results of observations and in-depth interview show that most coastal fishermen in North Minahasa Regency have no alternative livelihoods other than fishing, so their survival mechanism is highly dependent on the catch of fish and other marine products. This becomes a problem when they do not go to sea or face the fish famine. While the results of Focus Group Discussion and survey fishermen about the need for model of fishermen empowerment program based on economics education and off fishing concept show that 95% or almost all of the fishermen who become research samples wanted coastal fishermen empowerment model based on economics education and off fishing concept to be implemented to them to improve their standard of living. It is very possible to diversify livelihoods of fishermen based on coastal resources without having to rely on the results of fishing. From this research, fishermen are suggested to attend education and training on the utilization of other coastal resources such as coastal and marine ecotourism, agriculture, animal husbandry, and integrated coastal fish farming. In the integration of coastal fish farming, coastal farming, coastal animal husbandry, fishermen can farm while planting corn and cassava, and looking for rough fish as the fodder for their livestock. Corn flour, rough fish flour, and cassava can be processed into high-quality feed for the livestock such as poultry.
5. In this research, it was found that most coastal fishermen in North Minahasa Regency are not actively involved in the organization or group of fishermen. Most of them work individually in fishing activities. Although they are not involved in the group, they often help each other and have strong social interaction in their social life, both among fishermen and with other groups such as farmers, traders and other communities. In Focus Group Discussion, it was identified that fishermen lacked the ability to form a joint business group or fisheries economics institution. The results of survey after FGD found that 90% of fishermen expect training on how to form groups

and how the legalization mechanism of fishing groups. From this research, fishermen are suggested to form a group or organization of fishermen with the aim to help each other in marine activity and other productive activities around the coastal area. With group, fishermen can exchange information, knowledge, and energy in order to increase their productivity and income.

6. One of the findings of researchers from observation and in-depth interview which were conducted to fishermen is that some of them do not understand how to establish partnerships with fishermen business groups or larger business partners. Focus Group Discussion also identified the problem of the inability of fishermen in finding and establishing business partnerships with business partners that can support the fishing business. Survey on the need analysis of fishermen empowerment through education and training to establish business partnerships found that 80% of fishermen expect fishermen empowerment model through education and training on how to build business partnerships with other mutually beneficial parties. Fishermen are suggested to be able to build networks and partnerships with other parties, individuals, economic institutions and corporations in order to transfer knowledge, marketing network of fish and processed fish products, market information, business capital assistance, technology assistance, and various partnerships with fishermen's partners.
7. Research findings show that most fishermen do not have the ability to access capital. Fishermen are often in debt to usurers with high rates of interest. In covering the cost of family living and the cost of fishing, fishermen are often stuck in debt. The results of survey of the need analysis of coastal fishermen empowerment program through training on how to access capital through types of capital assistance for fishermen confirm that 92.5 percent of fishermen who became the study samples agreed that they want education and training on capital access procedures as well as types of capital which can be accessed by fishermen. It is suggested to conduct intensive training and assistance for fishermen groups, so that they have knowledge and ability in accessing fishery business capital and also other business capital in which fishermen are involved.
8. From the research findings, it can be seen that most fishermen do not understand family financial management mechanism. Most of the fisherman and their families do not have savings and do not know how to save in financial institution like bank. 85% of the research samples stated that they were in needof knowledge of financial management, planning for proper financial allocations, and saving procedures. There should be learning and training of family financial management for fishermen and their families.

7. References

[1] Maria CA,dkk. (2012). Model Kurikulum pemberdayaan Masayarakat Nelayan Pesisir berbasis ekonomi produktif di Kabupaten Cirebon dan Kabupaten Jeneponto.Lembaga Riset Kementerian Kelautan dan Perikanan.
[2] Nzmar, E. (2012). Model Pemberdayaan Ekonomi Rumah Tangga Nelayan Skala Kecil dengan Pengembangan Off Fhising di Kota Padang Sumatera Barat, Jurnal Ekonomi Manajemen dan Akuntansi Vol 18 (1) Juni 2018.
[3] Fauzi, S. (2000). Valuasi Ekonomi Sumberdaya Pesisir. Materi pada Seminar PengelolaanWilayah Pesisir dan Pulau-pulau Kecil Secara Berkelanjutan. Proyek Kerjasama IPB dengan The Papua New Guinea University of Technology.
[4] Elfindri. (2002). Ekonomi Patron Client. Fenomena Mikro Rumah Tangga Nelayan dan Kebijakan Makro. Andalas University Press.
[5] Kusnadi. (2009). Keberdayaan Nelayan Dan Dinamika Ekonomi Pesisir, Aruzz Media Yogyakarta.
[6] Kusnadi. (2005). Akar Kemiskinan Nelayan. Yogyakarta LKiS, Pelangi Aksara
[7] Harunisya, N. (2014). Pendidikan Ekonomi Masyarakat Melalui Program Pemberdayaan Nelayan Kabupaten Probolinggo.Jurnal Ecobus Vol 1 No.2 2014.
[8] Hikmat A. (2006). Strategi Pemberdayaan Masyarakat, Humaniora Utama Press Bandung.

[9] Evans, D.R. (1981). The Planning of nonformal education,Published in 1981 by The United Nations Educational, Scientific and Cultural Organization 7 Place de Fontenoy, 75700 Paris.
[10] Jogiyanto. (2007), Sistem Informasi Keprilakuan. Penerbit, Andi Yogyakarta.
[11] Cotton Wayne, et al. (2009). A Journey Through a Design Based Research Project. Research Online Faculty Of Education and Faculty Of Social Science University f Wollongong.

Salt Salinity Tolerance on Nursery of *Indigofera zollingeriana*

Marhamah Nadir [1]Muhammad Jabar Anugrah[1] Purnama Isti Khaerani[1,]

[1]Faculty of Animal Science, University of Hasanuddin, Indonesia
Email: marhamahnadir@unhas.ac.id
[1]Faculty of Animal Science, University of Hasanuddin, Indonesia
Email:Email: jabaranugrah@gmail.com
[2] Faculty of Animal Science, University of Hasanuddin, Indonesia
Email : puputkhaerani@gmail.com

Abstract: *Indigofera zollingeriana.*, is a legume plant that has a high protein content and is very potential to develop as one of the forage crops. This study was conducted to determine the tolerance level of *Indigofera zollingeriana*, concerning to the level of NaCl concentration in cultivation seedlings as an effort to develop these forages on a marginal land, especially on saline land, so that the potential of saline land can be utilized properly. This study used a randomized block design with 3 groups and 5 replications, maintained and observed intensively for 30 days. The first treatment (P1) was grown on soil and compost media (1:1) and given a water without NaCl concentration. The second treatment (P2) was grown on soil, sand and compost media (1:1:1) and given a water with a concentration NaCl of 50 mM/2.925 g/ l, and the third treatment (P3) was grown on soil, sand and compost media (1:1:1) and given a water with a concentration NaCl 100 mM/5.85 g/l. The results showed that Indigofera zollingeriana growth and develop on medium on the salinity of 9.9 indicated that the plant's tolerance high salinity levels despite the effect of salt stress affecting growth plant.

Keywords: *Indigofera zollingeriana.*, Legume, Salinity, NaCl, Forages

1. Introduction

Forage feed is the main source of feed for ruminant livestock with daily consumption reaches 70% of the total ration and plays a very important role because forage contains almost all the necessary substances for livestock. The presence of fiber in forage feeds such as cellulose and hemicellulose is a source of energy for rumen microbes, as well as minerals and proteins, especially those derived from leguminous as the source of N for rumen bacteria.

Legume has been known to have potential as a source of high-quality feed, especially during the dry season when the availability of forage grass declined sharply. Forage quality is reflected in the nutritional value contained in it. Forage generally contains crude protein, fat, crude fiber, and minerals. Crude protein is a very important nutrient for livestock. Forage that contains high crude protein is owned by most leguminous plants.

One type of leguminous plant that has not been explored much is *Indigofera zollingeriana*, The crude protein content and crude fat of *Indigofera sp.*, Were high, ie 24.2% and 6.2% respectively. With a high enough protein content, relatively low fiber content and high levels of digestibility, *Indigofera sp.*, Is potentially used as a source of forage and as a source of protein supplement feed [1].

In order to provide forage feed for sustainable livestock feed is needed that is tolerant of various environmental stress conditions. Amongst a variety of environmental stresses, salinity is one of the most common stresses [2]. Less fertile land use such as saline for feed crops is a very important alternative to deal with the declining productive lands that convert to residential functions. The use of saline also has great potential because Indonesia is an archipelagic country with long coastline dominated by saline field. [3] Land in Indonesia about 181.000 km2 located along the beach and has not been utilized. Strategies to deal with these marginal land issues agronomically can be done by utilizing plant cultivars that are tolerant to salinity stress [4]

Efforts to use cultivars of salinity tolerant food plants to date are still constrained by the limited availability of salinity tolerant plant feed cultivars. The problems of infertility of the land also need to be overcome by creating cultivars of feed plants that are tolerant to salinity by making alternative methods of testing fast and easy before direct planting to the field. *Indigofera zollingeriana.*, ones of forage animal feed that has tolerance to dry soil conditions [1]. *Indigofera zollingeriana*, still able to survive and produce at the level of severe drought stress (25% field capacity), despite decreasing productivity. With the ability of *Indigofera zollingeriana.*, To survive in drought conditions, it is expected that *Indigofera sp.*, Also able to survive in salinity condition, because the effect of drought stress is almost the same as salinity stress [5]

This study aimed to determine the tolerance level of salinity of *Indigofera zollingeriana.*, In the nursery which is expected to be the base of *Indigofera zollingeriana.* Plant development, on marginal land especially on the saline field, so that potential of *Indigofera zollingeriana.*, As forage feed and also the availability of saline area wide enough to be put to good use.

2. Method

The study was conducted at the experimental farm of Animal Science Faculty, Hasanuddin University, Makassar, Indonesia. At first, 12-d old seedlings were used for germination. The seeds were planted in the pot using medium compost: soil (1:1) until cotyledone leaves change to complete 2-3 leaves. Germinated seeds were transferred to poly-bag contain medium sand: soil: compost (1:1:1), each treatment contains 2 poly-bags. The study designed by Randomized Block Design with 3 treatment and 5 groups. The treatments were : P1 = water without NaCl control); P2 = water + NaCl concentration of 50 mM (2.925 g/l); P3 = water + NaCl concentration of 100 mM (5.85 g/l).

This experiment has been carried out in a randomized block design with 3 treatment and 5 blocks per treatment. Statistical significance between mean values was assessed using the analysis of variance and conventional Duncan's multiple range tests using SPSS Statistical version 17.0 software. A probability of $p<0.05$ was considered significant. The data obtained were analyzed by Variance (ANOVA).

3. Results

3.1. Measurement of Soil Salinity

Based on the result of soil analysis from Soil Chemistry and Soil Fertility, Faculty of Agriculture, Department of Soil Science, Hasanuddin University, as a medium of treatment of legumes of *Indigofera zollingeriana describe* in Table 1.

Table 1. Concentrations of Salinity of Medium *Indigofera zollingeriana*

No.	Treatments	Salinity level (dS/s) Extract 1:2,5
1	P1	0,54
2	P2	1,14
3	P3	9,99

Description: P1 (Without addition of NaCl), P2 (Addition of NaCl 50 mM / 2,925 g / litre of water), P3 (Added NaCl 100 mM / 5.85 g / litre water)

3.2. Morphology Indigofera zollingeriana

Morphology of plants can also show the effect of treatments on *Indigofera sp.* that can appear from the shape, color, and response of plants. More specifically, the morphology of *Indigofera sp.* in each treatment can be seen in figure 1.

Figure 1. *Indigofera sp. Morphology by salinity treatments*

4. Discussion

Based on soil analysis results it can be seen that in the treatment of P1 (without addition NaCl) the salinity rate of the media is 0.54, then at the treatment of P2 (Addition of NaCl 50 mM / 2,925 g / litre of water) the salinity level of 1.14 and P3 (NaCl addition of 100 mM / 5.85 g / litre of water) of media salinity level of 9.99. At treatment P3 with salinity level 9.99 classified as level with high salinity. This is consistent with the opinion [7] that salinity levels of 8-16 include high salinity levels and only for high tolerant plants. This indicates that the legumes of tilapia (*Indigofera sp.*) can survive with a high salinity level of 9.99 although it provides a salt stress effect with changes in plant height, leaf number, leaf area and leaf area index.

Reduced rate and quality of *Indigofera sp.* Growth in saline conditions can be due to the decreased potential of water from the substrate where it grows, increased absorption of Na and Cl, or both. The effects of water loss on plant tissue will decrease cell turgor, increase the macroeconomic concentration of molecules as well as compounds with low molecular weight, affect the cell membrane and the potential chemical activity of water in plants. It is supported by [8] that the role of water is very important for plants which can have consequences that directly or indirectly the water shortage in plants will affect all of its metabolic processes so as to decrease plant growth. Salinity causes a further effect on plant growth because it can reduce the potential of groundwater which further causes osmotic stress, affect the imbalances of ions in the cell, in particular, decreases the concentration of K^+, Ca^{2+}, and NO_3^-ions, and causes the poisoning of Na^+ and Cl^-. The disruption of plant growth occurs due to inhibition by both Na^+ and Cl^-, where Na^+ is the main causative agent that damages plant growth [9].

Potential soil saline osmotic soil solution is similar to that caused by drought (dry), then some symptoms due to salt stress are also seen in plants that experience drought causing barriers to the process of plant growth [10]. Based on research [11] the effect of salinity stress on rice crops is the decrease in plant height and number of tillers, root growth is inhibited, the weight loss of 1000 grains and the total protein content in the seeds due to excessive Na absorption, and reduced dry weight of the plant. [12] also reported that the high growth rates of rice and wheat crops decreased differently with increasing NaCl concentrations in plant growth mediums.

The main effect of salinity is the decrease in leaf growth which directly leads to reduced plant photosynthesis. It is supported by [10] that in saline conditions, growth and development of plants are inhibited due to excessive accumulation of Na and Cl in the cytoplasm, causing metabolic changes in the cells. Enzyme activity is also hampered by salt.

This is in accordance with the study [13] which states that the salinity level will affect the number of leaves from Rhizophora stylosa seedlings and Avicennia marina where the higher the salinity, the less number of leaves produced. NaCl salt can inhibit the balance of growth hormone. The addition of NaCl may increase ABA while the concentration of auxin, gibberellin, and cytokinin decreases dramatically. Therefore, the formation of new leaves will be inhibited by the decrease of cytokines and gibberellins that play a role in cell division and growth [14].

The decrease in total leaf area is the response of plants to the water supply. The response of the plants to the water supply is suspected because the leaf cells are still young and are holding a water stress experienced by salinity. This situation causes the enlargement and elongation of young cells that are not maximal plastic. When the cells become mature and the cell wall is not plastic anymore due to lignification, the leaf cell size remains small, this causes the leaf area becomes narrow. That the total leaf area is the most affected variable compared to other growth parameters, which can be judging by the drastic decline of leaf area indicating that high sensitivity of leaf growth to salinity [15].

Further explains that the reduction in leaf growth rate is a form of salt stress adaptation because increased soil salt levels inhibit the absorption of water by plants and to limit leaf transpiration. Smaller leaf sizes are essential for maintaining turgor [16]. While root lignification is necessary for the adjustment of osmosis which is essential for maintaining the necessary turgor for plant growth and normal activity. Reduced leaf area will affect photosynthesis. The area of the leaf plays an important role, as the rate of photosynthesis takes place following the broad development of the leaf [17].

Based on research [18] the broad growth of leaves of green mustard greens has been effected at salt treatment concentrations of 4000 ppm - 10,000 ppm. This condition is caused by the presence of excess salt that inhibits the process of cell division. According to [19], one of the negative effects of salt on the plant can be a decrease in the ability of plants to absorb water so that it affects the process of cell division and affect hormonal function. Excess salt may decrease the function of the hormone auxin. [20] Auxins play a role in spurring cell division and enlargement [21]

Salinity causes structural changes that improve the water balance of the plant so that the water potential in the plant can maintain turgor and all biochemical processes for normal growth and activity. Structural changes include smaller leaf sizes, smaller stomata per unit of leaf area, increased suction, cuticle thickness and wax coating on the leaf surface, and earlier root lignification [17].

Some researchers consider the reduction of cell size in stress conditions as an adaptation mechanism of extreme environmental conditions. Reduced cell size appears to be the primary responsibility of cells to water deficiency caused by drought or salt stress. The status of water is very sensitive to salinity and therefore greatly determines the response of plants to stress.

5. Conclusion

Adaptability of *Indigofera zollingeriana* growth on medium *NaCl* indicates these plants very tolerance to salt salinity with high salinity levels despite the effect of salt stress affecting plant growth, therefore these species potential developing forages in the coastal land.

6. Acknowledgment

This research was supported/partially supported by the University of Hasanuddin and The Ministry of Higher Education and Research Technology, as supporting institution who provided insight and expertise that greatly assisted the research. We would also like to show our gratitude to the researcher for sharing and working their pearls of wisdom during this research. We are also immensely grateful to Forum Dosen Indonesia (FDI) for their comments on an earlier version of the manuscript.

7. References

[1] Hassen, A., Rethman NFG, Van Niekerk, Tjelele TJ. 2007. Influence of season/year and species on chemical composition and in vitrodigestibility of five Indigofera accession. J Anim Feed Sci Technol. 136:312-322.

[2] Gedoan SP, Indradewa D dan Syukur A. 2004. Tanggapan varietas kacang tunggak terhadap cekaman salinitas. Fakultas Pertanian, Universitas Gadjah Mada. Jurnal Agrosains. 17(1):102-103.

[3] Suhardi. 2008. Pengembangan agroindustri berbasis pangan lokal untuk meningkatkan kedaulatan pangan. Pros. Semnas Pengembangan Produk Berbasis Pangan Lokal. Universitas Mercu Buana, Yogyakarta

[4] Mudgal, V. 2004. Physiological Studies On Growth And Nitrogen Metabolism In (*Cicer arietinum* L). Under saline Conditions. Ph.D. Thesis. Rohilkhand Univ., India.

[5] Haryadi, 2013. Pengukuran luas daun dengan metode simpson, Anterior.Vol.12, no.2:1-5p

[6] Sitompul, S.M. dan B. Guritno. 1995. Analisis Pertumbuhan Tanaman. Gadjah Mada University Press. Yogyakarta.

[7] Abrol, I. P., J. S. V. Yadaf, and F. I. Massaud.1988. Salt-Affected Soil and Their Management. Food And Agriculture Organization Of United Nations, Rome.

[8] Sinaga, S. 2002. Asam absisik sebuah mekanisme adaptasi tanaman terhadap cekaman kekeringan. Hal 1-6. Diakses dari http://www.daneprairie.com Diakses 13 November 2017.

[9] Tuteja, N. 2007. Mechanisms of high salinity tolerance in plants, chapter twenty- four. Methods in Enzymology. 428: 419-438.

[10] Yuniati. R. 2004. Penapisan galur kedelai *glycine max (l.) merrill* toleran terhadap NaCl untuk penanaman di lahan salin (screening of soybean cultivars *glycine max (l.) merrill* under sodium chloride stress condition). Departemen Biologi, FMIPA, Universitas Indonesia. Depok. Makara, sains. Vol. 8, no.1:21-24

[11] Rahmawati. 2006. Status perkembangan dan perbaikan genetik padi menggunakan teknik transformasi Agrobacterium. Agrobiogen 2:364-375.

[12] Zhou, J., X. Wang, Y. Jiao, Y. Qin, X. Liu, K. He, C. Chen, L. Ma, J. Wang, L. Xiong, Q. Zhang, L. Fan, and X.W. Deng. 2007. Global genome expression analysis of rice in respone to drought and high salinity stresses in shoot, flag, leaf, and panicle. Plant Mol Biol 63:591-608

[13] Basyuni., Lollie, A, P, P., Berliana, N dan Putri, E. 2014. Growth and biomass in response to salinity and subsequent fresh water in mangrove seedlings avicennia marina and rhizophora stylosa. JMHT.Vol XX (1): 17-25.

[14] Syamsuardi. 1992. Tanggapan Tanaman Bayam Cabut (*Amaranthus tricolor L.*) dan Bayam Duri (*Amaranthus spinosus L.)* terhadap Salinitas NaCl dan Pemberian Pupuk Nitrogen pada Tanah Regosol. Tesis. Prog Studi Biologi Jurusan Ilmu Matematika dan Pengetahuan Alam Universitas Gajah Mada.

[15] Bai, R., Z. Zhang, Y. Hu, M. Fan, U. Schmidhalter. 2011. Improving the salt tolerance of Chinese spring wheat through an evaluation of genotype genetic variation. Australian J. of Crop Sci. 5(10): 1173-1178

[16] Carillo, P., G. Mastrolonardo, F. Nacca and A. Fuggi. 2005. Nitratereductase in durum wheat seedlings as affected by nitrate nutrition and salinity. Funct. Plant Biol. 32:209–219.

[17] Harjadi, S. S. dan S. Yahya. 1988. Fisiologi Stress Lingkungan. PAU-IPB, Bogor

[18] Eti Dwi, Mukarlina, Irwan Lovadi., 2015. Toleransi tanaman sawi hijau *(*Brassica juncea *L.)* terhadap cekaman salinitas garam NaCl. J.Protobiont. Vol.4(1): 203-208

[19] Fitter, AH & Hay, RKM, 1994, Fisiologi Lingkungan Tanaman, Gadjah Mada University Press, Yogyakarta

[20] Pessarakli, M, & Tucker, TC, 1993. Dry matter yield and nitrogen-15 uptake by tomatoes under sodium chloride stress. Journal Soil Science.Vol.52:698-700

[21] Gardner, PF, Pearce, RB, & Mitchel, RL, 1991, Fisiologi Tanaman Budidaya, UI Press, Jakarta.

[22] Takagi, H dan S. Sumadi. 1984. The growth of soybean as affected by plant density. Penelitian Pertanian. 4 (2): 83-86.

Divorce and Family Role Based on *Slerek* as Traditional Fishing Method in Muncar Banyuwangi Coastal Area

Burhani Buhani[1], **Sri Surahmi**[2]

[1)] Universitas PGRI Banyuwangi, Jl Ikan Tongkol 22 Banyuwangi 68416
Email: buhani@uniba-bwi.ac.id

[2)] Universitas PGRI Banyuwangi, Jl Ikan Tongkol 22 Banyuwangi 68416
Email: surachmi@uniba-bwi.ac.id

Abstract *:* Muncar is a sub district in Banyuwangi which is a ocean fishery area. Banyuwangi has a big number of divorce in marriage. In Muncar there are many separation cases but the number of divorce is zero in 2012. This research is aimed to describe the causes of this phenomena. The research is done by field interview to local people. The interview results are used to describe the causes of the phenomena. The result shows that separation happen because economical and ethnical factor. There is also a unique understanding which come from the philosophy of traditional fishing method that called slerek. This method describe a couple of fishing boat or vessel which represent a role of marriage couple. Slerek did not effect the knowledge and attitude to prevent separation and divorce in Muncar district but can be used as premarital education material.

Keyword**:** divorce, slerek, Muncar, Banyuwangi

1. Introduction

Family is the basic and essential unit of life that gives safety and comfort to their members. Marriage is necessary for forming of family [1]. One of the family problems is divorce. Divorce is related with marriage and family, and is a social innovation; it had been used as an instrument exposing failure through marriage [2]. Society should make every effort to support healthy marriages and to discourage married couples from divorcing [3]. Divorce is a complex event that can be viewed from multiple perspectives. Sociological research has focused primarily on structural and life course predictors of marital disruption, such as social class, race, and age at first marriage [4].

Divorce has received significant attention in many developed countries because of its striking effects on individuals and society such in Bangladesh [5] and Iran [1, 2, 6]. Divorce has many effects due to the member of the family. Divorce causes personal, domestic and social disintegration, and in most cases, has greater banes for women compared to men [2]. Divorce has impact due to family structure on the health of children [3]. Former husbands and wives were more likely to blame their ex-spouses than themselves for the problems that led to the divorce [4].

Research in divorce topics has done for the reason [1, 2, 4, 6, 7], the impact [3], and the phenomenon of divorce [8]. There are a several factors that cause a divorce. A research result shows that the demographics of families are changing, and with that, the philosophical underpinnings of elationships are also changing [3].

Banyuwangi as a district has number of divorce which is the the biggest in 2009 [9] and second biggest in East Java Province [10]. Muncar is a sub district ocean with traditional fishing society which the second biggest in Indonesia. The number of population in Muncar is 64.254 persons with 13.810 persons working in fishery sector [11]. This research is aimed to describe the causes of the divorce phenomenon in Muncar.

2. Methodology

The research is done by field interview to local sources with snow ball method. The interviews done for local government, 10 family couples and 5 local fishermen. The interview is aimed to get the causes the tendency of separation rather than divorcing. The interview results are used to describe the causes of the phenomena.

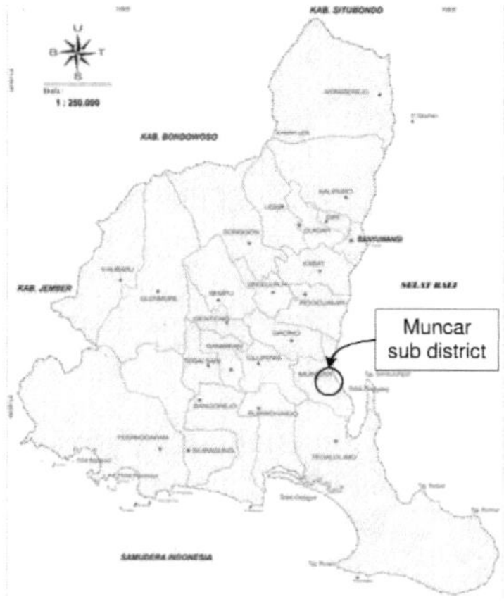

Fig. 1. Muncar sub district

3. Result and Discussion

There is a divorce phenomenon in this sub district. According to Badan Pusat Statistik Kabupaten Banyuwangi (Central of Statistical Bureau of Banyuwangi District) the number of marriage, separation and divorce is shown at table 1.

Table 1. Number of Marriage, Separation and Divorce 2010-2012 in Muncar

Year	Marriage	Separation	Divorce	Ref
2011	1.130	17	23	[12]
2012	1.294	43	85	[13]
2013	1.227	141	0	[14]

Table 1. shows that the number of separation tends to increase. Number of divorce also tends to increase during 2011-2012 but tends to decrease during 2012-2013. These numbers show that people who have problems with marriage tend to choose separation rather than divorce. Separation has a chance that the problem can be fix. The intensity of marital conflict would be higher if the wife lives with her husband [15].

Traditional fishing method in Muncar is called *slerek*. This method use 2 fishing boats or vessels that function as a couple [16]. People in Muncar called *slerek* as "husband and wife in a family" [17]. The first ship function as fish catcher and the second one as a provider for the need of the first one. It is similar with the role of husband and wife in Javanesse culture. The husband work for the family and the wife stay at home to provide the needs of the husband. This fishing method is known in people daily life that can be the foundation of family role in society.

The cause of divorce mostly is economical factor. Fishermen is known as relatively poor society. Muncar people mostly are Madura ethnic. Usually, the role of woman has been measured up by domestic jobs, it goes around bathroom, cooking room, and bed room. These roles are considered ideal for woman [18]. The development of the world has made changes in this paradigm. It is also drive by women emancipation movement. Studying historical trend of divorce phenomena, among contemporary societies, revealed that whatever we shift from feudal system to liberal and industrial society, the divorce possibility and frequency will increase [2].

An important factor that affects the quality of marital relationships can be values and belief systems of the partners and the similarities and differences of beliefs and values in the couple system [6]. The knowledge of *slerek* concept should be a marriage foundation in family daily life in Muncar.

The phenomenon shows that divorce number is still relatively high. This shows that *slerek* concept did not has effect due to the family role. This concept can be used as premarital education material in order to decrease the number of separation and divorce.

4. Conclusion

Slerek as a pair of traditional fishing boat has a concept of husband and wife role in a family. This concept did not effect the knowledge and attitude to prevent separation and divorce in Muncar district. This philosophical knowledge can be used as premarital education material.

5. Refrences

[1] Barikani, A., Ebrahim, SM., Navid, M. (2012). The Cause of Divorce among Men and Women Referred to Marriage and Legal Office in Qazvin, Iran, Global Journal of Health Science 4(5): 184-191

[2] Mohammadi, SMRG., Tafti, FC. (2014). The Main Causes of Divorce in Islamic Republic of Iran (Yazd), International Journal of Humanities and Social Science 4(9): 193-197

[3] Anderson, J. (2014). The impact of family structure on the health of children: Effects of divorce, The Linacre Quarterly 81(4): 378–387

[4] Amato, PR., Previti, D. (2003). People's Reasons for Divorcing: Gender, Social Class, the Life Course, and Adjustment, Journal Of Family Issues, 24(5): 602-626

[5] Parves, KN. (2011). Social Changes and Women-Initiated Divorce in Dhaka, Bangladesh: Gaining or Loosing Power?, Thesis, Master of Philosophy in Gender and Development, Department of Health Promotion and Development, Faculty of Psychology, Bergensis University

[6] Akbar, T., Hossein, V., Hengameh M. (2015). Evaluation Affecting Factors Of Emotional Divorce (Case Study: The perspective of Higher Education Married Staffs in Karaj Province in 2014), MAGNT Research Report (ISSN. 1444-8939) 3(3): 459-467

[7] Scott, SB., Rhoades GK., Stanley SM., Allen ES., Markman HJ. (2013). Reasons for Divorce and Recollections of Premarital Intervention: Implications for Improving Relationship Education, Couple Family Psychol 2(2): 1-20

[8] The University of Hong Kong. (2014). On the Study of the Phenomenon of Divorce in Hong Kong, Final Report, Central Policy Units, The Government of Hong Kong Special Administrative Region

[9] Ernawan, Y. (2016). Peran Kyai Pada Perceraian Masyarakat Migran Pulau Bawean, BioKultur V(2): 209-236

[10] Purnama, DYE., Santoso, B. (2013). Aspek-Aspek Penyebab Perceraian Gugat Di Desa Karangbendo Kecamatan Rogojampi Kabupaten Banyuwangi (Aspect Causes Of Divorce Contested In Countryside of Karangbendo Subdistric Rogojampi Regency of Banyuwangi), Artikel Ilmiah Hasil Penelitian Mahasiswa I (1): 1-7

[11] Pusat Statistik Kabupaten Banyuwangi. (2015). Kecamatan Muncar Dalam Angka 2015 (Muncar in Figures)

[12] Badan Pusat Statistik Kabupaten Banyuwangi. (2011). Banyuwangi Dalam Angka Tahun 2011 (Banyuwangi in Figures)

[13] Badan Pusat Statistik Kabupaten Banyuwangi. (2012). Banyuwangi Dalam Angka Tahun 2012 (Banyuwangi in Figures)

[14] Badan Pusat Statistik Kabupaten Banyuwangi. (2013). Banyuwangi Dalam Angka Tahun 2013 (Banyuwangi in Figures)

[15] Dewi, EMP., Basti. (2008). Konflik Perkawinan Dan Model Penyelesaian Konflik Pada Pasangan Suami Istri, Jurnal Psikologi 2(1): 42-51

[16] Perkasa, T., Wijayanto, D., Fitri, ADP. (2016). Analisis Produktivitas Purse Seine Gardan Dan Purse Seine Slerek Dengan Fishing Base Di Pelabuhan Perikanan Pantai (Ppp) Muncar

Kabupaten Banyuwangi Jawa Timur, Journal of Fisheries Resources Utilization Management and Technology 5(1): 102-110
[17] Rachmawati, I. (2016). Mengenal Kapal Slerek "Suami Istri" di Pelabuhan Ikan Muncar Banyuwangi, http://www.kompas.com, downloaded April 22nd, 2017
[18] Mulyadi, A. (2011). Perempuan Madura Pesisir Meretas Budaya Mode Produksi Patriarkat, Karsa 19(2): 199-213

Potency of Education Historical Tourismof World War II Japanese Cavesand Bunkersin Coastal Banyuwangi

Miftahul Rahmi[1], Ikhwanul Qiram[2]
1) Universitas PGRI Banyuwangi, , Jl Ikan Tongkol 22 Banyuwangi 68416
Email: ikhwanulqiram@gmail.com
[2] Universitas PGRI Banyuwangi, Jl Ikan Tongkol 22 Banyuwangi 68416
Email: miftahulrahmi70@gmail.com

Abstract: Banyuwangi district has some Japanese caves and bunkers of World War II. The location of the objects are along the Banyuwangi coast as a maritime defense during the war. This structures can be used as education historical tourism object. There are many similar structures in other area that have been neglected and do not get enough preservation attention. This research is aimed to identify the potency of education historical tourism of Japanese caves and bunker in Banyuwangi. The research is done by field research for the observation of objects physical condition. It is also done by interviewing local government, historical actors and surrounding community. The result shows that the caves and bunker have a great potency but have not been used as education historical object.

Keywords: Japanese, cave, bunker, education, history

1. Introduction

Indonesia is a country that is very rich in historic attractions that can be utilized as a tourist attraction concept education [1]. Tourism will usually be more developed or can be developed if in an area has more than one type of object and tourist attraction [2]. The implementation of tourism is a very important tool in regional development, to be able to create jobs, increase and level up the income of the community and introduce local cultural arts [3]. The main purpose of tourism development is for more tourists to come to a tourist area, stay longer, and spend more money in the tourist attractions they visit [4].

Research on the potential of historical tourism has been done, among others, historical objects in Ilir I Palembang [5], cultural tourism potential of the Majapahitkingdom heritage site at TrowulanMojokerto [6], tourism colonial tourism potential in Malang [7], Gilimanuk district site potential Jembrana Bali [8], historical objects in the city of Ternate [2] and the Van der WijckGombong Citadel of Kebumendistrictin Central Java [1].

One of the historical tourism potential is the cave, including caves and bunker forme of the Japanese colonial period. Many of Japan's caves and bunkers are not getting attention, such as cannons and World War II defense cave in Alak sub-district, Kupang city, Nusa Tenggara Timur province [9]. Bunker or arsenal weapon of Japanese army former located in Pangandaranvillage threatened dismantled [10]. Dozens of caves and bunker former of the Japanese occupation which condition are neglected, whereas the heritage buildings can be used as a tourist attraction [11]. In fact, Japanese bunker ruins are damaged and taken for sale, then some are also used for trash dumps [12].

Cave exploration systematically is not only related to physical aspects and findings but also includes systematic documentation. This field documentation shows cave exploration systematically covering the surrounding soil surface, cave surfaces, detailed observation records, rock and lime characteristics, and documentation photographs. This data is combined with maps and descriptions that can serve as a basis for similar research or subsequent research [13].

Cave research has been done a lot. The cave is examined in terms of lime distribution and its control in Japanese Yoron-Jima [14]. Caves in Scandinavia are extensively researched in terms of location, history and use [15]. The cave is also investigated in terms of exploration [16] and historical exploitation [17]. In Indonesia, research was conducted for the development of Kreo cave tourism object to socio-economic life of society [3].Cave research is documented and mapped to the location based on historical records related to the exploration, mining and development of cave sites with techniques developed from caving search hobbies. These data can be tested on related historical information. This could be the basis for the conclusion of the land conditions around the cave and the comparison with its regional territory [17].

The potential of Japanese caves and bunker in Banyuwangi district is still not developed as a historical tourism asset that can be beneficial to the world of education. As in other areas, Japanese caves and bunker tend to be dormant and do not get enough attention from the public and government. For this reason we need a study that examines the potency of this historical heritage.

2. Method

The research is done by field observation and interview method. The observation of objects physical condition is done by a camera. Surrounding location is observe for accesibilty and supporting facilities by photograph documentation. It is also done by interviewing local government, historical actors and surrounding community to get informations about the Japanese caves and bunkers in Banyuwangi district.

3. Result and Discussion

Japanese caves and bunkers in Banyuwangi district have characteristic such:

1. Relatively small in size which indicates the base for small army group.
2. Located in slope hill with footpath acces which gave advantage for defense.
3. The direction facing to the sea because the function as maritime defense.

The condition of the caves and bunkers are relatively good but have not been explore as tourism objects. There is a board sign in the side of the entrance of the cave. Children from surrounding community use the objects for as playing ground. Some teenagers and older people came to the object because of the view and the unique photograph background.

Figure 1. Japanese caves at Kalipuro subdistrict

Figure 2. Japanese bunkers at Kalipuro subdistrict

According to some sources, there are more than 25 Japanese caves and bunkers along the coastal line in Banyuwangi district. These locations are understandable because coastal line in Banyuwangi is part of Bali strait in east side which make the coast is strategic place for troop landing and Indonesian ocean in south side which have access to international territory. These Japanese caves and bunkers have not been identify and explore by researchers.

The Japanese caves and bunkers have potency as research and exploration. Systematic documentations can be done for objects characteristic such as measuring size, mapping, soil and rock, etc. It is also can be done for environmental study for further preservation program.

Figure 3. Coastal line of Banyuwangi district

The Japanese caves and bunker have not been used as historical education material, especially for elementary school children. There is no documented information about the objects. These objects can contibute to educate children about local history or even national and international history.

Utilization as tourism object is needed in order to explore the objects for history education. Tourism attractiveness will has effort due to preservation activities and maintain the existence of the object for the next generation. This activity will need a program which need cooperation between government and surrounding community.

4. Conclusion

Japanese caves and bunker in Banyuwangi district have own characteristic and a great potency for tourism objects. These structures have not explored and used as education historical object. The government and community can work together for conducting the preservation program.

5. Reference

[1] N.A. Kurniawan, G,. Sridiatmoko. (2015). *Pemanfaatan Benteng Van Der Wijck Gombong Kabupaten Kebumen Jawa Tengah Sebagai Sumber Belajar Sejarah dan Objek Wisata Pendidikan*. Skripsi. Prodi Sejarah Fakultas Keguruan dan Ilmu Pendidikan Universitas PGRI Yogyakarta

[2] M.Y. Fadlhy, L,. Tondobala, S,. Tilaar. (2015). *Permasalahan Pengembangan Objek Bersejarah Dalam Menunjang Wisata Kota di Ternate*. Hasil Penelitian. Program Studi Perencanaan Wilayah dan Kota. Universitas Sam Ratulanggi. Manado

[3] Y.A. Ginting. (2014). *Perkembangan Objek Wisata Gua Kreo Terhadap Kehidupan Sosial Ekonomi Masyarakat*. Jurnal Ilmiah Pendidikan Sejarah IKIP Veteran Semarang 02(1): 1-8
[4] G. Gunardi. (2010). *Identifikasi Potensi Kawasan Wisata Kali Pasir, Kota Tangerang*. Jurnal Planesa^{TM} 1(1): 28-35
[5] S. Nugroho. (2005). *Kajian Potensi Linkage Obyek Wisata Sejarah Budaya di Kelurahan I Ilir Palembang*. Laporan Penelitian. Fakultas Teknik. Universitas Sriwijaya
[6] K.. Anwar. (2009). *Potensi Wisata Budaya Situs Sejarah Peninggalan Kerajaan Majapahit di Trowulan Mojokerto*. Laporan Tugas Akhir. Fakultas Sastra dan Seni Rupa. Universitas Sebelas Maret. Surakarta
[7] D. Budiyono., R, Djoko. (2010). *Potensi Wisata Bangunan Kolonial di Kota Malang*. Buana Sains 10(1): 83-92
[8] F.D. Yanti. (2012). *Potensi Situs Gilimanuk Sebagai Objek Wisata Sejarah di Kecamatan Melaya Kabupaten Jembrana*. Skripsi. Program Studi Pendidikan Sejarah. Jurusan Pendidikan Ilmu Pengetahuan Sosial. Fakultas Keguruan dan Ilmu Pendidikan. Universitas Jember
[9] Balai Pelestarian Cagar Budaya Bali. (2014). *Konservasi Meriam dan Gua Pertahanan Perang Dunia Ke II di Kecamatan Alak, Kota Kupang, Provinsi Nusa Tenggara Timur*. http://www.kebudayaan.kemdikbud.go.id/bpcbgianyar.com. accesed May 25th, 2016
[10] S. Dadan. (2015). *Bungker Jepang Terancam Dibongkar*. http://www.radartasikmalaya.com. accesed May 25th, 2016
[11] V.S. Pranyoto. (2016). *Balai Arkeologi: Puluhan Gua Jepang Terbengkalai*. http://www.antarayogya.com. accesed May 25th, 2016
[12] W. Kharisma. (2016). *Bungker Jepang Dirusak, Dijual, dan Jadi Tempat Sampah*. http://www.pikiran-rakyat.com. accesed May 25th, 2016
[13] P. Kambesis,. (2007). *The Importance of Cave Exploration to Scientific Research*. Journal of Cave and Karst Studies, 69(1): 46-58
[14] J.P. Terry. (2005). *Karst Distribution and Controls on Yoron-Jima, an Emerged Reef Island in Sub-Tropical Japan*. Journal of Cave and Karst Studies, 67(1): 48-54
[15] K. Jennbert. (2011). *Scandinavian Cave Archaeology*. Journal of Cave and Karst Studies, 38(3): 144-150
[16] S.A. Craven. (2012). *A History of Cave Explorations on the Cape Peninsula Mountain Chain of South Africa: 1709 et seq*. Journal of Cave and Karst Studies, 39(3): 101-108
[17] S. Damm, A. Young. (2013). *Caving in: Archaeology of Historical Cave Exploration and Exploitation*. DigitalCommons@University of Nebraska–Lincoln

IOP Conf. Series: Earth and Environmental Science **156** (2018) 012030 doi:10.1088/1755-1315/156/1/012030

Magnifying Democracy and Sovereignty In Indonesian Maritime Governance Through Open Marine Spatial Data Practice

Adipandang Yudono[1]
[1]Dept. of Urban and Regional Planning
Brawijaya University, Jl. M.T. Haryono No.167, Malang, 65145, INDONESIA
Email: adipandang@ub.ac.id

Abstract. This research has attempted to discover a new approach in magnifying democracy and sovereignty in the Indonesian maritime governance through open marine spatial data practices between governments and citizens. The research has been done in order to fill in the gap of bridging marine spatial data or information at all government levels and between citizen and government. The research predominantly used qualitative methods with specifically approach was discourse analysis using legal document analysis and in-depth interview to elites, and local digital mapping communities. The coherence and synergy of maritime development can be achieved through dialogue between the elites and the public. A solution to bridge political communication between the elite and the public is sharing or open marine spatial data and information.

Keywords:Democracy, Sovereignty, Maritime Governance, Open Marine Spatial data

1. Introduction

Today, issues of maritime affairs and fisheries in Indonesia are the main targets to be achieved under JokoWidodo's cabinet, the current President of Indonesia, known as "*NawaCita*". *NawaCita* has meaning thenine priority agenda of JokoWidodo - JusufKalla government cabinet, which become the national flagship programmes to be realized during the term of office period. In an effort to strengthening Indonesian image as the maritime state, the development of maritime sector as well as sea-tolls infrastructure for inter-island logistics transportation and utilization of fishery and marine resources for the welfare of the Indonesian people are prioritized for immediate implementation. One of immediate actions to achieve *NawaCita* is marine open data practice between government levels and citizens.

Maritime data are central to the working of the Indonesian government to implement regular programmes, management and strategies, particularly in maritime governance. Governments at all levels (central, state/province and local level) often run their maritime programmes on the bedrock of accurate, seamless, reliable and up-to-date data which can be transformed into valid maritime information. The government usually createsthe information by compiling it from various authorised data to formulate policies and strategies that affect society. However, although in maritime governance, marine spatial data is crucial, often it is not given sufficient attention.

An effective communication amongst maritime governance stakeholders (i.e. governments, planners, contractors, citizens) becomes a fundamental factor in maritime governance issues. It not only allows transaction of maritime information but also avoids/minimizes maritime information gap (fragmentation) resulted form lack of coordination. The persons involved will get maritime information from the agencies and feedback from maritime data and information users. However, research in communication related to open marine spatial data at governance levels and the relationship with non-government spatial data providers in the crowdsourced information context is still in its infancy and invites a challenge for the new knowledge.

This research has been done in order to fill in the gap of bridging marine spatial data or information at all government levels and between citizen and government. The objective of this research is to discover a new approach to enhance more democratic maritime governance by implementing open marine spatial data practice involving central government, regional government, local government, andcitizens by investigating official spatial data development accommodating crowdsourced geographic information context in a nation.

2. Methods

This research predominantly used qualitative methods with specifically approach was discourse analysis using legal document analysis and in-depth interview to elites, and local digital mapping communities. The legal document collections were needed to gain a comprehensive understanding of maritime governance and marine spatial data infrastructure development. Furthermore, the interviews were used to understand the perspective of the government and citizens in terms of their current strategy and initiatives for promoting maritime governance through open marine spatial data practices from a political and social background.

An in-depth interview activity is undertaken to get information about the acquainted and substantial potential of official spatial data and crowdsourced geographic information integration in the maritime governance practice. In this case, the respondents explained their current working activities and experiences using Geographical Information Systems (GIS), including their expectancy and threat towards spatial data sharing amongst institutions. In order to facilitate the process of interview and to ensure all information are secured, note taking with the help of digital voice recording has been applied and later transcribed. Furthermore, the legal document collections were used to serve the purpose of providing a fundamental background of the policy context to gain a comprehensive understanding of the marine spatial data development and sharing in maritime governance applications that has been done and will do in the future.

The next step after accomplished interviews, legal document collections and field observation was doing coding. It was be done subjectively by applying the method utilised by Renner and Powell (2003)[1] to draw out the significance of the words or the topics, which are recognised as far as the thoughts or ideas. These were then sorted out into thematic classifications that will abridge and convey intending to the content. The coded information then investigated utilising a content analysis. In the content analysis, the examination moved from a generous viewpoint to look for a specific insight of the research topic. This diagnostic might be investigated with discoveries from documents and investigations from the respondents.

3. Result

3.1 The Open Data Practices in Indonesia

From 1966 to 1998, under the Suharto regime, the former 2nd President of Indonesia, the distribution of government data and information were very strictly controlled by the state. Access to data and information were subject to very strict regulations; there was no public access to official information, and even inter-governmental institutions found it extremely difficult to exchange data or information without obtaining recommendations from the heads of government departments [2][3][4].

The phenomenon of limited access to obtain data and information can be traced by examining The Law No. 7 of 1971 concerning principal filing provisions, especially Article 11, Paragraph 1 of the criminal provisions which states "*Whoever intentionally and unlawfully has official data and information without legal procedure may be liable to imprisonment for a maximum of 10 (ten) years.*" [5]. In addition, Paragraph 2 states that "*Whoever deliberately or unintentionally passes on the content of official data or information to third parties who are not entitled to know it, the person in charge of keeping the data and information safe may be liable to imprisonment for a life sentence or imprisonment for a maximum of 20 (twenty) years.*"[5]. A further provision is found in Government

Regulation No. 34 of 1979 on "Depreciation Archive" in the explanation of Article 15 which states that "*Depreciation archives referred to in this government regulation be implemented with due regard to the security and the confidential nature of the archive.*" [6].

Political reform in 1998 marked the collapse of the Suharto regime and brought about fundamental changes to the Indonesian Constitution. Indonesia ratified human rights covenants and amended the Indonesian Constitution 1945 in 2002. One of the amendments guarantees the rights of citizens to access information, as specified in Article 28F of the Amendment: "*Everyone has the right to communicate and obtain information to develop their personal and social environment, and the right to seek, obtain, possess, store, process and convey information by using all available channels*". The Article became the basis for constructing a law on public information disclosure [7].

The draft of the Indonesian Freedom of Information Act, *Rancangan Undang-Undang Kebebasan Memperoleh Informasi Publik* (RUU KMIP) was first put forward in 2000 by the Indonesian Centre for Environmental Law (ICEL). The difficult negotiation of the RUU KMIP took until 2007, due to tough negotiations between the Executive government under the President's control and the Legislature under The House of Representatives control [8].

In April 2008, RUU KMIP was finally passed into the law as the Indonesian Freedom of Information Act, *Undang-Undang Keterbukaan Informasi Publik* (UU KIP) (Law No.14 of 2008). Two years later, The Law. No.14 of 2008, has been run effectively upon issuance of Government Regulation No. 61 of 2010 of the Law No. 14 of 2008 guidelines. It took eight years for Indonesian Freedom of Information Act to be ratified, and based on the Indonesian law procedure, the law can be implemented when law derivative, in this context is the Indonesian Government Regulation for the law implementation guideline, is stipulated. Ordinarily, the government regulation enactment from the law ratification takes two years, The situation means to impose Freedom of Information Act in Indonesia needs a 10-year since the 2nd Constitution 1945 amendments in 2000 (see Figure 1)

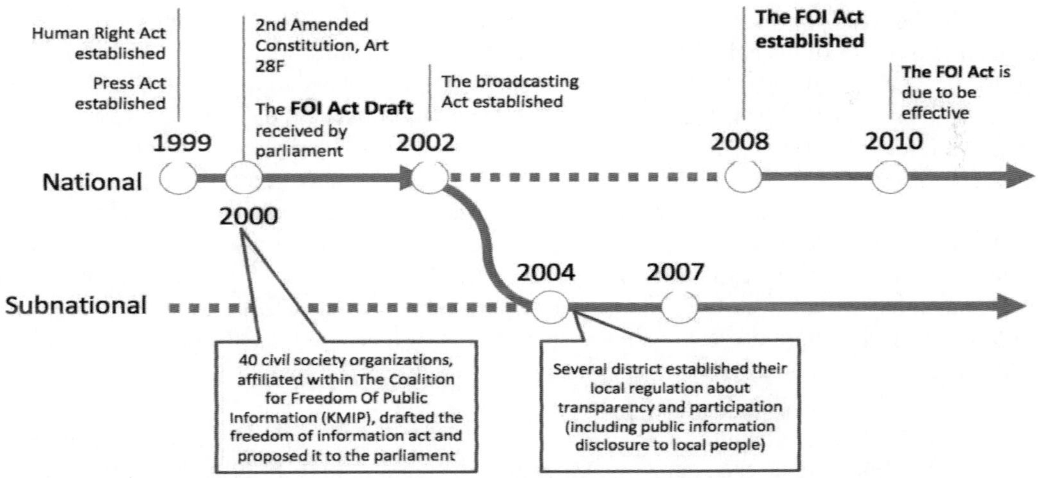

http://www.asef.org/images/stories/ccs4/ws%206_alamsyah%20saragih 20indonesia.pdf

Figure 1. Establishment of Freedom of Information Act in Indonesia

The ratification of the Freedom of Information Act and regulations was completed in 2010 officially beginning an era of open data in Indonesia. One of the substantial agendas of open government data in this research topic is open marine spatial data in maritime governance context.

IOP Conf. Series: Earth and Environmental Science **156** (2018) 012030 doi:10.1088/1755-1315/156/1/012030

3.2 The Indonesian Role in the Protection and Empowerment of Marine Resources in the International Arena

Today, The Indonesian Ministry of Maritime Affairs and Fisheries (KKP) in incorporate with other Association of South-East Asian Nasion (ASEAN) maritime/marine ministries under the coral triangle Initiative programme, which is focus in a program of protection and empowerment of marine resources living therein, details can be found at http://www.coraltriangleinitiative.org. This coral triangle area is very strategic as a food supplier and a place to live for fish, especially tuna and a variety of reef fish, where they spawn, lay eggs, and thrive. However, as adults grow, these fish communities migrate to other regions / countries outside the coral triangle area.

The countries included in the Coral Triangle Initiative programme agree to run the global fishery trade system fairly. One of the formation of this organisation is made label / brand for certification, or currently known as Eco-label, that products out of coral triangle area has been friendly, or have standardization, where when the country entering the coral triangle area is buying fish outside the country of the region, already has a bargaining price.

However, it is acknowledged that in reality, the implementation of the Coral Triangle Initiative is full of strong political content. There are some conflicts, such as the Australian State disagreeing with the concept of bargaining price, which is considered to be detrimental to it. Another case, where the country of Vietnam has a political repayment with America, all fisheries exports to America from Vietnam is made easier. While other member countries of the Coral Triangle Initiative, such as Indonesia, Malaysia and other countries that do not have political relations with America have difficulty to export fisheries. Another case is illegal fishing, where Indonesian fishery and marine products are stolen and processed in several ASEAN Member States and East Asian countries. Thus, the term is, Indonesia does not get any results from this fishery and marine resources, whereas Indonesia maintains its ecosystem, but no one pays for the services of fish breeding ecosystem for Indonesia, so this is the idea to increase open data / data sharing in the KKP environment, with the concept of the Payment Ecosystem Service, where the fee is paid using certification from the Coral Triangle Initiative organisation.

In the world of international fisheries, there is a division of fishing zones arranged in various regional authority groups called Regional Fisheries Management Organization (RFMO), based under FAO-UN, more details, can be found at http://www.fao.org/fishery/topic/166304/en . Indonesia itself is surrounded by 3 RFMOs, the Indian Ocean Tuna Commission (IOTC), The Convention for the Conservation of Southern Bluefin Tuna (CCSBT) and the Commision for the Conservation and Management of Highly Migratory Fish Stocks in the Western and Central Pacific Ocean (WCPFC). This means, Indonesia cannot be arbitrary to catch fish outside this region. However, there is an odd thing, for example the type of tuna fish, where when Indonesia has its own abundant tuna fishery resources, for its arrest regulation, Indonesia must have licenses. So that tuna has a quota of populations that are allowed to be fishing. The purpose of the tuna fishing quotas is to conserve the world's fish. Here, there are calculations, such as the number of tuna in an RFMO, which will be subdivided into each country. By:

- State Compliance.
- Capacity of the Country

It is felt another odd things with the quota of tuna in this RFMO, which obtained the largest quota is a country that lacks / lacks tuna resources as big as Indonesia. in the context of illegal (non-official) systems they contribute, where the system is as follows, the vessel for catching its tuna comes from countries where the marine territory has a small population of tuna while for licensing, in collaboration with person in Indonesia as maritime territory with abundant tuna population, then out permit for fishing from companies with name or belonging to Indonesia, but ship and its ship crews (*ABK*) is all stranger. The case of illegal fishing is also happening in Solomon Island, based on the study of World Wide Fund (WWF), neatness of fishing administration is not good too, so the case of illegal fishing is much more rampant there.

In relation to illegal fishing, when marine and fishery issues begin to lead to trade, superpowers, such as America and European countries, focus on certifying food for the safety of food products that enter the country. To that end, these superpower nations are assisting countries rich in fish and other marine products, to select fisheries and marine products toward certification and have security for consumption for their people. On the basis of that, fisheries management began to be tidied, ranging from the legal administration of fishing and the manner of his capture. Thus, fish and other marine products netted in the case of illegal fishing will be banned, or fish or seafood products have been blacklisted, so it cannot be accepted by the world fishery market.

At the current ASEAN level, there is cooperation with a programme called Regional Plan of Action Illegal, Unreported and Unregulated (RPOAIUU), a cooperation program involving 11 countries in ASEAN in implementing open data / data sharing related to illegal fishing indication. However, at the time of implementation, the regional secretariat of RPOAIUU, in this case Indonesia through the KKP with the executor of the Directorate General of Marine and Fishery Resources Control, said that when asked for data to countries that have political power with superpower countries such as Thailand and Vietnam, for data acquisition. When a forum of RPOAIUU meeting was held, delegates from countries with political power relations with the superpower were present, mostly not officials at decision level positions, but operational staff who lack knowledge of illegal fishing data and information. This is the obstacle in conducting data sharing illegal activities of fisheries and marine environment of ASEAN.

In fact, Indonesia itself is quite sophisticated in the acquisition of data and information related to fishing and other marine products, which is applied to the Vassel Monitoring system, a kind of GPS tracking that is used on large ships with a fishing range above 12 miles. These data are connected to the internet and can be accessed to global data fishing tracking such as marine traffic or ship finder that works like Flight Radar 24.

In multilateral cooperation, open data / data sharing system is required commitment and understanding and vision the same. In fact, for the ASEAN region, such as Indonesia, Malaysia, Thailand and the Philippines have good Human Resource capacity and management performance of fishery and marine resources compared to archipelagic countries such as Timor Leste, Solomon Islands and Vanuatu, the commitment to mutual prosperity can be realized in the effort of marine and fishery data sharing.

For the scope of Indonesia itself, in ensuring the transparency of data between institutions engaged in maritime, KKP, as the highest authority of marine sector has been bridging with the loosening of marine data sharing bureaucracy, the motivation is the understanding in various data, but sometimes, there are initiatives that are the same purpose, his political commitment is different, ultimately the data sharing is not running. Not focus on one purpose, because there are other activities done in fisheries management, so data sharing is ignored. Therefore, things can be taken, that leadership commitment has crucial action, and this is recognized very difficult to implement. For example, in government institutions, when there is a change / transfer of positions, communicating data sharing programs between the old and new officials is not necessarily done, because the different vision and mission carried by the new leadership, often not in line with the vision and mission of previous officials. So Leadership and Policy become the main aspect for successful implementation of open data system, moreover is component of commitment to willingness to do data sharing among institution that move in maritime sector.

3.3 The role of open marine spatial data in strengthening seafood security

In terms of the development issues for fulfillment of fishery and marine products, the national programme launched by the Indonesian government is food Security. In the exposure of food security in maritime sector, there are 3 main components that become the concentration of operations, namely:

1. Traceability. The focus of operations is the tracking of producers to end users, such as knowing the geographical location of fishing and other marine products, the location of the distribution of the harbor for the collection of fish, the location of fish processing and seafood, the location of the port / airport that becomes the destination of the processed fish and

seafood, export location to other fish and marine fishing destination countries, then detects the recipient of fish and other seafood in the country of export to detect the final consumer.
2. Affordability. The extent to which people are able to buy fishery products and other marine resources
3. Supply Stock (Availability), at what rate, the market is able to provide other fishery products and sea fisheries.

3.4 The Potential Strategy Creating Open Marine Spatial Data Alliances in The Context of Maritime Governance

Various Information systems scholars acknowledge that achieving good organisational performance requires equal vision and mission amongst members [9][10][11[[12]. Specifically, Budhatoki and Nedovic-Budic (2007) argue that the essential point of data or information sharing in an organisation performance factor is an inter-organisational Collaboration-Cooperation-Coordination (3C) element[13]. In relation to this, this section will examine the broader 3C concept and its relevance to spatial data development and sharing in spatial planning process.

3.4.1 Collaboration

Collaboration gives an opportunity for sharing learning, background and aptitudes with various individuals, keeping in mind the end goal of altering objectives and creating improvement. Within a specific end goal to effectively work together, there must be adequate assets, a society that energizes compelling collaboration and participation and in addition clearly understood obligations[14] [15].

Moreover, colleagues involved in collaboration must trust and respect each other. There must be open correspondence and readiness to acknowledge information from others. Since in all groups there are clashing objectives, decision-making must take place through cooperative methods. Collaboration guarantees that no one individual, gathering, association or foundation is completely accountable for anything; rather, all are included and bear a measure of obligation [16](Characteristics of the collaboration and the relationship with cooperation and coordination can be seen in Figure 2).

With relevance to the context of this research, collaboration amongst authorised and non-authorised inter-organisations that produce and manage spatial data can create willingness for each organisation to share their knowledge and assets in data and information in a commitment to the obligation of achieving a nation aim. Therefore, collaboration is one of the essential elements necessary to achieve successful potential integration of official open marine spatial data and VGI in supporting the maritime governance process.

3.4.2 Cooperation

Cooperation is depicted as a casual relationship without a typical mission in which data or information is shared on 'an as-required' premise, power stays with every association, there is little (or no) danger assets are kept separate. Furthermore, inter-organisational partnership includes assets, abilities and skills in the quest for common enthusiasm for the achievement of the organisations' goals [15] [16].

The amalgamation of data and information production with marketing or dissemination is a type of joint advantage, central to achievement [17]. In terms of this research, cooperation amongst authorised and non-authorised inter-organisations that produce and manage marine spatial data can create commitment to the obligation of achieving a nation aim, in this case, achieving coherent maritime governance programmes at different levels (Characteristics of cooperation and the relationship with collaboration and coordination can be seen in Figure 2).

3.4.3 Coordination

Coordination suggests understanding the dependencies between the responsibilities the various class representatives are performing and the way the class coordinates their duties [18]. Coordination ensures that all specialists and divisions recognise what and whenthey have to accomplish. Along these lines, work startswith one office then onto the next without hindrance.

In any association, nation or activity, all individuals must be organised in such a way as to guarantee that general vital aims are accomplished and every individual makes a commitment. Therefore, coordination has an essential point in an organisational performance. Every single departmental arrangement and spending plans must be facilitated to guarantee they are cooperating to accomplish agreed goals [16](Characteristics of the cooperation and the relationship with collaboration and coordination can be seen in Figure 2).

In this study, coordination amongst authorised and non-authorised inter-organisations that produce and manage marine spatial data can integrate, synchronize and simplify the different tasks on an ongoing basis to achieve effective and efficient maritime governance.

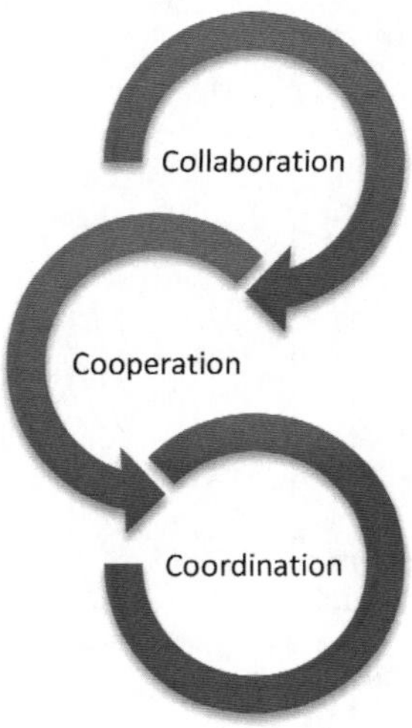

- Driven by mutual self-interest
- Requires high level of responsibility of each member
- Often requires expertise of each party
- Requires high level of trust about the safety of sharing proprietary data and information

- Addresses the substantial needs of each member
- Value often derivative (eg. Process improvement, efficiency initiatives)
- Value may directly accrue only to one party (or neither)

- Driven by order
- Team work has significant value in achieving success
- Typically effective in short-term goals

Source: Modified from The Economist Intteligence Unit 2008
https://www.trinityp3.com/2012/08/did-you-want-collaboration-cooperation-or-coordination-with-that-marketing-process/

Figure 2. Characteristics of Collaboration-Cooperation-Coordination (3C) and The Relationships of 3C Elements

The 3C concept is helpful in creating an inter-organisations (authorised and non-authorised) partnership to develop and open marine spatial data at the macro scale. However,this concept needs to understand the spatial data transaction relationship in terms of how inter-agency partnership can achieve consensus for sharing their marine data and information knowledge assets. The next session will discusses this issue.

4. Discussion

According to governance perspective, potential spatial data and information integration between official spatial data and crowd-sourced geographic information in the maritime governance context is determined by active communication between government and citizens. The researcher proposes a stepped model of community participation the government's agenda through open marine spatial data and information between official spatial data and crowd-sourced geographic information. Purpose of proposing this model is to identify the levels of public participation and government participation in creating collaboration, cooperation and coordination in the maritime governance circumstance (See Figure 3).

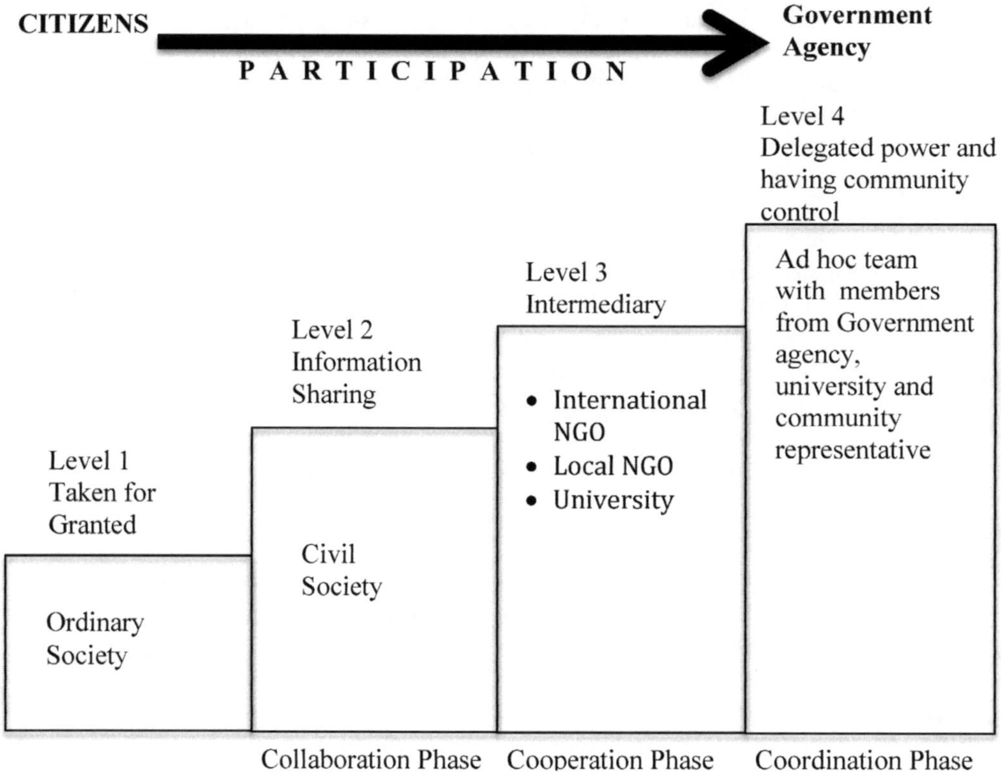

Figure 3. A Stepped Model for Community Participation in Government Agency Agendas

Level 1: Taken for granted

Citizen who live in coastal areas have passive attitudes. They take for granted that planning activities are developed and implemented by the government. Likewise, from the government's perspective, the community does not need to be involved in the maritime development agenda. At this stage, political communication between community and government in terms of maritime governance has not been achieved.

Level 2: Information Sharing

People are concerned about the area where they live and provide current information about conditions in their local region. Information can be delivered through *Musrenbang* (public hearings) or submitted online, including spatial data sharing in a user-friendly platform between the public and the government. The government tends to accept public input for improving the planning agenda. At this stage, collaboration between citizens and government has begun.

In terms of spatial data sharing, citizens as sensors have volunteered to be local mapping contributors by interpreting spatial objects on a user-friendly application or mobile digital mapping platform. It aims to inform other communities and reports current condition as input in spatial policy making.

Level 3: Intermediary

This stage bridges the gap between public demand for improving their living environments and the government agenda, which has been drawn up for maritime development implementation. In a development activity, political communication between a government agency and the public often reaches a deadlock. Thus, facilitating communication between the two parties needs an intermediary actor has been selected based on personal or community organisation experience with a good reputation trusted by the both sides. Frequently, the intermediary actors in planning activities are international or local NGOs and universities. At this stage, participation between citizens and government has moved from collaboration to cooperation.

In terms of spatial data sharing, high community interest in spatial data will need a local digital mapping organiser to control spatial data quality and reliability. Sometimes, the task of the local digital mapping organiser is as a substitute for local mapping contributors to the continuous updating of spatial data.

Level 4: Delegated power and community control

This stage is the ultimate step in creating coordination between government and the public. The members of the *ad hoc* team are respresentatives from government, universities and community organisations and are generally established in planning and development practice. This *ad hoc* team works to formulate spatial policy and determine maritime planning and development programme priorities in real development projects. In terms of spatial data sharing context, the local digital mapping organiser and government officers join the *ad hoc* team to manage spatial data sharing transactions and work to ensure reliable quality and quantity of official spatial data.

Overall, this research has attempted to discover a new approach in magnifying democracy and sovereignty in the Indonesian maritime governance through open marine spatial data practices between governments and citizens. The coherence and synergy of maritime development can be achieved through dialogue between the elites and the public. A solution to bridge political communication between the elite and the public is sharing or open marine spatial data and information.

5. Conclusion

The current Issues of maritime affairs and fisheries in Indonesia are the main agenda of the new President of Indonesia known as "*Nawa Cita*", the nine priority agenda of JokoWidodo – Jusuf Kalla government, which become the national flagship program to be realized during the term of office period. In an effort to strengthen Indonesia has the identity as a maritime State, the development of maritime sector as well as sea-tolls infrastructure for inter-island logistics transportation and utilization of fishery and marine resources for the welfare of the Indonesian people are prioritized for immediate implementation. One of immediate actions to achieve *NawaCita* is open marine spatial data between government levels and citizens.

Leadership and Policy become the main aspect for successful implementation of open data system, moreover is component of commitment to willingness to do data sharing among institution that move in maritime sector. Finally, The coherence and synergy of maritime development can be achieved through dialogue between the elites and the public. A solution to bridge political communication between the elite and the public is sharing or open marine spatial data and information.

6. Acknowledgement

The author would like to thanks to The Indonesian Endowment Fund for Education (LPDP Scholarship) to support this financial research, also thank you to my supervisor, Prof. Alasdair Rae, with his patient, always support and motivate me to do this research.

IOP Conf. Series: Earth and Environmental Science **156** (2018) 012030 doi:10.1088/1755-1315/156/1/012030

References

[1] Renner, M. and Powell, E.T., (2003) Analysing Qualitative Data, Programme Development & Evaluation, University of Wisconsin-Extension Cooperative Extension, Madison, Wisconsin, G3658-12.

[2] Lim, M., 2004, The Internet, Social Networks, and Reform in Indonesia. In Coultry, N., and Curran, J. (eds.), *Contesting media power: alternative media in a networked world.* Rowman and Littlefield, Lanham, MD, pp. 273-288.

[3] Kitley, P., 1994, Fine tuning control: commercial television in Indonesia, *Continuum: the Australian Journal of Media and Culture* Vol. 8, No.2.

[4] Shoesmith, B., 1994, Asia in Their Shadow: Satellites and Asia, *Southeast Asian Journal of Social Science*, Vol.22, pp. 125-141.

[5] Government of Indonesia, GoI, 1971, *The Law No. 7 of 1971* of Ketentuan-ketentuanpokokkearsipan (Principal terms in Indonesian official documents).

[6] Government of Indonesia, GoI, 1979, *The Government Regulation No. 34 of 1979* of PenyusutanArsip (The reduction of the number of volumes of official documents).

[7] Saragih, A., 2010, *SejarahKeterbukaanInformasiPublik di Indonesia* (The History of Freedom of the Information Act in Indonesia), at https://taskforce14.wordpress.com/2010/08/18/150/ at 1st March 2016.

[8]Sumrahyadi,2007,*Pemikiranketerbukaanarsipdinamisdalammenyongsongditetapkannyaruukebebasanmemperolehinformasi* ***public*** (The Idea of Dynamic Archives Openness in Welcoming the Enactment of Freedom of Information Act).National Archives of The Republic of Indonesia, Jakarta, Indonesia. Access in http://www.anri.go.id/assets/download/Artikel_Online_Pemikiran_Keterbukaan_Arsip_Dinamis_afc25d97ab774c7c.pdf accessed 1st March 2016

[9] Doherty, N, and King, M., 2001, An investigation of the factors affecting the successful treatment of organizational issues in systems development projects. *European Journal of Information Systems* Vol.10, No. 3,pp. 147–60.

[10] Lambert, R, and Peppard, J, 1993. Information Technology and New Organizational Forms: Destinations But No Road Map? *Journal of Strategic Information Systems*Vol. 2, No.3, pp. 180–205.

[11] Williams, T., 1997, Interorganisational information systems: Issues affecting interorganisational cooperation. *Journal of Strategic Information Systems,*Vol.6, No.3, pp.231–50.

[12] Suomi, R., 1994.What to take into account when building an inter-organizational information system.*Information Processing and Management,* Vol. 30, No. 1, pp. 151–59.

[13] Budhathoki, N.R., Nedovic-Budic, Z., 2007, Expanding the Spatial Data Infrastructure Knowledge Base. In Onsrud, H., (ed), 2007, *Research and Theory in Advancing Spatial Data Infrastructure Concepts*, ESRI Press, Redlands, USA

[14] O'Flynn, J., and Wanna., J., 2008, *Collaborative Governance: A New Era of Public Policy in Australia?*, ANU E-Press, Canberra, Australia

[15] Ramjit, D.M, 2011, *Strategic Planning: Collaboration, Cooperation, and Coordination* in https://publicpuzzle.wordpress.com/2011/02/18/strategic-planning-collaboration-cooperation-and-coordination/ accessed on 7th December 2015

[16] Denise, L, 1999, Collaboration vs. C-Three (Cooperation, Coordination, and Communication), *Innovating* Vol.7, No.3

[17] Gulati, R., Wohlgezogen, F., Zhelyazkov, P., 2012, The Two Facets of Collaboration: Cooperation and Coordination in Strategic Alliances.*Academy of Management Annals,* Vol. 6 pp. 531–583.

[18] Crowston, K., Rubleske, J., and Howison, J., 2006, Coordination theory: A ten-year retrospective. In Zhang, P, and Galletta, D, (eds.), *Human-Computer Interaction in Management Information Systems*: Foundation (Advances in Management Series), Routledge, London, UK.

Combining an Analytic Hierarchy Process and TOPSIS for Selecting Postharvest Technology Method for Selayar Citrus in Indonesia

Andi Dirpan[1]

[1] Department of Agricultural Technology, Faculty of Agriculture, Hasanuddin University, Makassar 90245, Indonesia

E-mail:dirpan@unhas.ac.id

Abstract. This research was intended to select the best handling methods or postharvest technologies that can be used to maintain the quality of citrus fruit in Selayar, South Sulawesi, Indonesia among (1) modified atmosphere packaging (MAP (2) Controlled atmosphere storage (CAS) (3) coatings (4) hot water treatment (5) Hot Calcium Dip (HCD) by using combination between an analytic hierarchy process (AHP) and TOPSIS. Improving quality, applicability, increasing shelf life and reducing cost are used as the criteria to determine the best postharvest technologies. The results show that the most important criteria for selecting postharvest technology is improving quality followed by increasing shelf life, reducing cost and applicability. Furthermore, by using TOPSIS, it is clear that the postharvest technology that had the lowest rangking is modified atmosphere packaging (MAP), followed by controlled atmosphere storage (CAS), coatings, hot calcium dip (HCD) and hot water treatment (HWT). Therefore, it can be concluded that the best postharvest technology method for Selayar citrus is modified atmosphere packaging (MAP).

1. Introduction

Selayar regency is one of the main citrus fruit producing regions in South Sulawesi, Indonesia. The harvest of citrus fruit in a large scale has brought the consequences in the postharvest deterioration[1,2]. Postharvest deterioration of citrus fruit could be caused by many factors, including metabolic changes, (biochemical changes associated with respiratory metabolism, ethylene biosynthesis and action, and compositional changes), growth and development (anatomical and morphological changes), physical injuries, water loss, physiological disorders, and pathological breakdown. On the other hand, citrus postharvest quality and shelf life are becoming increasingly aspects, as consumers expect the quality fruit to be available throughout the year [1]. Therefore, harvesting and handling hold the key to getting desired results from postharvest treatments.

Some handling methods or postharvest technologies can be used to maintain the quality of citrus fruit, such as: modified atmosphere packaging (MAP)[3–6] and controlled atmosphere storage (CAS) [7], coating or waxing[8–11], hot water treatment[12,13] and hot calcium dip [14]. However, every method has drawbacks and advantages in preserving citrus fruit. Therefore, the main objective of this research is to select the best postharvest technology by combining analytic hierarchy process (AHP) and TOPSIS.

Some studies on constructing and applying AHP or TOPSIS have been conducted. Alphonce (1997), have applied AHP models in agriculture in developing countries. In his study, decision hierarchies have been suggested for: (a) determination of farm portions to allocate to each of the food crops; (b) resource allocation to agricultural activities; (c) best location for a village store; (d) choice between subsistence and cash crops production and (e) determination of the crop production technology[15]. Chavez et al. [16] investigated criteria and farming activities for tobacco diversification by AHP. The results showed that livestock activities and spring–summer crops are important alternatives for tobacco production. Chen et al [17] used AHP to determine the optimum edible film for preserving kiwifruits. Amiri (2010) combined AHP and TOPSIS to select project for oil-fields development[18]. Another study by Berdie (2017) also combined AHP and TOPSIS in the field of integrated software systems[19]. From the above studies, it can be concluded that AHP and TOPSIS can be applied to the agriculture sector. However, little attention has been given so far to the application of the AHP to the postharvest technology, although the AHP seems to be suitable also to sector of postharvest technology [20].

Our previous study suggests that AHP is suitable for selecting postharvest technology method for citrus fruit in Japan[1]. However, in this study we only proposed AHP to determine the best postharvest technology for citrus in Japan. The present study was, therefore, conducted to select the best postharvest technology for Selayar Citrus by combining analytic hierarchy process (AHP) and TOPSIS. The AHP is used to score and set the share of importance of the relative weights for the decision criteria – sets and the TOPSIS method is used to obtain the final ranking.

2. Method

2.1. *The analytic hierarchy process*

This study was quantitative research which employed the application of AHP for selecting postharvest technology method for citrus fruit in Selayar regency, Indonesia. The AHP method is one of the multi-criteria decision analyses, and is applicable to solving problems containing more than one decision criterion. This method is a mathematically based, was introduced by Saaty (2008)[21]. It uses the pairwise comparison method to rank order alternatives of a problem that are formulated and solved in a hierarchical structure. Briefly, the step-by-step procedure in using AHP is the following [22]:

1. Define decision criteria in the form of a hierarchy of objectives. The hierarchy is structured in different levels: from the top (i.e. the goal) through intermediate levels (criteria and sub-criteria on which subsequent levels depend) to the lowest level (i.e. the alternatives). In this study, we are dealing with a multiple decision problem, the aim of which is to select postharvest technology for fruit citrus. The highest level of the hierarchy is the goal of the problem, the second level is the criteria as applied to the ranking of postharvest technology (improving quality, increasing shelf life, reducing cost and applicability), while the third one is the decision variants, i.e. modified atmosphere packaging (MAP), controlled atmosphere storage (CAS), coatings, hot water treatment (HWT), and hot calcium dip (HCD).
2. Weight the criteria, sub-criteria and alternatives as a function of their importance for the corresponding element of the higher level. For this purpose, AHP uses simple pairwise comparisons to determine weights and ratings so that the analyst can concentrate on just two factors at one time. One of the questions which could be asked in using a pairwise comparison is: how important is the "improving quality" factor with respect to the "increasing shelf life", in terms of the goal for "selecting postharvest technology method for Citrus fruit"?. The answer may be "equally important", "moderately more important", etc. The verbal responses are then quantified and translated into a score via the use of discrete 9-point scales that we can see in Table 1. In principle, the evaluators are expected to express their ratings in odd numbers (1, 3, 5, 7 and 9), while even numbers (2, 4, 6 and 8) are used where there is no consensus in the group of evaluators.

Table 1. Judgement scores in AHP.

Judgement	Explanation	Score
Equally	Two attribute contribute equally to the upper-level criteria	1
Between equally and moderately		2
Moderately	Experience and judgement slightly favour one attribute over another	3
Between moderately and strongly		4
Strongly	Experience and judgement strongly favour one attribute over another	5
Between strongly and very strongly		6
Very strongly	An attribute is strongly favoured and its dominance demonstrated in practice	7
Between very strongly and extremely		8
Extremely	The evidence favouring one attribute over another is of the highest possible order of affirmation	9

3. After a judgment matrix has been developed, a priority vector to weight the elements of the matrix is calculated. Priority vectors (w) are obtained from the pairwise comparison matrix (A) by solving an eigenvalue problem in the following relation (Eq. 1):

$$Aw = \lambda_{max} w \quad (1)$$

where λ_{max} is the highest matrix eigenvalue.

4. Evaluate the soundness of the judgements with the inconsistency ratio I_R. This is a peculiarity of the AHP technique. Before determining an inconsistency measurement, it is necessary to introduce the consistency index (CI) of an *n*×*n* matrix (of judgements) defined by the ratio (Eq. 2):

$$CI = \frac{\lambda_{max} - n}{n-1} \quad (2)$$

Then, I_R is defined as the ratio (Eq. 3):

$$I_R = \frac{CI}{RI} \quad (3)$$

where RI is the corresponding average random value of CI for an *n*×*n* matrix. Judgements can be considered acceptable if $I_R \leqslant 0.1$. In cases of inconsistency, the assessment process is immediately repeated for the inconsistent matrix. An inconsistency ratio of 0.1 or more may warrant further investigation. Analyses using the AHP method can be performed easily and quickly with the support of numerous available IT tools. In this study, all AHP analyses were performed using the Expert Choice ver 11.

2.2. *Topsis*

The final stage in order to achieve the proposed objectives is to apply the TOPSIS method. The basic principle of this method is that the optimal decision shall be as close as the most advantageous solution and as far away as the worst solution. This method is often criticized for its inability to adequately handle the inherent uncertainty and imprecision associated with the mapping of the decision-maker's perception to crisp values. In the traditional formulation of the TOPSIS, personal judgments are represented with crisp values. However, in many practical cases the human preference model is uncertain and decision makers might be reluctant or unable to assign crisp values to the comparison judgments[18]. The ideal solution should have a rank of one and the worst alternative should have a rank approaching 0 [19,21,23].

Through the proposed model, the TOPSIS method involves the following steps:

- Construction of the decision matrix. This step leads to the Table 1 whose configuration is the decisional matrix.
- Calculation of the normalized decision matrix. The goal of normalization is to transform different dimensional attributes into dimensionless attributes.

3. Results and discussion

There are two types of pairwise comparisons in this study. The first type describes pairwise comparisons of elements of the criteria with respect to the goal as shown in Table 2. One of the question being asked is this: "Comparing reducing cost and applicability, which one more dominates for the selecting postharvest technology method for Citrus fruit and by how much important?" The resulting priority vector, the maximum eigenvalue (λmax) and the inconsistency ratio (IR) are shown in Table 2.

Table 2. Pairwise comparisons of elements of the criteria with respect to the goal.

Goal	Local priority vector
Improving quality	0.636
Increasing shelf life	0.117
Reducing cost	0.117
Applicability	0.130

$\lambda_{max} = 4.112$, $I_R = 0.0039$

The second type describes pairwise comparisons of the elements of postharvest technology alternatives with respect to their parent criteria element as shown in Figure 1, 2, 3 and 4. For the example, the question being asked in Figure 1 is : "Comparing coatings/pelilinan and hot water treatment, which one more dominates the improving quality criteria, and by how much important?".

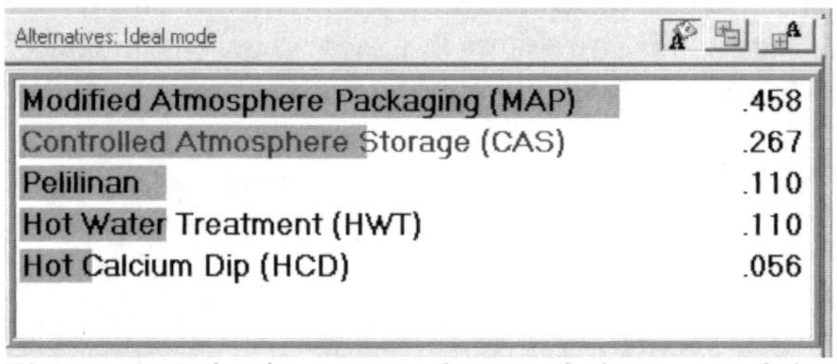

Figure 1. Pairwise comparisons of elements of the alternatives with respect to **the improving quality criteria** ($\lambda_{max} = 4.112$, $I_R = 0.0039$)

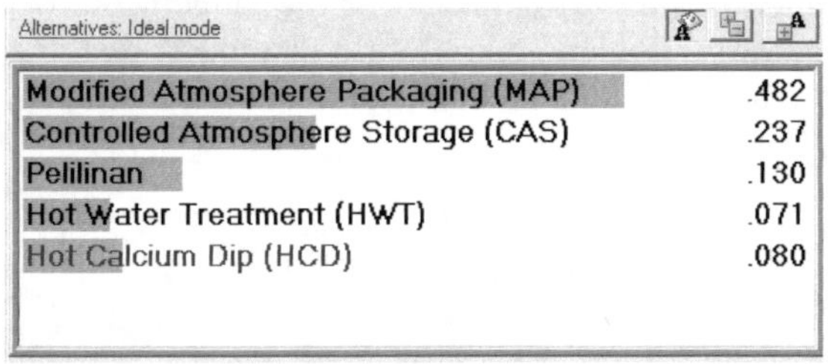

Figure 2. Pairwise comparisons of elements of the alternatives with respect to **the increasing shelf life criteria** ($\lambda_{max} = 5.012$, $I_R = 0.05$)

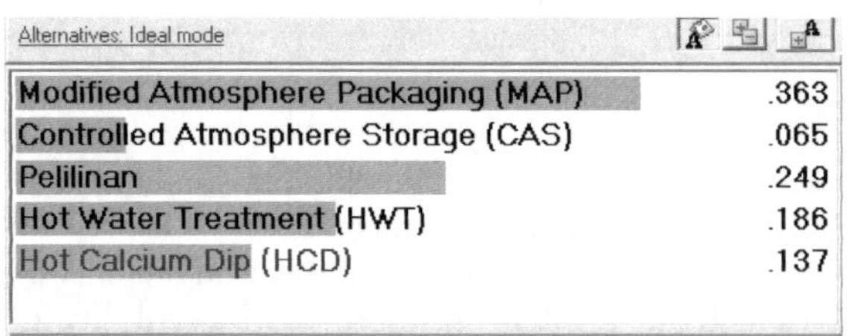

Figure 3. Pairwise comparisons of elements of the alternatives with respect to **the reducing cost criteria** ($\lambda_{max} = 5.574$, $I_R = 0.02$)

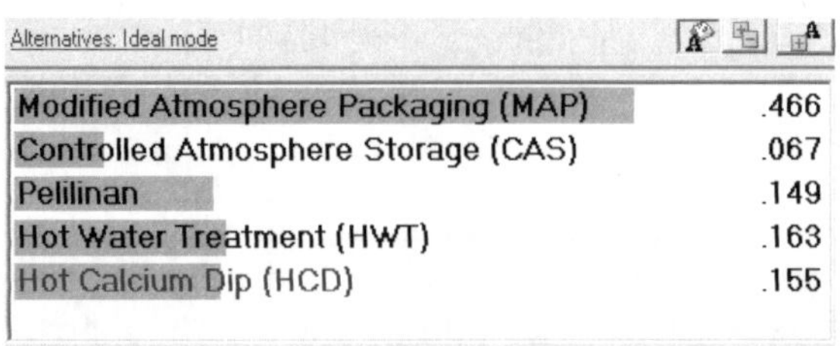

Figure 4. Pairwise comparisons of elements of the alternatives with respect to **the applicability** ($\lambda_{max} = 5.458$, $I_R = 0.049$)

The Figure 1, 2, 3, and 4 display the local priority vectors of each pairwise comparisons matrix with their corresponding maximum eigenvalues and inconsistency ratio (IR). IR values range from 0.0039 to 0.05 which satisfy the 0.10 threshold.

After applying the TOPSIS algorithm the preferred postharvest technology in terms of performance is the modified atmosphere packaging (MAP) (0.354), because she results with minimum score. This hierarchy is considered valid for the defined selection criteria and their corresponding weights.

4. Conclusion

The postharvest technology is method to maintain the quality of citrus fruit and to choose the best candidate by using a multi-criteria approach that using linguistic preferences can be very useful for uncertain situations. The AHP is used to analyze the structure of the best postharvest technology and to determine weights of the criteria, and fuzzy TOPSIS method is used to obtain final ranking. Proposed model results indicate that modified atmosphere packaging (MAP) is the best alternative with CC value of 0.354. In the application, it is shown that calculation of the criteria weights is important in fuzzy TOPSIS method and they could change the ranking for other project.

5. References

[1] Dirpan A and Slamet A S 2016 Selecting Postharvest Technology Method for Citrus Fruit using Analytic Hierarchy Process (AHP)Selecting Postharvest Technology Method for Citrus Fruit using Analytic Hierarchy Process (AHP) *KnE Life Sci.* **3** 77–82

[2] Dirpan A, Hikida Y, Chiba H, Aji G K and Rahman A N 2017 Improving the accuracy of intercellular space volume of citrus prediction by artificial neural networks *Acta Horticulturae* (ISHS) pp 37–44

[3] Zagory D and Kader A 1988 Modified atmosphere packaging of fresh produce *Food Technol.* **42** 70–7

[4] Kader A A and Zagory D 1989 Modified atmosphere packaging of fruits and vegetables *Crit. Rev. Food Sci. Nutr.* **28** 1–30

[5] Porat R, Weiss B, Cohen L, Daus A and Aharoni N 2004 Reduction of postharvest rind disorders in citrus fruit by modified atmosphere packaging *Postharvest Biol. Technol.* **33** 35–43

[6] Techavises N and Hikida Y 2008 Development of a mathematical model for simulating gas and water vapor exchanges in modified atmosphere packaging with macroscopic perforations *J. Food Eng.* **85** 94–104

[7] Dirpan A and Hikida Y 2015 Effect of various citrus sizes on the resistance to gas diffusion *Procedia Environ. Sci. – Elsevier* **28** 391–8

[8] Ben-Yehoshua S, Burg S P and Young R 1985 Resistance of citrus fruit to mass transport of water vapor and other gases *Plant Physiol.* **79** 1048–53

[9] Park H J 1999 Development of advanced edible coatings for fruits *Trends Food Sci. Technol.* **10** 254–60

[10] Hagenmaier R D and Baker R A 1993 Reduction in gas axchange of citrus fruit by wax coatings *J. Agric. Food Chem* **41** 283–7

[11] Chien P J, Sheu F and Lin H R 2007 Coating citrus (Murcott tangor) fruit with low molecular weight chitosan increases postharvest quality and shelf life *Food Chem.* **100** 1160–4

[12] Hong S I, Lee H H and Kim D 2007 Effects of hot water treatment on the storage stability of satsuma mandarin as a postharvest decay control *Postharvest Biol. Technol.* **43** 271–9

[13] Porat R, Daus A, Weiss B, Cohen L, Fallik E and Droby S 2000 Reduction of postharvest decay in organic citrus fruit by a short hot water brushing treatment *Postharvest Biol. Technol.* **18** 151–7

[14] Nutakorn T, Yoshio H and Toshio K 2011 A Comparative Study of the Effects of Postharvest Treatments on Occurrence of Kohansho and Quality of "Kiyomi" Tangor *Food Preserv. Sci.* **37** 167–72

[15] Alphonce C B 1997 Application of the analytic Hierarchy process in agriculture in developing countries *Agric. Syst.* **53** 97–112

[16] Chavez M D, Berentsen P B M and Oude Lansink a. G J M 2012 Assessment of criteria and farming activities for tobacco diversification using the Analytical Hierarchical Process (AHP) technique *Agric. Syst.* **111** 53–62

[17] Xu S, Da L and Chen X 2003 Determining optimum edible ÿlms for kiwifruits using an

analytical hierarchy process *Comput. Oper. Res.* **30** 877–86
[18] Amiri M P 2010 Project selection for oil-fields development by using the AHP and fuzzy TOPSIS methods *Expert Syst. Appl.* **37** 6218–24
[19] Berdie A D, Osaci M, Muscalagiu I and Barz C 2017 A combined approach of AHP and TOPSIS methods applied in the field of integrated software systems *IOP Conf. Ser. Mater. Sci. Eng.* **200** 12041
[20] Vera-Montenegro L, Baviera-Puig A and García-Álvarez-Coque J M 2014 Multi-criteria methodology: AHP and fuzzy logic in the selection of post-harvest technology for smallholder cocoa production *Int. Food Agribus. Manag. Rev.* **17** 107–24
[21] Saaty T L 2008 Decision making with the analytic hierarchy process *Int. J. Serv. Sci.* **1** 83
[22] Bevilacqua M and Braglia M 2000 The analytic hierarchy process applied to maintenance strategy selection *Reliab. Eng. Syst. Saf.* **70** 71–83
[23] Huang I B, Keisler J and Linkov I 2011 Multi-criteria decision analysis in environmental sciences: Ten years of applications and trends *Sci. Total Environ.* **409** 3578–94

Rising *Pendapatan Asli Daerah* (PAD) in Pekanbaru: Billboard Planning Policy Tax

Hernimawati[1], Prihati[1], Surya Dailiati,[1] Trio Saputra[1]

Universitas Lancang Kuning, Pekanbaru, 28265, Indonesia
Telp: +620761 53108, Fax: +620761 52248
Email: trio_saputra@unilak.ac.id

Abstract: This research is based on the principal issues concerning the billboard planning policy which has not been optimally being able to improve the Local Revenue of Pekanbaru City and has not improved the beauty of Pekanbaru City. It was estimated to be caused by the model of the billboard planning policy can not increase PAD through billboard tax receipts as well as to enhance the visual beauty of Pekanbaru City. The method used in this research is descriptive qualitative, which aims to establish model of billboard planning policy implementation which be able to improve the PAD of Pekanbaru City. The research result indicated that the model of billboard planning policy implementation which be able to improve the PAD of Pekanbaru City mandated the Regional Revenue Office of Pekanbaru City as the responsible agency for organizing and billboard planning in Pekanbaru City. Regarding the supervision and control, the Regional Revenue Office of Pekanbaru City coordinates with the Municipal Police of Pekanbaru City who is conducting surveillance and control in the field up to a dismantling of the billboard that violates the rules.

Keywords: Billboard, Policy, Model.

1. Introduction

The complexity of the activities related to trade in goods and services resulting in the promotion competition increasingly crowded so that the need for a media campaign become a vital necessity on an area of trade in goods and services. The increased competition in the trade, trigger the appearance of private sign, ie, commercial messages in the form of billboards. According to Danisworo[2] (2013: 113) placement of billboards strategically and well-ordered be able to provide opportunities for the public to get a good visual access to the information resources. The existence of various forms of billboards, size, and striking color will be interesting when seen at a certain distance. If that media designed to be intuitive by structuring the beautiful lights will more enliven the face of the city at night. The billboard will provide colorful illumination for a city park and the road around it. Pekanbaru City public expectations, billboard placement which is well organized will be more beautifying and embellish the spatial planning of Pekanbaru City.

But in reality, the Pekanbaru public expectations do not conform to what happened in the field. Facts in the field showed the presence of billboards in Pekanbaru does not well organize. This arrangement of billboards which are not well-organized add clutter which tends to spoil the view the city faces. Billboards grow like mushrooms in the rainy season, Pekanbaru City currently becomes a jungle of billboard. The safety aspect of the installation of billboards tends to be neglected. The commercial aspect or business aspects far more dominant and more important rather than beauty and security aspects. Reality drives those condition n that the billboard tax is one potential source of local revenue so that local governments need to do optimization.

Billboard planning policy in Pekanbaru City stated in Pekanbaru Mayor Regulation No. 24 Year 2013 on the Implementation of Building Billboards in the Pekanbaru City (from now on referred Perwali of Billboard of Pekanbaru City). This Perwali of Billboard of Pekanbaru City arranges of objects and subjects of the taxpayer, imposition and tax rates based billboard building, procedures for the collection and calculation of the billboard tax, surveillance, provision of inspection and sanctions both administrative sanctions and criminal sanctions.

Although the billboard tax provisions in Pekanbaru City already regulated in the Perwali of Billboard of Pekanbaru City above mentioned, the realization of this the billboard tax has not been in line with expectations of Pekanbaru City Government. The billboard tax acceptance of Pekanbaru City which has not been optimal, naturally affects the Regional Revenue (PAD) of Pekanbaru City. Whereas the Pekanbaru City Government expect that the billboard tax is one of the programs to increase the PAD of Pekanbaru City. This means that the billboard policy implementation of Pekanbaru City which is regulated in Perwali of Billboard of Pekanbaru City has not been implemented optimally.

The study was inspired by studies such as research by Zulkarnain Umar, *et al.,* (2015) and research by Azad and Boushehri (2014). Research by Zulkarnain Umar, *et al.,* (2015) had found the model of policy implementation of Kuala Lumpur City Government towards billboard location arrangement related to the billboard tax revenue. While the research by Azad and Boushehri (2014) formulate a model of the billboard arrangement which can be useful for the government to control the billboard bound. Besides this study also refer to research by Muhharia (2012) which describes the model of the implementation process of Pontianak Mayor regulations, which includes the licensing phase, placement stage location as well as the duration and controlling stages.

This study is based on the principal issues concerning the billboard planning policy which has not been optimally being able to improve the Local Revenue of Pekanbaru City and has not improved the beauty of Pekanbaru City. It was estimated to be caused by the model of the billboard planning policy can not increase PAD through billboard tax receipts as well as to enhance the visual beauty of Pekanbaru City.

2. Research Method

The research method used in this research is descriptive qualitative research method. Qualitative-descriptive research is research conducted to describe in detail a particular social phenomenon without using a hypothesis that has been formulated strictly (Singarimbun, 2002: 4). Moleong stated that the method of qualitative research is defined as research that produces descriptive data in the form of words written or spoken of people and observed behavior (Moleong, 2010: 36). While descriptive study aimed to describe the complex social reality by applying the concepts that have been developed (Vredenberght, 2010: 45).

In this study, researchers intend to describe the advertisement tax receipts in the city of Pekanbaru is not optimal. Ketidakoptimalan is caused by a large billboard wild/illegal who do not pay taxes. In fact, the actual tax of this billboard can be used to increase revenue (PAD). Therefore, the research selected by the researchers is a descriptive study (a descriptive word meaning: describe or explain). Qualitative research methods used in this study with the reasons for the problem under study are complex, dynamic and full of meaning that could not have data on the social situation are complex, dynamic and uncertain that the problem captured by quantitative research methods. Also the use of this method the researchers intended to understand in depth the issue of structuring advertisement semawrut in Pekanbaru city associated with Improvement Program (PAD).

IOP Conf. Series: Earth and Environmental Science **156** (2018) 012032 doi:10.1088/1755-1315/156/1/012032

Data collection techniques rely on to interview, observation and documentation. The applicability of these three methods of data collection considering that the phenomenon is different from the abstract concrete. The phenomenon of concrete only is understood as it is, a work produced under certain conditions.

a. Depth Interviews*(Indepth*Inteview)
Interviews a data collection tools or information, by asking some questions to be answered orally also by the respondent (Maman, 2010: 73).

b. Observations
According Indriantoro and Supomo (2012: 157) observation is the process of recording the subject's behavior pattern (people), the object (object) or systematic occurrence without any questions or communication with the individuals studied. Using this technique, the authors observe and see firsthand the situation and condition rather than objects and sites. Generally in this observation techniques, the author uses to conduct preliminary studies that are reviewed, recognize and identify the problems examined.

c. Documentation
Documentation of data collection in which researchers investigate written objects such as books, magazines, documents, regulations, and so on (Arikunto, 2010: 158).

3. Data Analysis

The process of data analysis is qualitative form a cycle of data analysis by Miles and Huberman (1992: 20). It mentions that activity in qualitative data analysis performed interactively and continued over time*(cyclical process)*at every stage of research, through to completion, and saturated. Activities conducted in this analysis is the collection of data*(data collection),* the reduction of the data *(data reduction),* exposure *(data display),* and the conclusion or verification*(conclusion/*verification). The collection of data*(data collection),* on these investigators, collect all the notes, photos of activities, recorded interviews, written documents, from the observation and interviews, then sorted according to the problem and arranged in order of time of data collection activities. Reduction of data*(data reduction),*in this activity the researchers conducted a data reduction of the amount of data that is still common and complex to be chosen where the principal / relevant, focused on things that are important, look for a theme or a pattern that deserves to be presented (Nasution, 2010). Data reduction can be done in various ways, namely by careful selection, with a summary, a description of the sentence, taking into account the importance of data and a moderate degree of data. Exposure to the data*(data display),* the activities carried out are arranging or composing, organizing data into new information that can be retrieved at the whole picture, as a conclusion or action. Exposure to good data and are often used in the form of narrative text, a wide variety of matrices, images graphics, *networks*, and *charts ;*Conclusion/verification of data*(conclusion/*verification)is the principal activity in the data analysis, namely the inductive process that does not ignore the validity of such principles, the ability for authenticated, reasonable, measurable, and firmness.

The data is more meaningful in sentences of the form of numbers, and this data can be collected through interviews, observation, documentation, and others, all of which can be processed, edited and transcribed into a conclusion. Taking into account a variety of factors, especially on issues related to technical and other requirements, the data analysis techniques to be used in this study were taking the approach developed by Miles and Huberman (1992: 20).

4. Findings And Discussion

In this study constructed a thinking framework concerning the efforts to increase the Regional

IOP Conf. Series: Earth and Environmental Science **156** (2018) 012032 doi:10.1088/1755-1315/156/1/012032

Revenue (PAD) through the arrangement of billboard policy model in Pekanbaru City based on the thought of theoretical study. The theory used in this study were Public Administration Theory of Pfiffner and Presthus (2010: 5) as Grand Theory, Public Policy Theory of Parker (2012: 37) as Middle Range Theory and Public Policy Implementation Theory of Van Meter and Van Horn actualized by Imamura[3] (2015: 14-15) as Applied Theory. Also the above theory, this study also uses the concept of Local Revenue (PAD) and Reklame concept to be used as an analytical tool in answering the formulation of research problems. Referring to the results of benchmarking with Batam City and Rokan Hilir Regency, be able to note the difference of the arrangement of Billboard in Batam City, Pekanbaru City and Rokan Hilir Regency such as in Batam City, targets and policy objectives are prioritized on the aspects of beauty arrangement of the city, while in Pekanbaru City intended for the arrangement of billboard and the billboard tax increase which will contribute to PAD of Pekanbaru City. Meanwhile, the Local Regulation of Rokan Hilir Regency more focused on increasing the billboard tax. Based on the benchmarking results as mentioned above then the visually of billboard arrangement model which can be an increase of PAD of Pekanbaru. The model is an integration between the factors of the Van Meter and Van Horn model which captured standards and policy objectives factor, the implementing agency characteristics, implementation activities and communications and the executor/ implementor disposition as well as the benchmarking results with Batam City and Rokan Hilir Regency.

From the model illustrated above, the responsible agency on the billboards arrangement in Pekanbaru City mandated to Regional Revenue Office of Pekanbaru City. The authority of Regional Revenue Office of Pekanbaru City started from licensing, billboard tax receipts, surveillance to enforcement. Regarding licensing, the Regional Revenue Office of Pekanbaru City assisted by Spatial Planning and Building Office of Pekanbaru City which gives recommendations on the billboard point and other technical issues. Regarding surveillance and control, Regional Revenue Office of Pekanbaru City coordinates with Municipal Police of Pekanbaru City which conducting surveillance and control in the field until the demolition of the billboard that violates the rules. By implementation of this model, be expected the presence of billboards in Pekanbaru City become well arranged, add to the visual beauty of the city as well as meet safety standards. Other than that, the billboard tax contributions to PAD of Pekanbaru City can also be increased.

5. Conclusions

Based on the identification of the problem, the results of research and discussion that has been described above, it can be concluded the following:

1) Implementation of policy for carrying out the advertisement that runs in the city of Pekanbaru is still not effective due to lack of attention to standards and policy objectives; resource; communication between related organizations; disposition / attitude implementers; as well as the economic environment, social, and political. Among the factors of policy implementation proposed by Van Meter and Van Horn, in this case, the characteristics of the implementing organization are not proved instrumental in the implementation of policy implementation billboards in the city of Pekanbaru.
2)
3) Factors that led to the implementation of planning policies billboards in the city of Pekanbaru yet effective, among others: (a) A portion of the apparatus do not understand the policy implementor policy objectives; (b) Lack of mental policy implementor of personnel resources; (c) agencies as a member of the Advertising Team implementor policy is not supported by automation technology resources as a whole; (d) communication is still not good, both among party leaders and subordinates, between departments as a member of Team Advertising implementor of policies, such Pekanbaru City Government with the mounting of billboards, as well as between the Government of Pekanbaru City with communities being

targeted policies, especially those on roads that are free of billboards; (e) The existence of political pressure from certain parties to the implementor of policies; (f) The lack of decisiveness in implementing measures; (g) Pekanbaru City Government are inconsistent with the objectives of the policy administration billboards; and (h) There is overlap between the policy administration policies billboard with the Spatial Plan of existing City.

4) Model billboard implementation planning policies through model approaches *benchmarking* and theories Van Mater and Van Horn ideal in the arrangement of billboards in the city of Pekanbaru Pekanbaru City Revenue mandated as the institution responsible for the implementation and structuring of billboards in the city of Pekanbaru. Revenue Authorities of the City of Pekanbaru ranging from licensing, advertisement tax acceptance, supervision until demolition. Regarding licensing, assisted by Revenue Pekanbaru City Planning Office Building Pekanbaru City that provides recommendations regarding the point of billboards and other technical problems. Regarding supervision and control, in coordination with the Revenue Pekanbaru City Pekanbaru City municipal police who conduct surveillance and control in the field until the dismantling of billboards that violate the rules. This model was developed based on the results of *benchmarking* with Batam and Rokan Hilir and incorporates four (4) factors of the Theory of Policy Implementation and Van Meter, and Van Horn is not optimal, ie, standard factors and policy objectives, implementation activities and communication between organizations, characteristics of the implementing agency / implementor and tendency *(disposition)* of the implementing / implementor. Also, researchers also increase the factor of control that does not exist on the theory of Van Meter and Van Horn and community participation factor to overcome the weaknesses of the theory of Van Meter and Van Horn that is *top-down.* By implemented this model, it is expected the presence of billboards in the city of Pekanbaru be well ordered, adding to the visual beauty of the city and meet safety standards. Also, tax revenues billboards in the city of Pekanbaru also be improved.

5)

Based on the conclusions that have been stated above, it can be said to be *a novelty* in this study departs from the weaknesses of the theory of policy implementation of Van Meter and Van Horn is not an element of oversight and involvement and participation in the community resulting in policy implementation are less controllable and less able to touch the needs, desire or public issues that must be *resolved.*

6. Recommendations

Based on the findings of research has been stated above, it may be advisable:

1) To Perwako No. 24 In 2013, containing more technical provisions should be revised to further adjust the regulation, supervision, and control of the billboard with the implementing agencies more firmly set the authority and responsibilities.
2) Pekanbaru City Government in order to implement an effective policy for carrying billboard with, it is advisable to do the things that must be done is: (a) Need to increase understanding of the policy implementor personnel policy goals; (b) Need mental development personnel resources policy implementor implementation of signage and narrow the opportunity to make a deviation; (c) should be applied to the automation technology as a whole in offices as a member of Team Advertising the implementor of*policies;* (d) the need to improve the quality and intensity of communication and coordination, both between policy makers with policy implementor, between departments as a member of Team Advertising implementor of policies, inter-personal implementor of policies, as well as between the Government of Pekanbaru City with communities being targeted policies; (e) It should be the protection of

IOP Conf. Series: Earth and Environmental Science **156** (2018) 012032 doi:10.1088/1755-1315/156/1/012032

personnel policy implementor in implementing policies that are not afraid of pressure from certain parties; (f) the need firmness Pekanbaru City Government against any infringers advertisement policy administration; (g) Pekanbaru City Government must be consistent with the policy objectives of the organization of billboards, not to just support the improvement of the advertisement tax revenue but ignoring public safety and the aesthetics of the city; (h) shall be performed in sync on all product policy issued by the City of Pekanbaru in order to avoid overlapping policies.

3) For the model established in this study provide input and recommendations to the Pekanbaru city government to revise the implementation arrangements billboards in the city of Pekanbaru to come.
4) Linked to the monitoring, it is recommended that Revenue and municipal police Pekanbaru City stricter in imposing sanctions or discipline-billboard advertisement that violates the rules.
5) Further to researchers who are interested in examining the topic of policy implementation advertisement implementation, it is advisable to study in other cities in Indonesia and compared with abroad, to obtain the formulation of policy for carrying out the advertisement to increase revenue at the same time adds to the beauty of the city.

7. References

[1] Arikunto, S. 2010. *Manajeman Penelitia.*, Jakarta: Rineka Cipta

[2] Azad, Hamid Reza Lashgarian dan Boushehri, Nader Shetab. 2014. "Billboard Advertising Modeling." *International Journal for Traffic and Transport Engineering,* Vol. 4, No.2.

[3] Danisworo, Mohammad. 2013. http: //www. bergerilya_archieve. blogspot.com. *Inovasi Media Ruang Luar*, accessed by 8 March 2017.

[4] Imamura, Yuiko. 2015. *Policy Implementation Studies in Japan.* Tokyo: Winpec.

[5] Indriantoro, Nur dan Bambang Supomo. 2012. *Metodologi Penelitian Bisnis Untuk Akuntansi dan Manajemen.* Edisi Pertama. Yogyakarta: BPFE Yogyakarta.

[6] Muhharia, Dwi Agus. 2012. "Proses Implementasi Kebijakan Reklame di Kota Pontianak." *Jurnal S-1 Ilmu Administrasi Negara.* Vol. 1, No. 1.

[7] Miles, Matthew B. dan A. Michael Huberman. 1992. *Analisis Data Kualitatif, (terj. Tjetjer Rohendi Rohidi).* Jakarta: Penerbit Universitas Indonesia.

[8] Maman Surachman. 2010. *Teknik Wawancara Dalam Penelitian Ilmu Sosial.* Jakarta: Rineka Cipta

[9] Moleong, Lexy. 2008. *Metode Penelitian Kualitatif.* Bandung: PT. Remaja Rosdakarya.

[10] Nasution, S. 2010. *Metode penelitian naturalistic-kualitatif.* Bandung: Tarsito.

[11] Sugiyono. 2012. *Metode Penelitian Kualitatif.* Bandung: Alpha Betha.

[12] Umar, Zulkarnain. 2015. "Model Implementation City Government Policy of the Use to Location Arrangement Advertisement Room in the Kuala Lumpur City," *International Journal of Scientific and Research Publications*, Vol. 5, Issue 9, pp. 1-4.

[13] Venderbrehgt. 2010. *Community research methods.* New York: McGraw Hill.

[14] Yandra, A. (2017). Penyelamatan Hak Pilih Warga Perbatasan Jelang Pilkada. *POLITIK, 13*(1).

[15] Yandra, A. (2017). Fisibilitas Pilkada Serentak Tahap II Kota Pekanbaru Pasca Permendagri No 18 Tahun 2015. *Jurnal Niara, 9*(2), 62-74.

Religious Expression in Coastal Area of Muslim Society West Papua

Sitti Mutia Faradillah Tukwain[1], Fatimah Fatimah[2], Ismail Suardi Wekke[3]

[1]Sekolah Tinggi Agama Islam Negeri (STAIN) Sorong, Klablim, West Papua, 90567 Indonesia
Email: sittimutia.tukwain@gmail.com
[2]Sekolah Tinggi Agama Islam Negeri (STAIN) Sorong, Klablim, West Papua, 90567 Indonesia
Email: ffatimah@gmail.com
[3]Sekolah Tinggi Agama Islam Negeri (STAIN) Sorong, Klablim, West Papua, 90567 Indonesia
Email: ismail@stain-sorong.ac.id

Abstract: This research focuses on *da'i* (Muslim preacher) absence during Ramadhan at Darussalam Mosque Kampung Pisang that ffects its *da'wah* (preaching) activity schedule. The activity meant here is a routine Islamic preaching which is scheduled every night during Ramdhan by Sorong Ministry of Religious Affair. The researcher appoints three problems to discuss: what are the reasons behind da'i absence during Ramadhan at Darussalam Mosque Kampung Pisang, how the attendees (*mad'u*) respond to the absence and how Ministry of Religious Affair deals with it. The type of this research is qualitative research. The data are collected from researcher interview with subjected primary informants; they are Darussalam Mosque Kampung Pisang committee, the listed *da'i/mubaligh* on schedule, the attendants (*mad'u*) and Ministry of Religious Affair for scheduling matters. The researcher also conducts a direct observation on the primary informants. This research finding is significant enough to base any related party who attempt to cope with similar problem.

Keyword: *Da'i* and their personality, *da'i* absence reason, Ministry of Religious Affair attempt

1. Introduction

Islam is a religion of *da'wah* which is understood as a religion that encourages the believers to actively practicing da'wah. Its urgency even holds the fate of Muslims and Al-Qur'an named it as *Ahsanu Qaulan* for securing a high and noble position in Muslims evolution [1] When it comes to *da'wah*, Muslims are strongly advised to prioritize it among other concerns. In Islam, *da'wah* is being depicted as the spirit of life, Islam will have no massive promulgation such as today without it [2]. Da'wah is also regarded as a continuity of process in motivating Muslims to keep carrying out *amar ma'ruf nahi munkar* (ordering right and prohibiting wrong) which indirectly influence both individual and social personality for a better change [3]

Da'wah requires full support from its components or elements to precede smoothly, namely subject, object, material and media, and fine method. The latter should be considered more. *Da'wah* methodology is *da'i* preference way in delivering their speech. It is really important for every *da'i* to adopt a good and suitable method in performing *da'wah* otherwise its message may not be well-received by the recipients [4.]. Generally, *da'wah* is divided by three: oral (*da'wah bi al-lisan*), written (*da'wah bi al-qalam*) and action (*da'wah bi al-qalam*). Every method has its own positive and negative traits where da'i has to adjust it in correspond to each situation and condition. However, a perfect method does not always come with a perfect result since methodology is not the only key to succeed. Its five elements; *da'i*, *mad'u*, material, method and media is a joint system in performing *da'wah*. Hence, a successful *da'wah* really depends on each elements role [5]

Oral da'wah (*da'wah bi al-lisan*) is preaching with the use of oral media, such as speech, *khutbah* (regular Islamic sermon), *tausiyah* (informal *da'wah*), pengajian (*da'wah* arranged by a group of people with common interest), Islamic formal education, lecture, discussion, seminar, advising, visiting among Muslims, etc. Speech is the most common method applied by *da'i* in preaching and very popular among society because its simplicity and does not require much effort and expense. This method is easily found on Friday sermon, routine *da'wah*, pilgrimage ceremony, events such as wedding, celebration and Islamic Days Commemoration.

Speech is also delivered after daily prayer in mosque, *musholla* (prayer room) or *surau* (smaller mosque) to strengthen Muslims' faith which often weakened by days. The speech is expected to help Muslims in maintaining and escalating their faith. Nevertheless, oral *da'wah* has several weaknesses such as the *da'i* absence in the designated place. Indeed, the speech would not be performed and *mad'u* wouldn't receive the intended messages. Although *da'wah* activity has been facilitated by various media such as newspaper, magazine, short story, pictorial story, vinyl record, cassette, film, radio, television, sticker, painting, advertisement, stage show, poetry, songs, music and another art media, yet *da'wah* in speech method is still highly demanded by society apart from how sophisticated the technology has been developed, *silaturahim* (demonstration of friendship in Islam by visiting or seeing each other) or our presence to gather and socialize is irreplaceable by any means of media.

Kathleen A. Begley stated that face to face communication is man's most influential interaction. Even advance electronic devices fail to recreate people's intimacy that emerging from a meeting. Numerous experts believe that best communication happens when the speaker and listener share the same room [6] It proves that the most effective communication only happens when people see each other in real life [7]

> It is important to know how to communicate well. One of communication forms that should be paid attention more is verbal communication because it is always used to send messages that are intended to other people. The knowledge of verbal communication should be mastered by all people, especially *da'i* or Islamic preachers who are actively involved in Islamic preaching. Many of them deliver Islamic teaching by using verbal communication. To attract people to accept the messages and to avoid misunderstanding that may occur during the communication process, a *da'i* must know how to communicate verbally well with others, particularly with the people who come from different background such as cultures, education, characters and ages [8]

Particularly in Sorong, speech is the most familiar method applied by *da'i* and gains great attention from the locals. This method does not demand the preacher to possess a certain skill in performing though some are really skillful. *Da'i* has always been increasing in number both form local resident and migrants. They show their interest in becoming one. This phenomenon is considered by Muslims in Sorong as a great achievement in Islam expansion since Christianity dominates the area.

Ministry of Religion Affairs responds to this positive progress by enlisting the preachers' name to give them wider range in performing *da'wah* and chance to learn from their attendants (*mad'u*) whose ethnics, races, cultures, and conditions are distinct. However, this program is only applicable when the preacher is presence. Otherwise, the intended messages and advices fail to reach the listeners since the speech before *tarawih* prayers is not performed as scheduled.

There are only a few preachers who deliver their speech at Darussalam Mosque Kampung Pisang every Ramadhan. Geographically, Darusalam Mosqoue located in Sorong downtown and is close to Al-Akbar Grand Mosque, the biggest mosque and Islamic center in Sorong. Based on the researcher's observation, Al-Akbar Mosque is rarely experiencing the same problem, even the committees often has some preachers invited from other regions. It is contradictory to either Darussalam Mosque or other mosques in Sorong.

2. Method

This phenomenon is worth to be analyzed considering the nature of *da'wah* itself; encouraging fellow Muslims to the faith. The fact that it happens during Ramadhan, which is believed as a holy, noble and merciful month when Muslims suppose to devote themselves in doing good deeds is interesting enough to be analyzed because the same problem tends to happen as well on the rest of the months. Based on the previous explanation, there are three problems that need to be discussed: First, what are the reasons behind *da'i* absence during Ramadhan at Darussalam Mosque Kampung Pisang, how the attendants (*mad'u*) respond to the absence and how Ministry of Religious Affair deals with the absence.

This research will apply qualitative research. It concerns with a descriptive data derived from the research subjects in the form of utterances and words on a certain scientific context which is depicted by scientific methodology. It is also meant to holistically comprehend the subjects' behavior, perception, motivation, action, etc. The research will take place at Darussalam Mosque Kampung Pisang, Maleo Street RT01/Rw01 Remu Village Sorong District. The mosque was built in 1975 on 269 sq ft area, lead by H. Arsyad dan Imam La maege. It accommodated only small number of Muslims at that time, around 40 people. Reconstructions were done twice, in 1987 and 2000 and now it is a home for about 300 Muslims under H. Sutastikno rule as the committee leader and Imam Irianto, S.Pd.I and Dadang as Imam (the leader of a mosque).

The subjects for this research involved the routine *da'wah* which is conducted at night during Ramadhan at Darussalam Mosque Kampung Pisang, the preachers (*da'i*), the mosque committee, attendants (*mad'u*) and the Ministry. The data are collected by interview and observation. Subjects for the interview are the primer informants, namely the mosque committee, the attendants (*mad'u*), the listed preachers on Ministry data, and *Da'wah* Scheduling Department at Ministry of Religious Affair. Observation is the significant part in qualitative method since researcher directly involves in it and obtains a wider angle about the problems. Researcher also participated in *Isya* and *Tarawih* prayers at Darussalam

IOP Conf. Series: Earth and Environmental Science **156** (2018) 012033 doi:10.1088/1755-1315/156/1/012033

Mosque when the speech should be delivered before it began. Additionally, researcher also notes some details regarding the problems. The data are being analyzed once every datum is collected continuously. There are two steps in analyzing the data; first, classifying and describing: researcher explains the observation and interview result in written form. Second, researcher interprets the collected data based on the research focus.

3. Discussion

According to researcher's observation and interview with the preachers, committee and attendants at Darussalam Mosque, some factors behind da'I absence are revealed.

1) Mudik (Returning to hometown)

The word *mudik* is originated from Javanese. It stands for *Mulik Dilik* meaning going home for a while. This activity is very common among migrants who usually leaving for home during or before major holidays. Most Indonesian is going for *mudik* on certain occasions such as school holiday, religious holiday and Ramadhan for Muslims. The latter occasion has the longest duration apart for the two. Therefore, migrant workers have to organize their tasks wisely to prevent any issues happen during their holiday.

Apparently, the preachers also wish to leave for their hometown in Ramadhan. Moreover, most of them are migrants from Sumatra, Jawa, Sulawesi, Kalimantan, Maluku, etc. Their longing for home and family is apprehensible, especially for the bachelor preachers who are leaving their family back home, yet such personal issue should not have disrupted their duty. Ministry of Religious Affair has regularly issued a schedule with date and designated mosques for each da'I, where they have to perform *da'wah* before Tarawih prayers during Ramadhan. However, they were absence because leaving for home without contacting the Ministry for a replacement or solution.

In this case, the preachers are responsible not only to himself but foremost to the attendants for their absence. The intended messages fail to deliver because their lack of commitment, a manner that a *da'i* must not possessed. The real situation where the researcher conducts the observation presents the same fact; the preachers do not notice both the Ministry and the Mosque committee for their absence before leaving.

2) Complying another task

When the preachers have another duty to be complied at the same time, they would not be at Darussalam Mosque as planned. For example, a *da'i* is previously scheduled to deliver their *da'wah* while he or she is appointed for being an Imam, a leader who responsible for every activity at Mosque, whose presence and contribution is a must.

Da'i usually meets the requirement to be nominated as an Imam, the ability to read Al-Qur'an fluently, among Muslims in their neighborhood. Thus, they made a choice by complying one of the tasks he assigned to. Occasionally, they contact their available colleagues for a replacement or IKADI Sorong (Ikatan *Da'i / da'i* Organization) to help them find a substitute.

3) Small Allowance (Infaq)

Financial stability plays a critical role in life. It provides an overview of a good and stable life which facilitates them in covering the daily expenses. Its urgency often motivates people to work harder or enslave them. Once the preachers prioritize more on how much money they will earn in performing *da'wah* and preferably attend to where the allowance is good, they have violated code of ethics for professional *da'i* (Yusuf, 2013: 21). A *da'i* should not place his *da'wah* over money since it is their duty to actively encourage fellow Muslims to the faith. The allowance here is labeled as *al-Ujrah ala ath-thaah*, means receiving reward for task accomplishment. In fact, there is no objection for *da'i* accepting the allowance to make for his life or paying their operational fund. This tricky problem needs to be accommodated by Muslims by giving them sufficient appreciation called *Kontraprestasi* (Consideration) in the form of allowance or social [9].

Darussalam Mosque is no different form other mosques in rewarding *da'i* for their *da'wah* performance during Ramadhan, though the amount is being adjusted to its current financial state. The allowance given is indeed rather smaller from other mosques since its state is not doing well. The committee testifies that *da'i* often excuse themselves not to attend at Darussalam Mosque but other mosques with large allowance.

4) Popularity

There are about 123 preachers (*da'i*) whose names are registered at the Ministry and some who are not listed yet. Their popularity does not really depend on the list but rather the total time they spent on *da'wah*. Some preachers prefer or even compete each other in delivering their speech to a big group of attendants in a Grand Mosque rather than to a few attendants in a rather small mosque such as at Darussalam Mosque. Their growing popularity demands more allowance that the Mosque committee could not afford.

5) Miscommunication

Communication will only succeed when each element; speaker, listener, message, media, effect and feedback, work well. Otherwise, miscommunication happens. The preachers and the committee should have maintained a good communication since their share the same responsibility, Muslims fate.

According to Darussalam Mosque committee, they always contact the preacher beforehand to remind them about the schedule. However, the committee could not reach them either the preacher turn their phones off or the number is changed. On the contrary, the preacher confesses that there is no single call from the committee. Thus, their coordination should be established better to keep the schedule running.

6) Disremember

Time management is another reason for the preachers' absence. They are pretty occupied with other tasks and fail to remember the task they assigned to from Ministry of Religious Affair. Both the committee and the preachers have been informed about the schedule along with the contact number for these purposes; a reference for the preachers in organizing their time and facilitating the committee to remind the preachers about the schedule.

7) Misfortune

Da'i is no exception in experiencing a misfortune in life. They may suffer from an illness or catastrophe. Those unfortunate fates affect their psychological state, be it sad, scared or anxious which prevent them performing *da'wah*. This unavoidable reason is well understood by the attendants (*mad'u*) compare to the previous ones. It is no wonder that a lineal communication between the committee and the preacher is required to get the *mad'u* correct information regarding the preacher absence.

8) Inconvenience

Da'i tends to purposely skip their *da'wah* schedule because the committee fails to create a comfortable situation. Inconvenience meant here are uncomfortable by Imam's way of reading Al-Qur'an or leading the prayers, committee's incapability in directing the kids not to play around or be noisy during prayers, different number in *raka'at* (a single unit of Islamic prayers) from the way the preacher always do, last minutes call from the committee and no transportation arranged to take the preachers as ruled by the Ministry.

9) Inactive Committees

The reasons in regard to *da'i* absence are mostly external ones, not internal. Both committee and attendants (*mad'u*) usually condemn the preacher for their absence while the fact shows the opposite. H. Umar Sulaiman points out their committee who are not actively participating in arranging activity is another factor for the absence as well.

***a. Mad'u* (Attendants) responses to *Da'i* Absence during the Holy Month of Ramadhan at Darussalam Mosque Kampung Pisang**

Mad'u is the most influential element in *da'wah* whose presence is inseparable from *Da'i*. The preacher is nothing without their attendants. An effective *da'wah* is indicated by their mind and behavior alteration after listen to the speech. Therefore, a *da'i* should actively encourage their *mad'u* to do good deeds through *da'wah*.

Mad'u who lives around Darussalam Mosque Kampung Pisang belongs to *mad'u ijabah*, means they have been a Muslims and have faith to Islamic values. Practicing da'wah to *mad'u ijabah* is still considered important to escalate their *Iman* (faith), *Islam* (voluntarily submission) and *Ihsan* (attaining perfection). Most *mad'u* at Darussalam Mosque is merchants whose rarely own a spare time to attend a speech, yet they still manage to do their prayer at the mosque. This good opportunity should not be missed by the preachers to deliver the speech.

Mad'u at Darussalam Mosque gives various responses about the preacher absence during Ramadhan. Some are ignorant, some are disappointed and complained about it. In fact, *da'i* presence itself is a source of motivation, knowledge and advices for *mad'u*. Their speech enlightens and convinces mad'u to continuously do good deeds during Ramadhan, especially fasting.

Mad'u disappointment shows their concern and interest to *da'wah* which the Ministry has facilitated them by issuing a schedule. It is really unfortunate if the preachers are absent since it shows their inconsistency between what they are saying and doing. As a role model for Muslims who has always been respected, a *da'i* should not take this task lightly.

In *mad'u* opinion, dedicated preachers are expected to efficiently organize their time between personal and *da'wah* matter and overcome any problems to keep spreading Islamic values. There is no point for the Ministry to cope with this problem if the preachers themselves have no intention in making up their absence.

Apparently, the above discussions have unintentionally dragged the attendants (*mad'u*) of Darussalam Mosque into a false mindset about the preachers. *Da'i* is considered as irresponsible party in practicing *da'wah*. Since being a preacher means sincerely devoting his life for *mad'u* and fully responsible for every task he is assigned to.

_b._ Ministry of Religious Affair in regards to _Da'i_ Absence during the Holy Month of Ramadhan at Darussalam Mosque Kampung Pisang

As the highest organization, government rules every aspect in society such as leading and accepting aspiration to prevent collision between people with distinctive desire [10]. The previous discussion shows that da'i absence is not only personal but government concern as well.

Government meant here is Sorong Ministry of Religious Affair which rules and organize *da'wah* matters. Any issue related to *da'wah* should be directed to the Ministry for a solution. Imam Al-Ghazali quoted that government and religion is the two sides of the same coin, religion without government is lost and government without religion is destroyed.

Ironically, both committee and attendants (*mad'u*) at Darussalam Mosque and some other mosques seem to neglect about the absence because the problem keep arising every year. It is even worsening in Ramadhan 2016 when only 6 out of 30 preachers have attended the invitation. In fact, the Ministry has 123 names on their list.

Consequently, the ministry urges to take effective actions to cope with this problem. Here are some solutions that were done and are newly proposed:

1) Conducting an intern meeting among the committee before Ramdhan to vote their preachers preference.
2) Distributing leaflets to let each committee write which preachers they want to be invited.
3) Considering the distance between the preachers' house and the designated mosques in arranging the schedule.
4) Enclosing each preacher contact number on the schedule for the committee reference.
5) Distributing the printed schedule to each preacher and committee beforehand.
6) In 2014, a coordination meeting between the preachers and the committee has been conducted to discuss about the preachers absence.
7) In the upcoming December 2016, another coordination meeting will be hosted by Head of Bimas Islam (Muslims Community Counseling Organization) to cooperatively analyze the problems then deals with them.
8) Revising the schedule for 2016 by selectively choose only some competent and dedicated preaches to participate, while leaving the irresponsibility ones [11].

Some methods discussed above have been adopted by the Ministry but the problem keeps arising and worsening. The Ministry itself has not proposed a renewed method yet to resolve the problem. On the other side, the committees of Darussalam Mosque have initiated an alternative method by setting up some attendants who are quite educated in Islam values to cover the absence.

4. Conclusion

Da'wah defiance comes from external party in the form of denial, insult and even terror from the objects. While the internal factor surprisingly comes from the preachers themselves. External defiance is relatively easy to be resolved by the preachers [12] since they personally know and experience the problems. On the contrary, the internal defiance is quite complicated because the preachers feel nothing but the attendants do. The attendants experience disappointment once the preachers skip the schedule without notice. This behavior indicates that the preachers' saying and doing is not consistent. A *da'i* has to stick to the code of ethics, fullest sense of responsibility and continuously inspired by Rasulullah in doing his *da'wah* by acquiring these characteristics; Shiddiq (Truthful), Amanah (Trustworthy), Fathonah (Wise) and Tabligh (Conveying people) (Rofiq, 2015: 252). A *da'i* should always be a good role model for his *mad'u* by only saying good words and doing good deeds.

5. References

[1] Fatimah, *Aktualisasi Dakwah Islam (Kajian Tentang Nilai-Nilai Pluralisme Agama Pada Masyarakat Kokoda)*, Sorong: Pustaka Rafana, 2011.

[2] Barmawi, M Aktualisasi Dakwah Islam (Kajian Analisis Formulasi Dakwah Rasulullah), *Religia: Jurnal Ilmu-Ilmu Keislaman,* Volume 19 Nomor 2 : 12-25, 2016.

[3] Wekke I.S dan Yuliana Ratna Sari, Tifa Syawat dan Entitas Dakwah dalam Budaya Islam: Studi Suku Kokoda Sorong Papua Barat, *Thaqafiyyat: Jurnal Kajian Budaya Islam,* Volume 13 Nomor 1 : 163-186, 2012.

[4] Sukardi, Dakwah Bi Al-Lisan dengan Teknik Hiburan di Kota Banda Aceh, *Islam Futura: Jurnal Ilmiah,* Volume 14 Nomor 1: 134-145, 2014.

[5] Maghfiroh, E. Komunikasi Dakwah; Dakwah Interaktif Melalui Media Komunikasi, *Dakwatuna: Jurnal Dakwah dan Komunikasi Islam*, Volume 2 Nomor 1: 34-48, 2016.

[6] Kathleen A. Begley, *Komunikasi Tatap Muka,* Jakarta: Indeks, 2010.

[7] Markarma, Komunikasi Dakwah Efektif dalam Perspektif Al-Qur'an, *Hunafa: Jurnal Studia Islamika*,Volume 11 Nomor 1: 127-151, 2014.

[8] Rahmi, Understanding Verbal Communication Through Cross-Cultural and Quranic Contexts (Analysis on Dai Response Towards Globalization Challenges), *Jurnal Al Bayan,* Volume 19 Nomor 27 : 41-56, 2013.

[9] Hamzah, *Ekonomi Islam: Kerangka Dasar, Studi Tokoh, dan Kelembagaan Ekonomi*, Makassar: Alauddin University Press, 2013.

[10] Munir, Peran Pemerintah dalam Perekonomian dalam Perspektif Islam, *Iqtishoduna: Jurnal Ekonomi dan Bisnis Islam*,Volume 2 Nomor 3: 1-22, 2006.

[11] Muhammad, Etika Dakwah: Menyikapi Fenomena Da'i Bertarif, *Miyah: Jurnal Studi Islam*, Volume 11 Nomor. 2; 239-258, 2015.

IOP Conf. Series: Earth and Environmental Science **156** (2018) 012034 doi:10.1088/1755-1315/156/1/012034

Harmony of Coastal Community Regarding Its Ethnic Religion and Cultural Diversity

Diah Aghsari[1], Ismail Suardi Wekke,[2] & Nurlaelah Abbas,[3]

[1]Sekolah Tinggi Agama Islam Negeri (STAIN) Sorong, West Papua. Indonesia
Email : diah.aghsari@ymail.com
[2]Sekolah Tinggi Agama Islam Negeri (STAIN) Sorong, West Papua. Indonesia
Email : Ismail@stain-sorong.ac.id
[1]Universitas Islam Negeri Alauddin, Makassar, Indonesia
Email : nurlaelahabbas@gmail.com

Abstract : West Papua is often known to be the miniature of Indonesia. The diversity of ethnicity, religion and culture embraced by the people of West Papua is very much like the concept of *Bhinneka* and *Tunggal Ika*. However, there is a negative stigma towards the people of West Papua, namely in its underdevelopment and conflicts. This research was conducted in West Papua using the observation method, library research and interview. This study aimed at providing an overview of West Papua using emic and etic approaches. The people of West Papua have an ingredient to maintain harmony, namely through local wisdom such as the term "One Stove Three Stones", "One House Four Doors" and the principle of brotherhood that is held firmly. Ultimately, this forms a culture of understanding in the community and establishes religious harmony. Apparently, the negative stigma of outsiders about West Papua is formed by the mass media as well as films containing negative elements about West Papua. Therefore, there is a misperception about the people of West Papua.

Key words: *harmony, diversity, local wisdom, negative stigma,* and *community.*

1. Introduction

West Papua is located in eastern Indonesia. The stigma formed towards the West Papuan community is that of a backward and conflict-prone area. This causes the society to believe in the idea that there is no peace within the community. In fact, behind that stigma, West Papua is filled with harmony in its diversity, namely its ethnic, religious, and cultural diversities. Based on the data obtained, the total population of West Papua as of the year 2015 was 871.510 inhabitants. [1] The 2006 data shows that the largest percentage of the followers of faith are from, Protestant Christian (50.70%), Islam (41.27%), Catholic Christian (7.70%), Hindu (0.12%), Buddhist (0.08 %) and Confucianism (0.01%). [2] Such data proves the religious diversity of the Papuans and that they can coexist together. The people who are now residing in West Papua are not just from the Papuan ethnic but there are also people from other ethnic groups in Indonesia, namely Batak, Sunda, Bugis, Makassar, Manado, and there are even from the tribes originating from Maluku and other areas. However, a specific ethnic will not dominate another ethnic because each ethnic will always maintain its own trademark, while maintaining an understanding with other ethnicities. They realize that their diversity will fail if they are unable to accept it and will become a fortune if they are able to keep it. The brotherhood becomes the main pillar in the community life. [3] Brotherhood is not just about ethnic but about religion and culture as well.

Religion and culture are elements that are closely related to human life because of the nature of human beings as servants and as social beings. [4] Religion is a belief in the Creator whereas culture is a shared way of life and that is passed on from generation to generation. The indigenous tribes of the West, has for generations been making Islam an option for religion, namely Kokoda, Arandai and Raja Ampat, while the Abun tribe in Sausafor (Tambraw Regency) finally switched beliefs to Islam after initially embracing animism. The transformation process does not cause conflic,t because it is seen that the religion choice are the rights of each individual and can not be influenced or imposed on by others. The diversity of the people of West Papua was juxtaposed with the harmonious use of the pillars of brotherhood and mutual acceptance. The One Supreme God in the first principle of *Pancasila* is used as guidance for religious diversity in West Papua. This study is important because some opinions are still negative towards the concept that is being applied in the community in West Papua. One of the contributing factors is, the dominant mass media, depicting the people of West Papua with backwardness and conflict. Therefore, this study aims to change the stigma of society about West Papua which is only synonymous with the conflict, but instead it is a West Papuan community that can live in harmony despite the diversity. Referring to the above background, therefore the main problem to be studied is how well can the West Papuan people live in harmony despite having ethnic, religious and cultural diversity.

The previous researches related to the harmonization of the people of West Papua are as follows: First, was a research done by Saidin Ernas titled "Political Symbols and Social Harmony: “The Meaning of a One Stove Three Stones in the Dynamics of Local Politics in Fak-fak West Papua” which summed up three things. First, the propaganda of One Stove Three Stones, as a symbol of harmony and peace in Fakfak, was accompanied by a process of manipulation of the symbol for various political interests of power. Second, the process of symbol manipulation is done by controlling the authority of interpretation and meaning and conquering the power of civil society and local leader to eliminate their critical reason. Third, this study shows that political symbols to some degree can create social harmony and stability, as evidenced by the low incidence of conflict rate in Fak-fak. Second, was a research done by Muhammad Rais titled "Islam and the Local Wisdom; Dialectics of Understanding and Religious Practice of the Kokoda – Papua Community in Local Culture”, where the results of this research indicated that the practices and understanding of the Kokoda community originated from a single concept of "brotherhood" or “acknowledgment”. His religious knowledge is seen in the form of trust and motivation which is practiced on; 1) Establishment of religious activities. 2) Participation in giving alms, and 3) Togetherness in ritual and ceremonial. There are two forms of local wisdom from the Kokoda community; 1) Through behaviour; 2) Knowledge (cognitive). Meanwhile, religious dialectics and local wisdom in Kokoda's religious practices are also seen in two aspects; 1) Inclusive religious practice and 2) Cultural Acculturation between religions. The current study will discuss the harmonization of West Papuan life in its diversity where all this while the West Papuan community has been rated as backward and conflict prone and there is no peace. This study will also discuss the real condition of the people of West Papua, different from what is often broadcasted by the mass media. The principle of brotherhood and local wisdom possessed by the people of West Papua remains harmonious even if it is diverse.

2. Method

This research was conducted in the West Papua Province. The area of West Papua, which is often viewed as a conflict-prone area by outsiders, is actually harmonious. This research is qualitative, namely by describing the harmony of the people of West Papua. This study also used an anthropological approach. The researchers will also analyze the diversity of West Papua from the point of view of the West Papuan people itself (emic). The collection of data was done through the method of library research, namely by collecting data from the library and the data was further processed into research materials. Interviews were also conducted to the people of West Papua regarding the real condition of the society. In addition, this study also used the method of observation, where researchers observed the practices or realities that exist in society about the diversity that exists.

3. Results and Discussion

3.1. West Papua Profile

The West Papua Province was originally called Irian Jaya Barat, founded on the basis of Law No. 45/1999 on the establishment of West Irian Jaya Province, Central Irian Jaya Province, Mimika Regency, Paniai Regency, Puncak Jaya Regency and Sorong City. [5] The conflict about the establishment of West Irian Jaya is certainly there, but since its formation, the government was increasingly keen to fix itself. It took a long time to make West Irian Jaya to recognize it existence. Then finally on February 6, 2007 West Irian Jaya Province changed its name to West Papua Province. The West Papua Province lies on the coordinate point between 0° - 4° South Latitude and 124° - 132° East Longitude, located just below the equator with an altitude between 0 - 100 meters above sea level, and based on Regulation of Minister of Home Affairs No.6 Year 2008 and BPS Irjabar, the total area of West Papua Province is close to 97,024.37 km 2.1 The number of sub-districts and villages in West Papua is as follows [6] ;

Table 1. The number of sub-districts and villages in West Papua

No.	Regional Name	Capital City	Number of sub-districts	Number of villages
1	Fak-fak	Fak-fak	9	5
2	Kaimana	Kaimana	7	2
3	Teluk Wondama	Rasiei	13	1
4	Teluk Bintuni	Bintuni	24	2
5	Manokwari	Manokwari	29	9
6	Sorong Selatan	Teminabuan	13	2
7	Sorong	Aimas	18	15
8	Raja Ampat	Waisai	17	4
9	Tambraw	Sausapor	7	0
10	Maybrat	Kumurkek	11	1
11	Sorong	Sorong	6	31

In addition to geographic information, this discussion will also inform about religion and tribes followed by the people of West Papua. Population in West Papua consists of various tribes with different languages. In fact, even in one tribe alone, there are many kinds of languages. In Papua and West Papua there are 310 languages spoken by the community and in West Papua alone there are 67 tribes inhabiting. In addition to diverse tribes and languages, religions embraced by the community are also diverse, namely Islam, Catholicism, Protestantism, Hinduism, Buddhism, Confucianism and others. It is clear that West Papua is made up of various ethnicities and religions, but lives side by side in harmony.

3.2. Mass Media as a Creator of Stigma

The stigma of the outsiders regarding West Papua is a conflict-prone and primitive region, is not excluded of the role of the mass media. Wherein the reporting or portrayal that is broadcasted is only an aspect of insurrection, conflict and primitiveness of West Papua. One of the news that caused a stir is of the Tolikara case. Tolikara is one of regencies located in the West Papua Province. It was rumored that the Tolikara case was a conflict that was triggered by different beliefs, in this case Muslims and Christians. Whereas the case of Tolikara has actually been designed by conflict actors, only so far with ethnic, racial and group identities have not worked, so experiments were carried out by prioritizing religious status. This case was deliberately designed by certain parties to divide the harmony of the Tolikara community. Even outsiders were provoked by the case, triggering religious sentiments.

In addition through the news, Papuan ethnics are often depicted through film. There are several Indonesian films that take Papuan ethnic as the main highlight. Among them are Denias and Lost in Papua. These films have a different story line and problems portrayed. However, behind these differences there is also the same role pattern which continues to be repeated among the films based on Papuan ethnics, which are, ethnic immigrants who are considered superior (in this case the Javanese ethnics) and come as the savior and are considered to make the Papuan ethnic to be considered as the more inferior ethnicity which makes them advanced and modern. [7] From the depiction of such a role it will enforce the thinking of the outside community about the primitive Papuan community and always rely on other ethnic groups to improve their living standards. In fact, it is not just people from ethnic immigrants who can become high-ranking officials in West Papua. As an illustration of the political conditions in Sorong (West Papua), members of the Regional House of Representatives elected for the period 2014-2019 consist of the composition of Ayamaru, Malamoi, Java, Sunda, Madura, Bugis, Makassar, Batak, Buton, Ambon, Kei, Toraja, Minahasa , Masohi, and Seram. [8] The indigenous Papuans included on the list are Ayamaru and Malamoi. In addition to proving the ethnic integrity of indigenous Papuans, the data can also depict West Papua which actually consists of ethnic diversity but can interact in harmony. Therefore, the term that West Papua is a miniature of Indonesia was spread. Thus, it is not only the bad image that dominates the social life of the people in West Papua. West Papua is also a peaceful and harmonious place. This is in line with the statement of one of the heads of urban villages in Sorong Regency, which said that:

There are about 1300 residents who live in the Klaru village, where Islam 70%, Christian 28% and Hindu 2% make up for the percentage of religion adopted by the people there. While based on ethnic or local origin makes up of Java 70%, Moi 15%, Maibrat 10%, Sulawesi 3%, and others about 2%. The conflicts that arise in the community are caused by problems of land or customary tradition. Regardless of its diversity, conflict is not about religion or ethnicity. In my opinion, harmony in the community can be achieved because good communication is always carried out and mutual understanding in the interaction of the community is maintained, so there is no gap between the differences [9]

From these interviews it can be concluded that ethnic and religious diversity is not the cause of conflict. Despite the conflict, there are other factors that cause it, one of which is the problem of customary tradition. Therefore, the news that West Papua is an area that often has conflict caused by differences of religion and ethnicity is not something absolute, because sometimes the conflict caused by other factors are misused by certain parties as a conflict of religion and ethnic. Thus, news or mass media is one of the formers of negative stigma towards the people of West Papua.

IOP Conf. Series: Earth and Environmental Science **156** (2018) 012034 doi:10.1088/1755-1315/156/1/012034

3.3. Local Wisdom as a Harmonious Adhesive

Local wisdom is the view of life and science and various life strategies that manifest the activities taken by the local communities in solving various problems regarding their needs [10]. Local wisdom or in the *fiqh* term is called *al-'urf*, is something that becomes custom and tradition for society either in the form of speech or deed. [11] By looking at this understanding then in this study, local wisdom concerns the tolerance and brotherly principles that gave birth to the harmonization in diversity in West Papua. West Papua is a province in Indonesia whose society consists of various ethnic, culture and even religion. West Papua is said to be a miniature of Indonesia, diverse but still one like the *Bhinneka Tunggal Ika.* The diversity can be unified with the principle of brotherhood, as what has happened in the West Papua region of Fak-Fak, Raja Ampat Islands, and the Kokoda tribe in Sorong. Fak-Fak is one of the regencies in the West Papua Province. The people in Fak-fak is not only dominated by one religion only, but there are three major religions embraced by the community there. For example in a family there are three religions, namely the family of the Mbaham tribe in one clan / fam there are three religions; H. Muhammad Temongmere (Islamic Religion), Demianus Temongmere (Catholic Religion), Yewisa Temongmere (Protestant Religion. [12] Therefore, the term "One Stove Three Stones" describes the religious condition of the people there.

The simple philosophy of "One Stove Three Stones" is a description of the brotherhood of the Fak-fak society across different religious and cultural identity. [13] The religious diversity in Fak-fak is integrated with the understanding of "One Stove Three Stones" is more motivated by the attitude of the public tolerance in all fields, so that there are no domination that can cause division among them. As which was disclosed by one of the Head of TR BAPPEDA of Fak-Fak Regency is that:
In applying the term "One Stove Three Stones", we not only regulate the issue of social life in the daily life in Fak-fak. However, religious issues are also applied in this tradition, so that there is no dispute of understanding and belief. If there is construction of a house of worship such as a mosque, then the chairman of the development committee is of the Catholic religion. Vice versa, if there is a church being built, then the chairman of the development committee is of the Muslim religion. Just an addition and information from me, that frankly I was once chairman of the committee of one of the church being built in Fak-fak Regency when I am a Muslim. In addition to the construction of houses of worship, the tradition of one stove three stones is also applied in securing religious rituals. For example, if Catholics are performing a mass or a Christmas celebration, then those who keep the worship safe are the Muslims. Meanwhile, if Muslims are organizing Eid prayers or other worship, then those who keep their security is from the Catholics. This is what makes Muslims and Catholics in the regency of Fak-fak always have cooperation in various activities. [14]

The "One Stove Three Stones" is like the "stove" as the family and "stone" as the religion. It is said that to cook on a stove, it requires at least three stones. The philosophy is that brothers must be one-hearted, whenever they are united, no force will be able to penetrate it. [15] Thus, it means that the people of Fak-Fak society can live in harmony even with the three major religions they have in their society are Christian, Catholic and Islamic. [16] The majority of people in Fak-fak are Islam which makes up of 53.80% of the local population as well as immigrants, while Protestant Christianity is embraced by 28.35% and Catholics 17.59%, other religions such as Hinduism, Buddhism and Confucianism are very few or below than 1%. [17] Such is the picture of religious diversity in Fak-Fak, West Papua Province. Unlike the case with Fak-fak, in the Raja Ampat Islands, there are four major religions embraced by the people, namely Islam, Catholicism, Protestantism and animism. [18] In this term, the people of Raja Ampat describes a family is a house which its family members can enter through any door. It is clear that what the community holds firmly is tolerance, where they will accept each other and respect whatever religion is followed by members of his family, instead of disowning a family member for having different beliefs.

Besides the people of Fak-fak and Raja Ampat Islands, the Kokoda society is also able to maintain the harmony by holding firmly to the principle of brotherhood. The harmony is inseparable from their belief in the 5 kings. There are three inherited messages of the five tribal kings who are always spoken by tribal chiefs (now) or Kokoda residents when talking about religion, firstly, maintaining religious harmonization among peers, secondly, no bonds should be broken because of religious belief, third, no matter wherever you go, everyone is your brother and sister. [19] It is clear that the principle of brotherhood is the key to Kokoda's harmony.

The three groups of society make the difference not as a breaker of unity but as a binder of harmony. Even when it comes to religion they will respect each other. For them the difference is not a barrier but a means to keep in touch with each other. This is obvious during religious festivals, where each religion will help each other.

Like when the celebration of *Eid al-Fitr* and *Eid al-Adha*, people from other religion will help. Similarly, with the celebration of Christmas, the Muslims will also lend a hand. When it comes to food, it is not an issue, as there is already an understanding of the host religion of what can and cannot be consumed by people from other religions. With such a small thing, religious differences are not a breaker, but can create a harmonious unity if coupled with tolerance and brotherly principles.

3.4. Harmonious Religious Diversity

The principle of brotherhood and tolerance demonstrated by the harmony of the people of West Papua can also be illustrated in the provincial Papuan *Musabaqah Tilawatil Quran* (MTQ) committees in West Papua. All the organizers are not only from the faith of Islam, but the pastor and church youths also joined in it. The celebration of *Musabaqah Tilawatil Quran* (MTQ) together with the brotherhood and togetherness is the key to its success. Religion is only a private matter and not an entity that has the right to influence its followers, and has nothing to do in communal affairs. Religion is not an excuse to create a division. Therefore, the implementation of religious rituals never gets in the way. The expansion of the *At-Taubah* Mosque in Malanu, Sorong City was supported by the willingness of the Protestant landowners to give their land. [20] Submission of this land is not free but the selling price was far more below from the market price. Meanwhile, the land owned by the State Islamic High School (STAIN) of Sorong was handed over to the church to build a house of worship. [21] This matter proves the religious harmony between both faiths.

3.5 Culture Strengthening

Culture is a way of life that is shared by a group of people and passed down from generation to generation. Culture is also a norm carried out by a group of people who are ultimately executing it on an ongoing basis. One culture that illustrates harmony among the diversity of the people of West Papua is a culture of friendship during a religious celebration. Observations made during the celebration of *Eid al-Fitr* and Christmas is that during the celebration of *Eid al-Fitr*, not only the Muslims visit each other, but the Christians will also come and join in the celebration. Similarly, when it is Christmas, the Muslims will visit and their Christian friends and join in the celebrations. This shows that the people of West Papua strongly holds firm to the belief of brotherhood and tolerance. Religious differences are not a cause for conflict but will encourage each people from different religious background to interact and acknowledge the diversity that exists. Conflicts arising on the basis of differences are only made by certain parties who use religion as an instrument to spread hatred. On the other hand, conflicts in the name of religion are carried out by those who have not understood or imposed things on what and how they interpret religion. The people of West Papua have made diversity as a place to strengthen the relationship, even gave birth to a new culture.

IOP Conf. Series: Earth and Environmental Science **156** (2018) 012034 doi:10.1088/1755-1315/156/1/012034

4. Conclusion

West Papua, although is a region rich in diversity but is still peaceful in harmony, however not only is it peaceful in religious diversity, but also ethnicity and culture. The diversity possessed by West Papua makes it known to be a miniature of Indonesia. Behind the diversity, the stigma of outsiders is still always negative. In this study, it was found that mass media was one of the origins of the negative stigma surrounding West Papua, where the dominant news gives a negative meaning to the West Papua Ethnic. The ethnic of West Papuans were described as, people who like rebellion and conflict, such as in the Tolikara case. In addition, the formation of a negative stigma towards the society was also supported by a couple of Indonesian films depicting primitive Papuan ethnic, one of them which was, the Lost In Papua film, in which the inferior Papuan ethnic needed the other superior ethnic in order to advance.

This study confirms that diversity is not a trigger for a conflict. In fact, diversity will contribute to the harmonization of life if built with the pillars of brotherhood and tolerance. The pillars of brotherhood and tolerance are reflected in the motto "One Stove Three Stones" and "One House Four Doors" which originated from the community residing in Fak-Fak and Raja Ampat Islands, West Papua, where in the family, religion is not a differentiator. The diversity of religion adopted is used as a means to strengthen the relationship. It is even entrenched in West Papuan society, that, when there is a religious festival it will provide opportunities for interfaith friendship and bond to be greater.

In addition, the harmony of the people of West Papua can also be seen in the implementation of the Musabaqah Tilawatil Quran (MTQ) which involves not only Muslims but also Catholics and Protestants. Infact, when it comes to places of worship, there is no difficulty in its construction. Brotherhood and tolerance among the diverse community is also clearly visible in binding the harmony of the people of West Papua amid their diversity of ethnicity, religion, and culture. Therefore, this research is expected to be able to change the stigma towards of the people of West Papua that is linked to conflict and backwardness, into a harmonious society despite its diversity.

5. References

[1] Official Website of the Province of Papua Barat through ://papuabaratprov.go.id/ sekilas-papua-barat/, accessed on February 23, 2017.

[3] Ismail Suardi Wekke, 2015. Agama, Persaudaraan, dan Ikatan Emosional: Harmoni Sosial Minoritas Muslim Papua Barat. Masika ICMI Orwil Sulawesi Selatan in Makassar.

[4] Diah Aghsari dan Ismail Suardi Wekke. 2015. Ritual Sasi Laut; Akulturasi Agama dan Budaya dalam Praktik Ritual Kebaharian Masyarakat Misool Raja Ampat. Jurnal Airaha, 4(1) : 4

[5] Firman Sujadi, A. Saibani and Junarsyah. 2013. Provinsi Papua Barat Cintaku Negeriku. Jakarta: Cita Insan Madani.

[6] Ismail Suardi Wekke, "Memahami Kasus Tolikara" (Makassar: Fajar, July 25, 2015).

[7] Cindy Erika Larasati, 2014. Representasi Identitas Etnis Papua pada Film Lost In Papua. Jurnal Unair Commonline Departemen Komunikasi, 3(3) : 492-493.

[8] Ismail Suardi Wekke, "Citra Sorong: Gambaran Model Pancasila", in Pancasila Rumah Kita Bersama. Editor Thamrin Sonata (Jakarta: Peniti Media, 2014).

[9] Kodhim, Head of Klaru Village (Sorong Regency), Interview, March 9, 2017, on Jalan Nusa Indah, Klaru Village (Sorong Regency).

[10] Ulfah Fajarini, 2014. Peranan Kearifan Lokal dalam Pendidikan Karakter. Sosio Didaktika: Sosio Science Education Journal 1(2) : 123.

[11] Sagaf S. Pettalongi, 2012. Local Wisdom dan Penetapan Hukum Islam di Indonesia, in Tsaqafah Jurnal Peradaban Islam 8(2) : 231.

[12] Abustani Ilyas, Surahman Amin and Rusyaid, "Satu Tungku Tiga Batu: Kearifan Lokal dalam Harmonisasi Beragam Masyarakat Fak-Fak Papua Barat", (Group research, State Islamic College (STAIN) of Sorong, Sorong, 2013), page 47.

[13] Saidin Ernas, Heru Nugroho and Zuly Qodir, 2014. Dinamika Integrasi Sosial di Papua Fenomena masyarakat Fak Fak di Provinsi Papua Barat. Kawistara, 4(1) : 69

[14] Abustani Ilyas, Surahman Amin and Rusyaid, "Satu Tungku Tiga Batu: Kearifan Lokal dalam Harmonisasi Beragam Masyarakat Fak-Fak Papua Barat", (Group research, State Islamic College (STAIN) of Sorong, Sorong, 2013), page 42-43.

[15] Idrus Al-Hamid, 2013. Islam Politik di Papua: Resistensi dan Tantangan Membangun Perdamaian. Millah, 12(2) : 459.

[16] Ismail Suardi Wekke, 2013. Islam di Papua Barat: Tradisi dan Keberagaman. Ulul Albab, 14(2).

[17] Saidin Ernas, 2013. Politik Simbol dan Harmoni Sosial: Makna Satu Tungku Tiga Batu dalam Dinamika Politik Lokal di Fakfak Papua Barat. Dialektika, 9(2)

[18] Ismail Suardi Wekke, 2016. Harmoni Sosial dalam Keberagaman dan Keberagamaan Masyarakat Minoritas Muslim Papua Barat. Kalam, 10 (2) : 299.

[19] Muhammad Rais, "Islam dan Kerifan Lokal; Dialektika Faham dan Praktik Keagamaan Kokoda-Papua dalam Budaya Lokal", Paper of Annual Conference on Islamic (ACIS) X in Banjarmasin, November 1-4, 2010, page 195

[20] Ismail Suardi Wekke, 2016 Harmoni Sosial dalam Keberagaman dan Keberagamaan Masyarakat Minoritas Muslim Papua Barat. Kalam, 10(2) : 306.

Islamic Education on Formation of Environmental Awareness in Pondok Pesantren Indonesia

Jumardin La Fua[1], Ratna Umi Nurlila[2], Fahmi Gunawan[3], Ismail Suardi Wekke[4]

[1]Institut Agama Islam Negeri Kendari, Jl. Sultan Qaimuddin No 17, Baruga, Kendari, Sulawesi Tengara 93563 Indonesia
Email : jumarddin81_stainkdi@yahoo.co.id
[2]Sekolah Tinggi Ilmu Kesehatan Mandala Waluya Kendari, Jl. Jend AH. Nasution Blok G-37, Poasia, Kambu, Kendari, Sulawesi Tenggara 93231 Indonesia
Email : rumi@gmail.com
[3] Institut Agama Islam Negeri Kendari, Jl. Sultan Qaimuddin No 17 Baruga, Kendari, Sulawesi Tenggara, 93563, Indonesia
Email : fgunawanp@gmail.com
[4] Sekolah Tinggi Agama Islam Negeri (STAIN) Sorong, West Papua. Indonesia
Email : ismail@stain-sorong.ac.id

Abstract. This study aimed at exploring Islamic education strategy in shaping environmental awareness in Islamic Boarding School Gontor Putra, Southeast Sulawesi, Indonesia. This is a model for actualizing environmental education through environmental hygiene management based on the values of the *Qur'an* and the *Hadisth*. This research was qualitative descriptive study examining the Islamic community by using an ethnographic approach. The results showed that those educational strategies were (1) conducting participatory activities, such as cleaning Friday, arrangement of garden and surrounding environment, creation of green open space, and (2) building collective awareness about the importance of environmental management through daily activities. The strategy can ultimately create students who have a sense of eco-spirituality in interacting with nature.

1. Introduction

Islamic Boarding School Gontor Putra is one of the religious institutions located in southern Konawe which concern the activity of spirituality and morality in shaping the behavior of students. One education model developed at the boarding school are education combines spirituality and environmental education. Through a model of integration between religious education and environmental education, it is expected that the relationship between humans and environment will harmonize [1−3], and can form human resources with the knowledge, attitudes, skills, motivation and commitment to participate in solving environmental problems and prevent exploitative behavior on the environment [4−5].

Islamic boarding school Gontor Putra which has spiritual teachings are expected to be able to produce students who can govern the relationship between human and environment, as well as being a catalyst for the formation of character in preserving the environment. In addition, it can produce individuals who have ecological piety in interacting with the environment.

The research on relationship between religion and environmental education to the formation of behavior has been widely conducted by many people using various perspectives. Religious and environmental education [6−7], strategy of religion in environmental education [8], religion and conservation [9], school environment [10], and eco-sprituality [11−12]. All these studies emphasize that religious values have relevance to environmental values, so religion is regarded as an effective approach to harmonize the relationship between human and environment. Therefore, the previous related studies indicate that no research yet relate the phenomena of formation of character environmental awareness through Islamic boarding school.

2. Method

This research was a study of a community that focuses on students at Islamic boarding school Gontor Putra. To collect data, interviews and observations by using instruments were conducted [13]. The interviews were perfomed by interviewing the leaders of boarding school, teachers, students and residents around. During the research, the researchers interact with students by following all the activities. The process of triangulation was carried out by examining subject of research at the boarding school, learning topics in different place. Then focus group discussion was performed by involving the researchers from IAIN Kendari and Halu Oleo University to ensure the accuracy of the data. Etnographic perspective was also conducted to analyze data obtained.

3. Results and Discussion

The combination of spiritual elements and religious themes in environmental education will provide empirical learning experience that can increase knowledge and awareness of students in managing the environment [7]. Swam [14] says that the religious movement would be able to inspire people to build harmony with nature since religious education is considered as a solution to minimize environmental damage [15−16].

Islamic boarding school Gontor Putra provides a religious approach to the students [17] and encourage them to behave well to the environment. The strategy developed by the boarding school to provide environmental awareness is to care environment. In addition, building a collective awareness of students are conducted through some activities performed in daily life, such as cleaning classrooms, cleaning the cabin environment, doing reforestation, disposing of waste and picking it up in place. Those collective activities are an effort to build community awareness of students through instilling the values of environment preservation. Then, it enhance students to become carefully discern and care for the environment. In addition, this strategy is expected to be one form of Islamic education as a node in the behavior of environmental awareness in students community.

Another strategy to foster environment awareness is the development of participatory activities carried out independently and supervised by religious leaders of boarding school. This strategy aims to habituate students in performing activities which are beneficial for them and is able to shape their personalities caring for environment. Those participatory activities are management of boarding school yard and cleaning Friday activities. These activities are expected to be able to strengthen knowledge, awareness and behavior of students community about the importance of preserving the environment. In addition, they can shape students' habit of environmentally sound. This strategy may further provide the training o establish students' wisdom to interact with the environment. Lijun and Ya [18] called it as ecological spirituality. It is the relationship between human and the environment which based on religious values [19−21]. Ecological spirituality will encourage students to have an attitude and behavior reflecting the character to preserve the environment.

To achieve the environmental education strategy requires a considerable time because the students build unity gradually and actively involved in environmental education activities. Each stage of the strategy requires commitment, consistent and high fighting spirit to build community awareness of students about the importance of maintaining and mestarikan environment. There are four stages conducted by religious leaders of Pesantren Gontor in building character of environmental education in students community, namely forcing the students to care on environment, familiarizing students to conduct clean-up activities, making the students be diligent to any existing program relating with preserving environment, and supervising the character formed through application of the rules since these activities can still run well. According to Oman [12] The internalization of the values can occur to students if they are willing to accept the influence and are willing to behave in accordance with what is affected because it can support the success of a program that has been planned.

Environmental education has been introduced to the students by their leaders since they were studying at boarding school. Its' implementation is done in such simple and extracurricular activities, such as involving students to clean their learning environment, planting and caring for trees and other activities. Through this simple activity, students are introduced on the attitude of maintaining and preserving the environment from an early age, which in turn can shape the character of caring environment in community of students in the future. Involving students from the beginning is also one of the strategies developed by boarding school in producing generation that has a positive attitude towards the environment. In addition, it is expected to produce the students who have high ecological sensitivity and can contribute to a variety of environmental problems that exist in society.

Environmental education implemented by Pondok Gontor Putra is a form of education based on the values of Islam with an emphasis on the concept of man as earth hastask prosperity of the earth and do not make mischief on earth. This concept is in line with Kula [22] and schultz et al [23] stating that the relationship between man and nature in environmental management is not a relationship between conquerors and conquered, but a relationship of togetherness under obedience to God. The role of religious institutions in environmental education have also been conducted by several researchers such as Sheikh [24] in Pakistan, Almeida and Cutter [25] in India, schultz et al [23] that environmental education will be successful when they involve religious institutions, because culturally religious institutions have a charismatic religious leader and highly respected.

Various environmental education strategies that have been introduced by Islamic Boarding School Gontor Putra in implementing environmental education on students has played a very significant role in changing the perspective of students to the environment by shaking the ecological consciousness of students to get involved in the behavior and lifestyles that are environmentally friendly. In addition, environmental education strategies that have been implemented seeks to touch the cognitive, psychomotor and affective so that students might be actively involved in ecological activities in boarding school environment.

4 Conclusion

This study confirms that religious education values is the most ideal form of education and is assumed to be a very powerful force in supporting the development of environmental education. This study also proves that religious values have given some contributions to increase the capacity, knowledge, and understanding of management and environmental protection. This provides a compelling argument for increasing the role of religion in environmental education as conducted by Islamic boarding school Gontor Putra.

References

[1] Bekalo S, Bangay C 2002 Towards effective environmental education in Ethiopia: problems and prospects in responding to the environment—poverty challenge *International Journal of Educational Development,* 22(1) pp. 35-46.

[2] Hitzhusen G E 2007 Judeo-Christian theology and the environment: moving beyond scepticism to new sources for environmental education in the United States *Environmental Education Research,* 13(1) pp. 55-74.

[3] Guth J L, Green JC, Kellstedt LA, Smidt CE 1995 Faith and the environment: Religious beliefs and attitudes on environmental policy *American Journal of Political Science*, pp. 364-382.

[4] Rasmussen J 1983 Skills, rules, and knowledge; signals, signs, and symbols, and other distinctions in human performance models *IEEE transactions on systems, man, and cybernetics,* (3) pp. 257-266.

[5] Roome N 1992 Developing environmental management strategies *Business strategy and the environment,* 1(1) pp. 11-24.
[6] Hitzhusen G E 2006 Religion and Environmental Education: Building on Common Ground *Canadian Journal of Environmental Education,* 11 pp. 9-25.
[7] Crowe J L 2013 Transforming Environmental Attitudes and Behaviours through Eco-spirituality and Religion, *International Electronic Journal of Environmental Education,* 3 (1) pp. 75-88.
[8] Haigh M 2010 Education for a Sustainable Future: Strategies of the New Hindu Religious Movements *Sustainability,* 2 pp. 3500-3519.
[9] Frascaroli F 2013 Catholicism and Conservation: The Potential of Sacred Natural Sites for Biodiversity Management in Central Italy *Human Ecology,* 41 pp. 587–601.
[10] Wekke IS, Sahlan A 2014 Strategy in Creating School Environment: Lessons from High Schools in Indonesia *Procedia-Social and Behavioral Sciences 143* pp. 112-116.
[11] Van Schalkwyk A 2011 Sacredness and sustainability: Searching for a practical eco-spirituality *Religion and Theology,* 18(1-2) pp. 77-92.
[12] Oman M S 2014 The Effect of Educating Environmental Ethics on Behavior and Attitude to Environment Protection *European Online Journal of Natural and Social Sciences,* 3 (3) pp. 141-150.
[13] Burke J R 1997 Examining the validity structure of qualitative research *Education,* 118(2) pp. 282-293.
[14] Swam JA 1978 Enviromental Education: a New Religion *Journal of Environmental Education* 10 (1).
[15] Tilbury D 1995 Environmental education for sustainability: Defining the new focus of environmental education in the 1990s *Environmental education research,* 1(2) pp. 195-212.
[16] Littledyke* M 2004 Primary children's views on science and environmental issues: examples of environmental cognitive and moral development *Environmental Education Research,* 10(2) pp. 217-235.
[17] Wekke I S, Andriansyah A 2016 From Gontor to Sorong: Muslim Minority Practices on Arabic Teaching and Learning *Sosiohumanika, 9*(1).
[18] Lijun W, Ya Li 2013 Enviromental Design and Cultural Spritual Ecology *Journal Landscape Research,* 34 (5) pp. 29-30.
[19] Kinsley D R 1995 *Ecology and religion: ecological spirituality in cross-cultural perspective* vol 8 Englewood Cliffs, (NJ: Prentice hall).
[20] Bruteau B 1990 Eucharistic ecology and ecological spirituality *CrossCurrents* pp. 499-514.
[21] Rícan P, Janosová P 2005 Spirituality: its psychological operationalization via measurement of individual differences: A Czech perspective *Studia psychologica*, 47(2) pp. 157.
[22] Kula E 2001 Islam and Environmental Conservation *Environmental Conservation* 28 (1) pp. 1–9.
[23] Schultz P W, Zelezny L, Dalrymple N J 2000 A multinational perspective on the relation between Judeo-Christian religious beliefs and attitudes of environmental concern *Environment and Behavior,* 32(4) pp. 576-591.
[24] Sheikh K M 2006 Involving Religious Leaders in Conservation Education in the Western Karakorum Pakistan *Mountain Research and Development,* 26 (4) pp. 319-322.
[25] Almeida S, Cutter A M 2011 the historical, present and future ness of environmental education in india *Australian Journal Of Environmental Education* 27 (1) PP. 122-133.

IOP Conf. Series: Earth and Environmental Science **156** (2018) 012036 doi:10.1088/1755-1315/156/1/012036

BOD and DO Identification of Jeneberrang-River Water as Water Source

Basir Basir[1] Ibrahim Abdul Haris[2]
[1]Environmental Health Depatment, Faculty of Public Health, Hasanuddin University Makassar. Indonesia
Email: baz.rasyid@gmail.com
[2] Environmental Health Depatment, Faculty of Public Health, Hasanuddin University Makassar. Indonesia
Email: ibrahimmharis@gmail.com, abdulharisibrahim@yahoo.co.id

Abstract: Water is a part of life on the surface of the earth. It is not a new substance which no life on earth can survive without water. This study aims to identify the quality of river water of jeneberang as municipal waterworks (PDAM) raw regarding to turbidity parameters, BOD, and DO. The used methodology is observational with descriptive approach. The sampling technique was done by grasping each sample for turbidity parameter, BOD and DO in four IPAM water inlet of Makassar City. This sampling was conducted at Laboratory of Environmental Health Engineering Center (BTKL) and Disease Control using Nephelometric Turbidity Unit, Titrimetry, and Spectrophotometric method. The results of the examination showed that the turbidity level> 5 NTU, so it is not eligible to be drunk directly. The levels for BOD and eligibles for not exceeding the specified limits are <3 mg / l and <25 mg / l and in DO for Maccini Sombala <4 mg / l region, so it is not eligible. The conclusion in this research that turbidity parameter, BOD and DO qualify as the raw water of PDAM have to be processed so that water can be drunk by society.

***Keywords*: Turbidity, BOD, and DO**

[1] Introduction

The provision of drinking water to public has to meet with the quality requirements including physical, biological, chemical and radiological. Nowadays, surface water (rivers, springs, reservoirs and so on) is still becoming a source of raw water to meet human demands, whether processed through a government-run or managed private water supply company. Biological and radioactive is necessary to control the quality of raw water to produce drinking water to public that meets the requirements because its water is easily contaminated by chemicals [1].

National Socioeconomic Survey Data (SUSENAS) (2010) revealed that Indonesians with access to clean water only 44.19% and 55.54% who had no access to basic sanitation facilities. The total population of Indonesia in 2015 was about 248 million people based on the results of the 2010 Basic Health Research as many as 110 million people (44.5%) had not access to sanitation and 55 million people (22.1%) did not possess access to drinking water , rural population was estimated at 153 million people (61.5%) who had no encroach to sanitation and 77 million (31%) who did not have infiltrate to drinking water [2].

According to Minister of Health Regulation (Permenkes) no. 416 / MENKES / PER / IX / 1990 referred to as "clean water" is water used for daily purposes whose quality meets health requirements and can be drunk when cooked. Clean water has to obviously own high quality physically, chemically and biologically to prevent disease [3].

Clean water supply in Indonesia especially for large scale is concentrated in urban areas, and managed by the Municipal Waterworks (PDAM) concerned. One of the Providers of Water Supply in South Sulawesi province, PDAM Kota Makassar. PDAM Makassar is one of 12 drinking water companies entering the big group in Indonesia. It has a customer above 100,000 customers. Raw water source is used by PDAM Kota Makassar to serve the provision of drinking water mostly derived from surface water (river), the Jeneberang River and Maros River [4].

Water quality can be determined by testing the water. The tests are often in chemical, and biological (smell and color) tests. It can be expressed by several parameters, for instance, physical parameters (temperature, turbidity, dissolved solids, etc.), chemical parameters (pH, dissolved oxygen, BOD, metal content, etc.) and biological parameters (presence of plankton, bacteria and so on) [5].

Regarding to data obtained from South Sulawesi BLHD in the end of 2013, Jeneberang River water quality was at severe pollutant level. The parameters refered to the TSS, BOD, NO2, H2S, and CI test results that were above the standard. This condition was exacerbated worsened by forest land around the Jeneberang River which is about 2,462 hectares, classified as critical [6].

Apriyantomo and Nurbowo stated that Jeneberang River as the raw water source of PDAM Makassar which was the supplier of urban water supply in Makassar must meet the criteria and requirements of the parameters. When the content of BOD and DO in water was high, it produced a high organic material so it could be a growth medium of various types of microbes causing unpleasant odors. To reduce the odor required activated carbon by utilizing halalan thayyiban materials in accordance with Islamic Shari'a. Therefore, determining the level of dissolved oxygen contained in water should be examined the levels of oxygen [7].

Based on the results of research which was conducted by Malokoudi River Ereniku, one of the rivers on the Continental Europe based on testing for physical and chemical parameters. There were several increases in temperature, pH, turbidity and dissolved oxygen (DO), relating to the analysis. The temperature as the most important parameter showed that there was an increase in temperature of each point of 7.9 ° at the first chemical point to 19.5 ° at the second point, turbidity also increased from 5, 9 to 25.5, while the pH value at the first point of 8.08 increased to 9.82 in the second point and DO at the first point 6.8mg / L increased to 10.8 mg / L [15].

[2] Method

Sampling Technique used for Ratulangi PDAM (IPA) is Graping Samples method. Temporary Example represents the state of water at any point from a place when a water source has characteristics that have not changed much in a short-time period. The characteristics of river water within a period or within a certain distance limit then the momentary example simply represents the time and place. Generally, a momentary example uses a 24-hour sampling period. The sample is taken from the point before entering into the inlet from the upstream of the river at the sampling frequency of three times in the morning at around 09:00 to 10:00 pm, at noon around 13:00 to 14:30 pm and in the afternoon at around 16:00-17:00 pm. Primary data was obtained from the results of the examination at the Laboratory of Environmental Health Engineering (BTKL) and the Control of Banta-Bantaeng Disease of South Sulawesi Province of Makassar City.

[3] Result

The results of examination from Laboratory of BTKL and Disease Control showed that the level of turbidity inlet IPA Makassar PDAM did not fulfill the requirement as raw drinking water based on second class because of the turbidity level of each inlets if related to PP. 82 of 2001 and Permenkes 416 of 1990 were in the range of 5 NTU and above while the maximum limit of turbidity is 5 NTU (Table 1).

The examined result of BOD inlet parameter of PDAM Makassar in Laboratory of BTKL and Disease Control showed that BOD content qualified as raw drinking water based on class II because BOD content in IPAM Inlet IPA Makassar was in the range <3 mg / l. Hence, it could be concluded that BOD content qualified as raw drinking water in accordance with PP. 82 of 2001 and South Sulawesi Provincial Governor Regulation Number 69 of 2010 (Table 2).

The examination results of the parameters in DO inlet IPA PDAM Makassar from BTKL Laboratory and Disease Control Center indicated that the level of DO there were qualified as raw drinking water based on class II. Some results were not eligible if associated with PP. 82 Year 2001 and South Sulawesi Provincial Governor Regulation No. 69 of 2010. IPA V Inlet Somba Opu, IPAT Ratulangi inlet and IPAK Malengkeri inlet of its DO content qualify as raw drinking water for the level of DO in the three inlets IPA Makassar PDAM was located in range> 4 mg / l. The level of DO did not meet the requirements of DO levels in IPA IV Inlet Maccini Sombala in the morning due the fact that its DO levels were in the range of 2.49 mg / l. Thus, it did not meet the requirements as raw materials of drinking water class of second class because <4 mg / l (Table 3).

[4] Discussion

The turbidity parameter was measured by using a Turbiquant 3000 IR instrument. The Turbiquant 3000 IR tool will read the turbidity scale measurement results (SNI.06-6989.25-2005). The all results which were read by the tool are not qualified as raw materials of drinking water class II in accordance with PP. 82 year 2001 and Permenkes 416 year 1990 because turbidity level is above 5 NTU with average pH is in the range of 6,4-6,8 [8]. Therefore, the water turbidity level of Jeneberang River as raw water of PDAM was not feasible to be drunk directly but it must be firstly neutralized. The PDAM should perform several clarification stages, for example, the coagulation stage which was the phase of charge neutralization with the addition of certain chemicals to ensure maximum contact between the suspended solids and the added chemicals.

The result of turbidity parameter research was in line with research conducted by Makassar PDAM side, with turbidity level in each IPAM IPA inlet ranged from 21,40 NTU-120 NTU. This indicated that the turbidity level in each IPAM inlet of PDAM Makassar originating from Jeneberang River was not qualified as PDAM raw water and as raw material of clean water and drinking water as contained in Permenkes 416 of 1990 that limit the maximum level for turbidity is 5 NTU.

BOD parameters are measured by titrimetric method. BOD is one of pollutant indicator parameters in water caused by organic waste. Its presence in the environment is largely determined by organic waste, whether from domestic waste or from industrial waste [9].

Based on the laboratory test result data, BOD inlet level of IPA PDAM Makassar if related to PP. 82 of 2001, and South Sulawesi Provincial Governor Regulation No. 69 of 2010 qualifies as raw drinking water regarding to class II because BOD content in IPAM inlet of Makassar PDAM is in the range <3 mg / l therefore it can be argued that the lower the BOD, the higher DO [10]. BOD in the Makassar PDAM inlet can be decomposed aerobically to the maximum because the organic matter contained in the waste water is low. Thus, the oxygen supply increases and goes on continuously [11].

Other research results were regarding to this research conducted by Fadhillah Jamal, which showed the measurement of raw water quality of PDAM on IPA transmission channel Panaikang. For BOD parameters, there were eligible whilst others were not qualified [12]. For example, point 1,3,4,5,6, and 7 qualified each ranged between 2.96 mg / l, 2.32 mg / l, 2.28 mg / l, 2.64 mg / l, 2.96 mg / l and 2.32 mg / l and all of these were <3 mg / l according to PP. 82 for 2001 and South Sulawesi Provincial Governor Regulation No. 69 of 2010. As for point 2 and point 8 BOD levels ranged between 3.28 mg / l and 3.92 mg / l and both were> 3 mg / l , hence, that they did not meet the requirements as the water of PDAM in accordance with PP. 82 of 2001 and South Sulawesi Provincial Governor Regulation No. 69 of 2010. This was because at point 2 that in the village of Kabung District Tanralili Maros regency there were activities of citizens who used water for bathing and washing, land use consisted of settlements, rice fields, gardens , poultry, cattle and buffalo. Point 8 that was in the IPA Panaikang Jl. Abd. Daeng Sirua was a dense end-use density intake, close to markets and shops, there were several garbage containers at the edge of the canal, lots of piles of scattered waste.

Similar to BOD research, DO research was also measured by titrimetric method using a winkler bottle. According to Government Regulation on Water Quality Management and Water Pollution Control and South Sulawesi Provincial Governor Regulation No. 69 of 2010 that the minimum dissolved Oxygen (DO) limit for class II water quality is 4 mg / l, based on the calculation of DOs obtained dissolved oxygen for three IPA IPAM inlets were IPA V Somba Opu inlet, IPAL Malengkeri inlet and IPA Ratulangi inlet are in the range of 5.3 - 8.5 mg / l, the dissolved oxygen in Jeneberang river water was below the minimum limit so that it qualified as raw drinking water.

The highly oxygen content dissolved in water at the IPA V inlet Somba Opu, IPAT inlet Ratulangi and Inlet IPA Malengkeri showed that the degree of water contamination in the inlet was small realtif. This was caused by the absence of settlements around the inlet, so that the contamination by the waste and domestic waste did not exist then caused the process of decomposition in the water was relatively low. It also witnessed that the Jeneberang River was still safe for the life of aquatic biota such as fish and others.

However, it could not be denied that the current levels of DO would decline someday due to human activity and the existence of the industry along the Jeneberang River such as tofu factories in Pangkabinanga and the Sungguminasa mains market. The difference of dissolved oxygen content in water at each sampling point was influenced by the difference of sampling location so that the effect on the temperature that affects the dissolved oxygen content in the water. As well as other factors such as biological activity, photosynthesis activities, and waste or the influence of terrestrial activity.

Uncomplicated DO levels whereas DO levels in IPA IV Inlet Maccini Sombala in the morning due to its DO levels were in the range of 2.49 mg / l so as not to qualify as raw materials for second class drinking water because <4 mg / l . The low content of dissolved oxygen at the sample point was due to the habit of disposing of waste by residents around the river into the river where most of the waste is organic waste. In addition, the researchers observed that the low dissolved oxygen content in the Maccini Sombala area in the morning was due to the fact that at night trucks picking up sand and sand dredging could increase the turbidity of raw water so that dissolved oxygen content was low. This was also because of the Mayor Regulation no. 94 of 2013 concerning the ban on 10-wheeler trucks operating during the day[13].

The waste will be decomposed in water by microorganisms so that the decomposition of compounds by microorganisms utilized oxygen so that the dissolved oxygen (DO) decreased. The higher the dissolved oxygen in the water, the easier the life of the water biota to live, the lower the oxygen was dissolved in the water, the more difficult life of the water biota to live[14].

Table 1
Results of Turbidity Parameter Examination on Jeneberang River Water in Every Inlet Municipal Waterworks (PDAM) Makassar City

Times	Results of Parameter Turbidity (NTU)			
	Inlet IPA Somba Opu	Inlet IPA Ratulangi	Inlet IPA Maccini Sombala	Inlet IPA Malengkeri
Morning	78,59	38,85	11,45	101,39
Noon	102,20	24,39	10,61	88,76
Afternoon	79,16	21,34	8,01	1972,8

Table 2
Results of BOD (Bissolved Oxygen Demand) Parameter on Jeneberang River Water in Every Inlet Municipal Waterworks (PDAM) Makassar City

Times	Results of Parameter BOD (Bissolved Oxygen Demand) (mg/l)

	Inlet IPA Somba Opu	Inlet IPA Ratulangi	Inlet IPA Maccini Sombala	Inlet IPA Malengkeri
Morning	0.92	0.62	1,28	0.92
Noon	1,24	0.51	1,60	1,08
Afternoon	1,72	0,58	1,40	0,63

Table 3
Results of DO (Dissolved Oxygen) Parameter on Jeneberang River Water in Every Inlet Municipal Waterworks (PDAM) Makassar City

Times	Results of DO (Dissolved Oxygen) (mg/l)			
	Inlet IPA Somba Opu	Inlet IPA Ratulangi	Inlet IPA Maccini Sombala	Inlet IPA Malengkeri
Morning	8,62	7,59	2,49	7,55
Noon	6,93	7,51	4,98	7,02
Afternoon	7,45	7,18	5,35	7,47

Table 4
Result of Measurement and Observation of Environmental Condition on Jeneberang River Water around Makassar IPAL Inlet Science

Times	Environment Condition			
	pH	Temperature (°C)	Current Velocity (m/s)	Weather
Inlet IPA V Somba Opu				
Morning	6.9	28	0.1	Sunny
Noon	6.9	31	0.15	Sunny
Afternoon	6.8	29	0.1	Sunny
Inlet Sungguminasa IPA Ratulangi				
Morning	7,1	27,5	0,20	Sunny
Noon	7,1	29	0,20	Sunny
Afternoon	7,1	28	0,26	Cloudy
Inlet Malengkeri IPA IV Maccini Sombala				
Morning	7,0	27	0,12	Sunny
Noon	6,8	28	0,12	Sunny
Afternoon	6,8	27	0,00	Sunny
Inlet IPA IV Maccini Sombala				
Morning	6,5	29,5	0,00	Sunny
Noon	6,4	32	0,00	Sunny

Afternoon	6,5	30	0,00	Sunny

[5] References

[1] Andhika, et al (2013). Kadar Sisa Chlor dan Kandungan Bakteri E. Coli Perusahaan Air Minum Tirta Moedal Semarang Sebelum dan Sesudah Pengolahan.
[2] Data Survei Sosial Ekonomi Nasional. (2007). Pedoman Kor. Jakarta.
[3] Menteri Kesehatan RI.(1990). Peraturan Menteri Kesehatan Nomor 416 Tahun 1990 tentang Syarat-Syarat Dan Pengawasan Kualitas Air. Jakarta.
[4] Kabupaten/ Kota Makassar Sulawesi Selatan. (2004).
[5] Yuliastuti, Etik. (2011). Kajian Kualitas Air Sungai Ngringo Karanganyar dalam Upaya Pengendalian Pencemaran Air.
[6] Halide, Jumaidil. (2014). Air Sungai Jeneberang Tercemar Berat. Sindonews.
[7] Apriyantomo dan Nurbowo.(2003). Panduan Belanja dan Konsumsi Halal. (19-24).
[8] Widayat, et al.(2010). Penyisihan Amoniak Dalam Upaya Meningkatkan Kualitas Air Baku PDAM-IPA Bojong Renged Dengan Proses Biofiltrasi Menggunakan Media Plastik Tipe Sarang Tawon.JAI. 6(1): 64-74.
[9] Peraturan Pemerintah Republik Indonesia Nomor 82. (2001). tentang Pengelolaan Kualitas Air dan Pengendalian Pencemaran Air.
[10] Peraturan Gubernur Sulawesi Selatan Nomor 69. (2010). tentang Baku Mutu dan Kriteria Kerusakan Lingkungan Hidup.
[11] Akbar dan Sudarmaji.(2010). Efektifitas Sistem Pengolahan Limbah Cair dan Keluhan Kesehatan Pada Petugas IPAL di RSUD Dr.M.Soewandhie Surabaya.
[12] Jamal, dkk.(2014). Analisis Kualitas Air Baku PDAM Pada Saluran Transmisi IPA Panaikang.
[13] Peraturan Walikota Nomor 94. (2013). Mengenai Larangan Truk 10 Roda Beroperasi di Siang hari.
[14] Manihuruk, Elperida.(2013). Pengaruh Waktu Pengambilan Sampel Terhadap Kadar BOD, COD dan Minyak/Lemak Dari Air Sungai Silau Kabupaten Asahan[Skripsi]. Medan:Universitas Sumatera Utara.
[15] Maloku, Faton. (2014). Defining Of Physical-Chemical Parameters Of Krena River As The Polluter Of Enriku River.European Scintific Journal.ISSN.3(1): 1857-7881.

Cultural Harmony Between Islam and Local Traditions of *Caramseguguk* in Rengas Payaraman Ogan Ilir Indonesia

Muhammad Syawaludin

Universitas Islam Negeri Raden Fatah, Jl.Prof. K.H. Zainal Abidin Fikri, KM.3.5, Kemuning, Pahlawan, Kemuning, Kota Palembang, Sumatera Selatan 30126, Indonesia
Email: syawalibrahim71@gmail.com

Abstract: Islam as a religion which teaches various aspects of normative and humanities must undergo a process of acceptance and rejection in society. Otherwise, traditions of a particular community identity also endure a process of eclectic adjustment along with Islamic teachings. It proves that Islam is not only understood as a doctrine but also as a result of interaction among cultural values, norms and morals that are present in the forms of religious practices that tolerate local values. The meeting of Islamic values and local traditions has maintained and formed Islam narrated activities as a form of internalisation of the teachings and behaviour. Such as *Caramseguguk* tradition in South Sumatera, that is the result of the harmonisation encountered between Islam and local traditions in everyday behaviour.

Keywords: Cultural Harmony, Islam, Local Traditions, Caramseguguk

1. Introduction

Islamic studies and the tradition continue to evolve to be eclectic studies that try to combine the tradition of Islamic studies with the tradition of social science studies and anthropology. In addition, there are also Islamic studies that look at social interactions and patterns in a Muslim society. According to this view, Islam developed in the life of society is inseparable from culture. Further, the combination of these can also be seen in a particular event or ceremony. It is difficult to say that the local tradition is not synonymous with Islam or vice versa in which Islam is an entity formed by local traditions. But, civilisation of religions in Indonesia archipelago is a series of process of formation of periodic religious life. Islamic historical process cannot be separated from spiritual civilisation that existed before Islam was present in this archipelago. Therefore, in the practice of Islam in Indonesia can cultural values of the religion be found. Islam in this sense might be not the one maintained and practised by Islam entities, e.g. practices of giving a dowry in a wedding ceremony or reading history haul someone who is considered "sacred". These are a tradition common among Hindus Mataram and Bikhsu.

This process lasts a long time and continues to be the entity and identity of Islam in Indonesia. It lives in the civilisation of religions network. Thus, having a Muslim identity means they have become part of the historical Islam Nusantara. This fact draws a new pattern of emergence of Islam Nusantara which serves as a social agent for Islam and the archipelago. It forms an episode of a slab of civilisation in Indonesia, which has an open and dynamic character that blends Islam and local traditions value. It is entirely possible for the meeting between Islamic values and local traditions maintained, and Islam narrated activities to exist in circumstances and daily habits. This meeting point is referred as local wisdom. This term refers to the notion that human beings and others are connected because human needs other humans, not because of the functions of the logic of work. In addition, *Caramseguguk* tradition still exists and takes place in public life. In Rengas Payaraman South Sumatera, this fact is interesting to be studied more in-depth empirically and holistically.

Furthermore describing the clan system governing family relationships and relationships between women and men in Simboer Tjahaya is derived from Islamic teachings; Nurdin, suggests that it is the culture of the indigenous people, pepadun Muakhi, in Lampung [1]. In his research, Fauzie sees Muakhi culture, indigenous people living in pepadun Lampung, is accommodated from various values, norms and philosophy of life which is continuously running. This experience is internalisation process in pracpepadun indigenous culture. Interestingly Muakhi was derived from the Arabic language which is often delivered by Islamic religious leaders in his message. It was where narratives of Islam became the source of the local traditions of indigenous peoples in Lampung. Research on social capital in the great tradition Agung Gawemeranjat II Ogan Ilir village is able to describe the glorious tradition as a tradition that contribute to social change by teaching about social capital [2]. According to her, one-on-one social value in *agenggawe* refers to an adaptation of the teachings of Islam embraced by the citizens of Meranjat II. Even according to the teachings of Islam it is able to bind a variety of local cultural differences [2].

2. Method

Through a case study approach, the data is derived from rationalism and the situation of socials, politics, and culture. The data is dynamic, open, and occurs in social relationships. The research seeks the real social formation and its dynamics. The data was captured through observations and interviews. The data was then analysed with interpretative method to understand the events and their relation to the historical period in certain past situations.

The theoretical approach is structural functionalism [3] as George Ritzer dan Ritzer points out that this approach provides explanation of how collective cultural accumulation occurs and how it can be interpreted as a form of local tradition [4]. The culture is reflected by event of *caramseguguk* in Rengas. While Islam becoming adhesives and evaluation tools values occur. Core concepts of structural functionalism theory lies [4], in the full understanding of the process of convergence of cultures, traditions, systems, materials in the form of traditions of everyday life and the events in *caramseguguk* tradition performed by Rengas society will impact on the development of Islam and social change [5].

3. Findings and Discussion

One of local wisdom values that continue to be maintained in society in South Sumatera is the behavior of *caramseguguk* build unity and mutual cooperation. The traditional social system was maintained in culture of Ogan Ilir and became the motto of this district. Ogan Ilir has Ogan Ilir indigenous peoples and tribes that migrate from Java and Sunda. The indigenous population of Ogan Ilir consists of (1) Ogan Ethnic: include residents along the river Ogan from the village Munggu to Embacang Muara Kuang and Lubuk Keliat. The language used is the language Ogan. (2) Pegagan Ethnic: include residents in the district of Tanjung Raja, Rantau Panjang, Sungai Pinang, Rantau Alai, Kandis, Pemulutan, Pemulutan West, South Pemulutan, Indralaya and South Indralaya. English is the language Pegagan famous. (3) Penesak Ethnic: or called Meranjat parts, covering a population in the district of Tanjung Batu and Payaraman and partly District of Lubuk Keliat (villages ex district of Tanjung Batu) in Malay Palembang known as Meranjat.

One of the villages that preserve *caramseguguk* is Rengas village Payaraman-of Ogan Ilir South Sumatera. Rengas village was originally entered in the district of Tanjung Batu, since 2005 divided into two, namely Tanjung Batu sub-district and District Payaraman, Rengas village included in the Payaraman. In this region, tribe Penesak, is one of the tribes that dominate in Rengas and other areas (Tebedak, Lubuk Bandung, Betung, Ketiau, TanjungDayang, Series Bandung, Seri Kembang, Payaraman, TalangSeleman, PayaBesar, Tanjung Lalang, Seri Tanjung ,Bangun Jaya, Tanjung Tambak, Tanjung Baru Petai, Cape roof, Limbang Jaya, TanjungLaut, Meranjat, Meranjat II, Meranjat Ilir, Beti, Tanjung Pinang, Tanjung Batu, Tanjung Batu Seberang, Pajar Bulan, Senuro,

Sentul, Burai, Tanjung Burai. Rengas village itself is the oldest village in the district of Ogan Ilir Payaraman which consists of two regions, Rengas 1 and Rengas 2. The word comes from the shanty payopandan (now located in out- plantation PTPN VII), while partly indeed have lived and settled in Rengas village. Because access in Rengas village more easily affordable then the previous residents decided not to leave their inpayopandan and moved to the village of Rengas.

Rengas word itself is taken from the word meaning Ronges cleft. Rengas village was once led by a Kerio but since 1989 the village leadership led by the village head. Land in the village Rengas when viewed from the ownership of land is divided into two types namely Personal Land and Land Garden Village called *Umohor Duson*. As for the ownership of land in the village Rengas based largely on land legacy for generations without a land administration, even if it is only in the form of a Certificate of Business Rights (SKHU) on land. Continuities *caramseguguk*system inRengas society has a role as a social and cultural institutionalization of the family and relatives. In a common type of family unit, usually to be seen through several aspects, (1) Kinship system, (2) Line of descent, (3) Inheritance system

Kinship is a social unit whose members have blood ties, while the kinship system in question is the customs, norms, values, knowledge, human behavior is bound by ties of blood or marriage. In a pluralistic society kinship Ogan Ilir embraced *patrilineal* system [6]. This system involves the type of family unit in a *patrilineal* society. This was evident in the system of local daily life, for example in dealing with jobs that require a lot of people, the feeling of collateral (*seduluran*) is very prominent at all. They are collateral will face severe problems or even work with full responsibility, feeling embarrassed when they can not participate in any employment or collateral relatives they did not succeed satisfactorily. For example, in the marriage ceremony, moving house or called *Sedekah*, stricken. The nature of family rose to be the basis of resolving conflicts among citizens in *Panesak* clan, has been known forms of communication between people who are rooted and frequently used media for conflict resolution among residents and conflicts with communities beyond them. The mechanism is Flour Fresh we called *Are Basare*. Fresh flour is a custom event that is performed to remove the brotherhood between people or between family members when they disagree and intend to finish it. Such disputes are generally resolved and facilitated by the adat. Whereas Are Basare is customary mechanism that is intended to lift the brotherhood between family through weddings, circumcisions, and other parties. This is done when the accidental guest name (either a husband, wife or child) from outside the indigenous community together with host names have urination. In the presence of many people present then citizens of the same name was sworn in customary manner. Rapturing brother like this is also known as *Angkanan*.

To maintain continuity between clans with *caramseguguk* no maintenance business patterns that motivated the desired behavior, so that the ideas of cultural system makes cultural attributes inherent in social systems are maintained, such as: the mating of a Pasirah (village head) is very much appreciated often asked to give advice called peradatan. Patterns in *caramseguguk*territorial relationship is symbolized in the person's status and role in the local community. This pattern is conserved terms like Pesirah, headmen, traditional leaders, Kerio, *pembarab*, saplings. Each of them in certain events kinship or cultural attributes are still used. To clarify their identities typically use oversized clothes each. A skullcap or cap *Pesirah* using Air-mas and a red umbrella, while *pembarap* or Kerio generally only half waist wear gloves and headband called pose. The nuclear family (family nucleus) with an extension (extended family), with the levels of age and profession with field-field and courtesy of family relationships horizontally turns tied to territorial factors and genealogy. Nuclear family is the core family formed through marriage, the elements and all the requirements refer to the customary rules and *Shari'a*. Nuclear family consisting of father, mother and children each serves as a protector and breadwinner and the successor generation in which there is the responsibility to continue the tradition [7]. Children receive and perform the role of tradition in the nursery-courteous both in the context of differences in age, sex and how to behave with others. Law and the law is the identification of parent-child in the extended family for the marriage relationship.

IOP Conf. Series: Earth and Environmental Science **156** (2018) 012037 doi:10.1088/1755-1315/156/1/012037

This leads to expansion of the functions of the role, rights and responsibilities of its members. Kinship that created a strong position and influence each other even become a solid foothold for various activities. Important underlined entire extended family members included in the circle of unrelated and indeed refer to *Shari'a.* Expansion of the nuclear family is generally obtained through the blood relationship of grandparents and obtained through called *besanan* relationship with your wife or husband. Norms applied in social interactions both within the family and the environment is an extension of the existing norms in the nuclear family. But the polite more showed a pattern firmer and tighter. With regard to Islamic procession in *caram seguguk* no longer need to be emphasized that local cultural elements that can or should be used as a source of law is that at least does not contradict the principles of Islam. The elements that are contrary to the principles of Islam itself must be removed and replaced. This is the meaning of the Islamic presence in a place or country. In the science of *Usul Al-Fiqh*, the local culture in the form of customs is often called *'Urf* (derived from the same root word with *al- Ma'ruf*). Because the *'urf* one that comes from people's behavior, and contains elements that are wrong and right, then naturally the Muslims must look more widely and critically. Ease the views of the socio-economic fulfillment so get "respect" from other social layers [8]. But there are also some people in the village and in the District PayaramanRengas different view associated social status of "elite" based wealth and jobs. The results of in-depth observation and analysis of public perception and understanding of Rengas generally that the "rich" will be deemed to have a social status as "elite" who respected if the person has a behavioral and life order in accordance with the values and norms in society in general.

As "elite" are people who have wealth and do not behave in violation of values and norms in society *Rengas* village. If a rich man, but do not behave according to the norms and values, it is considered that the person is not "elite" and unworthy of treatment is respected. Values and norms in public life generally *Rengas* village strongly influenced by socio-religious values of Islam and Malay customary norms. For example, if a rich man and had a job but the behavior of life often do cockfights or often harvest the pineapple on top of fairness, then that person is considered to be elite and does not deserve respect. From perspective *Rengas* villagers a social status will be more dignified and respectable if followed by the appropriate behavior to social norms that are loaded with the values of Islam. Social status is often based on the type of job one, it is also the case rural communities ratings in *Rengas*. For the people perceived as "elite" not only have the wealth but also determine the type of work. Of course such case the suitability norms and everyday behavior. This type of work is considered as having the social status of "elite" by the public *Rengas* is formal employment such as civil servants, military/police and non-formal employment such as fruit or vegetable agents, teachers, clerics, pesantren. Their tradition in *caramseguguk* are driven by four factors, namely (1) residents or citizens concerned are equally attached to the area occupied, (2) there is a feeling bonded to each other by reason of *puyang*(descent), (3) due to the incorporation of territorial factors and genealogy, (4) The similarity of the Islamic faith.

Aspects that stand in talks *caram seguguk* tradition is the aspect of confidence and trust of the people's behavior. *Caram seguguk* not just a rule of customary norms, however, abortion is closely related to the Islamic religious behavior that reach a good behavior is individual, group and society, applicable both in daily situations and incidental. In the tradition of tribal *caram seguguk Rengas panesak* can guarantee kinship families remain intact and in or in the neighborhood, where a certain family grove in habit a place, and makes the area as a place of their group settlement. Over time, the number of community members in the group one family grove was gradually increased. Key actors in the implementation of this *caram seguguk* is *Pasirah*, for a fixed *Pasirah* customarily have power and have the authority under the supervision of resources such as land and resource ownership hamlets. Pasirah also reserves the right to mobilize the labor of the population matagawe to various official or private interests. Peonage for example *wang* clan porters flunky, *kemit marga*, *kemit hamlets*, and *gawe road*. While the social structure is also a key caramsuguguk is Marga. Becouse *marga* is functionally plays a very important role for the life and the history of civilization in South Sumatera.

IOP Conf. Series: Earth and Environmental Science **156** (2018) 012037 doi:10.1088/1755-1315/156/1/012037

Traditionally, the clan is the highest institution of society after the institution of the family, village and hamlet. Clan headed by a figure who is generally known as *Pasirah*. With certain qualifications, clan leader also referred to as Depati and Prince. A head of the clan, to be called the *Depati*, if he has successfully been selected to assume the position of Chief of the clan at least two times in a row, while the Prince is to have at least five times in a row. Besides *Pesirah* as a key actor in *caram seguguk*, there are other actors who control norms and values of the organization of *caram seguguk* namely the clerics or religious leaders us. As the highest institution acting as custodian of the clan social order (social piety). Measurability Islamic values in the tradition *seguguk caram* very visible in the process of organizing itself *caramseguguk*, like berasan, bereguk, their leader, news, nolonggawe and cash clan. Relevance is charity *ma'ruf Islamic teachings, al-wakalah, at-ta'awwun, as-syirkah and al-busroh.* Waking patterns of Islamic values in the procession *caram seguguk* done with the involvement of *Kyai or Ustaz* as normative and ideological leader. Social processes that occur within each *caramseguguk* shows local actors in the social space that is dynamic and organic. As the following table;

Table 1. Islam and Local Traditions of Caramseguguk

NO	Caram Seguguk	Communicative action	Islamic Values
1	Bereguk	Inviting Consultative	Working together
2	Berasan	Delivering the will and intent	Please help
3	Nambah Keluargo	Build sympathetic	Kinship
4	Nyirenke Gawe	Invite and encourage residents to do something	Copyright Concerns
5	Nerimo sando	Provide assistance to residents' Trust	the use of the grant

Source: 2016, study of data processed

Social functions in the tradition of *caram seguguk* an organic functions directly felt by local communities such as; Solidarity called *pengemin* is a social solidarity is *pengemin* have awareness of and concern for the general public condition, characterized by invitation at the time of mutual cooperation to carry out or assist together. In addition *caram* locality *caram seguguk* is one model that puts mutual trust and maintain the trust on the basis of honesty everyday. For residents who have enough capital, then he could do auctions of personal care or be an actor capable of communicating to other residents voluntarily. This form is often known as resiprocity. The process of forming resipro city in *caram seguguk* not done mobilization or social structure, but run as everyday life alone or organic, not done by an agency or social engineering. It is unavoidable role in the mobilization of agents and actions encourage citizens to take action concern, but the process is not to create a mobilization structure and social structure, because the tradition of empirical *caram seguguk* an event that is always run by private citizens *Ogan Ilir Rengas*. In addition, real *seguguk caram* attitudes and behaviors of high culture which is managed by the residents themselves and the implications for the lives of the citizens themselves.

IOP Conf. Series: Earth and Environmental Science **156** (2018) 012037 doi:10.1088/1755-1315/156/1/012037

It may be said that caram seguguk one-on-one form of civilization locality is able to live and survive in the social system. When the local people do not do things that generally make good habits, then other citizens with the norms and values will provide corrections to citizens who do not perform the behavior.

Correction delivered culturally and maintain feelings of other citizens. What are termed *Tepo Selero* and *Sedulang Seperahuan.* In *caramseguguk* auction process involving various actors concerned in Rengas village. *Caramseguguk* actors are villagers from all social classes, professions and religions. Starting from the village up to the farmer stakeholders, starting from the midwife to the farmer workers. Social collective action is actually a manifestation of the teachings of *al-ihsan* and *tuminubillah.* This occurs because of the institutionalization of Islam in social institutions and institutional in nature. *Kyai* or stakeholders generally like *pesantren* in *Rengas*. While people *Rengas* send their children to pesantren or other Islamic institutions. In this way, the process of Islam in the local tadition *caramseguguk* able to co-exist and even complement as the norm and the value of life.

According to the above opinion, the encounter Islam in *caram seguguk* tradition is helping others by others to participate overcoming economic, cultural, social harmony and local residents. In this arrangement space *caram seguguk* promote social values that are not limited by the definition of humanity. Because for residents *caram seguguk* implemented is very important, so anybody community members who find it difficult to live should be assisted by the residents. Internalization of Islamic values recorded in *caram seguguk* community OganIlir Rengas including the following:

Table 2. Caramseguguk: Participation and Mutual Help

	Participation	**Mutual help**
Caramseguguk	Resiprocity	Cooperation
	Solidarity	Kinship,
	Sympathetic	Kinship sympathetic
	Mutual trust cooperation	Togetherness

Source: 2016, study of data processed

Caram Seguguk has its own values is something hereditary ideas that have been considered true and important members of society. Value always contain the consequences. Value always played an important role in people's lives and change sosial. Dua important value in *caram seguguk* so that it can serve as a model of natural assimilation of Islam and local traditions are organic community that is the value of togetherness and familiarity because the attitude of helping each other bring between members of the community with other community members help and help members of the public.

Saw the forms of social relations in the form of kinship, exists between members of society, which is the common goal. While other values are kinship is with the implementation of *caram seguguk* means merging two or more families bound together by mutual sympathy committed by family members each so that these families grow and establish family relationships. In the context of social relationships *caram seguguk* which is the motto of *Ogan Ilir* implies solidarity or mutual aid with harmonious , democratic and uphold human rights in the welfare and prosperity of society *Ogan Ilir.* When back at the encounter of Islam with local tradition *caram seguguk* in community life *Rengas Ogan Ilir* appears once the shifting that starts from the belief in animism-dynamism and deep thinking of previous generations, in the form of customary in the direction of traditional Malay culture which has come into contact with Islam.

IOP Conf. Series: Earth and Environmental Science **156** (2018) 012037 doi:10.1088/1755-1315/156/1/012037

Although initially not been acculturated intensively, because dealing with Hinduism-Buddhism that is still strong and the counter action from China. Intensive contiguity are still at the end of 19th century AD or centuries VII / VIII Hijria. During this period the process of acculturation of Islam and traditional Malay culture Palembang, can be said really happened and managed to transform the traditional Malay culture into Malay culture that is based on Islam. The success of traditional Malay culture to transform Islam into the Malay culture that is based on Islam is a form of acculturation of Islam and Malay culture is capable of receiving various aspects of culture; (1) Transformation of the belief system of animism-dynamism Malays to Islam derived from revelation; (2) Custom transformation (the rites of the life cycle, the system of government and social system); (3) Transformation tradition of myth and purpose to the tradition as a means of socialization and solidarity values.

One manifestation of that transformation is *caram seguguk* tradition itself. *Caram Seguguk* traditions are cultural creations and Malay reconstituted with Islam. This is possible because the appeal of Islam that has the character and spirit of egalitarian. Here the role of the presence of preachers and religious teachers who then creates social houses of worship, *madrasah* and in study greatly contributed to the process of internalization of Islam into the personality of the local culture itself.

4. Conclusion

The findings of this study provide a wide range of both academic contributions, empirical and practical operations. Academically, the contribution of this research are: 1) In terms of anthropology, the values of Islam will dominate and deeply rooted in the cultural system of a society where the values of Islam acculturated into the culture of the community through an intensive process, gradual, accommodating, empathy, and sustained, not frontal and confrontational; 2) In terms of sociology, Islamic acculturation into a society can make Islam as an identity and solidarity of a community binder (spirit de corps), because the identity and solidarity of a community is not absolutely based on ethnic unity. It also can also be formed on the unity of aqidah. Social unity is called the *ummah*; Empirically, the Islamic acculturation into the Malay culture in Ogan Ilir, has made Islam a person surnamed *Panesak* Malay identity, so that the Malay identity is not always based on genetic factors, but also can be formed on the basis of *aqidah*. Thus, the "Malay" is an open concept that can be entered anyone through the corridors of Islam.

5. References

[1] Nurdin A, Fauzie A 2009 Culture Muakhi And Regional Devt Towards a dignified. *Gama Media*,

[2] Irasti F 2013 Social Capital In The Great Tradition Gawe On Meranjat II Villagers Idralaya Ogan Ilir District, *Fisipol Unsri*.

[3] Hamilton P 1991 *Reading From Talcott Parsons,* Translation by Hartono Hadikusumo. *Tiara Wacana.*

[4] Ritzer G, Douglas JG 2005 Translation by Alimandan, 6th edition, third, *Pranada Media.*

[5] Alexander JC, Mast JL 2006 Introduction: Symbolic action in theory and practice. The cultural pragmatics of symbolic action. In JC Alexander, B Giesen & JL Mast (Eds.), Social performance. Symbolic action, cultural pragmatics, and ritual. *Cambridge University Press.*

[6] Berlian S 2002 Traditional Management Study of Islamic Gender On The Draft Simboer Tjahaya. *Millenium Press.*

[8] Radcliffe-Brown AR 1952 Structure and Function in Primitive Society. *Free Press.*

[9] Takari M 1998 Ronggeng Malay North Sumatera : History, Function and structure, Gadjah Mada University, Yogyakarta.

Environmental Conservation of Muslim Minorities in Raja Ampat: Sasi, Mosque and Customs

Ismail Suardi Wekke,[1] Boby Samra[2], Nurlaela Abbas[3], & Nurlaila Harun[4]

[1]Sekolah Tinggi Agama Islam Negeri (STAIN) Sorong, West Papua. Indonesia
E-mail: iswekke@gmail.com
[2]Department of Architecture, Universitas Lancang Kuning, Pekanbaru.Riau. Indonesia
E-mail: boby@unilak.ac.id
[3]Universitas Islam Negeri Alauddin, Makassar, Indonesia
E-mail: nurlaelahabbas@gmail.com
[4]Faculty of Law, Institut Agama Islam Negeri Manado, North Sulawesi. Indonesia
E-mail: laila.harun@iain-manado.ac.id

Abstract: The form of symbolic understanding of religion and the rules reflected in symbols for their own circles. Three terms used in this research are sasi, custom, and conservation. Sasi is a practice of adaptation to the nature that is run as part of interaction between Raja Ampat society and their environment. This qualitative research using in-depth interviews and non-participant observations were employ to assits the data collection. Interviews were conducted on a series of subjects consisting of Muslim community leaders, mosque imams, mosque administrators, village heads, members of political parties, and district governments in Raja Ampat Regency, West Papua Indonesia. Meanwhile, the observations were carried out in several opening and closing activities and attending several deliberations in mosques. The study lasted for six months, starting from April to October 2014. To meet the objective of data triangulation, the time was extended until December 2014. This study also conducting directed discussions with experts and researchers in the college environment as much as three series. The series of discussions involved experts from the Polytechnic of Maritime and Fisheries, State Islamic High School (STAIN) Sorong, and Muhammadiyah University of Sorong. These three forms were repeated and also with sufficient frequency to ensure the reliability and accuracy of data already obtained. This article shows that there is a dynamicization between religious practices and more technical applications in the form of environmental conservation, thus giving birth to sasimosque which has uniqueness in vernacular architectural design as well as an institution to leads the native ritual. In the end this study also found Sasi Mosque serve as institutions to support the preserving of the environment.

Keywords: Sasi, Mosque design, Mosque function, Customs, and Raja Ampat

1. Introduction

An effort to maintain the harmonious relationships between human beings andnature or the wisdom of being one with the nature or creater it seems existed in almost every ethnic including the native Raja Ampatin West Papua. Many of the native in Raja Ampat Island were Moslem that tried to preserve this philosophy of life. The Raja Ampatlocation on the bird's head of the island of Papua, with Waisasi was chosen as the center of government located on the island of Waigeo. The Moslem community of Raja Ampat occupies a coastal area. This geographic location of sea and mountain becomes very influnces with the native activities. The study of Islam in Papua is a necessity to explain that Islam actually has experienced spreading and encounter with many environments, not only in large islands in the western part of Indonesia. This article tried to discuss the existence of sasi, mosque and native custome in relation to the actualization of environmental conservation by Muslim minorities in Raja Ampat.

Further, studies on Islam and ethnicity customs (henceforth custom) have been described many scholars including Taufik Abdullah, that explained that they are mutually reinforcing among custom and religions (Islam) or streghtheningeach other. However, there also a potential for conflict that is always a constraint in the relationship of both. Moreover, the relationship between Customs and Islam has its ups and downs and there is also a dynamic transformation to form its own dynamics relations to sustain the social order [9].

Other studies that showing Islam and Customs can have a dynamicly relationship were explainedby Abdullah, Martinez and Radzi, showing the integration among it into the legal system that adopts the 'urf and subsequently become positive law in the legislation (Malay marriage case) [13]. This reiterates that customs are in fact the answer that the formation of the law can begin by identifying the prevailing in certain environments. Thus, this study see itsrelevance to put these study as basic thingking and distinguish there has been no specific research on Customs and Islam in the context of the Muslim minorities of Raja Ampat. In other word, previous researches have specifically examined the relationship between customs and conflict, values, and marriage, while customs can be used as a source of law based on the understanding of the local community while this article specifically examines the position of customs as a local wisdom that is driven by the existence of the mosque.The research gap proposed by this study is specifically explores Indonesian custom and Islam in the context of Muslim minorities in Raja Ampat. Therefore, three focus of this study that is going to be explorer are *sasi*, native custom (adat) and Mosque conservation.

Sasi is a ritual or practice of adaptation to the nature that is run as part of interaction between native Raja Ampatand their environment. With varying understanding and religious expression, each group treats the environment in accordance with religious rituals they follow [2]. Sasi means also witnesses, intended as a testimony of residents of a region to abstain from certain behavior with regard to the environment. Thus there will be no one who commits the offense and then argues for a reason of not knowing the time and the area subject to *sasi* status. In the context of a witness, when opening and closing an area or a certain period as a sasi, a ceremony is performed. This is to give testimony to every citizen that the place is subject to *sasi*, including timeswhere they are prohibited to perform any activity in the area that has been set as*sasi*.

As for adat or custom, is a legal instrument although it is sometimes not institutionalized in writing in each community [5]. It is a tradition that is run as a social demand and a part of social agreement that must be executed. When executing it, someone will get peace. Along with the demands of modernity, implementation of customs is a form of community resilience where in the midst of life with other communities they are still trying to carry out activities believed to be part of the ancestral heritage [3]. Meanwhile, conservation is part of efforts to maintain environmental excellences. The desire to always maintain the inhabited area is managed using local understanding [4]. This serves as a means to bridge the understanding of society with the goal to be achieved.

While mosques not only serves as place to pray but also an institutions that hold the authority to regulate environmental management including the execution date of sasi ritual. It should be based on the agreement between the mosques congregation, to decided the timing of harvesting, maintenance, and the activity of the relation between human and the centralized environment in communication are facilitated by mosques leaders. The selection of mosques leaders tends to be based on seniority and descent. The duties and responsibilities of mosques leaders are carried throughout life. When a leader of a mosque died, re-election is done. With strong belief in the existence of mosques leaders and their election agreed in all villages, then the community also gives greater confidence, not only in regulating the affairs of worship within the scope of the mosque building, but also reach out to the common affairs among Muslims [9] [3].

2. Method

This research used qualitative approach. The design of this study used two main characteristics to perform the analysis, namely a constant comparison of data and emerging categories and theoretical sampling of different groups to maximize the similarities and differences of information. With these categories forms of findings were found later. From the findings, the data theoretization was formed to answer the research question. The research approach was chosen based on the suitability between the research objectives and the research subjects so as to facilitate the collection of data in the field. In-depth interviews and non-participant observations were made throughout the data collection. The data were collected by using interview list instruments and observation sheets. Interviews were conducted on a series of subjects consisting of Muslim community leaders, mosque imams, mosque administrators, village heads, members of political parties, and district governments. Meanwhile, the observations were carried out in several opening and closing activities and attending several deliberations in mosques. During the course of the research, the researcher's presence was endeavored not to be realized by the community so that the activity takes place naturally. Preoservations also conducting by, the researcher sought to interact with the community for three weeks to minimize alienation and also bring about acceptance within the mosque's congregation.

The study lasted for six months, starting from April to October 2014. To meet the objective of data triangulation, the time was extended until December 2014. This three-month opportunity was an opportunity to re-check data repeatedly. Triangulation process was done in three forms, namely, first, to check the subjects of research in other villages where resources were tracked to various corners of the Raja Ampat with different subjects but the same topic. Then, checking the data in the same study area with the research subjects had diverse professional backgrounds. Finally, conducting directed discussions with experts and researchers in the college environment as much as three series. The series of discussions involved experts from the Polytechnic of Maritime and Fisheries, State Islamic High School (STAIN) Sorong, and Muhammadiyah University of Sorong. These three forms were repeated and also with sufficient frequency to ensure the reliability and accuracy of data already obtained.

This research was conducted in Raja Ampat Regency, West Papua. The research was limited to coastal villages, while villages located in the mountains were not selected as research sites. This was carried out as an attempt to impose limits on the study of the relation between Customs and Islam in a more specific context, namely the factor of sasi and mosque. The limitations of this study are only because it was implemented on large islands and did not reach some small areas.

3. Result and Discussion

With several limitation this study comprehensively describe relations among; sasi, Islamic religion representing by mosque that serve as instituions and the native custom in Raja Ampat. The discussion on findings presented in three main section below, it tries to cover the concept, the role and acculturaion of sasi ritual, mosque philosophy as well its function and native custom dynamic acculturation .

3. 1 The Concept: SASI MOSQUE

The existence of mosque is very important for the Muslims' as it is the place to worship Allah. Mosques are one type of the buildings which symbolise Islamic architecture. In Arabic language, this word is known as 'masjid' coming from the word 'sajd'. In architecture terminology, the word 'mosque' refers to a building used for the prayer activities. The mosque design therefore should portray symbol of prostration, which reflects to the meaning of 'masjid' as well as a time marking that connected past, present and future [10]. "In architecture, the measurable factors of the building design therefore should be symbolised by concept of prostration to His creations, which reflect to 'how the mosque is designed' in relation to the existing surrounding environment. Islamic architecture co-relates to an expression with an act of devotion in the mosque design. An expression of its physical form and space should reflect to the surrounding environment as its prostration values.

The geographical condition in Raja Ampat region that surrounded by sea and spaced by mountain becomes a reference for the native Raja Ampattraditional mosque designed. As a result, itswere created a mosque that carry not just a concrete building for praying but also as an institution that leads and direct the native. It was found also as a place that accommodate the traditional customs, tradition and ritual including sasiit self.Eventhough a classical vernacular design of mosque in Raja Ampat were also found has a space function that quite similar with others mosque aroud the globe such as: building orientation to qiblat direction, entrance gate, Ablution (wuduk) area, veranda area (serambi), Prayer hall, Niche area (mihrab), Sermon podium (mimbar), main roof design and minaret. Moreover, it is shaped most strongly by the regional traditions of the time and place where it was built. As a result, style, layout, and decoration can vary greatly. Nevertheless, because of the common function of the mosque as a place of congregational prayer, certain architectural features appear adopt a native philosophical in its vernacular design. Mosques for the native seems not only a place to do a worship, it is also transform become an institutions that hold the authority to regulate environmental management for certain occupation such as traditional ritual such as sasi. Based on the agreement between the mosques congregation, the timing of harvesting, maintenance, and the activity of the relation between human and the centralized environment in communication are facilitated by mosques.

In connection with mosque, the existence of *sasi* is integrated with the mosque, both in physical and spiritual forms, so there is attachment between Muslims' hearts with their places of worship. Mosque is not only a place of ritual and worship of *mahd'ah*, but also reaching a wider scope in the field of society. Mosquebecame a pioneer of pilgrims in connection with the efforts of environmental conservation. It helps maintain the continuity of living things, in the form of plants and animals. This is believed to be the manifestation of the task of preserving life on earth.Long before Islam was accepted by the native, the *Sasiritual* is divided into two types, namely intentional *sasi* and unintentional *sasi,* after Islam came and attach in their society, the sasi rituals were stil maintain however some changes were made e.g the spell, time and people who leads the ritual ceremony. Mosque in the end has role to manage and preserve the tradition. For example, *sasi* ritual both opening and closing ceremonies becomemosque responsibility to conduct in the end. Moreover, the unintentional *sasi* is the rule of law that applies throughout the time for certain behavior and treatments such as the ban on catching fish with bombs and potas, including the use of trawlswre monitored by the mosque as well as church and tribes courts. In other case, such as when a family will perform a wedding ceremony or a certain celebration approved by the leader of the mosque, they may take anything from the sea and the forest according to the purposes and time specified by the mosque leader. While unconditional sanctions are applied throughout the term, this is unconditional so that there is no reason for anyone to take anything from either region.

The opening and closing of *sasi* are held in ceremonies at the mosque. The closing is in the form of *zikr*togetherly and performed simply, followed by a prayer then asking to the God the blessings of nature in the form of His grace abundantly for human prosperity. In contrast to the closing, the opening was done lively in a joint meal. The mosque is decorated with yellow coconut leaves shaped like sea animals, prepared food to eat together, and then continued with the announcement of the start of harvesting fish, sea cucumbers, *lola*, and lobsters. It could be in one opening period there are certain animals that should not be captured at all. Thus, at the opening of *sasi,* shall be informed in the opening time what sea animals may be taken. Sometimes the sea in the cultivation area may be harvested, but some marine biota species should not be touched at all. All the devices established in the mosque are an integral part of the community's activities. The function of the mosque is in an essentially expanded environment, no longer limited to the physical function only. With the authority to manage, held and preserve of *sasi*, the mosque becomes the essential for the society, so that the emotional attachment is not limited to the aspect of spirit alone. Moreover, the mosque also concerns the livelihood, the existence of the environment, and the unification of the community for a wider and tangible role.The decision of the *sasi* starts from the proposal of the pilgrim or the observation of the leader of the mosque in a certain period of time so that a *sasi*is needed to give the abundance of fruit and also to maintain the certainty of the harvest.

3.2 Environmental Conservation: The Role of Mosque

The role of mosque is more solid than just being a place of prayer and recitation. Strengthening the capacity is needed to make mosque as the center of all the activities of Muslims in their respective regions. The combination of *sasi* with mosque becomes a form of how Islam is integrated with Customs. Both are actually strengthening each other. The mosque becomes a place of prostration, at the same time becomes a public building used together for the ceremony, ritual or other functions that involves the natives. Mosque holds a central position as the existence of village hall or church for the natives. Together, all these three bildingbecame the center of communication among their respective citizens. If mosque is worshiplace for Muslims, church is dedicated to Catholics and Protestants, while village hall facilitatesall of them. The mosque institution becomes an integral part of the life of the village community. The mosque becomes a facilitator of all the needs of its citizens. Not only in matters of worship but also in daily needs, including negotiations on mutual cooperation, deliberations to determine the right days to start construction of public facilities, and to help people in need of financial assistance in times of disaster or family celebrations. In matters concerning the harvest, as part of the needs of people's lives, mosque becomes the center for deciding what is considered strategic and generally applicable to all worshipers.

Many activities can take a place in it from meetings to discuss mutual interests can be conducted in the mosque. Especially since the beginning of a tradition that carried on from generation to generation, the mosque was built with one of the goals is to accommodate the need to accommodate large numbers of people accommodate participants. Population growth continues to evolve over time, while settlements are not increasing, so some newly established citizens choose to live in other areas. One of the main considerations is the limited land.

IOP Conf. Series: Earth and Environmental Science **156** (2018) 012038 doi:10.1088/1755-1315/156/1/012038

Mosque as a religious institution as well as a pillar of customs is given the confirmation of the role it has played for centuries. Through the enactment of the Raja AmpatRegency regulation, the mosque is getting stronger as the organizer of interests of the pilgrims in their respective neighborhoods. Despite the publication of the local regulation nothing has changed, the participation of the government and politicians has increased, followed by the increased portion of campaigns and more socialization has raised the awareness to maintain the existence of *sasi*ritual itself.Through local government policy, supported by community boards, mosques were given an authority to play as a main key to strengthen common interests, and to provide a place for environmental protectionactivities.

3.3 The Dynamic Acculturation of Islam and Customs

This study shows that the practice of *sasi* implemented in Raja Ampat is actually adapted from the Maluku and North Maluku region, because Raja Ampat is geographically and culturally very close to the kingdoms of Ternate and as well as Bacan and Jailolo the origin four kingdoms or Raja Ampat were this region attached its history in it. in the ancient times Tidore. As such, the practice of *sasi* is developed and adapted to each environment in accordance with the state of the territory, although the meaning used constextually may undergo changes and developments.The came of Islam dynamically acculturate with the local custom, since Islam coming in peace in this region, most of the principles as long it not forbid by the syariah Law in Islam will be accepted in some modification. From time to time since Islam came the tradition, ritual or community function were discussdtogether among them

The religious leaders and traditional/customary leaders sat together and deliberated the interests of the community by using the mosque institution as the starting location to discuss the solution and solve the problems that arose. Likewise with the customary institutions already present first before the arrival of Islam in the land of Bugis. With the arrival of Islam, it is precisely that customs have adapted and maintained a tradition that is not contradictory to the religious philosophy[15], so that Customs are preserved are Islam-based Customs [14]. The existence of mosque *sasi* becomes part of the legal establishment necessary for the setting up of the Raja Ampat community. This phenomenon occurs because of a unique interpretation in the context of the interpretation of Islamic teachings [16]. Islam is no longer seen merely as a teleogical aspect which then stops at the process of sharia containing the teachings of worship only. But more than that, the spirit of Islam is grown to help to preserve the environment. Religion serves as a pillar to foster awareness in protecting the environment, as in conservation practices undertaken. *Sasi* may differ in meaning. For example, in research on the religious nature of Raja Ampat fishermen, *sasi* is interpreted as oath. Although different in terms of meaning, but the practices that run in every region have the same concept. This is sustained by the attachment between Religion and Customs so that it can be productively encouraged also by the presence of customary leaders who are also religious leaders. *Sasi* can survive because it gets religious justification, of which comes from the spirit of Islam.

However, their research was limited to the study of *sasi*, mosque as an instituionsin relation to marine conservation. Thus, the particularity of this study emphasizes the study between Islam and Customs in relation to environmental conservation. Yet, the variables studied is the existence of mosque which is the main support capacity of the implementation of *sasi*.

4. Conclusions

This article shows that there is a dynamicization between religious practices and more technical applications in the form of environmental conservation, thus giving birth to mosque *sasi*. By using mosque instrument to institutionalize *sasi*, the residents gave the environmental management authority to the mosque leaders of their own choosing. *Sasi* means two things, namely testimony and prohibition. In these two meanings, coupled with the meaning of *sasi* as an oath, it becomes a tool to make certain areas or goods to be protected from excessive exploitation. Meanwhile mosque *sasi*is interpreted as a *sasi* organized by mosque institution inMuslims congregation and residences. Moreover, Mosque institutions serve as a support for preserving the environment. The existence of a religion symbolized by mosque is part of the interaction of Muslims. The encounter of people with different faiths provides an opportunity for dialogue and then together to translate religious messages in practical form. This is done as a form of demands. It also explains that religion actually goes hand in hand with its adherents. For that reason, religion is no longer in abstract norms, but it is managed in the form of beliefs that can be applied more operationally in everyday life.

this study main finding some how showed the encounter of Islam and daily practice in the form of local wisdom were realized and implemented in the *form* of their *sasi* which serves to maintain the continuity of nature which is carried out with the spirit of religious and keeping the tradition. Meanwhile, mosque as concrete buiding and institutions becomes a means to run the objective and It was found it were streghtheningeach others

5. References

[1] Pandamdari, Endang. "Dinamika Pengakuan Eksistensi Hak Ulayat Masyarakat Hukum Adat ".*Disertasi*. Jakarta: Program Pascasarjana Universitas Trisakti,2011.

[2] Hughes,Joanne."Contact and Context: Sharing Education and Building Relationships in a DividedS ociety".*Research Papers in Education*. 2 (Maret2014).

[3] Dayaram, Kandy dan Pick, David. "Entangled Between Tradition and Modernity: TheExperiences of Bhutanese Working Women". *Society and Business Review*. 2, 2012.

[4] Rahadiansyah dan APrayitno. *Transformasi Nilai Kearifan Lokal Dalam Pendidikan Bangsa: Dialektika Pentingnya Pendidikan Berbasis Lokal Genius*.Jakarta: Penerbit UniversitasTrisakti,2011.

[5] Lukito, Ratno. *Pergumulan Antara Hukum Islam dan Adat di Indonesia*. Jakarta:INIS, 1998.

[6] Belge, Cerendan Lisa Blaydes. "SocialCapitaland Dispute Resolution in Informal Areas ofC airo and Istanbul ".*Studiesin Comparative International Developmen*. 4 (Desember 2014).

[7] Rodemeier,S."Islam in the Protestant Environment of the Alor and Pantar Island".*IndonesiaandtheMalayWorld*.110(Maret, 2010).

[8] Mashad, Dhurorudin. *Muslim Bali Mencari Kembali Harmoni yang Hilang*. Jakarta: PustakaAl-Kautsar,2014.

[9] Abdullah, Taufik."Adat and Islam: AnExaminationofConflictin Minangkabau". *Indonesia*. 2, 1966.

[10] Samra B. Konsep Ruang Dalam Rumah Lama di Kawasan Senapelan Pekanbaru. JurnalArsitektur: Arsitektur Melayu dan Lingkungan. 2015;2(1).

[11] Lateff,E.E. A; Palil,M.R; Hassan,M.S. "FinancialandNon- Financial Distribution Efficiency Perfomance Among Zakat Institutions in Malaysia". *Jurnal Ekonomi Malaysia*. 2, 2014.

[12] Roth,DikdanGedeSedana."Reframing Tri Hita Karana: From Balinese Culture to Politics".*Asia Pacific Journal of Anthropology*.2 (Maret2015).

[13] Khurshid,A."Islamic Traditions of Modernity: Gender, Class,and Islamina Transnational Women's Education Project".*Gender and Society*. 17, (Februari 2015).

[14] Abdullah, Raihanah; Martinez, Patricia; dan Radzi, Wirdarti Mohd. "Islam and Adat".*Indonesia and the MalayWorld*. Vol.38No. 111, 2010.

[15] Zadrak, YanceRumahuru, Islam Syariahdan Islam Adat (Jakarta: R & D and Training Agency of the Ministry of Religious Affairs of the Republic of Indonesia, 2012), 156.

[16] Wekke, Ismail Suardi."Masjid diPapua Barat: Tinjauan Ekspresi Keberagamaan Minoritas Muslim dalam Arsitektur". *JurnalBudaya Islam Elharakah*. 2 (Juli-Desember, 2013).

[17] Brünig, B; dan Fleischmann, F. "UnderstandingtheVeiling of Muslim Women in the Netherlands ".*Journal for the Scientific Study of Religion*. 1 (1 March 2015)

[18] Benda-Beckman,F.Von.danBenda-Beckman,K.Von."Islamic Law in a Plural Context: The Struggle Over Inheritance Law in Colonial West Sumatra".*Journal of the Economic and Social History of the Orient*. Vol.55,No.4-5,2012.

[19] Curteis, Tobit dan Luxford Naomi. "Royston Cave: An Holistic Approach to Conservation ".*Journal of Architectural Conservation*. 3(September2014).

[20] Dsouli, Quarda Khan, Nadeem danKakadbase Nada K."SpiritualCapital,TheCo-evolutionof an EthicalFramework Based on AbrahamocReligiousValuesinTheIslamicTradition".*Journal of Management Development*. 10, 2012.

[21] Karpov, Y.Y. "TheDagestaniMountainVillage:FromtheTraditional Jamaatto Its Present Social Character ".*Antropology and Archeology of Eurasia*. 4 (April, 2010).

[22] Kato,T."Unbreakable Lance vs. Impenetrable Shield?–on The Relationship Between Islam and Matriliny in Minangkabau Society".*Southeast Asian Studies*. 2, 1980.

[23] McLeod, Elizabethetal. "Sasi and Marine Conservation in Raja Ampat Indonesia". *Coastal Management*. 6, 2009.

IOP Conf. Series: Earth and Environmental Science **156** (2018) 012038 doi:10.1088/1755-1315/156/1/012038

[24] Rashkow, Ezra. "Resistance to Hunting in Pre-independence India: Religious Environmentalism, Ecological Nationalismor Cultural Conservation?".*Modern Asian Studies.*2(Maret,2015).

[25] Rumahuru, Yance Zadrak. *Islam Syariah dan Islam Adat.*Jakarta: Badan Litbang dan Diklat Kementerian Agama RI, 2012.

[26] Strauss, Sophie. "Alliances Across Ideologies: Networking withNGOsina Tourism Dispute in Northern Bali ".*Asia Pacific Journal of Anthropology.* 2 (Maret2015).

[27] Wekke,IsmailSuardi dan Yuliana Ratna Sari."Tifa Syawat dan Entitas Dakwah dalam Budaya Islam: Studi Suku Kokoda Sorong Papua Barat".*Jurnal Kajian Budaya Islam Thaqafiyyat.*1(Juni,2012).

[28] Wekke, Ismail Suardi. et.al. "Identitas dan Ekspresi Keberagamaan Nelayan Muslim: Masyarakat Pesisir Raja Ampat Papua Barat". *Prosiding Seminar Nasional Kelautan dan Perikanan.*10Maret2015. Sorong: Politeknik Kelautan dan Perikanan.

[29] Samra B. The Characteristics of Malay House Spatial Layout of Pekanbaru in Accordance with Islamic Values. In IOP Conference Series: Earth and Environmental Science 2017 Dec (Vol. 97, No. 1, p. 012049). IOP Publishing.

Religion and Culture Encounters in Misool Raja Ampat: Marine Ritual Practice of Sasi Laut

Ismail Suardi Wekke[1], Diah Aghsari[2], Evizariza Evizariza[3], Junaidi Junaidi[4], & Nurlaila Harun[5]

[1]Sekolah Tinggi Agama Islam Negeri (STAIN) Sorong, West Papua
E-mail: ismail@stain-sorong.ac.id
[2] Sekolah Tinggi Agama Islam Negeri (STAIN) Sorong, West Papua
E-mail: daghsari@stain-sorong.ac.id
[3]Department of Indonesian Studies, Faculty of Cultural Science, Universitas Lancang Kuning, Pekanbaru. Indonesia
E-mail: evizariza@unilak.ac.id
[4]Department of Indonesian Studies, Faculty of Cultural Science, Universitas Lancang Kuning, Pekanbaru. Indonesia
E-mail: junaidi@unilak.ac.id
[5]Faculty of Law, Institut Agama Islam Negeri Manado, North Sulawesi. Indonesia
E-mail: laila.harun@iain-manado.ac.id

Abstract: The acculturation of religion and culture in such a way are gradually accepted without causing the loss of the native culture in one community will influence the social process. When religion and culture arise and merged into one, it can be clear to be found in their rituals. This ethnography study on qualitative in nature was conducted to see how the existence, practices, and acculturation between local culture and Islam in, *Sasi* marine rituals in Misool Raja Ampat in West Papua Indonesia. Data collection was done through observation, in-depth interview including library research on community manuscript and local government database. Using snowball sampling technique, this study is able to gain solid information from native people in Misoo that willing to take part as the informants. The informants in this study were found as influences native people in Misool community e.g; the head of Fafanlap village, community figures, female figures and youth figures. This study considers Misool community which still strongly maintains their cultures including sasi laut as a sign to start and stop harvesting the specific sea resources to maintain the sustainability of nature. The acculturation between the culture of Islam runs smoothly in changing the native mindset to eliminate the things that can cause *syirik* but did not eliminate the characteristics and the main part of the ritual of *sasi laut*.

Keywords: *Sasi, Misool, Religion, and Culture*

1. Introduction

Bhinneka Tunggal Ika or unity in diversity is the official motto to describe the chronicles of multicultural and religions that varies and ranging from Aceh to Papua which then united as the main part of the Republic of Indonesia. Hundreds of Ethnicities and thousands of sub-ethnics were recognized as Indigenous ethnics lived in Indonesia. Nevertheless, Islam, Protestant, Catholic, Hindu, Buddhist and Confucian religions are also the element that enriching the state. These religions were believed as the way to bring them to the happy life, therefore these religions were officially recognized and protected in Indonesia.

IOP Conf. Series: Earth and Environmental Science **156** (2018) 012039 doi:10.1088/1755-1315/156/1/012039

None doubt that, religion and culture are elements that are closely related to human life because of the nature of human beings as servants and as social beings. Religion acts as a belief held by the man and governs his relationship with God. While culture is a way of life that develops, shared by a group of people, and pass from generation to generation. Seeing this as big picture Indonesia is famous for its religious and cultural diversity.

Religion and culture might be two different things however, they are strongly interrelated. The merging of these two things into one is called acculturation. Acculturation is a social process that arises when a certain group of human beings is confronted with elements of foreign culture in such a way that the elements of foreign culture are gradually accepted without causing the loss of the cultural nature itself. When religion and culture have merged into one, we can see them in their rituals.

The population of Indonesia according was estimated to reach 262 million in 2017. Among other islands and ethnicities in Indonesia, Papua is a unique island it was multi-ethnic, multi-linguistic, and multi-religious region with a high influx of migration with, nevertheless Papua also recognized as an island with high frequently inter-ethnic conflict and demand for separation from the Republic of Indonesia. The thing that we also need to consider is the land of Papua seems one of the largest number of local ethnic groups in Indonesia, at least 261 ethnic groups, among all big islands in Indonesia [1].

Among places in Papua, Raja Ampat Regency seems one of the most interesting places to visit not only because its beautiful natural scenery but Raja Ampat also known as one of the richest places of cultural heritage. Located on the north-west tip of bird's head peninsula on the island of West Papua province, Raja Ampat, or the Four Kings, is an archipelago consisted of 1,500 small islands, shoals surrounding the five main islands of Waigeo, Kofiau, Salawati, Batanta, and Misool. The Raja Ampat archipelago is the part of impressive coral triangle which contains the richest marine biodiversity on earth. It covers more than 40,000 km^2 of land and sea, including in CenderawasihBay, the largest marine national park in Indonesia [1]. The population of Raja Ampat regency the Regency was more than 50.000 people, however, in the native who lives in Misool, which focus on this study only around 8.729 people, the distribution can be seen on the table below:

Table 1. Population Density of Misool Raja Ampat

No	District	Width (km^2)	Population	Density
1	Misool	318.7	1761	5.51
2	South Misool	469.1	3026	6.45
3	West Misool	203.1	1291	6.36
4	East Misool	403.1	2651	6.58

The majority of native Misool population live in a community that spreads in four districts as shown in the table below:

Table 2. District at Misool Raja Ampat

No	District	Capital of District
1	Misool	Waigama
2	South Misool	Dabatan
3	West Misool	Lilintah
4	East Misool	Foley

2. Misool Island Natives

Misool, meaning 'a house on the sea', is one of four large islands of Raja Ampat, West Papua, Indonesia. Its area covers 2,034 km^2. Its highest point is on 535 meters, its main city is Waigama, and it lies on the coordinates of 1°53'41"S 130°5'1"E [2]. Misool island is located in Misool district, Raja Ampat, and directly adjacent to the Seram Sea. This area lies in the bird's head of West Papua and is close to the city of Sorong. Misool is divided into two parts, namely South-east Misool and West Misool.

Misool Island is best known for its unique culture, customs, marine, culinary and terrestrial diversity. This area is included in the world's coral triangle having a variety of ornamental fish found in the ocean in which about 75% of ornamental fish and the world's coral triangle are also found in Misool area. Misool area also has a very large open sea that becomes the path of large animals, such as whales and octopus [3].

Even though, Misool Island also has potential on coal mines but mostly the worker was not from native Misool. Economic activities done by the community are livelihood-oriented to support per capita income. Socio-economic activities managed by the community leader were dominated by fishing activities where this area is rich with fishery resources. In addition, agro-tourism business is also developed in Misool since it is one of a series of islands in Raja Ampat that has beautiful sea and one of the best diving spots in the world.

The habit of *gotong-royong* or works as a together until now is still maintain in various aspects of their community life. The habit can be seen in activities such as in the activities of cleaning the village, processing of agricultural land, customary marriage party, circumcision, and so forth. A very prominent habit is also shown in the lives of local people, namely in receiving guests or if there are visiting immigrants. The attitude shown by the local community characterizes a strong togetherness and kinship in giving appreciation to the outsider that attached to their community. This attitude still maintained, considering the tribes and customs that have been mingled in Misool certainly can provide opportunities for information and communication. To enter that can directly or indirectly influence the behavior of the community without leaving the existing cultural values.

Due to its geographical conditions of Misool, sea transportation become the native favorite options to reach the place just as marine life to support their needs. Therefore rituals relating to marine somehow bond their daily activities. As a result, cultural and religions have contributed a significantly on their marine ritual on how, why and when they do the rituals. The native people of Misool Island are predominantly Muslims and still perform rituals related to religious, cultural and maritime affairs that performed in a festive season in specific time such as *sasi laut* festive season.

In the *sasi laut*, festive season, performed by Misool native people every year is a marine ritual which conducted in two sessions namely *buka sasi* and *tutup sasi*. *Buka sasi* is held when the west wind blows around November, while *tutup sasi* is done around May. (Kartini Limalol/personal interview: if-KL.12':53"). *Buka sasi laut* is held before the harvest of the seafood, that is, around November or December. Meanwhile, a ritual of *tutup sasi laut* is held about 6 months after the ritual of *buka sasi laut*, which is, around May every year. During every session many rituals performed by the native including katinting and oar competitions which make the *sasi laut* get their crowd.

According to Misool community leader, Sasi in Misool means oath to not to take the sea resources for sustainability. This is an ancient way as well as their local genius to support the sustainable use of their natural resources in Raja Ampat territory. In the culture of *sasi*, there is a customary ban for catching and fishing seafood for a year. The community is then allowed to collectively catch the seafood after the break. So there is a shared commitment to not catching certain species of marine resources and giving them time to increase their quantities and flourish, for examples, *bia kapa*, *lola*, natural snails, lobsters, *batu laga*, and sea cucumbers. (Abdul Rasyid Wainsyaf/ Personal Interview: im-ARW.2':11").

IOP Conf. Series: Earth and Environmental Science **156** (2018) 012039 doi:10.1088/1755-1315/156/1/012039

Therefore, this study considers it the main issues of interest to be studied in this research, to see how the existence of marine rituals in the community of Misool? The focus of this study consisting three main issues: First, how is the history of marine rituals in Misool Raja Ampat society? Second, how is the practice of marine rituals in Misool Raja Ampat society? Third, how are Islamic values acculturated with local culture in their ritual practice?

3. Method

This ethnography study in qualitative research was conducted at Misool Raja Ampat in West-Papua Indonesia. This location is a typical region where *sasi* becomes part of rituals that influences their everyday life. Data collection was done through observation, in-depth interview including library research on community manuscript and local government database. Using snowball sampling technique, this study is able to gain solid information from native people in Misoo that willing to be the informants. The informants in this study were found as influences people in Misool community e.g; a head of Misool Fafanlap village, Misool community figures, female figures in Misool, and Misool youth figures.

4. Results and Discussion

This section will present the results of the analysis found in the study. In this section, the first section starts out with the history, and philosophy beyond Sasi marine rituals in Misool Raja Ampat as this is the focus of the study. Then, it moves on to the discussion of the ritual which covers the concept of *Sasi* itself, and finally, discussing how Islamic values acculturated with local culture in their ritual practice.

4.1 The Philosophy behind Sasi

The meaning of *sasi laut*, according to Misool society, *sasi* means oath. Further, *sasi* which contains two meanings, namely ritual practice for the sea and as an application for permit or legality of the area that is sustained by the custom of the local community. *Sasi laut* is a ritual performed by the Misool community as a sign to start or stop the fishing session. *Sasi laut* is done with the intent and purpose to get the results of their fishing is abundant and in line with expectations. Misool community itself believes when they go to sea, the success depends on a ritual of *sasi laut* itself. (Munawir Haris /im-MH. 4':11")

Furthermore, according to Abdul Rasyid Wainsyaf to one of the leading figures in Misool, *Sasi laut* is derived from the language of Misool, means oath (witness). All the people in Misool do *sasi laut* when going to sea. *Sasi laut* is done when the west wind blows, around November, and ends when the south wind blows around May". *Sasi laut* is something that has become a mandatory ritual for the Misool community before harvesting the sea. The intended sea products e.g: *bia kapa*, *lola*, natural snails, *batu laga*, lobsters, and sea cucumbers. (im-ARW.17':12"). The same data also found a similar from Agung Saleh Bahalle a youth figure in Kampung Fafanlap however, this study also seems that it was also strategy to prevent an illegal fishing trip due to its fantastic price, stated that *bia kapa* grows at a depth of 15 meters, grows in the roots of bahar, and the texture is very sharp. The selling price of *bia kapa* is about IDR 15,000 per kilogram. *Lola* is a conical shell. *Lola* is sold at the price of IDR 40,000 to IDR 80,000 per kilogram. Moreover, a natural snail is a type of snail that within a period of 5 (five) years can produce natural pearls. The price range for natural snails is ranging from IDR 80,000 to over IDR 200,000 per kilogram. *Batu laga* is stone-like shells. These shells can be camouflaged with the surrounding rocks making it difficult to distinguish. *Batu laga* can be sold for IDR 120,000 to IDR 200,000 per kilogram. Lobster is a sea product that can be sold for around IDR 250,000 per kilogram. Meanwhile, sea cucumbers are oval-shaped marine products such as plantains and are valued at very high, namely around IDR 180,000 to IDR 1,000,000 per kilogram. (im-ASB. 24':21").

IOP Conf. Series: Earth and Environmental Science **156** (2018) 012039 doi:10.1088/1755-1315/156/1/012039

Approximately 85% of the total Raja Ampat Regency is the sea, especially Misool area [4]. Much of this area is covered by seawater, consequently many native people to have livelihoods as fishermen. Yet when many go to sea, a way is needed to limit society to not do it excessively. The data in the study underlined that the need for regeneration of abundant biodiversity has led to *sasi laut*. The philosophy merely on try to respecting and asking for permission from the Creator through *sasi laut* ritual as well as natural sustainability. So, *sasi laut* is also aimed at preserving the seafood, giving time for the seafood to grow larger and the result will be more. (Irsyah/ im-I.14':21")

4.2 The concept of Sasi: The ritual

Ritual of *sasi laut* becomes sign and symbol of the start of the seafaring season in Misool area. The native believes that by performing *sasi laut*, their marine products will increase as well as prosperity and serenity will always cover them. Therefore, Misool community will not start the activity to go to sea before *sasi laut* has been performed by traditional/ religion leaders. *Sasi laut* is divided into two stages, namely ritual of *buka sasi laut* and ritual of *tutup sasi laut*.

These *sasi laut* rituals fall into the calendrical rites category. Calendrical rites are periodic and predictable rituals based on weather or air changes, such as agricultural, marine, and other social works [5]. *Sasi laut* is carried out when the rainy season starts. There is an element of adjustment to weather conditions or seasons in the implementation of the ritual. Typically, the implementation of *sasi laut* is referring to the lunar calendar system according to the habits of the Misool community. *Sasi laut* is also a ritual that is strongly associated with agricultural and marine activities.

Sasi laut can also be classified into categories of political rites, rituals constructed, exhibited, and promoted by political institutions, such as kings, states, and village heads. Thus, *sasilaut*is not merely regarded as calenderical rites, but can also be categorized as politic rites. Further, Ritual of *sasi laut* can be predicted, precisely when the rainy season arrives. Although frequent weather changes, the rainy season always falls around November each year and *sasi laut* ritual is also held around November. This calendar is based on the Misool calendar system calculated according to the state or shape of the moon.

In ancient times, long before Islamic values attached to them, the timing of *sasi laut* was determined by customary leaders themselves, without any intervention from others, even from the government. Customary leaders first interact with gods, while asking the blessing. However, the current schedule of *sasi laut* adjusted to the rainy season is sometimes uncertain and the time is set unilaterally by the local government. The ritual seems to involving many people, *sasilaut* ritual and also requires funds that are not small in number, to buy materials, equipment, and ritual needs. The ritual materials and tools used in *sasi* laut are, among others, *sasi* rope, colorful flags, drums or tifa (traditional rectangular instruments made of special wood and animal skin), bond-shaped rice, young coconut flowers, coconut water, banana, sticky rice and all kinds of fruits, and others.

All kinds of ritual equipment and materials have different meanings and functions, such as *sasi* flags, strings, and betel nuts. Flags are symbols of protection against all dangers that are or will threaten the seafood, as well as a symbol of God's protection for the welfare and prosperity of the community. A round *sasi* rope is used with a length of 5 meters and golden color, while the betel nut is believed to be a sacred plant brought by their parents earlier. In addition, betel leaves become a complementary material must available in the ritual. In addition to the materials and tools that must exist in *sasi laut* ritual, Misool community also handed out artifacts and ritual attractions as a form of entertainment and tourism for the community [5]. The usual artifacts are human-shaped stone, gold-plated ropes, spear guns, and so on. Whereas one type of ritual attraction that is considered interesting and inviting admiration in the ritual of *sasi laut* is a trance dance. Artifacts and attractions performed initially have only magical and religious values, but now it can serve as the entertainment and tourism aspect for the newcomers.

4.3 Local Culture and Islamic Acculturation

Sasi laut has existed long ago, no one knows exactly when it was the first time the native perform it. Some said that *sasi laut* has existed for thousands of years ago, and some said since when native Misool began to go to sea. But there is one thing for sure, namely that *sasi laut* already existed before Islam entered in Misool area. *Sasi laut,* before the native accepted Islam, was implemented in order to honor the leaders and God(s) and then also to ask permission to the residents or unseen creatures residing in the area. This ritual was led by the customary heads and all aspects of society were involved in it. They believed that if they have asked for permission when going to sea, they will be protected.

Sasi laut ritual was originally practiced by native Misool who still believed that their creation are gods and everything had to be dedicated to gods. After Islam entered Papua around the 15th century, *sasi laut* then changed in its designation, process, and function. The point is, in the ancient times *sasi laut* was addressed to Gods, but after the entry of Islam, *sasi laut* was addressed to the Almighty, Allah SWT. Then, there were also some changes in the practice of this ritual. In the past, this ritual was supplemented with offerings for the Gods and the sea. Meanwhile, now it only uses a dry wooden tree tied with betel nut with colorful clothes, and a flag to indicate that the ritual has been opened. In addition, the function of *sasi laut* somehow has been changing. Before Islam came in, there had been pros and cons in the community about the ritual. Islamthen came to merging ~~with~~ the marine rituals by straightening and at the same time changing the mindset of the native without attached to any element of polytheism. Changes in mindset merely caused by the level of understanding and knowledge of the public who began to open with issues of difference and tolerance. Thus, Islam entered and instilled Islamic values, such as salawat to Prophet Muhammad SAW, prayer to facilitate good fortune, and prayer for salvation.

In practice, *sasi laut* ritual at Misool is headed by a Misool tribal chief and is attended also by all administrators of customary institutions, government, and local communities. Yasin Wainsyaf (head of *Kampung Fafanlap*) recounted that in the implementation of *sasi laut* several years ago, he came to the ritual venue with the Vice Regent of Raja Ampat and the community, both those who participated and who just came to witnessed the ritual procession. The government and the public must be present because the effects and benefits of *sasi laut* ritual will return to them in many ways, (im-YW. 4':1").

When *the sasi laut* ritual is performed, the descendants of traditional leaders make contact and direct interaction with God(s), so that they can know all the wishes and desires of God. Before dealing with or interacting with Him, the necessary conditions must be fulfilled, such as offerings of fruits, betel leaves, incense, coconut flowers and leaves, and much more to be provided. All that preparation is for God [5]. The existence of *sasi laut* in Misool community became a symbol of prosperity for the people. *Sasi* is considered as one of ritual worships to God, as well as a sign of respect and gratitude of society for the blessings revealed by.

As a form of grateful expression and public thanks to the creator, *sasilaut* ritual, the preparation of ferings foods and fruits addressed to God(s) who live on land and at sea. This is long before Islam accepted by the native in Misool the sasi performing starts in a specific day according to the nature sign. However, according to Yasin Wainsyaf (head of Misool village) said that *sasi laut* ritual, after Islam came, is held on Friday, but begins with a joint prayer on Thursday night. After the Maghrib prayer on Thursday night, the drums (Tifa) will be beaten repeatedly as a marker to the community when a joint prayer will be performed. Each family will bring 1 (one) food tray to the mosque or village hall, then hold a prayer together led by the village leader. After that, the whole community will eat together in the mosque. Then on the next day, the ritual *sasi laut* will be carried out by placing *timai* on the beach, then praying. After that, everybody allowed to harvest sea resources that had been banned before *buka sasi laut*, namely *bia kapa*, *lola*, natural snail, *batu laga*, lobster, and sea cucumber. After 6 months, *tutup sasi laut* will be held, which is around May. *Tutup sasi laut* is characterized by sticking coconut leaves on the beach. This *sasi laut* ritual procedure is not directed to gods, but the hope is directed to Allah SWT. (im-YW.15':25")

Furthermore, Abdul Rasyid Wainsyaf stated that the ritual of *buka sasi laut* is marked with betel nut tied with colorful cloth and then planted in. Such a ritual does not contain offerings but is done just by praying together for ease of fortune and for salvation. The changes that occur in *sasi laut* ritual is an acculturation between the culture of Misool community with Islam. Islam entered and changed the mindset of the people, eliminated the things that can cause *shirk*, but did not eliminate the characteristics and the main part of the ritual of *sasi laut*. (im-ARW.17':27")

5. Conclusion

Misool is an exotic island located in Raja Ampat, West Papua, Indonesia. This area is best known as a tourism destination, it a rich in fishery resources which strongly influence the Socio-economic activities of the naïve Misool. The native was dominated by fishermen with their marine activities. Further, agro-tourism in business scale recently is also developed in Misool since it is one of a series of islands in Raja Ampat the best diving spots in the world. The native with their unique culture was found also attracted the international and local tourist to come and experience the native cultural chronicles. The native hospitality and their culture seems supports the tourism campaign so far. Their unique cultures including *gotong-royong* and welcoming guests which they preserve until now are still inherent in various aspects of community life attract the crowd. The native were remarkable, their habits can be seen in many activities such as in the activities of cleaning the village, processing of agricultural land, customary marriage party, circumcision, including their famous marine ritual *Sasi* which they held every year.

Misool community seems still strongly holding cultures they have, including *sasi laut. Sasi laut* is a ritual performed by Misool community as a sign to start and stop harvesting the specific sea resources. Ritual of *sasi laut* in Misool has existed long ago, even before Islam entered in Misool area. The need for regeneration of abundant biodiversity led to this *sasi laut* ritual. In addition, respecting and asking permission from the Creator is a philosophy of this marine ritual. As for some of the seafood referred to in this ritual are *bia kapa*, *lola*, natural snail, *batu laga*, lobster and sea cucumber.

Sasi laut before the coming of Islam was performed in order to respect their leaders and their Gods. After Islam entered Papua around the 15th century, *sasi laut* then changed in "designation" and "process", where in the ancient times *sasi laut* was addressed to Gods, then it changed to address to the Almighty, Allah SWT. Further, some changes in the practice of this ritual. In the past, it was supplemented with offerings for the Gods and the sea, nowadays, it only uses a dry wooden tree tied with betel nut and colorful cloth, and a flag to indicate that the *sasi laut* has been opened.

Moreover, instilled Islamic rituals and values, such as salawat to Prophet Muhammad SAW, prayer facilitate good fortune, and prayer for salvation. The ritual was found no longer containing offerings and sacrifice but is done just by praying together for ease of fortune and for salvation. The changes that occur in *sasi laut* ritual is an acculturation between the culture of Misool community with Islam. The acculturation between the culture of Islam runs smoothly in changing the native mindset to eliminate the things that can cause syirik, however it does not eliminate the characteristics and the main part of the *sasi laut.*

6. References

[1] Ananta, A., Arifin, E.N., Hasbullah, M.S., Handayani, N.B. and Pramono, A., (2015). *Demography of Indonesia's ethnicity*. Institute of Southeast Asian Studies.
[2] Sujadi, Firman. (2013). *Provinsi Papua Barat Cintaku Negeriku,* Cita Insan Madani,Jakarta.
[3] E.Rusmana,dkk. (1993). *PetaGeologiPulauMisool,Irian Jaya,*Puslitbang Geologi Bandung, Bandung. Kamus Besar Bahasa Indonesia.
[4] Purwati. (2009). *Selayang Panvdang Papua Barat,*Intan Pariwara,Klaten.

[5] Rahman, Abdul. (2013). *Geliat Ritual Traditional Kasasyi; Rekontruksi Nilai-Nilai Lokal Suku Misol Raja Ampat.* Pusat Penelitian dan Pengabdian Masyarakat (P3M) Sekolah Tinggi Agama Islam Negeri (STAIN) Sorong, Sorong.
[6] Kurniawan, Beni. (2007). *Pendidikan Agama Islam untuk Perguruan Tinggi,* PT.Gramedia Widiasarana Indonesia,Jakarta.
[7] Rais,Muhammad, dkk. (2013). *Laporan Penelitian Kelompok Islam Rahma dan Toleran dalam Masyarakat Multikultural di Kabupaten raja Ampat,* Pusat Penelitian dan Pengabdian Masyarakat (P3M) Sekolah Tinggi Agama IslamNegeri(STAIN) Sorong,Sorong.
[8] Tp. (2012). *Buku Produk Domestik Regional Bruto, Kabupaten Raja Ampat Menurut Lapangan Usaha,* BPS, Raja Ampat.
[9] Waluya,Bagja. (2007). *Sosiologi: Meneyelami Fenomena Sosial di Masyarakat,* PT. Setia Purna Inves: Bandung.
[10] Evizariza, E. and Idayanti, I., (2017). Pendataan dan Digitalisasi Naskah Kuno Melayu di Kabupaten Kampar. *Jurnal Ilmu Budaya*, *14*(1).
[11] Junaidi, J., (2014). Islam dalam Jagad Pikir Melayu. *Buletin Al-Turas*, *20*(2), pp.46-56.
[12] Wekke, I.S., (2016). Religious education and empowerment: study on pesantren in muslim minority West Papua. *MIQOT: Jurnal Ilmu-ilmu Keislaman*, *37*(2).
[13] Wekke, I.S., (2015). Antara Tradisionalisme dan Kemodernan: Pembelajaran Bahasa Arab Madrasah Minoritas Muslim Papua Barat. *TSAQAFAH*, *11*(2), pp.313-332.
[14] Wekke, I.S., (2013). Islam di Papua Barat: tradisi dan keberagaman. *ULUL ALBAB Jurnal Studi Islam*, *14*(2).

Paddy Chlorophyll Concentrations in Drought Stress Condition and Endophytic Fungi Application

Syamsia [1], Abubakar Idhan[2], Noerfitryani[3], Marhamah Nadir[4], Reta[5], Muhammad Kadir [6]

[1]Muhammadiyah University of Makassar, South Sulawesi, 90221 Indonesia
Email : :syamsiatayibe@unismuh.ac.id
[2]Muhammadiyah University of Makassar, South Sulawesi, 90221 Indonesia
Email :
[3] Muhammadiyah University of Makassar, South Sulawesi, 90221 Indonesia
Email :
[4]Hasanuddin University, Makassar, South Sulawesi, 90234 Indonesia
Email : marhamahnadir@unhas.ac.id
[5]Plantation Department, Politechnic of Agriculture and Fisheries, Pangkajene and Kepulauan
Email :
[6] Plantation Department, Politechnic of Agriculture and Fisheries, Pangkajene and Kepulauan
Email: mkadir@gmail.com

Abstract - Drought is one of the major constraints in rainfed paddyfield production. Drought stress condition caused by long drought that often happens lately andit was major problem causing decreased paddy productivity. The research aimed to determine the chlorophyll concentration of paddy plants treated with endophytic fungi and drought. The research was conducted on farmer's paddy field in Galesong Sub-district of Gowa Regency from August to November 2017. The research consisted of two treatment that puddle height and fungi application. The puddle height treatment consists of two levels i.e. puddle5 cm height (normal) and 1 cm (drought). Endophytic fungi treatment was applied to paddy seeds before planting. Leaf chlorophyll content was measured on 52 days after planting by using SPAD-502 Plus chlorophyll meter. Chlorophyll-a content, chlorophyll-b content and total of chlorophyll contentwas measured by using extraction method in laboratory. The result showed that leaf chlorophyll concentration of paddy drought was low than the waterlogged conditions. Endophytic fungi application increasedchlorophyll-a, chlorophyll-bconcentration, total chlorophyll of paddy leaf on drought condition.

Keywords*:* chlorophyll a, chlorophyll b, total chlorophyll, SPAD-502 Plus

1. Introduction

Drought is major abiotic stress that greatly affect food crops in worldwide. The agronomic and physiological associated with drought tolerance are good indicators for drought tolerance of genotype selection to reduce the impact of drought stress condition onharvest in breeding programs [1]. Drought is one of important factor of abiotic stress [2];[3]. The measurement of physiological character such as chlorophyll concentration is one

IOP Conf. Series: Earth and Environmental Science **156** (2018) 012040 doi:10.1088/1755-1315/156/1/012040

of approach to study effect of drought condition on growth and production for this parameter is closely related to rate of photosynthesis [1]);[4]. Drought stress affects the biochemical processes in plant cells [5].

High chlorophyll and carotenoid were associated with plant tolerance to stress [6];[7];[8]. Measurement of chlorophyll fluorescence is a tool that can detect photosynthesis interference caused by environmental stresses [9];[10]. Chlorophylllossess is a negative consequence of environmental stress; However, it has been considered an adaptive feature in plants growth under drought stress condition [11];[12].

Endophytic fungi increasing host tolerance in stress conditions [13]. Endophytic fungi also produce diverse ofbiochemical important metabolites [14]. [15] was reported that inoculated *Trichodermahamatum* on cocoa plantswith endophytic fungi resistant in drought conditions. There were six genera of endophytic fungi were isolated from healthy tissue in different part of paddy plants, they were *Fusarium,Aspergillus*, *Curvularia*, *Penicillium*, *Gilmaniella* and *A. foliicola* [16]. [17] also were reported that there were 27 endophytic fungi isolated from Kuruluthuda rice variety, themost isolated genera were*Acremonium, Arthrobotrys, Colletotrichum*and *Humicol.*

[4] reported that the difference in total chlorophyll concentration, chlorophyll a, and chlorophyll b on cultivar Serayu and IR 64 paddy leaves treated with PEG with PA 0; -0.5 and -0.1 MPa. [18] showed that total leaf chlorophyll on ginger plants decreased 8% due to drought for 7 days, while total leaf chlorophyll content in control increased 9% to day 7.In drought stress condition also significantly decreased the chlorophyll-a, chlorophyll-b and total chlorophyllcontent at the vegetative and flowering stagesthree cultivars of chickpea [19]. Thusthis study to find outthe total chlorophyll concentration, chlorophyll-a, chlorophyll-b and carotenoid of paddy leaf treated by endophytic fungi application in drought condition.

2. Method

The research was conducted in the rainfed paddy field in Galesong Sub-district of Gowa Regency in August to November 2017. Irrigationpaddy field using a water pump powered by a generator.Paddy seedlings IMPARI 20 that are 21 days after planting were treated with immersion of endophytic fungi and without immersion (control). Drought stress treatment performed after first fertilization, that is 14 days after planting until harvest by adjusting puddle is 1 cmheight (drought) and puddle 5 cm height (normal). Analysis of chlorophyll content using ChlorophyMeter Soil Plant Analysis Development (SPAD) 502 Plus which has been calibrated. The measure leafwere cleaned using tissue paper. To find out the chlorophyll content, green leaf were clipped with SPAD Chlorophyll Meter sensor. Value seen on the screen SPAD chlorophyll meter chorophyll. Value of total leaf chlorophyll was calculated using the formula $Y = 0.0007x - 0.0059$, Y = chlorophyll content and x = value of the measurement results-502 Plus SPAD chlorophyll meter [20]; [21] . The concentration of chlorophyll a, chlorophyll b, carotenoid and total chlorophyll of paddy leaves was determined by leaf blade (lamina) sample for each treatment was taken 1 g, mashed with mortar and pestel, then extracted with 95% alcohol until all chlorophyll dissolved. Total chlorophyll concentrations, chlorophyll a and chlorophyll b were calculated by Wintermans & de Mots (1965) in Sasmitamihardja (1990) as follows: Total chlorophyll (mg / L) = 20.0 OD649 + 6.1 OD665 Chlorophyll a (mg / L) = 13.7 OD665 - 5,76 OD649 Chlorophyll b (mg / L) = 25.8 OD649 - 7.7 OD665 (OD = Optical density);[4].

3. Results and Discussion

Leaf greenishness level was measured using a chlorophyll meter SPAD 502 Plus. The principle of this tool is to record the greenishness level of leaves and relative total of chlorophyll molecules present in leaf in a single value based on the amount of light leaf transmitted (Konica Minolta, 1989).Result of measurement total chlorophyll of paddy leaf on drought treatment (puddle 1 cmheight) is lower than normal treatment (puddle 5 cmheight).While total chlorophyll content in treatment of endophytic fungi application showed that higher value both in droughttreatment and normal treatment (Figure 1). The results showed that increase in amount of chlorophyll with endophytic fungi treatment. Refer to [20], increased in the greenishness of leaf showed that increase of total chlorophyll in leaves. therefore, more of total chlorophyll in leaves, hence increasing the rate of photosynthesis. The leaves are yellow and old because of chlorophyllwas lost, the power of photosynthesis has decreased (Ahmadi, 1985; [3].Differences in physiological and biochemical responses in plants can be caused by drought condition in crops [4].

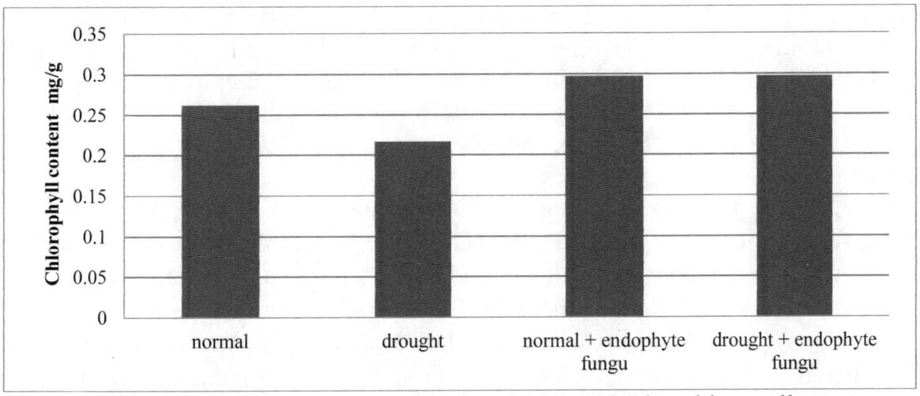

Figure 1. The measurement result using SPAD 502 Plus chloropyll meter

Result analysis chlorophyll content of paddy leaf showed that the highest value on puddle treatment 5 cm and fungi application that is 0.069 (Figure 2). Chlorophyll on paddy leaves treated by drought (puddle 1 cm height) showed that thevalue was low that was 0.058. Treatment of endophytic fungi application increased chlorophyll content of paddy leaf both in drought treatment(puddle 1 cm height) as well as control (puddle 5 cm height).This suggests that paddy plants treated with endophytic fungi were more resistant to drought treatment.According to [8];[3] states that the genotypes of wheat and corn resistant to oxidative stress have higher chlorophyll content than sensitive genotypes.Similarly also stated by [22]; [23] , drought condition in Brazilian green dwarf coconut (*Cocos nucifera L.* nana) resulting in decreased leaf chlorophyll concentration per unit leaf area. Effect of drought stress on plant chlorophyll content depends on plant genotype and environmental conditions, some varieties reduced chlorophyll content and some varieties increased [24].

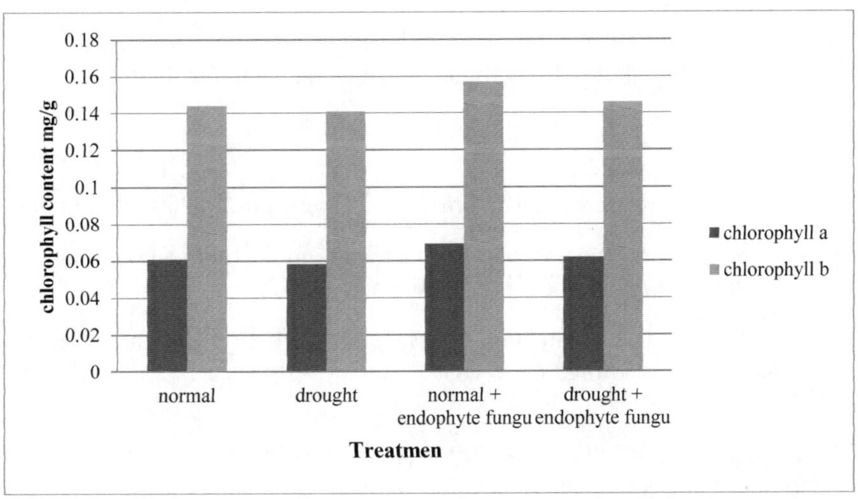

Figure 2. The content of chlorophyll a and b of paddy leaf on drought treatment and endophytic fungiapplication

Chlorophyll-b content showed that higher value than chlorophyll-a content in all treatments.Value of chlorophyll-b content onpaddy leaf is lowest on droughttreatmentthat is 0.141 compared with other treatments. According to [3] synthesis of chlorophyll is very important,after heavy rains total of chlorophyll increased,but in dry time its value decreased. Water stress condition also reduced chlorophyll content compared to well irigated plants. While in low water stress chlorophyll-a, chlorophyll-b and totalchlorophyll contentincreased by increasing stress pressure [25]. Chlorophyll-a and chlorophyll-b attributes for accumulation of solutes in cell sap through passive accumulation resulting from reduced cell size (Morgan, 1984); [26] which significantly did osmotic adjustment. [13] were reported that endophytic-infected plants were significantly higher chlorophyll content than non-inoculated endophytic fungi in stressful conditions.*Trichoderma harzianum* endophytic fungi inoculated in tomatoe increasing root, shoots growth and chlorophyll pigments compared with control treatment uninoculated of endophytic fungi, in addition plant resistant in drought stress condition [27].

Amount of chlorophyll content in plant will decrease in drought stress condition, endophytic fungi apllication affect to plant growth especiallyroots growth betterfor water absorb and water retention than without endophytic fungi application.Endophytic fungi significantly increased amount of chlorophyll in drought treatment as well as in control treatment. Refer to [23] showed that amount of chlorophyll a and chlorophyll b content and total chlorophyll in leaf was increased under drought stress condition.[28]. There was a symbiotic relationship between endophytic fungi and plant growth, endophytic fungi produce secondary metabolites and have a positive effect on plant growth under field conditions, they were indicated that *Penicillium citrinum* endophytic fungi produce secondary metabolites that incease in physiological characteristics such as chlorophyll content, especially chlorophyll b content, carotenoid content, phytosynthesis rate and tolerance to under hight-light and drought condition.

4. Conclusion

Paddy leaf chlorophyll concentration decreased during drought condition. Decrease of total chlorophyll content showed that response of paddy plants to drought treatment. Endophyticfungi application increasedchlorophyll-a, chlorophyll-b concentration, total chlorophyll of paddy leafon drought stress condition.

5. Acknowledgement

Authors would like to thank Ministry of Research, Technology and Higher Education for providing financial through Applied Product Research scheme for fiscal year 2017.

6. References

[1] LI, R. hua., Pei-guo, GUO., Baum, M., Grando, S., Ceccarelli, S, 2006. Evaluation of Chlorophyll Content and Fluorescence Parameters as Indicators of Drought Tolerance in Barley. *Agricultural Sciences in China*, 5(10), pp.751–757.

[2] Dash S, Mohanty N (2001). Evaluation of assays for the analysis of thermo tolerance and recovery potentials of seedlings of wheat (*Triticum aestivum* L.). J. Plant Physiol., 158: 1153–1165.

[3] Majid Khayatnezhad, 2012. The effect of drought stress on leaf chlorophyll content and stress resistance in maize cultivars (*Zea mays*). *African Journal of Microbiology Research*, 6(12): 2844–2848.

[4] Banyo, Y.E., Ai, N.S., Siahaan, P., Tangapo, A.M., 2013. The Chlorophyll Concentration in Rice Leaves Under Polyethykebe Glycol Induced Water Deficit. *Jurnal Ilmiah Sains*, 13(1), pp.1–8.

[5] Ai, N.S, 2011.Biomass and Concentration of Total Chlorophyll On Ginger (*Zingiber Officinale* L.) Leaf Under Water Deficit. *Jurnal Ilmiah Sains*, 11(1).

[6] Sairam RK (1994). Effect of moisture stress on physiological activities of two contrasting wheat genotypes. Indian J. Exp. Biol., 32: 594–597

[7] Kraus TE, Mckersie BD, Fletcher RA (1995). Paclobutrazole induced tolerance of wheat leaves to paraquat may involve antioxidant enzyme activity. J. Plant Physiol., 145: 570–576.

[8] Pastori GM, Trippi VS (1992). Oxidative stress induces high rate of glutathione reductase synthesis in a drought-resistant maize strain. Plant Cell Physiol., 33: 957–961.

[9] M. Bresti cand M. Zivˇak, "PSII fluorescence techniques for c´measurement of drought and high temperature stress signal in crop plants: protocols and applications," in Molecular Stress Physiology of Plants,G.R.Routand A. B. Das, Eds.,pp. 87–131, Springer, New Delhi, India, 2013.

[10] Jedmowski, C. Ashoub, A., Momtaz, O., Brüggemann' W., 2015. Impact of Drought , Heat , and Their Combination on Chlorophyll Fluorescence and Yield of Wild Barley (Hordeum spontaneum). *Journal of Botani* (2015) 2015

[11] Munne-Bosch S and Alegre L. 1999. Role of dew on the recovery of water-stressed Melissa officinalis L. plants. J. Plant Physiol 154: 759-766.

[12] Sarani, M. et al., 2014. The effect of drought stress on chlorophyll content , root growth , glucosinolate and proline in crop plants. *International Journal of Farming and Allied Sciences*, 3(9), pp.1997–2000.

[13]. Waqas, M., Khan, A.L., Kamran, M., Hamayun, M., Kang, SM., Kim, YH., and Lee, IJ., 2012. Endophytic Fungi Produce Gibberellins and Indoleacetic Acid and Promotes Host-Plant Growth during Stress. *Molecules* 17: 10754-10773.

[14] Khan, Latif, A., Hamayun, M. , Ahmad, N., Hussain, J., Kang, SM. , Kim, YH., Muhammad Adnan, M., Tang, DS., Waqas, M., Radhakrishnan, R., Hwang, YY., and Lee, IJ., 2011. Salinity Stress Resistance Offered by Endophytic Fungal Interaction Between Penicillium minioluteum LHL09 and Glycine max. L. *Journal Microbiology and Biotechnology*, 21(9): 893–902

[15] .Bae, H., Sicher, R.C., Kim, M.S., Kim, SH., Strem. D.M. Melnick, R.L. and Bailey, B.A., 2009. The beneficial endophyte *Trichoderma hamatum* isolate DIS 219b promotes growth and delays the onset of the drought response in *Theobroma cacao*. *Journal of Experimental Botany*,60(11): 3279–3295.

[16] Zakaria, L., Yaakop A.S., Salleh, B. and Zakaria, M., 2010. Short communication: Endophytic Fungi from Paddy. *Tropical Life Sciences Research* 21: 101–107.

[17]. Wijesooriya, W.A.D.K., and Deshappriya, N., 2016. An inoculum of Endophytic Fungi for Improved Growth of aTraditional Ricevariety in Sri Lanka. *Tropical Plant Research*. 3(3): 470 – 480

[18] Ai, N.S, 2011.Biomass and Concentration of Total Chlorophyll On Ginger (*Zingiber Officinale* L.) Leaf Under Water Deficit. *Jurnal Ilmiah Sains*, 11(1).

[19] Mafakheri, A., Siosemardeh, A., Bahramnejad, B., Struik, P.C., Sohrabi, E., 2010. Effect of drought stress on yield, proline and chlorophyll contents in three chickpea cultivars . *Australian Journal of Crop Science*. 4(8):580-585.

[20] Farhana, M.A, M.R Yusop, M.H. Harun, and A.K. Din. 2007. Performance of Tenera population for the chlorophyll contents and yield component. Proceedings of The PIPOC 2007 International Palm Oil Congress(Agriculture,Biotechnology & Sustainability). Malaysia Palm Oil Board. Malaysia. Vol.2:701-705

[21]Sudradjat and Siagian, N.A., 2014. Effect of Phosphorus and Potassium Fertilization on Oil Palm Growth (Elaeis Guineensis Jacq.) on Primer Plants Nursery . *Agrovigor*, 7(2), pp.105–116.

[22] Gomes, F.B., M.A. Olivia, M.S. Nielke, A.F. de Almeida, H.G. Leite, L.A. Aquine. 2008. Photosynthetic Limitations in Leaves of Young Brazilian Green Dwarf Coconut (*Cocos nucifera* L. 'nana') Palm under Well-Watered Conditions and Recovering from Drought Stress. Environmental and Experimental Botany 62: 195-204

[23] Song, N. and Banyo, Y., 2010. Water on Plants the Concentration of Leaf Chlorophyll As Water-Deficit Indicator in Plants. *Jurnal Ilmiah Sains*, 11(2), pp.166–173

[24] Sepehri, A. and Golparvar, A.R., 2011. The Effect of Drought Stress on Water Relations, Chlorophyll Content and Leaf Area in Canola Cultivars (*Brassica napus* L.). *Electronic Journal of Biology* 7(3): 49-53

[25] Pirzad, A., Shakiba, M. R., Zehtab-Salmasi S.Z., Mohammadi, S.A., Reza Darvishzadeh, R and Samadi, A., 2011. Effect of Water Stress on Leaf Relative Water Content, Chlorophyll, Proline and Soluble Carbohydrates in *Matricaria chamomilla* L. *Journal of Medicinal Plants Research* 5(12): 2483-2488.

[26].Chutia, J. and Borah, S.P., 2012. Water Stress Effects on Leaf Growth and Chlorophyll Content but Not the Grain Yield in Traditional Rice (Oryza sativa Linn.) Genotypes of Assam, India II. Protein and Proline Status in Seedlings under PEG Induced Water Stress. *American Journal of Plant Sciences*, 3(7), pp.971–980. Available at: http://www.scirp.org/journal/PaperDownload.aspx?DOI=10.4236/ajps.2012.37115.

[27] Mona, S.A., Hashem, A., Abd_Allah, E. F., Alqarawi, A.A., Soliman D.W.K., Wirth,

S. and Egamberdieva, D., 2017. Increased Resistance Of Drought By *Trichoderma Harzianum*fungal Treatment Correlates With Increased Secondary Metabolites And Proline Content. *Journal of Integrative Agriculture* 16(8): 1751-1757 (Abstract).

[28] Hwang, JS., You, YH., Bae, JJ., Khan, S.A., Kim, JG., and Choo, YS., 2011. Effects of Endophytic Fungal Secondary Metabolites on the Growth and Physiological Response of Carex kobomugi Ohwi. *Journal of Coastal Research* 27(3):544-548.

IOP Conf. Series: Earth and Environmental Science **156** (2018) 012041 doi:10.1088/1755-1315/156/1/012041

Religious Values in Environmental Conservation of Traditional Society in Wakatobi Island

***Sumiman Udu[1]**

[1]Universitas Haluoleo Kendari, Kampus Hijau Bumi Tridharma, Anduonou, Kambu, Kendari, Kambu, Kota Kendari, Sulawesi Tenggara 93132, Indonesia

Email: sumimanudu75@gmail.com

Abstract: Wakatobi is one of the *Bharata* that belongs to the governance system of Buton Sultanate. Therefore, as an ex-Islamic kingdom, certainly, Buton has various local wisdom in the life its society. One of the local wisdom can be seen in the system of their environmental conservation. This study made use of ethnographic approach; so, the researcher would like to expect that society might access multiple information related to the values that had become the foundation of environmental conservation in Wakatobi through this study. With this approach, the researcher would also like to expect that he gathered the data regarding how the society of Wakatobi viewed the Islamic values in their environmental conservation. The results of the study show that the biosphere heritage in Wakatobi land has been a genuine artwork by the traditional society of Wakatobi, which has been based on the Islamic values. Some of these values are internalized through *kangkilo* doctrine that includes: (1) *kangkiloawalu* (early holiness); and (2) *kangkiloakhiri* (final holiness). These values then have implications for their actions toward the environment. Several conservation concepts that have been based on these Islamic values can be seen in the forest, the *kaindea*, the *motika*, the *menani* (Napoleon) fish, the environmental management, and the shell (*rondora, untu, watu*). Also, the values of *rahmatan lil alamin* can also be found in the prayers that these traditional society utter when they are planting a tree.

1. Introduction

Wakatobi has been one of the *bharata* that belongs to the governance system of Buton Sultanate. In 2003, Wakatobi was expanded into an autonomous region in the Unified Nations of the Republic of Indonesia. Therefore, as an-ex Islamic kingdom, Wakatobi certainly has various local wisdom in their life that stems from the Islamic values. One of the local wisdoms can be seen in the system of environmental conservation that has been applied by the traditional society there [1]. In 2012, Wakatobi was appointed as the earth's biosphere heritage by UNESCO.

One of the Islamic values that can be found in the life of Wakatobi society is the concept of *thaharah* or holiness. This concept then becomes known by the local society as the concept of *kangkilo* or *kabusa* or *kili* [2]. It is this concept that then becomes the basis of various actions, both in the context of tradition and in the context of a family. It is these Islamic values that become the fundamental values of multiple activities performed by Wakatobi society, including in conducting the environmental conservation in their culture. However, these values then start fading along with the development of science and technology in the Wakatobi generation – Buton [3]. As a consequence, the Islamic values that have been inserted in all norms within the traditional society are considered irrational. Various myths that contribute to the traditional society-based environmental conservation are destroyed by the Muslim society there alone, because they consider that every single aspect should be Arabian; whereas, the local values that have been perfectly acculturated into the Islamic values are considered as something that breaks the *tauhid* and as a blasphemy. This generation can only retrieve the aspects of *syariat* while their Elders in Wakatobi have packaged the Islamic values into various myths [3] that have been in acculturation with the Islamic values. This matter can be seen from the local society's belief on *imbu* (sea ghost), a myth that has been turned into a story with *waliullah* meaning.

Therefore, a study about the Islamic values in the environmental conservation within the Wakatobi traditional society is an effort to uncover multiple discourses of traditional society-based environmental conservation on the one hand and an effort to eliminate these values due to the increasingly rational Wakatobi society. In addition, the presence of several religious schools which have been more radical and does not have tolerance to the acculturation between Islamic values and the local culture Puts the space of environmental conservation in Wakatobi in a worse situation [4].

2. Method

This study makes use of ethnographic approach; as a result, it is expected that the researcher may access various points of view among the local society regarding the values that the Wakatobi-Buton society hold in conducting the environmental conservation within the traditional society of Wakatobi. Radcliffe Brown and B. Malinowski stated that modern ethnography is differentiated from the preliminary ethnography based on the important characteristics, namely that these characteristics do not heavily pay attention to the matters that have been related to the cultural history of a community group [5]. Their main attention is in the present life, namely about the way of life that the related society hold. According to the view of the two anthropologists, the objective of ethnography is describing and establishing a social and cultural structure of society. Therefore, the researcher perceives that conducting interview will be insufficient; he would like to do a participatory observation along with the informants in this study. Then, the paradigm development in this study has focused his efforts to find how the society organize the culture in their mind and use the culture in their life [5]. The analysis in this study is not merely based on the researcher's interpretation; instead, the analysis has been a structure of mind from the traditional society that the researcher has brought about. Since the objective of this study is finding and describing the organization of ideas from society, the researcher's understanding of the language study becomes highly important in the method of this study. "By gathering the history of life and by using a mixed strategy, a language will appear in each phase of the process in this study."

In relation to this statement, Spardley defined culture as something under observation within the ethnography. In addition, culture also serves a learning process that has been implemented for interpreting the world around them and for designing behavioral strategies in dealing with the world. In his view, Spardley did not consider ethnography as a method for studying "other culture," a small isolated society; instead, he considered ethnography as a method for studying our own society, the multicultural society all around the world. Then he summarized his ideas into "Progressing Research Path" that consists of five principles namely: (1) a researcher is suggested to use one data gathering only; (2) a researcher is suggested to identify the main steps in the data gathering technique; (3) for example, a researcher should identify 12 main steps in an ethnographic interview from Spardley; (4) every main step should be performed sequentially; (5) a researcher should provide problem solving as part of his social responsibility; and (6) this research should not be intended to science-to-science anymore. The core of "New Ethnography" by Spardley is the effort to pay attention to the meaning of an action that impacts individuals whom we would like to understand through their culture. In performing the field work, an ethnographer concludes human culture from three sources: (1) what society says; (2) how society act; and (3) artefacts that have been used. However, in this article, Spardley focuses specifically on the conclusion drawing from what society have said. An ethnographic

IOP Conf. Series: Earth and Environmental Science **156** (2018) 012041 doi:10.1088/1755-1315/156/1/012041

interview is considered more able to explore the structure of mind within the society under observation.

Thereby, the results of this study are expected to disassemble the mind regarding the basic values of environmental conservation that lies in the awareness of Wakatobi traditional society. If within the last several years Wakatobi has been awarded by UNESCO as the earth's biosphere heritage [6], then actually Wakatobi is a representation of Islamic values that have been in acculturation with the values of the local culture [7]. Therefore, the basic values of environmental conservancy in the Wakatobi traditional society might be explained in the following sessions.

3. Findings and Discussion

3.1 Islamic Values in the Environmental Conservation of Wakatobi Traditional Society

Based on the results of the study, the researcher has found several fundamental values that become the foundation of Wakatobi society's culture in conducting the environmental conservation within each sub-ethnic of the traditional society. The achievement that the traditional society have in preserving the environmental preservation then has been appreciated by the world as the Wakatobi earth biosphere heritage. All of these are the authentic work of Wakatobi traditional society, which has been based on the Islamic values.

Several cultural values that hold important role which influences their paradigm and action in conducting the management of environment within the Wakatobi traditional society have been constructed by the doctrine of *kangkilo* (holiness). This doctrine includes: (1) *kangkilo awalu* (early holiness); and (2) *kangkilo akhir* (final holiness). These values then have implications for the traditional society's paradigm and action in dealing with the environment. Several concepts of conservation that have been based on the Islamic values can be seen in the forest, the *kaindea*, the *motika*, the *menani* (napoleon) fish, the shell, and the coral (*tondora, untu,* and *watu*).

3.1.1 Kangkilo as the Fundamental Value

Wakatobi society are a society that consists of multiple ethnics from all over Archipelago. Some society says that Wakatobisociety comes from Adonara in East Nusa Tenggara (Tomiasociety) [8] And some say that Wakatobi society are related to Wangi-Wangi society who come from Kei in Molluccas [1]; then, there are also traces from Chinese society in Binongko, Malay society in Tomia [9], and Javanese society in Liya [10]. Some other society also says that Wakatobi society has originally come from the remaining Tulukabesi troopers under the reign of King Hitu [10–12]. All of these statements lead to the long history of Wakatobisociety, who geographically settle on the east-west and south-north sailing direction. The trail of that long history can be found from multiple artefacts that have been found in Wakatobi. From these artifacts, one may find the process of acculturation between the local society's (Melanesian) culture and the other cultures such as Islam, Malay [9][13], China, and India. The arrival of Indian society (Hinduism and Buddhism) certainly has played its role in the early cultural formation in this region; it is certain that the arrival of these societies would influence the local society's knowledge in managing the environment [4].

The arrival of Islam in this region can be traced through the story of Tindoi in Wangi-Wangi Island [14], the story of Moori in Tomia Island [8], and the story of Culada Tape-Tape In Binongko Island [15]. The story of Islam in converting the society of Tukang Besi Island into Moslem cannot be separated from the role of Buton Sultanate in the past. From this long period of Islam, Wakatobi then have Islamic heritage in the form of noble values which become the doctrines [16] In their life management, including the environmental management. This statement is apparent from each of the traditional conservations that always have myths related to holy society or *waliullah*.

Understanding the *kangkilo* deeper, the Wakatobi traditional society recall *kangkilo* [2] As a ritual that includes *fudhu*, *mandi*, and *junub*. However, for Wakatobisociety holiness or *thaharah* does not only take the ritual form but also the doctrine form in their social dimension. It is in this social dimension that *kangkilo* has become the values that base the traditional society's mind and behaviors in establishing the relationship between the human beings and the deity, among the fellow human beings, and between the human beings and the environment.

The doctrine of *kangkilo* then is divided into two categories by Wakatobi traditional society namely *kangkiloawalu* (early holiness) and *kangkiloakhiri* (final holiness). According to la Musidu (40), *kangkiloawal* is a doctrine that sets the regulations for maintaining the self-holiness, which includes physics, mind, and sense, among Wakatobi traditional society. In the concept of understanding among the traditional society, this early holiness stage urges that all food and beverage that comes into the body should be in a holy condition. The food and beverages should not be *haram* due to uncleanliness or inappropriate belonging [17].

With this doctrine, Wakatobi traditional society decides to be hungry than to eat the food that they do not deserve. This doctrine then becomes the fundamental value of the belief that Wakatobi traditional society holds; they will not take the goods that do not belong to them and this also includes the environmental management. A fisherman who has internalized this doctrine will never catch napoleon fish because the fish does not belong to the society since they regard this fish as *sara* fish (*teika nu sara*). This decision is certainly based on the knowledge regarding *sara* [14] Regarding the Napoleon fish that has been very vulnerable to extinction if they are massively exploited. Therefore, the traditional society will let go the Napoleon fish back to the environment because *sara* has required them to return it before the *sara* [14]. If the fisherman eats the Napoleon fish, then he will just deface his physics, mind, and sense. According to the doctrine of *kangkilo*, this is a form of mind and action that might deface one's self. The traditional society will not destroy the forest due to the values of *kangkiloawal*. They will protect the forest, and they will not take away what does not belong to them. If they take the woods out of the forest without having the permission of *sara*, then they might deface themselves.

Kangkiloakhiri (final holiness) is a doctrine that Wakatobi traditional society believes to be the form of final holiness. According to their understanding, *kangkiloakhiri* is a doctrine that has been related to all the things that come out from themselves, and these things might take the physical form (action), manure, mind, and sense. These things may not degrade the sense (toward the deity, the human beings, and the environment). In the context of environmental preservation, Wakatobi traditional society lives this doctrine by not destroying the forest (*motika* And *kaindea*) and the coral. Even thinking about destroying the forest and the coral has been considered as disturbing and destroying their holiness. These values then influence the Wakatobi traditional society's paradigm and action in conducting the environmental conservation. Several regions that become the conservation area of Wakatobi traditional society are Tindoi traditional forest, Wabue-bua forest, Bungi Longa forest, and Kaidea Teo forest that is under the traditional territory of *kadhia* Wanse. Then, Malarau forest and Lagiampa forest are the conservation area under *Kaindea*, Mandati [1]. Next, Kapota forest becomes the conservation area of Kapota traditional society while Mangrow forest in Binongko Island becomes the conservation area of *kadhia* Wali traditional society. In Tomia Island, by the way, there is a conservation forest that still survives up to date and this forest is located in Suo-Suo Fortress under the conservation area of Kawati Tonggano.

Then, the traditional sea conservation areas are located in *sangiatondo* sea, *sangia tanjung patipelong* sea (Tongano traditional society), and *ujutondoa/ pasikeawa'altadhupalea*sea (let's get drunk) that belongs to the traditional area of *kawati* Waha. Then, there is also another sea conservation area in front of the eastern part of Onemai, known as Nuko. This area is considered sacred by the local society. Another region under the area of KawatiWaha is PasiTumondu or the coral that has been sunk in front of Tomia Harbor. However, these conservation areas are not considered sacred anymore because of the educational modernization; as a result, nowadays society believes their logical thinking more. The shift from the world of mythology to the world of science (Western style), and the arrival of radical Islam that highly pays attention to the Arabian symbols have influenced the paradigm of Wakatobi society in recent years.

The efforts to not degrading the surrounding environment can be traced from the prohibition to insult these sacred places. Tomiasociety knows a term *sakito*, which has been a prohibition to insult;

IOP Conf. Series: Earth and Environmental Science **156** (2018) 012041 doi:10.1088/1755-1315/156/1/012041

according to this doctrine, these sacred places should not be degraded with inappropriate utterances, behaviors, and actions. Sahruddin in an interviewing said that, "*Karang koko*h as many meanings; they can be *Homali*, *karamah*, and *sangia*". He said that *karangkoko* has been one of the traditional areas in Wali, which myth is that this place has many guardian spirits; they often see something strange in this place along with the abundant number of sunken ship. Society should respect Karang Koko by not saying inappropriate words around this place. The fisherman, especially, should accept the amount of their catch; however, if they have good catch then they should be careful because this area belongs to *waliullah*. This myth becomes a concept of continuous fishery activities within Wakatobi traditional society.

Moreover, in the region of *Kawati* Timu, there is Liang Kuri-Kuri which has become the myth of *waliullah* (holy society) settlement. In this area, society can see *sangia* (sacred traditional forest). In this cape as well, there are canon and a plate-like stone disc that has been used for performing rituals. Society will come here for praying and this is the reason why that belief still survives in the life of Wakatobi traditional society. One of the rituals that are usually performed here is the ritual of throwing away *loloa* for casting out evil spirits. In this ritual, society will read prayers for Allah from the evening until the dawn. It is very important to throw the *loloa* appropriately; otherwise, a huge catastrophe might fall upon the village. This society cast out the *loloa* with Islamic prayers and expectation that the village will be avoided from the big catastrophe.

Other areas of sea conservation can be found in several *untu* such as UntuWa Ode, UntuMelambi, and also in several Tondora. Although these conservation areas are protected by myths, there is a structure of thinking that lies within each member of traditional society there. The myth of *Untu Wa Ode* and *Untu Melambi* says that the two *untu* have been appointed as the region of *waliullah* city and this appoint has put aside the myth of *imbu*. In other words, the myth of *imbu* has been reconstructed into the myth of holy society or *waliullah*. Therefore, destroying the environment here will mean destroying the settlement of the holy society. The same paradigm also applies to WaBue-Bue Forest and Bungi Forest, which have also been considered as the city of holy society or *waliullah*. The traditional society there believe that they should not disturb other living creatures and this belief also includes the doctrine of defecation in *thaharah* manner from Islamic concept within Wakatobi society; this doctrine has been effective in protecting their environment [14].

In the context of Tindoi Forest, for instance, there is a myth says that society should not pick leaves and crack twigs. The reason is that Wakatobi traditional society believes that Tindoi has been the core area in the conservation of consumption water within Wangi-Wangi Island. Such myth, along with the other typical myths in this area, has been the representation of *kangkilo*. This traditional society will not pick the leaves or crack the twigs because they understand highly the doctrine *kangkilo*. The myth that has been developing among Tindoi society (WutaMohute and Ta'IBhete) shows that Tindoi region is a region of *karomah*. They believe that the spirits of their elders who dwell in Tindoi are society who has high religious knowledge. WaYai said that their elders who dwell in the region live straight. They maintain their holiness; as a consequence, the traditional society there is not allowed to pick leaves, crack twigs, and event cut trees. They believe that if they leave this belief, then they will suffer from the curse of *karomah* in Tindoi because they have been considered destroying the holiness that becomes the elders' way of life.

It is this doctrine that has been maintaining the Tindoi sustainability up to date. Despite the fact, Wakatobi has changed into a tourism destination and the lack of society understanding due to the shift in their awareness, the transition from the myth to the rationality-based science has caused many guests litter around the environment. These guests have lost the cultural values that frame Tindoi Forest as a bio-sphere heritage and the center of consumption water conservation in the same time, especially *uwedhi* Posalu or Posalu water. NurArafah [1] stated that the transition of traditional society institution in Wanci and Mandati has rendered a strong pressure to the environment within Wanse and Mandati traditional area [1]. Thereby, there has been a change of traditional society orientation in which *kadhia* institution has turned into a village, and this turning is heavily affecting the value sets in Wakatobi traditional society.

IOP Conf. Series: Earth and Environmental Science **156** (2018) 012041 doi:10.1088/1755-1315/156/1/012041

3.2 The Form of Local Conservation a Model of Acculturation between the Local Culture and Islam in Wakatobi

Among Wakatobi traditional society, there have been many models of acculturation between the local culture and Islam; these models are apparent in any form of conservation, both in the land one and in the sea one. Almost all of these acculturation models are framed in the concept of Islamic-style mythology because every conservation area is associated with *waliullah* or holy society. Such mythology displays that there has been an acculturation between the local culture (Hinduism, Buddhism, animism, and dynamism) and the Islam-based holy values.

The presence of *kangkilo* doctrine has changed the fundamental values of each local conservation area in Wakatobi. *Kangkilo* values, both in the initial holiness and the final holiness, provide an important meaning for the environmental conservation in Wakatobi. Multiples norms that have been developed by Wakatobi traditional society, for example, urge the society not to say inappropriate words or to do inappropriate actions in a conservation area; this is an effort of maintaining the holiness. The society is not allowed to say inappropriate words, to have an inappropriate mind, and to have negative feelings toward other society, the environment, and Lord. In several cases, the traditional society believes that they should have a *wudhu* before entering the area of *karomah* or otherwise they will be damned. Another example is that fishermen should not say swearing words whenever they enter Pasi Koko or they will not have huge catch. However, they should also be careful when they have huge catch. This is related to the values or the principles of *kangkilo* doctrine. The main point is that society should submit them to Allah when they do have bad fortune, and they should not be arrogant when they have good fortune. It is in this situation that the Islamic values become the fundamental values within the continuous fishery development in Wakatobi.

By understanding these Islamic values, a collective awareness among the traditional society might be born in order to maintain the environmental sustainability in Wakatobi. By doing so, it is expected that this society will not bomb the coral, which is destructive and even will not say swearing words in this conservation area. They should not need to be guarded by the forest rangers or the coast guards as well. These Islamic values should be internalized in their mind and soul so that this traditional society will always be guided by those values in their daily life as farmers or fishermen.

3.3 The Fading *Kangkilo* and the Threat of Environmental Damages in Wakatobi

The fading *kangkilo* values as the Islamic values within Wakatobi society leads to the increasing number of environmental damage in Wakatobi traditional area. La Rumadi said that the amount of environmental damage in Wakatobi has been caused by the decreasing number of the people who practice *kangkilo*. According to La Rumadi, it has almost ten years in Bau-Bau civil society that *kangkilo* has been practiced by two societies as a peculiar characteristic of Buton society. The implication is that nowadays there have been many societies who do not care anymore about Buton peculiar characteristics. These societies tend to do anything, and that includes destroying the environment. In other words, *kangkilo* doctrine has started to fade from Buton society.

There are several cases of environmental damage in Wakatobi. The perpetrators do not feel guilty when they destroy the environment. The reason is that they do not hold onto *kangkilo* doctrine anymore. They are willing to destroy the environment because they do not hold onto the Islamic values anymore. Many other acts of crime have been caused by the perpetrators' indifference toward the deity teachings, and they are even willing to hurt other society. They have abandoned their humanity. They have crossed the limit or *sabharagau* or the moral limits that have been set the by the traditional society.

In addition, the shift of social and political system has caused the fading Islamic values among Wakatobi traditional society, and this affects their context of environmental management. Wakatobi traditional society seems to be helpless in dealing with the social change both politically and culturally. The birth of young generation who are not familiar with *kangkilo* doctrine has affected the environmental sustainability in Wakatobi. The government's efforts in maintaining the environmental sustainability, including the implementation of multiple programs with NGOs, have not been able to give birth to successful environmental conservation like wha have been done by Wakatobi traditional society in their life. The reason is that the culture of establishing values has been internalized in these society's awareness. This condition enables every person to have some software that controls his or her mind regarding the environmental damage. *Kangkilo* doctrine will lead to the awareness that taking away what does not belong to an individual will ruin the individual himself, his relationship to the deity, and his relationship to the surrounding environment. La Nia said that that the society of kadhia Wanse has been very afraid of exploiting the forest heavily in the past especially around the part of Tindoi forest [14].

On the other hand, the conservation programs that the government has been pursuing are always based on the threat of positive and negative law. In the case of 2008 around Tindoi forest, one person was almost put into a traditional trial by the local society [14]. The reason is that Tindoi society had been aware that cracking a small twig might lead to the death of one person in that region while cracking a big twig might lead to the death of the prominent local figure. This understanding triggered Posalusociety in Tindoi area to almost kill the perpetrator. Fortunately, the security officers directly evacuated the perpetrator to the nearby police office. Meanwhile, there have been cases in which the local society have exploited the forest and have turned the forest in dry farming areas; such cases can be found in Bungi Longa forest, WaBue-Bue forest, and KaindeaTeo forest. In the case of Longa Patuno forest, KaideaTeo, for example, the local society do not understand their *kangkilo*. They deface their physics, mind, and soul by destroying the forest, whereas the elders of this society in the past would not do that. As long the as the software (namely, the Islamic values) still occupies these society's supra-structure, their awareness will lead them to establish *khilafah* on the earth.

Wakatobi traditional society in Buton who live the *kangkilo* doctrine will always hold onto their principles firmly as the *khalifah* on the earth. Under this principle, they will carry out their action as the form of *rahmatanlilalamin*. They will not destroy the environment, the relation among fellow human beings, and their relationship to the deity. It is here that Islamic values play their role in establishing the conservation area within Wakatobi. Thereby, the fading *kangkilo* doctrine within Wakatobi traditional society can be a serious threat to the environmental damage in Wakatobi. Moreover, in the social life, the fading *kangkilo* doctrine might lead to the birth of opportunistic society which prioritizes their own needs above the needs of other society.

4. Conclusion

The Islamic values in *kangkilo* have become the fundamental values that construct the paradigm of Wakatobi traditional society in performing the environmental conservation around Wakatobi. They who hold onto the *kangkilo* doctrine will always identify themselves with the maintenance of their holiness, which includes the dimension of physics, mind, and sense. Through the *kangkiloawalu* doctrine, Wakatobi traditional society will never eat something that might deface their physics, mind, and sense. While through the *kangkiloakhiri* doctrine they will always maintain what they do to the environment in order not to deface the physics, the mind, and the sense of both the human beings and the other creatures. Thereby, Wakatobi traditional society will always strive to maintain himself, the environment, and the deity. Through the concept of *kangkilo*, which has been the fundamental Islamic values, Wakatobi traditional society has been able to maintain the sustainability of earth bio-sphere heritage since from hundred years ago. For the society who lives with the *kangkilo* concept in their life, they will affect their life by not defacing their physic, mind, and humanity. By doing so, *kangkilo* can be a space of mental revolution within the environmental management among traditional society throughout the world.

References

[1] Arafah N 2015 *Kaindea: Adaptasi Masyarakat dalam Pengelolaan Hutan di Pulau-Pulau Kecil.* Kendari, Penerbit Oceania Press.

[2] Udu H 2017 *Tradisi Lisan Kangkilo: Refleksi Paham Tasauw dan Politik Kekuasaan dalam Masyarakat Buton.* Denpasar, Disertasi Program PascasaranaUniversitasUdayana.

[3] Udu H 2015 "Kultur Modern dan Kearifan Masyarakat Buton dalam Tradisi Lisan Kangkilo" dalam *Folklore and Folklife dalam Kehidupan Dunia Modern: Kesatuan dan Keberagaman* (ed. Suwardi Endaswara dkk). Yogyakarta, Penerbit Ombak.

[4] Alifuddin M 2007 Islam Buton: Interaksi Islam denganBudayaLokal. Yogyakarta, Disertasi Program Pascasarajana UIN Sunan Kalijaga.

[5] Spradley J 1997 *Metode Etnografi.* Yogyakarta, Tiara Wacana.

[6] Udu S 2016 *Sastra Lisan: Konservasi Lokal Cagar Biosfer Bumi Wakatobi.* Yogyakarta, Makalah Simposium Internasional Himpunan Sarjana Kesusastraan Indonesia Bekerja Sama dengan Fakultas Bahasa dan Seni Universitas Negeri Yogyakarta.

[7] Zaenu LO 1985 *Buton dalam Sejarah Kebudayaan.* Surabaya, Penerbit Suradipa.

[8] Sabaruddin LO 2002 *Cerita Rakyat Tomia.* Kendari, *Skripsi* FKIP Universitas Haluoleo.

[9] Taalami LO et al 2016 *Historiografi Benteng-Benteng di Wakatobi.* Wangi-Wangi, Laporan Penelitian yang dilaksanakan oleh Dinas Pariwisata dan Ekonomi Kreatif bekerja sama dengan Pusat Studi Sains, Budaya, Pariwisata dan Humaniora Indonesia (PUSBUDPAR-I)

[10] Udu S 2016 *Tradisi Lisan Bhanti-bhanti Wakatobi: Pementasan, Formula dan Komposisi Skematinya.* Yogyakarta, Disertasi Program pascasaraja Fakultas Ilmu Budaya Universitas Gadjah Mada.

[11] Zuhdi S 1999 *Labu Wana Labu Rope: Sejarah Buton Abad XVII-XVIII.* Jakarta, Disertasi Program Pascasarjana Universitas Indonesia.

[12] Hamid AR 2010 *Spirit Bahari Orang Buton.* Makassar, Rayhan Intermedia.

[13] Lapian AB 2009 *Orang Laut, BajakLaut, Raja Laut: SejarahKawasanLaut Sulawesi Abad XIX.* Jakarta, KomunitasBambu.

[14] Udu S 2015 "Tradisi Lisan sebagai Media Konservasi Lingkungan di dalam Masyarakat Wakatobi" dalam *Folklore and Folklife dalam Kehidupan Dunia Modern: Kesatuan dan Keberagaman* (ed. Suwardi Endaswara dkk.). Yogyakarta, Penerbit Ombak.

[15] Mbaru, LR 2016 *Culada Tape-Tape: Peradaban Binongko, WakatobiButon.* Kendari, Penerbit Oceania Press.

[16] Rosdin A 2015 *Nilai-Nilai Kehidupan Masyarakat Buton: Sumbangan Kabanti Kabhanti Anjonga Yinda Malusa untuk Revolusi Mental Indonesia.* Kendari, Penerbit Oceania Press.

[17] Taalami LO 2011 Hikayat Negeri Buton: *Analisis Jalinan Fakta dan Fiksi dalam Struktur Hikayat dan Fungsinya Serta Edisi Teks.* Bandung, Disertasi Program Pascasajana Universitas Padjajaran.

IOP Conf. Series: Earth and Environmental Science **156** (2018) 012042 doi:10.1088/1755-1315/156/1/012042

Environmental Health Risk Assesement in Flood-prone Area in *Tamangapa* Sub-District Makassar

Ibrahim Abdul Haris[1], Basir Basir[2]
[1]Environmental Health Department of Public Health Faculty of Hasanuddin University
Email: ibrahimmharis@gmail.com
[2]Environmental Health Department of Public Health Faculty of Hasanuddin University
Email: baz.rasyid@gmail.com

Abstract: Environmental health in Indonesia is still caution to concern, poor sanitation in Indonesia is characterized by the high incidence of infectious diseases in society. The society in flood-prone area has a high-risk exposure on the disease based on the environment because they live in disaster-prone area. This research aimed to describe the condition of sanitary facilities and risky behavior on public health in flood-prone areas in Manggala district particularly in *Tamangapa* sub-district of Makassar. This reserach uses an observation method with a descriptive approach. The data is processed by using SPSS and Arc View GIS applications. Environmental risk category is determined by the approach of Environmental Health Risk Assessment (EHRA). The results showed that the flood-prone area in RT 04 RW 06 was included in very high-risk category at 229 with an index value of environmental health risks 212-229. Meanwhile, RT 04 RW 05 was in the category of low risk in the amount of 155 with an index of 155-173. Environmental health hazards identified in *Tamangapa* flood-prone areas sub-district includes domestic sources of clean water, domestic wastewater, and household garbage.
***Keywords*: Risk, environmental health, flood**

[1] Introduction

Republic of Indonesia is widely located in prone areas of natural disasters such as landslides, earthquakes, volcanic eruptions, tsunamis, floods and others. Indonesia ranks second in the list of highest deaths due to natural disasters in Asia-Pacific. Over the past 20 years, various natural disasters in the country have also caused economic losses of at least US $ 22.5 billion. This data is available in the 2010 Asia Pacific Disaster Report prepared by the Economic and Social Commission for Asia and the Pacific (ESCAP) and the UN International Strategy for Disaster Reduction (UNISDR) [1,2].

People living in flood-prone areas are at risk of exposure to environmental-based diseases, as they live in hazard-prone environments. Flood-prone areas are the areas that affect the health of population who live in the area, influenced by the management of the living habitat and the management of the disturbed environment. One of the problems that often happen in the city of Makassar is the problem of flooding. Almost every year the flood in Makassar occurs on every coming of the rainy season. A total of 24 urban villages in 6 sub-districts with a total area of 2761.84 Ha often become flood subscriptions in the rainy season [3,4].

Environmental Health Risk Assessment (EHRA) is a study to understand the condition of sanitation facilities and behaviors that pose a risk to public health, facilities which are observed includes the Clean Water Source, a healthy latrine facility, household waste, and sewerage of households. For behavior, Hand-washing with Soap and disposing of garbage and Bleeding Behavior are learned, in this research will also be done distribution or assessment of environmental health risk [5].

According to the Sanitation Handbook of Makassar City 2011, sanitation is a deliberate behavior in the culture of clean living with the intention of preventing humans from direct contact with dirt and other hazardous waste materials in the hope that this effort will maintain and improve human health. Sanitation hazards may be physical, microbiological, and chemical or biological agents of related illnesses. The waste material that can cause health problems comprises human or animal feces, solid waste residue, domestic waste water (laundry, urine, bath or laundry materials), industrial waste, and agricultural waste materials [6,7].

Globally, 19% of deaths were due to infectious diseases related to basic sanitation and hygiene or hygiene risk factors. For Indonesia, environmental health is still a concern. The lack of optimal sanitation in Indonesia is characterized by the high incidence of infectious diseases in the community. Data from the Ministry of Health of the Republic of Indonesia shows that the presentation of the population with adequate sanitation facilities in 2011 was only 54.99%. This indicates that less than half of Indonesia's population still has inadequate sanitation facilities [8,9].

Historically, there are spot of flood areas in Romang Tangngaya RT 04, RW 06 and *Bontoa* RT 04, RW 05 *Tamangapa* Sub-district in Makassar City which topographically is in the lower plain of the surrounding area beside the River Kajenjeng flowing from upstream that is located in the administrative area of Gowa Regency. Hence, while the rain with high intensity occuring, it will cause floods that inundate not only settlements, but soaking the rice fields and destroying rice crops and crops. The flood often happens for two to three months each year [10].

[2] Method

The type of research was using observation with descriptive approach that aimed to identify the condition of sanitation facilities and behaviors which were at risk to public health in flood-prone area in *Tamangapa* Sub-District, Manggala, Makassar. The research was conducted on December 8th, 2015 to January 8th, 2016. The population in this study was all households in RW. 5, RT. 4 (*Bontoa*) and RW.6, RT. 4 (*Romang Tangaya*) flood-prone areas of *Tamangapa*, 129 houses, each 72 houses in *Bontoa* and 57 houses in *Romang Tangaya*. The samples in this research were all households in flood-prone areas of *Tamangapa* Sub-District, Manggala district, Makassar. Respondents in this study were household members found when conducting research, allowing for interviews. Sampling was done by exhaustive sampling. The data analysis used was descriptive by making interpretation and narration of each variable and presentation of data presented in the form of narration, tables of environmental health risks.

[3] Result

The result shows that most of the respondents occupy the house which is owned by the respondent themselves as many as 115 or 89,1% that is each spread in RW 05 RT 04 as many as 58 houses and at RW 06 RT 04 counted 57 houses. Of the 2 research sites in *Tamangapa* Village, the majority of the population owns 74 stage house houses or 57.4%, each spread in RW 05 RT 04 with 27 houses and RW 06 RT 04 with 47 houses.

Distribution by source of clean water and drinking water treatment was found that for drinking purposes most used by respondents was unavoidable dug well water of 39 or 30.2% and the least was water from unprotected springs as much as 1 or 0.8% , for the most commonly used water cooking purposes is well protected water wells as much as 39 or 30.2% and the least is water from unprotected springs as much as 1 or 0.8%, for the purposes of MCK as much as 40 or 31% , and the least is water from unprotected springs as much as 1 or 0.8% (Table 1).

Private latrines owned by respondents are much more around 92 or 71.3% while those who do not have 37 or 28.7%. For bowel space there are 92 or 71.3% using private latrines and 1 or 0.8% using excavation pits (Table 2). The respondents who have a waste bin of 29 or 22.5% and the number of respondents who do not have more than 100 or 96.1%. For the respondents who dispose of waste before disposal as much as 5 or 3.9% while that does not sort the waste more by 124 or 96.1%.

The incidence of environment-based disease in flood-prone areas of *Tamangapa* Sub-district, Manggala District, Makassar City is still dominated by diarrhea disease with the highest percentage of 69.8% of total respondents. While the lowest incidences of disease are eye disease and vector-based disease. Based on 4 variables which are used to determine the table of environmental health risk index. The cumulative results show that RW 06 RT 04 with the environmental health risk index value of 229, becoming the very high-risk category, while RW 05 RT 04 with an environmental health risk index score at 155, falling into the less risky category (Table 3). Then the results are inserted at the point of coordinates in pairs using the Arc View Gis computer program and performed digitization and marking at each coordinate point used to obtain an environmental health risk map (Picture. 1, Picture. 2).

[4] Discussion

It was found in this study that most households utilize well dug water is not protected in meeting their daily needs of 30.2%. On the other hand that many also used by the respondents is the source of water coming from the well water drill. The shape of topography in a region can affect the groundwater in the area. In the lowland area the frequency of groundwater extraction is relatively large because in this area the population growth is growing rapidly while the plateau area, this area is located on the slopes of the mountain foot [11].

The practice of bowel movements can be one of the risk factors for contamination of the environment including water sources, especially when open-air practices such as in river / gutters / gardens, but can also include comfortable latrine facilities at home. Although defecation is done at home with convenient toilet advice, but if the shelter and treatment facilities are not adequate, for example because it is not water-resistant, the risk of pathogen contamination will remain high. For people who have a habit of defecating carelessly obtained data of 23.2% done by family members who are not known clearly who the person. It may be influenced by the embarrassment by the respondent when questioned about it [6,12].

In this study also asked about the intensity of people's waste, the data show 45% of respondents throw their garbage every day. The longer the garbage is stacked at home, the higher risk of environmental health, because the garbage that builds up in the house will be the source of the smell, the source of the bacteria, and the source of the disease-carrying vector. Based on survey results of 129 respondents in the data, there are 70 respondents or 54.3% who do not have sewerage of households. Some studies suggest that untreated or well managed waste water can be harmful to human health as it can serve as a disease-carrying medium, especially infectious diseases that are transmitted through contaminated water. Seeing the condition of the location of research that when entering the rainy season will experience flooding, so it is possible to occur at the location of the study. As for handwashing variables are as many as 83 or 64.3% who have done handwashing with soap. Judging from the 5 important time to wash the hands, it is concluded that no one performs CTPS at a critical time. This is due to the lack of public knowledge about the importance of handwashing with soap.

Diarrhea is the highest disease that is 69,8%, this is because almost all household area use dug well water is not protected and also research location is flood-prone area. Several studies have revealed that the incidence of diarrhea is largely determined by the frequency of a polluted area caused by floods carrying bacteria and pathogens. The process of transmission of the disease is influenced by the characteristics of the host (immunity, nutritional status, health status, age, and sex) and host behavior (personal hygiene and food hygiene) [13].

Each value of environmental health risk index obtained then accumulated at each research location and obtained value at RW 05 RT 04 equal to 155 while at RW 06 RT 04 equal to 229. From that value to determine 4 categories of risky area that is less risk, medium risk, high risk, and the very high-risk. So that the results obtained less risky category entry on the vulnerable 155-173, the risk is susceptible value 174-192, high risk categories exist in value 193-211 and the last risk is too high between 212-229 value.

If it is reviewed overall, the risk picture in the flood-prone areas of *Tamangapa* Sub-district, Manggala district, Makassar. *Romang Tangaya* or RW 06 RT 04 area is very high. iI is because the access of facilities and infrastructure is far from the standard of sanitation and behavior which is still low and also locations that are difficult to access if the rainy season becomes one of the factors that affect it. In terms of population precisely *Romang Tangaya* is less than *Bontoa* (RW 05 RT 04) but the risk in *Bontoa* is lower than in *Romang Tangaya*, so the population does not guarantee high risk or

IOP Conf. Series: Earth and Environmental Science **156** (2018) 012042 doi:10.1088/1755-1315/156/1/012042

less risk, but depends on people's behavior. If community behavior is healthy then it will be able to reduce environmental health risks and opportunities [6,12].

Table 1. Distribution of Cleaned Water Source in Flood-prone Area *Tamangapa* Sub District, Makassar

Distribution of Cleaned Water Source	RW/RT				Total	
	05/04		06/04			
	n	%	n	%	n	%
Drinking Water						
Mineral Bottle Water	5	6,9	0	0	5	3,9
Refill Water	27	37,5	0	0	27	20,9
Well Water Drill	7	9,7	14	24,6	21	16,3
Well Water Dug Protected	11	15,3	8	14	19	14,7
Well Water Dug Unprotected	5	6,9	34	59,6	39	30,2
Protected Water Spring	0	0	0	0	0	0
Unprotected Water Spring	0	0	1	1,8	1	0,8
Rainwater	0	0	0	0	0	0
Water Work Municipal	17	23,6	0	0	17	13,2
Cooking Purposes						
Mineral Bottle Water	1	1,4	0	0	1	0,8
Refill Water	5	6,9	0	0	5	3,9
Well Water Drill	18	25	14	24,6	32	24,8
Well Water Dug Protected	15	20,8	8	14	23	17,8
Well Water Dug Unprotected	5	6,9	34	59,6	39	30,2
Protected Water Spring	0	0	0	0	0	0
Unprotected Water Spring	0	0	1	1,8	1	0,8
Rainwater	0	0	0	0	0	0
Water Work Municipal	28	38,9	0	0	28	21,7
Extensive Purposes (Bathing,Washing, Etc)						
Mineral Bottle Water	0	0	0	0	0	0
Refill Water	4	5,6	0	0	4	3,1
Well Water Drill	21	29,2	14	24,6	35	27,1
Well Water Dug Protected	16	22,2	8	14	24	18,6
Well Water Dug Unprotected	5	6,9	34	59,6	39	31
Protected Water Spring	0	0	0	0	0	0
Unprotected Water Spring	0	0	1	1,8	1	0,8
Rainwater	0	0	0	0	0	0
Water Work Municipal	26	36,1	0	0	26	20,2
Total	**72**	**100**	**57**	**100**	**129**	**100**

Table 2. Distribution of Toilet Ownership, Toilet and Type of Toilet in Flood-prone Area *Tamangapa* Sub District, Makassar

Distribution of Toilet Ownership, Toilet and Type Of Toilet	RW/RT				Total	
	05/04		06/04			
	n	%	n	%	n	%
Toilet Ownership						
Yes	64	88,9	28	49,1	92	71,3
No	8	11,1	29	50,9	37	28,7
Toilet						
Private Toilet	64	88,9	28	49,1	92	71,3
General Water Closet	8	11,1	23	40,1	31	24
Garden	0	0	3	5,3	3	2,3
Digging Hole	0	0	1	1,8	1	0,8
Neighbhour Toilet	0	0	2	3,5	2	1,6
Type of Toilet						
Swan Squat Closet	62	86,1	17	29,8	79	61,2
No Closet	8	11,1	29	50,9	37	28,7
Old-style squat toile	0	0	8	14	8	6,2
Swan Sit Closet	2	2,8	1	1,8	3	2,3
Garderobes	0	0	2	3,5	2	1,6
Total	**72**	**100**	**57**	**100**	**129**	**100**

Table 3. Score of Environmental Health Risk in Flood-prone Area *Tamangapa* Sub District, Makassar

RW/RT	IEHR Score	Risk Category
05/04	155	Low Risk
06/04	299	Very high-risk

[5] References

[1] Departemen Kesehatan RI.Keputusan menteri Kesehatan Republik Indonesia Nomor 145/Menkes/SK/I/2007. [2007]. Pedoman Bencana Bidang Kesehatan. Jakarta: Pusat Penanggulangan Krisis.

[2] Chazainul, M. [2013]. Governance and Capacity Bulding dalam Manajemen Bencana Banjir Indonesia. Jurnal Penanggulangan Bencana. 4(2):5-12.

[3] World Health Organization .(2004). Environmental Health Criteria XXX : Principles for Modelling, Dose Response for The Risk Assess-ment of Chemicals. Jenewa: IPCS.

[4] Rauf, S, Samang, L. (2012). Analisis dan Pemetaan Daerah Rawan Banjir di Kota Makassar Berbasis Spasial [Skripsi].Makassar: Universitas Hasanuddin.

[5] ISSDP.(2013). Penilaian Risiko Kesehatan Lingkungan Kota Blitar. Jakarta: Indonesia Sanitation Sector Development Program.

[6] ISSDP.(2011). Penilaian Risiko Kesehatan Lingkungan Kota Makassar. Makassar: Pokja Sanitasi Kota Makassar.

[7] Daud, A. Aspek Kesehatan Penyediaan Air Bersih.(2007). Makassar: CV. Healthy and Sanitation.

[8] Badu, A.(2016). Gambaran Sanitasi Dasar pada Masyarakat Nelayan di Kelurahan Pohe Kecamatan Hulonthalangi Kota Gorontalo

[9] Departemen Kesehatan RI. (2012). Persentasi Penduduk yang Memiliki Sanitasi yang Layak. Jakarta: Depkes RI.
[10] Badan Penanggulangan Bencana Daerah.Rencana Kontinjensi Bencana Banjir Kota Makassar. (2014). Makassar: Badan Pusat Bencana Daerah.
[11] Putranto, T, Kusuma, K. (2009). Permasalahan Air Tanah pada Daerah Urban. Jurnal Teknik. 30 (1):48-55.
[12] Simanjuntak, C.(2009). Demam Tifoid, Epidemiologi dan Perkembangan Penelitian. Jakarta: Cermin Dunia Kedokteran.
[13] Carr, R.(2001). Excreta-related Infections and the Role of Sanitation in the Control of Tranmission. In: Bartram LfaJ. Water Quality: Guidelines, Standards and Health. London: IWA Publishing; 90-107
[14] Chicken, Posner.(1998). The Philosophy of Risk. P.7. London : Tomas Telford.
[15] Marwah. (2014). Penilaian Risiko Kesehatan Lingkungan di Pulau Bonetambung Kota Makassar [Skripsi]. Makassar: Universitas Hasanuddin.

Implementation of Zakat on Income Management for Coastal Community Empowerment

Nur Mohamad Kasim

Faculty of Law, Universitas Negeri Gorontalo, 96128, Indonesia
Email: nurkasim76@yahoo.co.id

Abstract: The purpose of the research is to explain the substance of zakat on income regulation, to examine the government role in the management, utilization, and the importance of zakat on income management. Furthermore, the study aims to determine and analyze the constraints that obstruct the management of zakat on income in Gorontalo. The research is a socio-legal research, which uses a qualitative analysis technique. The data is comprised of primary and secondary data. This research used the legislation, law concepts or theories, and case approach. Moreover, the research employed literature review, observation, and interview as the data collecting techniques. The results showed that the management of the zakat on income in Gorontalo is still low. There is a substantial weakness in the management of zakat on income, as the existing regulation has not accommodated the interest of Islamic community, especially those who are obligated in giving zakat. Meanwhile, there are a lot of professional groups who are not aware on the obligation of giving their zakat on income. Hence, there is a necessity to penalize those who did not obey the regulation to raise awareness in the importance of paying zakat. Other factors that affect such phenomena are the lack of reliable and professional human resources as well as less supervision from the government. In addition, lack of community awareness of zakat on income culture and government role in zakat distribution also result in low attention on the management of zakat on income. Zakat on income can empower the less fortunate people by underlying the importance of zakat on income management and by doing a proper administration of the process of collecting, distributing, utilizing, as well as enhancing the process of zakat on income. In addition, the management of zakat on income must be done as professional, transparent, and proportional as possible. Therefore, it can give a positive contribution to the local government especially National Zakah Board or *Badan Amil Zakat Nasional (BAZNAS)* of Gorontalo Province.

Keyword: Zakat on income; regulation; management; professional; empowerment.

1. Introduction

The management of zakat is based on the Act No. 38 of 1999 about the Management of Zakat which is considered to be no longer appropriate in accordance with the times and legal needs in society and therefore needs to be replaced. Act No. 23 of 2011 about the Management of Zakat which replaces the former act has become a good basis for managing zakat in Indonesia. Today, the development of zakat is quite dynamic and hopefully zakat can play a role in increasing the economy of the society and the poverty alleviation in the future.

In the modern era, the progress of human civilization has given diverse impact on the human livelihood. Agricultural occupations were abandoned and considered less attractive. The most visible form of today's income is the income that is obtained through one's professions. The professions may be in the forms of employment (private or public), or occupations that do not depend on others (professionals), such as doctors, advocates, tailors, artists, and so on. This type of jobs earns income either in the form of salary, wage, or even honorarium.

IOP Conf. Series: Earth and Environmental Science **156** (2018) 012043 doi:10.1088/1755-1315/156/1/012043

Zakat on income or sometime also known as profession zakat as well as zakat on salary is a term that emerges nowadays which the *Salaf cleric*[1] calls by the term *al-mal al-mustafad*, which is zakat included *inal-mal al-mustafad*. This type of zakat is taken from income generated from professions such as salaries of employees, consultants, doctors, and others or unexpectedly generated sustenance such as sweepstakes and a quiz prize. Zakat on income has been a debate since a long time. Those who oppose zakat on income argued that there is no clear or vigorous argument and no specific concrete example from the Sunnah of the Prophet Muhammad SAW. However, those who agree and support the zakat on income assume that the actual zakat on income is not a new thing in Islam, in which the source is so clear and vigorous from the Quran[1].

In the early twentieth century, the existence of zakat on income has become widely recognized. This is marked by the emergence of a great contemporary scholar and highly prolific writer of *Egyptian masyhurasal*, Yusuf al-Qaradawi [2]. His books became the relevant references of Muslims in the world and gave a significant influence on the development of discipline related to zakat, including zakat on income.

The distribution of zakat property can be done privately or regularly through the aid of amil zakat institutions that have the authority to conduct activities in collecting and distributing the zakat property to the groups who deserve to receive it. The use of institutional zakat authority is to achieve its ethical objectives for social justice purpose based on the ideals of spirituality[3].

The potential of zakat will continue to grow along with the increasing number of employees of state-owned enterprises, along with private and civil servants, so that the benefits of zakat can be felt by the poor community and thus the economic independence can be raised. The point is that zakat can help the government in alleviating poverty. This is aligned with the purpose of zakat management that is professional, responsible, and legal conducted by the government with the community.

The zakat is also supported by the Act Number 23 of 2011 on Zakat Management and Presidential Instruction No. 3 of 2014 on Optimizing Zakat Collection. Currently, some ministry employees have implemented their zakat obligations according to Presidential Instruction No. 3 of 2014. The zakat of the employees is paid by deducting 2.5% of their monthly salary[4]. This is also followed by most of the civil servants in Gorontalo. The zakat requirement is also included in Gorontalo Governor Regulation No. 7 of 2014 on the Collection of Zakat in the Government of Gorontalo Province. Due to the importance of the regulation on the management, the government of Gorontalo province issued a decree in the form of governor's regulations that are binding to the people of Gorontalo, especially the *muzakki.*

It is also explained in the governor regulation that the subject and object of zakat collection is the *muzaki* which consists of the civil servants candidates and officials within the scope of the Government of Gorontalo province. Their calculation of income has fulfilled the obligation to pay zakat, and that the object of zakat is the revenue derived from the profession as civil servant candidates and officials[5].

The zakat collection case is done within the Government of Gorontalo Province and conducted by Local Government agencies spending treasurer with the calculation of 2.5% from the net salary of civil servants candidates and officers' allowance received every month and has been registered on payroll list [5].

The problems often arise in collecting the zakat on income in the Gorontalo Provincial Government and the Regency/ City Government. The problems are resulted from the implementation of collecting the zakat on income of civil servants/ civil servant candidates and officials who have become fixed *muzaki*, which slow down the distribution of zakat funds from local government agencies. The zakat from every institution is forwarded to the Institution of Amil Zakat and distribute it to the *mustahik* or the recipients who must receive the zakat. Along with the increase of poverty rate, the number of zakat or *mustahik* recipients is also increased. The data obtained by the researchers in the field, for Gorontalo Province with a population of 1.2 million, the *mustahik* number of zakat recipients is 611.274 people.[2] This is a serious issue that the local government must address. Without the help of zakat on income, the poor society (underprivileged) will not be able to scope with their everyday life. In this case, zakat is essential to help them.

2. Method

This research will review and analyze the legal issues related to the research on the urgency of on zakat on income management for Gorontalo community empowerment. This study used a statutory, conceptual, and legal theory as well as case approach. The research utilized literature, observation, and interviews as the data collection techniques.

The approach of this research is normative, sociological and philosophical approach. The approaches used in this research are:

First, the statute approach. The use of this approach in normative research is imperative, because by logical law, the focus of normative research is on research on written legal materials (written law).

Second, the conceptual approach. The relevance of this approach is important as it concerns with the regulation, policy strategy, and perception of the zakat on income. Policy-makers need to update the concept, so it will have consequences on the expansion of new theories in Islamic law in order to effectively support the implementation zakat on income in Gorontalo.

Third, the philosophical approach of law. The use of this approach is to seek and analyze the legal principles implicit in the legal arrangement of zakat on income and philosophy which will become the foundation of the scholars in giving the fatwa to realize the welfare of a just and equitable society.

3. Results and discussion

3.1 Zakat Management Based On Act Regulation

It has been discussed previously that zakat is included in a group of religious laws or is called the *diyani* law. However, based on the reality in Indonesia, the regulation does not only regulate the management of zakat, but also enters the substance of the law of zakat itself. This means that zakat worship is not only normative/ serviced, but also formally juridical/ *qadha'i*[3] as the law regulate it. The intended Act is Act No. 23 of 2011 on Management of Zakat.

Act No. 23 of 2011 is the replacement of Act No. 38 of 1999 on the Management of Zakat which is already in effect. In consideration of the Act. No. 23 of 2011 article states that Act No. 38 of 1999 on the Management of Zakat is not in accordance with the development of legal needs in society, so it needs to be replaced. This is the government's consideration to replace the Act No. 38 of 1999 with the Act No. 23 of 2011, to adjust based on the condition and the situation of the people of Indonesia which is always dynamic and evolving along with the times, as a result of globalization and modernization. Therefore, the zakat act must also be changed and updated according to the needs of the Indonesian people nowadays.

Article 1 Sub-Article a of Act No. 23 of 2011 states that: "the management of zakat is the activity of planning, executing and coordinating in the collection, distribution, and the utilization of zakat." While the purpose in article 3 of Act No. 23 of 2011 is as follows:

a. Improving the effectiveness and efficiency of services in the management of zakat, and
b. Increasing the benefits of zakat to create the welfare of society and poverty alleviation.

IOP Conf. Series: Earth and Environmental Science **156** (2018) 012043 doi:10.1088/1755-1315/156/1/012043

3.2 The Implementation Of Zakat On Income For The Community Empowerment

The ups and downs of the implementation of zakat on income in Gorontalo are due to the policy on zakat management regulation that has not been maximized, thus affecting the implementation of its management. The implementation of zakat based on a clear regulation is essential to place public or private employees as the subjects. Civil servants or other professions, who have high income as those who are obliged in the payment of zakat, can be used for the benefit of other people. This is aimed to improve the welfare of the poor [6].

The presence of Act No. 23 of 2011 on the Management of Zakat has given widespread implications for its management institutions. The payment of Zakat is expected to be entrenched in a capable Muslim society and can be functioned optimally as the most important instrument of economic empowerment of people and nation [7]. Similarly, Act No. 38 of 1999 on the Management of Zakat is very relevant to the provisions of Islamic law, where the management of zakat is entirely submitted to the government. Nevertheless, in the implementation of this zakat management, the government is not doing this alone. The government has the power and has the authority to compel the obligatory zakat to give their property and set aside some of their income for the benefit of many people as done by Prophet Muhammad's companions.

Such practices should be undertaken and followed by most civil servants or private employees or people with other professions in Gorontalo so that the management of zakat is more leverage and professional. This is similar to the results of the interview with the Head of Economic Section Zulbahri Yusuf, ST., MT. on April 14, 2017, that the zakat deductions by the local government have been made well. Though there are still obstacles encountered, especially in terms of understanding the obligation to give zakat which is still personal, and the report on zakat funds distribution has not run optimally. The circumstances have encouraged regional Baznas institutions to be more efficient in managing zakat, so there will be no more competition in collecting zakat, and there will be no more overlaps in zakat distribution. Furthermore, there will be no different information to the *muzakki* in depositing their zakat. This will be done as an effort to grow trust and curb the data as the form of accountability.

As the effort for actualizing the proper and professional management of zakat in accordance with the provisions of religion, it requires good and perfect law enforcement with law enforcement officers who have sufficient moral or faith quality, are willing to sacrifice with high dedication, and have a great responsibility to the interests of many people. According to Jeremy Bentham, the purpose of the law is to create the happiness for many. The existence of the state and the law is solely for the sake of the true benefit of people's happiness [8]. Firmly, according to this theory, the ideal society is a society that tries to increase happiness and minimize unhappiness. It also means a society that attempts to give happiness to the people in general, so that the unhappiness can be kept to a minimum. Happiness means pleasure or lack of misery; unhappiness means misery and lack of pleasure.

Therefore, to achieve happiness for many people, it is essential is to fulfill the obligation of zakat for those who are capable. The zakat management should also be supported by a good and professional procedure based on the provisions of sharia. In order to avoid inequality on the implementation of zakat management function, it is expected that National Zakah Board of Gorontalo or *Badan Amil Zakat Nasional (BAZNAS)* institution can channel the right program to the needy and give information related to zakat management, with the hopes that the management of zakat in the future will be improved to fulfill the prosperity of society. In addition, zakat management should be equipped with zakat financial management regulation as well. If the zakat financial management mechanism is guided by the regional financial management model, the financial management can be carried out in proper order. Zakat funds can be managed properly if the managing agency has a solid funding mechanism.

Meanwhile, related to the management of zakat fund, there should be a synergy between the central government, local government, National Zakah Board in advancing and developing the management of this charity from a social institution into a sharia financial institution. This can be done solely to make the management of zakat funds to be more transparent and accountable in accordance with the provisions of sharia.

4. Conclusion

The problem of zakat management especially zakat on income is quite complex, in which it has become the problem faced by all Muslims nowadays, especially for the *muzakki* as the zakat payer. The importance of building Muslim solidarity through zakat on income as its strength in promoting and improving the welfare of society to create awareness that zakat on income is efficient and useful for the community.

The realization of zakat management in Gorontalo does not seem to run effectively because the open management of zakat management is still experiencing a number of difficulties, particularly on the mechanism of collecting and distributing the zakat. Also, those who are obliged of zakat on income appear to be reluctant to pay zakat through National Zakah Board. There are still concerns of the *muzakki* regarding to the distribution of zakat and the lack of information to the public.

5. Acknowledgments

In conducting this research, the research team got help from various parties, and also overcome obstacles in the implementation of research activities. For that, the researchers express deep gratitude to the parties as follows:

a. Gorontalo Provincial Government (Gorontalo Regency, Boalemo Regency and Gorontalo City) who have helped and participated in this research activity;
b. Rector of State University of Gorontalo and Chairman of Institute of Research and Community Service State University of Gorontalo that has agreed and facilitate and facilitate the process of implementation of this research activity;
c. National Zakah Board or *Badan Amil Zakat Nasional (BAZNAS)* of Gorontalo Province;
d. National Zakah Board or *Badan Amil Zakat Nasional (BAZNAS)* of District Boalemo.
e. All parties who have assisted this research activity, so that this research can be implemented properly.

6. References

[1] Fakhruddin 2008 *Fiqh dan Manajemen Zakat di Indonesia* (Malang: UIN Malang) p135-136
[2] Hafifuddin D 2007 *Agar Harta Berkah dan Bertambah* (Jakarta: Gema Insani Press) p 96
[3] Mas'udi MF 1993 *Risalah Zakat Dalam Islam* (Jakarta: Pustaka Firdaus) p 126
[4] Ali HMT 2016 *Peran Strategis Zakat Dalam Masterplan Arsitektur Keuangan Syariah Indonesia. In: National seminar National Zakah Board, at Museum Kebangkitan Nasional* (Jakarta: National Zakah Board) p 6
[5] *Governor Regulation No 7* 2014.
[6] Muhammad Hadi 2010 *Problematika Zakat Profesi dan Solusinya Sebuah Tinjauan Sosiologi Hukum Islam* (Yogyakarta: Pustaka Pelajar) p 204
[7] Nur Mohamad Kasim 2014 *Membangun Rasa Nasionalisme Melalui Zakat Profesi* (Yogyakarta: Pohon Cahaya) p 70
[8] Achmad Ali 2009 *Menguak Tabir Hukum Dan Teori Peradilan Termasuk Interpretasi Undang-Undang* (Jakarta: Kencana) p 273

IOP Conf. Series: Earth and Environmental Science **156** (2018) 012044 doi:10.1088/1755-1315/156/1/012044

Implementation Study of Restorative Justice for Juvenile Criminal Justice System by Customary Court in Mainland Sulawesi

Nur Rochaeti[1] and Pujiyono Pujiyono[2]

[1] Faculty of Law, Universitas Diponegoro Semarang
Email: iyenk283@yahoo.co.id
[2]Faculty of Law, Universitas Diponegoro Semarang
Email: pujifhundip@yahoo.com

Abstract: Children's rights are specifically different from adult human rights. In all circumstances the rights of children must take precedence over other interests, namely the existence of legal protection for them and the existence of special protection aimed at children who violate the law. Act No.11 of 2012 on the Juvenile Criminal Justice System requires restorative justice efforts as a form of protection for children, which emphasizes restoring the harm caused or incurred by criminal acts. Restorative justice has long been practiced in indigenous community of Indonesia, i.e. where a criminal act occurs by its citizens (including juvenile delinquency), the dispute settlement is conducted within the indigenous community internally without involving the State apparatus. The formulation of the problem is, firstly, how is the policy of restorative justice formulation of juvenile criminal justice system in handling the juvenile with conflict of law, secondly how is implementation of restorative justice of juvenile justice system by customary court in Makassar. The results of the study indicate that customary court in Makassar has legal and cultural characteristics, moral values that can solve problems by deliberation, involving various parties in litigation such as in restorative justice mechanism in the juvenile criminal justice system.

Keyword: Implementation Study, Restorative Justice, Customary Court

1. Introduction

Every child is protected from inside of the womb to become adult. Children also have specific rights different to adult people. Physically, the condition of children and their mental states are unstable, making their ability to act and be responsible to their action different to the adults. Court process can cause psychological impact of harming children. They will get pressure and stigmatization during the process. Therefore, the actions in juvenile court conducted by policemen, attorneys, judges, or other officers should be based on the principle of children's welfare and interests. However, the court process often takes a long time. The treatment of the officers to the offenders of juvenile delinquency is full of pressure which can give bad experience to children that will be remembered and recorded to their life forever.

Thereby, they need different treatment to those adults in getting the process of juvenile criminal justice system, making their need special concerns, particularly for juvenile court justice system. The settlement of criminal act through restorative justice is deemed as one choice to cover the weakness and disappointment to retributive justice system and rehabilitation in general criminal justice system. According to Sudarto, "juvenile justice system includes all activities including investigation and verdicts on children's behalf". Based on the Regulation of Minimum Standard of United Nation regarding Administration of Juvenile Court (Beijing Rules), Article 1 section 4 regulates that juvenile court should be deemed as the integral part of the process of national development in every country in a comprehensive framework for social justice to all children; thus, it gives contribution to young generation and maintain the order of peace in the society".

IOP Conf. Series: Earth and Environmental Science **156** (2018) 012044 doi:10.1088/1755-1315/156/1/012044

In the Convention on The Children's Rights, it is regulated that legal process should be done as the last resort in a short time. There should be an alternative effort or the best solution which does not make a stigma to children. Therefore, the legal action to children should be balanced with integrated program related to education, health, psychological guidance, and the involvement of community which will avoid the children from repeating the same crimes. Mechanisms in the judicial system process can be done as a means to create justice that emphasizes the recovery of dangers and relationships due to abuses committed by adolescents. Both parties must reach agreement without any pressure.

In his article, Tyler said that the goal of this review is to point to common elements shared by this procedural justice approach, the restorative justice movement, and a focus on morality and moral development. These approaches are united by their common goal: to encourage the development or activation of people's internal values so that they will become more motivated to engage in self-regulatory behavior. It is suggested that a civil society is most effective when its members buy into commonly held values and follow rules because they are motivated to do so, rather than because they are motivated by incentives or sanctions. [1]

According to Llewellyn and Howse (1998): The main elements of the restorative process involve voluntariness, truth telling, and a face-to-face encounter. Consequently, the process should be completely voluntary for all participants; the offender needs to accept responsibility for the harm and be willing to openly and honestly discuss the criminal behaviour; and the participants should meet in a safe and organized setting to collectively agree on an appropriate method of repairing the harm. [2]

The next statement of John Braitwhite states that : The types of restorative justice standards are articulated: limiting, maximizing, and enabling standards. They are developed as multidimensional criteria for evaluating restorative justice programs. A way of summarizing the long list of standards is that they define ways of securing the republican freedom (dominion) of citizens through repair, transformation, empowerment with others, and limiting the exercise of power over others. A defense of the list is also articulated in terms of values that can be found in consensus UN Human Rights agreements and from what we know empirically about what citizens seek from restorative justice. Ultimately, such top-down lists motivated by UN instruments or the ruminations of intellectuals are only important for supplying a provisional and revisable agenda for bottom-up deliberation on restorative justice standards appropriate to distinctively local anxieties about in justice. A method is outlined for moving bottom-up from standards citizens settle for evaluating their local program to aggregating these into national and international standards.[3]. This is defined in the mechanism of justice recovery, and the method of governance is the need of citizens to also regulate their local programs for national and international standards.

According to Sidney W.A. Dekker and Hugh Breakey, culture needs to be considered, including in substantive justice that regulates how rules and procedures are fair and legitimate; and restorative justice that aims to restore the status of individuals involved and heal the relations and injuries of victims and the wider community after the occurrence of ethical violations. His opinion is as stated below:

Culture needs to be considered. These include substantive justice which prescribes how regulations, rules and procedures themselves are fair and legitimate; procedural justice which sets down processes for determining rule-breaches, offers protections for the accused, and governs who should make such determinations; and restorative justice which aims to restore the status of the individual involved and heal relationships and injuries of victims and the wider community in the wake of an ethical breach. Just culture approaches need to provide such foundations for a genuinely just safety culture, to be conducive to reporting, engagement, and safety improvement.[4]

The structure of society in Indonesia is assigned by two unique characters of horizontal and vertical. Horizontally, it is assigned by the existence of social unities based on differences of tribes, religions, culture, and regions. Indonesian society, in Furnivall, is mentioned as plural societies (Nasikun,1974). Social plurality also shows cultural diversification. This difference of culture causes many contradictions.

On the one hand, there is a local principle which should be kept. Meanwhile, on the other hand, people are demanded to adapt with global life. According to Van Ness, customary justice system is significantly a part of restorative justice in three different forms with two characters of restorative justice program adapted from customary practices: conference system (found in the traditional practices of Maori tribes in New Zealand) and round-table system (in the practice of American first civilization). Second, the philosophical basis of custom process is the court tries to repair the criminal structures in the society which injures them and give information regarding restorative justice. Third, some forms of customary court is included in formal efforts in the process of criminal handling.[5].

Vertically, the structure of Indonesians is assigned by the differences of upper and lower layers of agrarians and industries. It causes the disparities in expanding the development; since, society is still doing agrarian activities, while the others have stepped to industries or technology of information fields. Fred W. Riggs says it as prismatic society. [6]. In Indonesia the existence of customary court in some areas is still considered, making the restorative justice principle not become a new thing. Marc Levin states that the old and traditional approach is recently used as progressive approach. [7]

2. The Problems

Naturally, the settlement of juvenile cases is not simple. It needs equal perception and vision to give the children protection and fulfill their rights with good commitment in the form of ratification of international convention, law, and local wisdom for children's behalf. Related to this statement, there are some questions raised in this research: How is the formulation of restorative justice system for juvenile crimes or children with law problems? and How is the implementation of restorative justice system for juvenile crimes by customary court in Makassar?

3. Findings

3.1 The Formulation Policy of Restorative Justice for The Juvenile Criminal Justice System

The results show that various policies in international legal instruments have been established in the legal protection effort for child offenders, including the criminal justice system of children. It is needed to prioritize then on-formal settlement process to avoid the stigma for children. Children who was in custody before the court will be housed separately from adults and shall be detained in separate institutions from institutions convicting adults, receive cares, protection, and all necessary assistance for children, like social, educational, skills, psychological, and physical treatment they need according to age, gender, and personality.

The protection of juvenile against the law has been set up in several international legal instruments, including UN Standard Minimum Rules on the Administration of Justice for Juvenile ("Beijing Rules") by the United Nations Assembly Resolution No. 43/33 November 29, 1985, and UN Guidelines order Crime Prevention Centre ("Riyadh Guidelines") authorized and declared in the UN General Assembly Resolution No. 45/112 December 14, 1990, with the assertion that the UN resolution is the minimum standards for the protection of juvenile from all forms of deprivation of liberty, which is based on human rights and prevent the child from the side effects of all forms of detention in order to achieve the integration of children into society. In addition, the Convention on the Rights of the Child (Convention on the Rights of the Child) is acting as international law, which began on 2 September 1990, seeing the need to perform systematic and comprehensive steps to provide protection to children who are in conflict with the law.

Law No. 11 year 2012 regarding juvenile court justice system explains the process of justice system to children with lawful conflicts. The attempts of intervention have been done by Government and all stakeholders to fulfill the rights of the child with lawful conflicts. Likewise, Law No 35 year 2014 as the amendment of Law No.23 year 2002 regarding children's protection also explains that children should get

IOP Conf. Series: Earth and Environmental Science **156** (2018) 012044 doi:10.1088/1755-1315/156/1/012044

protection and "special" treatment if they are in emergency situation, like suspected as criminal offenders, involved in violence, etc. It is affirmed by the law and international law of the rights of the child convention by explaining the existence of protection and "special" treatment to children with law conflict. In Indonesia, the Act No. 11 of 2012 regarding Juvenile Criminal Justice System regulates the entire process of settling disputes concerning Juvenile against the law, starting from the investigation phase to the stage after a sentence of supervision.

People are expected to participate and be involved in the children case through Diversion and Restorative Justice. Furthermore, under Section 28 I of the Constitution of 1945, in clause (3), cultural identities and rights of traditional communities should be respected along with the times and civilization. This means that any activity in the community with regard to the values of traditional/ local knowledge is respected and recognized in accordance with the times and civilization.

In international instruments and national instruments, it has been arranged that the legal mechanisms as well as law enforcers and citizens are connected and responsible for the implementation of the juvenile criminal justice system and prefer the informal approach as preventing juvenile against the law.

3.2. The Implementation of Restorative Justice In Juvenile Court Justice System In Makassar

South Sulawesi is located in 0°12′ – 8° South Latitude and 116°48′ – 122°36′ East Longitude. Its area coverage is 62.482,54 km². This province is bordering with Central Sulawesi and West Sulawesi in its North, Bone Bay and Southeast Sulawesi on the east, Makassar Strait on the west, and Flores Sea on the south. South Sulawesi's civilization consists of four main tribes, Toraja, Bugis, Makassar, and Mandar.

The latest data on July 2017 showed that 55 criminal offenders, like violence and vehicle thefts, were arrested by the policemen of city police of Makassar, South Sulawesi. The offenders were dominated by dropout underage teenagers who repeatedly convicted. 21 of them did the thefts of vehicles, 22 of them did thefts with violence, and 12 of them did thefts of weighted items.

Police officers had confiscated some evidences including electronic devices, such as television, monitor with CPU, 2 units of smartphone, and fabric bridal accessories. Next, these 52 offenders will be suspected by Article 363 of Book of Law with five years of imprisonment.

According to A.A.G. Peters, law should not be used to reach arbitrary decision, even if it is attached in written or not written rules. Law should be used as the director of justice and legitimation which is oriented to law principles and law values based on living law of the society. Therefore, it is not wrong if there is an attempt of reactualization and revitalization of custom law in the society. [8]

Indonesia is a country with rich varieties of cultures, with diverse social order, and has a process that is based on the values of local wisdom without involving the legal process in an approach that is based on the principles of decency and fairness in the mechanism agreed upon jointly. Human involvement in the implementation of the law shows the relationship between culture and law. Legal culture determines individual attitudes, ideas, and values against the law in society. Friedman states that differences in education, gender, ethnicity, nationality, and other income may be a factor affecting a person's legal culture. Legal culture is the key to understanding the differences contained in one legal system to another system. In the restorative justice system, the role of the community is very important, since the understanding of the justice system is crucial for implementation. The restorative justice system will fail if people do not participate in the implementation process.

As with other kingdoms in South Sulawesi, every noble is written in his birth in "*panguriseng*" (genealogy chapter) by specifying the time of birth, the special event of the day and the degree of blood of his nobility. Determining degree of nobility is always based on the formula: Dad's Degree + Degrees Mother = X / 2. This is based on a provision in Luwu, namely: "*AmbE 'pawata' cappa 'passapu, Indo' parenreng ikko 'simpolong*" (the father pulls with the tip of the destar, the mother heaved with a bun). Then at other times clarified with the motto too, namely: "*AmbE 'mappabbati, Indo' mappattase* "' (the father lowered the nazab, mother made the blood cook). It is clear that both provisions constitute an oversight that the people of Luwu embrace the bilateral system, not patrimonial or matrimonial. Thus, a Datu Luwu is the absolute owner of the purest blood of the nobleman who is weighed from his father's blood and mother's blood by lontara experts coordinated by AnrEguru Anakarung.

Bugis tribe as a regional community of Sulawesi is using the language Bugis (talk Ogi) is a language that has a script, namely: *Sure' / Hurupu' Ugi* (Lontara: Mak.). This language has 4 strata, namely: *Talk DEwata* (I la Galigo or Bissu Talk), *Talk Ade 'Baruga*, *Speak Lalo* and *Speak Ammeng*. But then, when spoken everyday according to the user community in its communication function, Bugis language can be stratified into 6 levels as well as called "Wari Ada" (language strata) which is often abbreviated as "WErEkkada". Then if according to the function classification, Wari Ada is termed "PasilElE Ada" (mutual switching) which is often referred to as "EXE Ada" with the limits of simple meaning, namely: *pinutur*.

Based on interviews with Mr. Andi Oddang, the source persons as Luwu Royal People's People are as follows:

A sustainable customary formation since 36 Generation TopapoataE Datu / Pajung Luwu to this day. Tana Luwu the only "nation" in the history of the beloved Republic of Indonesia is the former Kingdom in South Sulawesi that is sustainable and its Adat Council still exists to this day and Insya Allah throughout the world's future life. Luwu) actually existed since the oldest and largest kingdom in Eastern Indonesia was founded by TopapoataE Bataraguru. But then it is always perfected in every era of leadership Datu / Pajung Luwu next where the completion of the formation according to the needs of the time (period).

Tana Luwu was established and organized by the government with love, dignity and justice, "tessibEtta-bEtta olo, tessiala akka" (not stepping over authority, not taking honor each other) or in other words: Professional Managerial. All matters concerning the policy in the organization of the kingdom institute is implemented with the principle of "appadaEloreng" (consensus to reach consensus), as in Wajo called it "assamaturuseng" (agreement). So with this mechanized system, a Datu Luwu is glorified by a special treatment, as it is called: "Naiyya Datu E mattuku ulu mattukku ajE, angel arie tenna, tennawellang tikka, iyapa napasedding, ade'pa teddu'i" (actually Datu Luwu covering the head, covering the legs, not blowing the wind, and not being exposed to the sun, he will wake up, if the customs stir).

In Luwu society Pangadereng custom is interpreted as a custom, then it must stand on the values of goodness. Say good words, be nice and always try to think well and then applied to the behavior in the framework of communication among human beings, according to the ability of our customs performance. The custom is placed in the deepest niches of culture. Until then the form of custom is emitted with language and behavior. So the content of the true customs is the culture of mutual human-language and behaving according to their respective pranatanya, especially in terms of language culture, which customs call it "wari ada" (grammar) until it is known as "werekkada". Khazanah pangadereng reveals: "mappabati sadda exists, there mappabati gau ', gau' mappabati tau" (the voice that gave birth to the word, the word that gave birth to deeds, deeds that gave birth to human).

Words heard or read by others are human reflections of how to declare or write them down, according to the judgment of their fellow human beings. As for the underlying intentions, it is none other than the only known to be the Most Goib, then it is fitting that the words are guarded and should be used with great respect to uphold the dignity of humanity, for in fact only the human beings themselves can recognize and glorify the dignity his own humanity. Thus, the teachings and customs exist in the Luwu community as human beings that are part of a cultured society.

The solution of cases in Toraja is also based on its historical values. Whereas, all Tana Toraja area respects and obedient to the justice system of *Tarian pitu*, since it works under the teaching of *Aluk Todolo* which states that justice is united to its custom. According to local myth, *pitu* justice system has been assigned to Tana Toraja since the first civilian comes to Earth.

IOP Conf. Series: Earth and Environmental Science **156** (2018) 012044 doi:10.1088/1755-1315/156/1/012044

Then, Bugis Bone tribe of Makassar acknowledges the custom of *Malaweng*. Based on different sources, *Malaweng* has three levels of rule:

1) First class *Malaweng* (*Malaweng Pakkita*), means people who offend crimes through their vision. For example, by seeing other people cynically, seeing other gender coming from different *mahram* (blood relation), and etc.
2) Second class *Malaweng* (*Malaweng Ada-ada*), means people who conduct crimes by their spoken words. For example, swearing, disgracing people, being arrogant, says harsh words, and etc.
3) Third class *Malaweng* (*Malaweng Kedo-kedo*), means people who do crimes because of their action. For example, people who do consensual sexual intercourse to their relatives, eloping, conducting incest, murders, stealing, and etc.[9]

Regarding the development of culture, Bugis tribe tries to nurture local values as the part of national culture based on the application of noble value and local wisdoms of Bugis people. One of the forms is the facilitation of Culture Institution of "Saoraja" as the partner of custom and culture preservation as well as cultural development. Bugis tribe in South Sulawesi also upholds *pancanorma* as five fundamental living norms, which differentiates this tribe from other tribes.

In the justice system of Bugis, specifically in Kajangese custom, they also know customary justice. The customary justice in this area is known as *A'borong* (sitting together in a round of discussion). *A'borong* or Bugis- Makasssarese is known as *Tudang Sipulung*or the process of presenting all customary apparatuses to certain cases.

Restorative justice is required to replace Retributive Justice. Crime in restorative justice is not seen as a violation of the state's interest, but is considered infringement of a person by another person. In this case, the restitution is a means of improvement of the parties and the conciliation, with restoration as the ultimate goal. The victims and offender of criminal acts are recognized, both in the problems and in the settlement. The rights and needs of victims are recognized; while criminals are offenders who need to take responsibility.

In the context of human rights, restorative justice takes on a very important role to ensure that children's rights are protected. Today, Indonesia has implemented restorative justice as a way of resolving juvenile delinquency conflicts. Indonesia has a model of deliberation, which has the same idea as restorative justice as an attempt to create and utilize alternative measures to handling juvenile delinquency cases. Restorative justice in Indonesia encourages a better criminal justice system. Restorative justice resolves the delinquency case by fixing the damage or loss caused by the delinquency behavior. It is a process of familial settlement of delinquency cases by children by involving victims and perpetrators in a deliberation to get the best agreement for both parties.

4. Conclusion

In international instruments and national instruments, it has been arranged that the legal mechanisms as well as law enforcers and citizens connected and responsible for the implementation of the juvenile criminal justice system prefer the informal approach as preventing juvenile against the law.

The customary law in Toraja Land is the customary law of *Kombongan* which has become an institution that has principles for the people of Tanah Toraja from *Kombonan*. Customary Law can also direct the Tanah Toraja people to become better and there is no division between them.

For the Bugis, Kajao Lalliddong teaches the law to society, both individually and noble groups plus taboos to make it perfect and complete. Kajao then became the guide of the Bugis kingdom in South Sulawesi. In fact, Bugis society has the strength of the bonds of values, morals, culture and local wisdom in solving the problems that occur in society. The joint participation of the community in the criminal justice of children is expected to realize a condition that supports all interests to realize the future of a better child.

5. References

[1] Tyler TR 2006 Restorative Justice and Prosedural Justice : Dealing with Rule Breaking *J Soc Issues* 62(2).

[2] Dowden JL; C 2005 The Effectiveness Of Restorative Justice Practices : A Meta-Analysis. *Prison J* 85(2):127–44.

[3] Braithwite J 2002 Setting Standars For restorative Justice. *Br J Criminol* 42(3).

[4] Sidney W.A. Dekker; Hugh Breakey. Just Culture: *Improving Safety By Achieving Substantive. Proced Restor Justice* [Internet]. Available from: www.elsevier.com/locate/ssci Accepted 26 January 2016

[5] Van Ness, D. W 2010 *Strong K. Restoring justice: An introduction to restorative justice* 4th ed. (New Jersey: Matthew Bender & Company, Inc)

[6] Ronny Hanitijo Soemitro 1985 *Law and Society Studies* (Bandung: Alumni)

[7] Zulfa, Eva Achjani 2009 *Restorative Justice Indonesia, Study of the Possibility of Implementation of Restorative Justice Approach in Criminal Law Enforcement Practice), Summary of Dissertation, University of Indonesia* (Fakultas Hukum Program Studi Ilmu Hukum Kekhususan Sistem Peradilan Pidana)

[8] AA. G. Peters dan Koesriani Siswosoebroto 1990 *Hukum dan Perkembangan Sosial, Buku Teks Sosiologi Hukum. Buku III* , Jakarta: Sinar Harapan,.

[9] Nasir, Muhammad *Norma dan Hukum Adat Suku Bugis*

[10] Abdullah. Hamid, 1985. *Manusia Bugis Makassar*. Jakarta: Inti Idayu Press

[11] Jeff Latimer 2005 The Effectiveness Of Restorative Justice Practices : A Meta-Analysis *The Prison Journal* 85(2) 127-144

[12] Soemitro, Ronny Hanitijo 1985 *Law and SocietyStudies*, (Bandung: Alumni)

[13] Van Ness, Daniel W, Jihnstone, Gerry 2007 *Hand Book of Restorative Justice* (Portland:Willan Publishing).

IOP Conf. Series: Earth and Environmental Science **156** (2018) 012045 doi:10.1088/1755-1315/156/1/012045

Legal Design of Domestic Workers Protection Based on Gorontalo Community Local Culture

Mutia Cherawaty Thalib
Faculty of Law, Universitas Negeri Gorontalo (UNG), 96128, Indonesia
Email: mutiathalib_barryf2@yahoo.com

Abstract. This study was conducted with an empirical juridical approach. The juridical approach was done by identifying community norms and legal policy related to the domestic workers existence, while the empirical approach was done by observing social phenomenon of housemaid and local culture that underlies the working relationship between employer and domestic workers (housemaid). In-depth interviews and group discussions were done to obtain the data. The result shows that the domestic workers existence in Gorontalo cannot be relied upon the domestic service market because it is increasingly eroded by socio-cultural changes that evolve in the rapid rise society awareness of human rights and technological development. Huyula's culture values, timoa, ambu, bilohe, and tolianga remain as survival strategies for some domestic workers who last longer with their work. For new domestic workers, the bargaining position is increasingly high with the poor quality of work. Some of the rural workers who still hold the principle of "dila biasa" (uncustomary principle), moomu (unwilling), moolito/ moqolito (shame), affect their resilience in working as domestic workers. On the other hand, domestic work relations as a social institution is not supported by strong instruments like the government. Consequently, it needs an integrated thinking and step by step designing of the form of protection for domestic workers based on the local culture values of Gorontalo people.

Keyword: Legal Design; Protection; Domestic Worker; Local Culture

1. Introduction

The existence of domestic workers is still a polemic, since there is no clear domestic workers concrete arrangement, as other workers have already arranged through Act No. 13 of 2003 on Employment. Nevertheless, as a precautionary measure, the Employment Minister for the Republic of Indonesia has issued Decree of the Minister of Employment of the Indonesia Republic No. 2 of 2015 on the Protection of Domestic Workers, pending the draft law on the Protection of Domestic Workers discussed and endorsed by the Indonesia legislative council into the law.

Regarding to the previous research result conducted by the researcher, it can be generally described that the domestic workers rights fulfillment in reality had not been met well, and it even caused many problems. Similarly, the domestic workers obligation fulfillment, revolving around domestic workers quality on the process and their low achievement, also caused many problems [1]. In particular, the existence of housemaid as domestic workers in the local labor market has a downward trend in the last 3 to 5 years. It is unbalanced with the higher need of housemaid service. Meanwhile, the local housemaid employment agencies or institutions officially operating in Gorontalo have not existed. The right protection and supervision for domestic workers has not been protected by the regional regulation, since there is no domestic workers regulation, as it is currently undergoing bill discussion

IOP Conf. Series: Earth and Environmental Science **156** (2018) 012045 doi:10.1088/1755-1315/156/1/012045

and endorsement by the parliament. In early 2015, the Employment Minister of Republic of Indonesia issued Employment Minister Regulation No. 2 of 2015 on the Protection of Domestic Workers as a Guideline for the Domestic Workers Minimum Protection, which regulates the employment relationship between individual service users and domestic workers as service providers through working agreements, both orally and written with the local Government involvement such as the chairman of the residents in the residence area of service user. The other reason is the low quality of domestic workers both from aspects of work skills and work culture[2].

According to the matters explained above, research questions arise as follows: (1) How is the Domestic Workers Protection in Gorontalo?, (2) What is the perspective of local culture viewed from historical aspects and its influence on the existence of Domestic Workers in Gorontalo?

2. Method

This study employed a juridical-empirical approach. Identifying the norms, laws, and policies for the domestic workers is the notion of juridical approach, while observing phenomena, such as social events and scopes of local culture, is the focus of empirical approach. The idea of observing such phenomena is based on the relationship between the employer and the workers. Interviews and group discussions with selected respondents are used to obtain more data and explore the issue further.

3. Result and Discussion

3.1. Understanding Domestic Workers

Domestic workers refer to people who work for an individual in a household; their jobs are mainly about helping domestic works. Their income is not only monthly salary in the form of money, but also other types of reward. This applies to domestic workers who either stay or not stay in the house of their employer in carrying out their job.

Domestic Workers should be the persons work for an individual in a household to carry out the domestic work by receiving wages and/ or other forms of remuneration including those who are staying and who are not staying (Article 1 Point 1 Labor Ministerial Regulation No. 02 of 2015).

3.2. The Aspects of the Domestic Workers Protection

3.2.1. Protection of Rights in CEDAW. Women's Rights as workers are generally contained in the CEDAW (Convention on the Elimination of All Forms of Discrimination against Women) that are relevant to the rules of law for development programs, which have been used as normative standard by the women's rights observers. One of them is the protection for the obstacles that women face concerning the dominant human rights, such as the state's attention to human rights is more focused on the public sphere than on the private sphere. It is therefore necessary to specifically formulate the regulation of human rights in private life.

In contrast to the instruments within the laws protecting human rights, CEDAW concerns more on the women at disadvantage. They suffer from oppression due to external factors, such as gender discrimination from societies or cultures, particularly inequality of the women's rights within the public domain (e.g. common beliefs about women in society) and the private domain (e.g. women's position in their family and work place). CEDAW also sets some instruments for the country regarding to the obligation to ensure the instruments of human rights which encompasses the rights to be respected, fulfillments of the rights, and protecting the rights and freedom through establishing a situation, including institutional, to include normative principles and effective procedures as an effort to protect these policies without delay. This is in accordance with the Article 2:

a. "To take all appropriate measures, including legislation, to modify or abolish existing laws, regulations, customs and practices which constitute discrimination against women.
b. To establish legal protection of the rights of women."

Further measurements are included in the Article 11 of CEDAW which states that "state parties shall take all appropriate measures to eliminate discrimination against women in the field of employment in order to ensure, on the basis of equality of men and women, the same rights, in particular [3]:

> "The right to equal remuneration, including benefits, and to equal treatment in respect of work of equal value, as well as equality of treatment in the evaluation of the quality of work; The right to social security, particularly in cases of retirement, unemployment, sickness, invalidity and old age and other incapacity to work, as well as the right to paid leave; and The right to protection of health and to safety in working conditions, including the safeguarding of the function of reproduction".

3.2.2. Legal protection. The concept of Indonesian law of protection has Pancasila, the foundational and philosophical ideology of the life of the nation, as its grounding. The principle of legal protection in Indonesia that recognizes and protects the dignity of human being is originated from Pancasila. On the other hand, the tenet of legal protection against the government act is originated from the recognition and protection of human rights, as this legal system aims at setting and restricting the obligation of the citizen and the government. The legal protection plays a major role in a nation of law; this is because all the people within such a nation have an obligation to uphold the law. It is common for a nation to have a reciprocal relationship between all the civilians. This leads to the duties and rights of those people; as a result, each citizen has his or her rights to ask for legal protection. The concept of legal protection is to protect the human rights from the miscarriages of justice so that the rights of the citizens are being fulfilled.

The specific arrangements regulate the domestic workers existence, which so far are not yet existed, due to the domestic workers characteristics, which have not been accommodated in Law no. 13 of 2003 on Employment. Practically, what has been done in the working relationship is to make a contract of work or a written agreement. In the Indonesia Codification of Civil Law, it is set forth in Article 1338. This method can be found in modern societies, and even then is performed only between housemaid service users with housemaid distributors. In Gorontalo, there has been such practice, but because it is not supported by the existence of the operational permit from the regional government, the agency is no longer operating[2].

3.3. The Domestic Workers Existence and Local Culture

The research result shows, in terms of sociological aspect, there are still phenomenon of working relationship which lead to the unequal rights and obligations that are motivated by knowledge, skill, culture, social economy, and housemaid's own view and society toward domestic workers position.

Discussing the socio-cultural life of Domestic Workers in particular is not separated from the cultural essence of Gorontalo society in general, which has values that are always inherited, interpreted, and implemented in line with the process of social change. The invasion impacted in Gorontalo on the ideology and local wisdom of the people led to a massive change in the government and economic system. An authoritarian kingdom system was inevitable because some lower-position chiefs called Baate was not regarded nonetheless. On top of that, some people escaped to forests (become Polahi) considering that Dutch soldiers were supportive to authoritarian kingdoms. People were also obliged to give some gold for their taxes that forced them to be a slave if they could not pay. Human trafficking became a common phenomenon and preceded with a significant increase of Wato or slaves working under landlords. This signifies that imperialism economic system and feudalism were present at the time[4].

One of the local cultures of the people of Gorontalo is the *Huyula* Culture[1] , which contains the noble values of *Pancasila* [2] and describes the life of a community full of spirit of mutual cooperation

IOP Conf. Series: Earth and Environmental Science **156** (2018) 012045 doi:10.1088/1755-1315/156/1/012045

and togetherness. However, some people start abandoning this belief. In the development, Huyula culture began to be abandoned by society. The presence of globalization of local wisdom of Gorontalo is increasingly marginalized, exposed to the behavior of Gorontalo people who have begun to ignore Huyula culture which has the values of cooperation, responsibility, and tolerance [5]. Alim S. Niode, a sociologist/ humanist, stated that the socio-cultural capital of other Gorontalo people, not only Huyula, but also the *timoa, ambu*[3], *bilohe* [4], *tolianga* [5], which remain a strategy for survival in marginal position, no longer serves as a social spirit that guides the discovery and creation of culture. Mental cognitive culture and socio-cultural formations of people fell into the pit, so there was a crisis on the system of social and cultural meaning. For the job-seekers who are "forced to work as housemaids", they work with "trial and error", because there is no other choice. When the work atmosphere became "uncomfortable", they will abandon their job and move to another workplace.

In the last few years, the regional development has been getting better. It is characterized by increasing economic growth and better community civilization, shown by population data aged 15 years and over by activity types during the past week in Gorontalo. In 2015, the force labor consists of Working Population which amounted to 92.195 persons with 6.031 in unemployment. There were 20.826 Non-Work Force Junior High Schoolers, 28.129 household members, and 4.306 others. Further described, labor force participation rate is 64.83% while the unemployment rate is 6.14% [6].

The domestic workers existence is different in the past until the present. The domestic workers' bargaining position is increasingly high with poor work quality; the household work culture is very low, as some rural workers still hold the hold the principle of "dilabiasa" (uncostumary principle), moomu (unwilling), moolito/ moqolito (shame). Another picture shows, there are relationships based on family relationships, in which working relationships are maintained, even for children. In general, families with a high level of education able to understand and place the position of housemaid as part of human rights, which dignity needs to be maintained. Even the employer's attention to the quality of housemaids dropping out of school is often done by including housemaids on the B and C equivalent study. This is shown from previous research results, in which housemaids are empowered through skills courses to continue their education to universities.

3.4. Model of Protection for Domestic Workers in Gorontalo City

As the social situation change, norms in family customs and beliefs are also gradually shifted, particularly in the existence of household workers. The differences between the existence household employees in the past, present, and future urge a set of laws protecting these workers to be established. However, the instruments considering the issue are still in process since the government is already busy in handling other priorities, so the task is left the local government to design measurements concerning the rights of household workers. This is closely related to the legal awareness of people. The notion of legal awareness refers to the act of a community rather than an individual since its grounding originates from social structures, customs, and functions within the law concerning a particular society [6]. Indonesia employs positivistic-oriented law system in which the problems regarding the professional relationship in a household as a part of family custom heavily focus on domestic norms. On the contrary, the family customs are developed globally according to the dynamics of study on laws, and it does originate from only one perspective.

Protection Forms against the domestic workers can be done by preventive and repressive efforts. The preventive protection model intended in this case is a Mutualistic Partnership Model with social engineering. Through the establishment of "Sahabat Pramuwisma" container, in which there are training programs that include knowledge and skills. Mutualistic partnership is a partnership of two or more parties who are equally aware of the importance of doing partnerships, namely to mutually

---

benefit and gain more benefits, so that will achieve goals more optimally, making it easier for each party in realizing its goals. The repressive way of partnership model is a linear "collaborative of partnership" that is not to distinguish the size, status/ legality, or strength of the partners. The main emphasis is on the vision, the mission of mutual filling with each other, and in handling the problems of domestic workers, whether included in the realm of Criminal Law, Civil Law, Human Rights, and others related to the function of local government institutions Province, namely Social Service and Local Government of Gorontalo City that is on the Department of Population Control, Protection of Women and Children and Police, Non Governmental Organization (NGO), and Legal Aid Institution through mentoring program.

4. Conclusion

4.1. Conclusion
The Domestic Worker's existence changes as social change takes place, from the past to the present, which cannot be avoided. One of the problems is the female workers declining presence as domestic workers both in terms of quantity and quality aspects. Among others, it is caused by the society civilization development, the increasing understanding on human rights, technological growth, economy, as an impact on the local culture decreasing values of society in the form of Huyula culture.

4.2. Suggestions
The domestic workers rights protection needs to be done regarding to the equal standing between the Procuring Entity and the Service Provider needs as domestic workers, by seeking solutions to all aspects of domestic workers' problems, especially to housemaids. One of them is the local culture of society, which has not been accommodated as part of the employment agreement, along with a process of awareness of the reality of community change. So it is necessary for the cooperation of all elements; government, NGO's, employers, Service Providers, and housemaids, to revive the domestic work market by promoting local cultural values in the working relationship, lawful, justice because it is based on cultural values as the character of Indonesian nation.

5. Acknowledgments
The authors would like to thank the Gorontalo State University Research Institute (LEMLIT UNG) for the opportunity in National Competitive Research, ICOLEG (International Conference of Law, Economic and Governance) 2017 that has facilitated the delivery and publishing of this Journal. Parties who have provided much input to this research, from Gorontalo Provincial Government elements through the Employment Sector, Employment section, Section of Empowerment and Protection of Women and Children of Gorontalo City, Central Bureau of Statistics of Gorontalo City and Legal Division of City Local Government which helped with the secondary data. Infinite remarks to Mr. Alim S. Niode, as sociologist, as well as Chairman of Ombustman Institute of Gorontalo Province and Mr. SuwardiBai, who have spent much time to be interviewed, peer reviewers of the problem of Domestic Workers as well as Service Users, as well as my friends, the Housemaid partner, namely Domestic Workers; who persistently struggle the family economy, which during this time convey many information related to their existence in the world of domestic work.

Hopefully this journal will be useful for the development of law and social-culture science.

6. References

[1] Thalib M C and Ibrahim S 2012 *Fundamental Research: Perlindungan Hak Pekerja Rumah Tangga dalam Pemenuhan Hak dan Kewajiban Bekerja* (Gorontalo: Lemlit UNG Gorontalo)

[2] Thalib M C and Junus N 2015 *Fundamental Research "Kajian Sosio Yuridis terhadap Eksistensi Pembantu Rumah Tangga (Domestic Worker) di Kota Gorontalo* (Gorontalo: UNG Research Institute)

[3] Partner For Law in Development United Nation Development Fund for Women 2007 *CEDAW (Convention on the All Forms Elimination of Discrimination against Women)* (Jakarta: SMK Grafika Desa Putra)
[4] Niode A S 2006 *Gorontalo, Perubahan Nilai-nilai Budaya dan Pranata Sosial* (Jakarta: Pustaka Indonesia Press) p 4-5
[5] Yunus R 2014 *Nilai-Nilai Kearifan Lokal (Local Genius) Sebagai Penguat Karakter Bangsa: Studi Empiris tentang Huyula* (Yogyakarta: Penerbit Deepublish) p 13
[6] Kharlie A T 2013 *Hukum Keluarga Indonesia* (Jakarta: Sinar Grafika) p 81-82

IOP Conf. Series: Earth and Environmental Science **156** (2018) 012046 doi:10.1088/1755-1315/156/1/012046

Tolerance Attitude among Religious People in Marine Envronment: Case Study of Mosque of Ex-Kampong Texas

Faradila Hasan[1], Rosdalina Bukido[2], Ismail Suardi Wekke[3], Rahman Mantu[4]

[1]Faculty of Sharia, IAIN Manado, Sulawesi Utara, Indonesia
Email :faradila.hasan@iain-manado.ac.id

[2]Faculty of Law, IAIN Manado, Sulawesi Utara, Indonesia
Email : rosdalina.bukido@iain-manado.ac.id

[3]Sekolah Tinggi Agama Islam Negeri (STAIN) Sorong, West Papua. Indonesia
Email : ismail@stain-sorong.ac.id

[4]Faculty of Ushuluddin Adab and Da'wah, IAIN Manado, Sulawesi Utara, Indonesia
Email : rahmanmantu530@gmail.com

Abstract. This paper raises the issue of the conflict that occurred between religious people in ex-kampong Texas. The conflict was marked by an action promoted by the Tolerance Care Community Alliance (Makapetor) Tou Minahasa. The method used in this research is qualitative method using multicultural approach. This study focuses on the attitude of tolerance among religious people, in this case, that are Muslims and Christians. From some informant's statement, the writer can draw the conclusion that the tolerance among religious people in Manado City is still available, although it can not be denied that there are still suspicions between each party that there are people who benefit from this conflict and also critic the government of Manado City.

1. Introduction

When hearing about the term 'multicultural', this can not be separated from the attitude of tolerance among religious people, although multicultural is not only always associated with inter-religious tolerance, but also the attitude of tolerance among tribes, races, groups and others. In communities, inter-religious conflict often happens. Thus, the word tolerance is often juxtaposed with the word 'religion'. The discourse of multiculturalism for the context in Indonesia found its momentum when the authoritarian-militaristic national system collapsed with the fall of the Soeharto regime. At that time, various conflicts between ethnic groups and between groups happened. Such conditions made various parties increasingly question what kind of national system was suitable for a change for Indonesia, as well as what system could make the people of Indonesia live peacefully by minimizing the potential for conflict (Masykur, 2016).

Related to tolerance, Baidi (2010) revealed that religious tolerance talks are still important in the context of the plural Indonesian nation. Multiplicity in language, religion, ethnicity, and custom requires a sophistication of conflict management, so that conflict with tension on an ongoing basis can be well managed.

From the image of tolerance among religious people, especially in Manado city whose population consists of several tribes such as the Minahasa, Bolaang Mongondow, Sangihe, Gorontalo, Makassar, Java and many other tribes, we know that the tribes also represent the majority religion that exist in the origin of each and all of these tribes coexist without any meaningful conflict.

Another case of religion is that there are several assumptions of religious leaders and people of Manado city that the actual conflict that occured was not on the inter-ethnic, but on religion. Their assumption is that there were some minority religions in the city of Manado who felt disturbed, from the case of worship to the case of permissions for the worship houses construction. Of course, this can not be ignored.

However, by looking at the condition of the community, both minority and majority in Manado city, there is still no significant conflict threatening harmony among religious people. This is evidenced by the existence of inter-religious activities where there was a diversity festival held in the Hall of DPRD Building of North Sulawesi Province on December 10, 2016 ago. The festival presented a mix of inter-religions in North Sulawesi, but it caused controversy in the community because in one appearance, shalawat gloria was played, it was shalawat with tones of gloria hum. There was also training activities about interfaith communities, New Generation Initiative and Engagement (INGAGE), in September 2016 in the city of Manado, and many other interfaith activities held in Manado.

From interfaith activities and the absence of meaningful conflicts between religious people and the inter-ethnic cohesion in the city of Manado, it is quite natural that the author says that Manado is a tolerant multicultural city. It is also agreed by Manado State Institute of Islamic Studies (IAIN) by seeking human resources of multicultural insight as it becomes the vision and mission of IAIN Manado in 2020.

We should keep the tolerance that has been established so that all can live peacefully in diversity. However, some time ago exactly on October 26, 2016, the Alliance of Kawanua Community Care Tolerance (MAKAPETOR) Tou Minahasa took action in downtown Manado precisely in District Wenang ex-kampong Texas. This action was allegedly as an intolerant act against Muslims. There were three demands from the action made by the MAKAPETOR Alliance that were: first, the return of the function of ex-kampong Texas as a religious tourism park. Second, the acceleration of the construction of religious tourism parks. Third, the dismantling of buildings that did not fit allotment in Ex-kampong Texas land.

The action is a concern in various circles, there are some that link the action as a backlash of various actions that occurred in the capital of Jakarta, the case of alleged blasphemy by the governor of Jakarta Basuki Tjahaya Purnama (Ahok) which led to the establishment of Ahok as a suspect. Some say that this action is an intolerant action among religious people and there are some that also link this action related to some political interests until certain elements can be benefited.

In fact, there are also some that link this problem with the discriminating efforts by the city government that is intolerant behind and is tolerant in front of the public. We need the best solution on this case so that no party will feel aggrieved if the problem is not protracted. So that in the end, there will be no clash between religious people.

Apart from these views, we need to pay attention to not to do so because the things mentioned above can be our reason for divisiveness. Harmony must be maintained although intolerant behavior is emerging outside the North Sulawasi.

This paper focuses on the attitude of tolerance among religious people who reflected on the conflict between Christians and Muslims that occurred in ex-kampong Texas Masjid Al-Khairiyah in District Wenang Manado City.

2. Research methods

This research includes the type of qualitative research using a multicultural approach which aims to explore information in depth about the reality of tolerance attitude among religious communities in the city of Manado. The author assessed the form of tolerance through the opinions expressed by some informants who are successfully interviewed. The successfully interviewed informant are 6 people.

Informants in this research are six people. Among them are informants who were directly involved with the actions of the Tolerance Care Community Alliance (MAKAPETOR) Tou Minahasa on Wednesday, October 26, 2016 and also some Christian religious leaders and Imam Masjid Al-Khairiyah ex-kampong of Texas Manado.

Sources of data in this study are obtained from the results of interviews with six informants as well as some scientific literature from books and articles related to the concept of tolerance and multicultural.

3. Results of Research and Discussion

No society can escape from conflict, at least the so-called non-existence or unreal conflict. In contrast to realistic conflicts that really become empirical facts and can be seen by naked eye, unreal conflict is a misunderstanding of other groups contained in human thought. But if there is a trigger, the conflict will be a realistic conflict (Arifin, 2015).

Conflict refers to the relationship between the conflicting individual and / or group. In the social life of friction, conflicts and disputes among the citizens of society is not impossible to be caused by various factors such as social, economic, political, cultural, and so forth. However, conflicts can also be caused by wider issues than these (Maulana, 2016).

Religion is not infrequently used as a "tool" and is accused as the cause of every riot or conflict in society. The problem of differences between religious groups, in this case are Islam and Christianity, is not rarely raised on the surface by the religious elite so that the phenomenon that appears in every religious conflict is more religious jihad form of "holy war" to fight for and defend religion. The use of religious labels has been used as a means of dispute, causing hostilities and worsening the climate of religious harmony. There is a religious tendency to be used as a tool to "raise" and "justify" disputes (Maulana, 2016).

Conflict is created by the desire of a group of people to gain a momentary advantage that causes many victims, but it needs to be understood, not all dissent can mean conflict. Competition is the same as conflict, but it is easy to lead to conflict, especially when there is competition that uses ways that are contrary to agreed rules (Artis, 2011).

According to Artis, the things that cause conflict in the midst of pluralist society are: first, the difference of race and religion. Second, differences in culture. Third, majority and minority issues where the majority feel belonging and power in an area so that the minority feel depressed (Artist, 2011).

The term tolerance comes from Latin "tolerare" which means being patient with something. So tolerance is a behavior of people who follow the rules, in which one can appreciate, respect for the behavior of others. The term tolerance in the socio-cultural and religious contexts means attitudes and actions that prohibit discrimination against different groups or classes within a society, such as religious tolerance, in which the majority religious group within a society provides a place for other religious groups to live in its environment (Bakar, 2015).

Nevertheless, the word tolerance is still controversial and criticized from various quarters, on the principles of tolerance, both from liberals and conservatives. However, religious tolerance is an attitude of respect and respect for other religious groups (Bakar, 2015).

Tolerance is a concept that describes mutual respect and cooperation between different ethnic, linguistic, cultural, political and religious groups. The concept of tolerance is a great and noble concept that fully becomes an organic part of the teachings of religions, including Islam (Hertina, 2010).

Thus, tolerance leads to a willingness to accept the reality of the differences that others have. Tolerance can be interpreted as giving a place to a different opinion, accompanied by an attitude, restraint or patience. Therefore, among people of different opinion, we must pay attention to the same attitude of mutual respect by considering the unity of the nation and the state (Artist, 2011).

According to Hertina (2010), tolerant attitude does not mean to justify the views allowed, but to recognize the freedom and rights of its adherents. There are three kinds of tolerance are as follows:

3.1. *Negative Tolerance*
The content of the teachings and adherents is not appreciated. The content of the teachings and adherents is only left alone because it is in a state of necessity.

3.2. *Positive Tolerance*
The content of the teaching is rejected, but the adherents are accepted and respected. For example, you are Muslim must obligate the law to reject the teachings of other religions based on belief in your religious teachings, but you cherish its adherents or human beings.

3.3. *Ecumenical tolerance*
The content of the teachings and adherents is appreciated, because in their teachings there are elements of truth that are useful to deepen their own beliefs. For example, you and your friends follow the same religion Islam or Christian but different streams or understanding.

In the context of inter-religious tolerance, Islam has a clear concept. "There is no compulsion in religion", "for you your religion and for us our religion" is a popular example of tolerance in Islam. In addition to those verses, many other verses are scattered in various Surahs and also a number of traditions and practices of tolerance in Islamic history. Historical facts show that the issue of tolerance in Islam is not a foreign concept. Tolerance is an integral part of Islam itself whose details are then formulated by scholars in their tafsir works. Then these formulations were perfected by the scholars with new enrichments that eventually became a historical practice in the Islamic community (Aslati, 2012).

Tolerance among religious people in Islam receives important and serious attention. Moreover, religious tolerance is a matter that concerns the existence of human belief in Allah swt. He is so sensitive, primodial and easy to burn conflict so as to absorb great attention from Islam (Bakar, 2015).

But nowadays Islam faces a lot of criticism coming from outside Islam. Islam is considered an intolerant religion that is often discriminatory and extreme. Islam is considered a religion that does not consider religious freedom. Such assumptions are rooted, moreover proliferation of terror in the name of Islamic sharia enforcement by the elements of radical understanding is growing. Muslims must work hard to restore the image of Islam that becomes a blessing for the whole of nature, the attitudes of tolerance exemplified by the Messenger of Allah should be more in da'wah.

Indeed Islam is present as a grace of lil'alamin for the universe. Becoming grace in the sense that the presence of Islam brings peace and avoids conflicts, both vertical and horizontal conflicts. In Islam, correct understanding always leads to good and moderate. There are some circles doing the actions on behalf of Islam that cause horizontal conflict can not necessarily be a reason to blame Islam. Usually such acts occur because of the misconceptions of the teachings of Islam or of emotional factors, such as the consequences of non-Muslim crimes committed in Muslim countries, as happened in the Middle East (Bakar, 2015).

Islam as a religion teaches mankind to always respect and tolerate others and keep the holiness and truth of Islamic teachings. Historical events have proven that Islam is a religion that teaches tolerance to all religions. Islam teaches its people the importance of maintaining unity and harmony, both internally and internationally. Islam also teaches its people to always tolerance to fellow believers and inter-religious people, and love each other among followers of religion. Furthermore, Islam also instills the values of patience and freedom of expression. One of the verses in the Qur'an from the many theorem that enjoins Muslims to respect other religions is "*God does not forbid you to do good and do justice to those who have not fought you for religion and have not expelled you from your country. Allah loves those who are just. Allah only forbids you to make your friends those who fight you because of religion and drive you out of your country, and help (others) to drive you out. Whoever* *makes them friends, they are the wrongdoers (*Surah al-Mumtahanah: 28). Similarly, the hadith of the Prophet who teaches Muslims to respect other religions is *"There is actually a body that passes before the Messenger of Allah, then the Apostle stood in his honor. Then, it was said to him: Surely the body is Jewish. " The apostle replied: Is not he also a man*" (Bakar, 2015).

The above verses and traditions affirm that there is no prohibition to do good and to be fair to people outside of Islam. Muslims must respect people outside of Islam as long as they do not fight against Muslims.

The standard of tolerance and justice in Islam, in which Islam recommends not to respond to the low and despicable charges of opponents, doing so will make Islam itself cruel. Conversely, forgiveness is a better act and if it is required to retaliate, the reply with the note does not exceed the limits that have been incurred in Islam. That is, if they are powerless and give up, then do not do excessive actions (Burn, 2015).

Tolerance in Islam is authentic. It means familiar and even it has existed since Islam existed. Because of its organic nature, tolerance within Islam is just a matter of implementation and a commitment to practice it consistently. However, religious tolerance according to Islam is not merged into beliefs nor it is to exchange confidence among religious groups of different religions. Tolerance in question is in the sense muamalah (social interaction). So, there are shared limits allowed and should not be violated. This is the essence of tolerance in which each party has to control themselves and provides a space to respect each other's uniqueness without feeling that the belief or rights is threatened (Aslati, 2012).

From the concept of tolerance that has been described above, it can be understood that the concept of tolerance is actually respecting the followers of religion as part of the community but do not share faith in his religion. It is appreciating people of other faiths, but not recognizing or justifying other religious beliefs. In other words, do not interfere with each other's faith, but keep each other's harmony well maintained and coexist peacefully.

To maintain harmony among religious people, multicultural insight becomes very important to be grown and developed as an effort to reduce conflict that often occur in the society. Growing good social values amid a multicultural environment does not come by itself but should be started from family, school and community environment. Multicultural approach in the study of religions has become a demand amid diversity and has a great role in nation building. Indonesia as a country that stands on the diversity of culture requires the importance of multiculturalism in nation building. With this multiculturalism, the principle of "bhineka single ika" as listed in the state base will be realized. The cultural diversity possessed by the Indonesian nation will be an inspiration and potential for the nation's development so that the ideals to bring about a just and prosperous Indonesia society as stated in the preamble of the 1945 Constitution can be achieved (Aslati, 2012).

It has been suggested that multicultural is closely related to tolerance. Because tolerance means accepting the culture of tribe, race, class and even other religions of our own. Characteristics of Indonesian society, especially Manado city people are very plural and the plurality make the people live side by side with peace without any significant conflict.

During this time, this is what is understood by many people. So, the question arises is: is the people of the city of Manado a tolerant multicultural society? Thus, it is necessary to conduct in-depth research on this subject. Do not let, tolerance in the city of Manado is just a false tolerance as some have alleged. Do not let tolerance only be highlighted by leaders who are forced to be tolerant, while the attitude of tolerance that highlighted the leader is not applied to the grassroots. Leaders are tolerant while the people are intolerant.

This tolerance was questioned when a community action calling itself with the Tolerance Care Community Alliance (Makapetor) Tou Minahasa on 26 October 2016 in the ex-kampong area of Texas Al-Khairiyah Mosque, by some allegedly an intolerant act of Christians as a majority to Muslims as minorities in the city of Manado.

For that, the authors interviewed six informants who are considered to represent the city of Manado and the authors shared opinions expressed by the informant are as follows:

IOP Conf. Series: Earth and Environmental Science **156** (2018) 012046 doi:10.1088/1755-1315/156/1/012046

Tolerance in the city of Manado has run well and is important to be taken care of. In relation to the action of the Makapetor alliance, one of the informants said that the action was directed to Manado city government, it just happened to Manado Police office which became the point of action adjacent to the house of worship (Mosque) which was in the ex-kampong of Texas. The action done has nothing to do with the demolition of houses of worship (mosque). But according to the informants, this action can provoke misunderstanding and familiarize the perspective of everyone. Even an equally Christian informant criticized the actions of the Makapetor alliance that invite misunderstandings and could harm religious harmony in the city of Manado. This action is not a concern of Minahasa Christians it is marked by the Christian Group of Minahasa Christian Churches (GMIM) that are large enough not to give statements, even GMIM priests continue to preach and urge their congregation not to be provoked by religious issues , because tolerance is a priority in this area, and it has been intertwined and treated for so long.

Related to Minahasa customary values, of course, nothing contradicts with the construction of houses of worship this can be proved by the existence of culture where there are kampong locations that háve several groups, such as the Arab village, the village of China, the village of Java Tondano. If this problem continues to be allowed and the government does not take steps to explain the problem, then things that lead to potential conflicts will occur due to small movements. Tolerance is mutual respect, respect between religious communities. Muslims always value other religions, but there are boundaries that should not be exceeded.

Relation with action conducted by Makapetor alliance, some opinion from informant said that action alliance of Makapetor Tou Minahasa which have done since 2015 has counted 4 times. In the action held last year, before going to the ex-kampong texas land they came to the municipal office and the city council (DPRD) Manado city. In contrast to this last action, the action point was directly to the Police office Manado without visiting the office of city and office (DPRD) of Manado City. This was twisted by the people that our action was leading to the house of worship (Mosque), but it was not like that, the mosque was not guilty but was the new building temporarily built at that location.

The action has drawn criticism from other Christian groups, previously arguing that the action could invite misunderstandings and could harm harmony among people. If the lawsuit concerns illegal buildings that do not have Building Permit (IMB) which is the working area of the city government, why action in front of Manado Police office is not in the Mayor's office? Moreover, the office is adjacent to the ex-kampong area of Texas in which there is a house of worship of the ummah, in this case the action can disrupt the harmony among religious people who are the construction of society and religious leaders for a long time in this area (Manado). Further criticism that Makapetor used the tolerance lebel in naming it, is that this is a way of showing to the public that the area of North Sulawesi is a tolerance, or a way to show that this alliance movement is not captured as an intolerant movement. Because in reality, the actions taken by the Makapetor alliance can trigger intolerance. Sometimes this kind of action is spontaneous, or just seeking attention. After the request has been followed by the government then this action will be silent again. The positive thing about this action by Makapetor alliance is that it is re-awakening the people of North Sulawesi that the harmony of our religion is dynamic.

This group is only spontaneous, there is no special agenda, but the indication of riding by the beneficiaries and so forth may be true. There are some things that have been cut first, this issue is not a concern of Christians. For example, if you see a small group of Christians of the Minahasa Christian Church (GMIM) do not give statements, on the contrary, GMIM priests continue to preach and encourage their congregation not to be provoked by religious issues, for tolerance is preferred, which has been interwoven and treated for so long. Secondly, sometimes such actions are only spontaneous, or merely seeking attention. After the request has been followed by the government then this action will be silent again.

If the problem is related to the IMB, then what should be done is to urge the government and find out what the target was, instead of taking a government function expert to do the curbing.

IOP Conf. Series: Earth and Environmental Science **156** (2018) 012046 doi:10.1088/1755-1315/156/1/012046

There were also Christian informants who were invited to join the Makapetor alliance but refused that, first of all, the demands and targets of the action were not appropriate, which was less precise is this alliance demanding to eliminate the addition of new buildings around the mosque in the area of Ex-kampong Texas, especially if the target was the mosque administrator. Yet from the information obtained in social media that the time of laying the first stone of renovation of the construction of the mosque, the city government was present at that time, other information according to the board of the mosque itself they were entitled because it has long occupied the land. Makapetor Alliance demanded to stop the construction of the new building because there is an agreement from the municipal government that in the land of Texas ex-village will be established as Religious Tourism Park. Although the construction of Taman Wisata Religi itself is not appropriate because considering the area of land that is not possible, for there is no location for parking. In addition, what he questioned was "why is the target of action (demonstration) not directed to the city government as policy maker?"

In an action held by the Makapetor alliance according to the informant, there are some ambiguous things such as there are some banners of inappropriate demands that he witnessed himself when he accidentally passed the action mobs in front of the Manado Police office, he saw a banner saying Reject and Disband FPI, similar information he also got from some friends who were involved in the action according to them even there were banners demands related to "Terrorism" and "Refuse ISIS". Which he thinks is inappropriate and deviates from the main demands, "what municipal policy issues have to do with terrorism, ISIS and FPI!"

Secondly, theer was permissiveness of the 'mass organizations' leaders gathered in the Alliance, for example when there were skewed and 'unethical' remarks in the discussions raised by the members, the leaders did not quickly correct them.

In line with the above opinions, the imam of Al-Khairiyah Mosque also revealed that indeed the ex-kampong of Texas land is not eligible to be made Religious Park because of its narrow land. He also suggested, if it will be built as religious tourism park. Instead, it will be moved in the wax monument land next to the Soekarno Bridge. The land is larger and the location is also still adjacent to the ex-kampong area of Texas.

Manado city government in implementing its policies relating to the case of land ex-kampong Texas reap criticism and also advice from each informant. Namely as follows:

First, the city government in managing inter-religious harmony in the city of Manado in order to immediately resolve the problems that exist in the land of Ex-kampong Texas and the government should be more observant to see the problems made by certain elements. Because it has disturbed the people in Manado city, the government in this case the security forces must be firm not only because some people (provocateurs) will lead to chaos. Although the problems that exist in the city of Manado is the responsibility of the municipal government itself, it's just how they live how the city government work together or partner with the community to solve the problems that exist.

Second, the birth of such movements could be due to the lack of government control through the role of Forum Kerukunan Umat Beragama (FKUB) and the Inter-Religious Cooperation Agency (BKSAUA). For that reason, there must be an immediate meeting between religious leaders in this area to deal with this, and especially with the problems that exist in the ex-Kampung Texas land there must be a coordination of the related parties, among others the policy makers of the government to conduct a comprehensive study after that the results are socialized to grass root level.

Third, it is the government's weakness in managing religious symbols in the public sphere. Instead, I see it is the weakness of the state, whether he is in favor of the state ideology or the anger of the masses, in favor of the constitution or the ideology of a particular group.

Fourth, the city government who, when planning to establish a religious tourism park at that time, was one of the promises of the campaign, as if the elected government is now a hero for siding with all elements of society, one of them Muslims. This is just a political game of the city government now, they just want to find a voice from the Muslim community. Once traced, it turns out the development of religious tourism park until this religious building has no budget. This plan has been in talk since 2007 but until now it has not been realized.

The alleged existence of individuals who use this case and seek profit, the informants argue that this action can be ridden by certain parties such as investors who use labels and symbols of religion, culture and custom. In the end the struggle is not for the benefit of many people, but only for the sake of a few people, whether it is the investor or one or two people in the organization. Because if it is seen from the demands whether when the government realize the religious tourism park then the affairs are done? What are their advantages of the affair? Or after a religious tourism park is established they will get a job from there? And in the end also many groups will be harmed over such movements. In addition there are also informants who do not associate the action with the elements who take advantage of the action made alliance Makapetor.

All informants said the tolerance in Manado city was well maintained. This can be proved in the SETARA Institute's release results on November 16, 2015 which empirically placed Manado as one of the 10 highest tolerant cities in Indonesia. And the issue of SARA that developed in the community is only a psychological implication of the community from the issue that developed because of the conflict that occurred in Jakarta about the blasphemy case by the Governor of DKI Jakarta (Ahok).

According to the authors, so far the tolerance between religious people in Manado City is quite awake and run well. It is marked by no significant inter-religious conflicts that occurred in Manado City. In line with the informants' opinion about the existence of villages with multiple identity groups and also about the appeal of religious leaders and community leaders to keep restraint and maintain harmony among religious communities. Do not, just because the case of Ex-kampong Texas land and the intolerant issues that exist in the center causing harmony that has been awakened, make the people of Manado plural city directly divide.

Furthermore, for the municipality of Manado, they better as soon as possible solve the existing problems so as not to affect the issue of intolerance. Certainly the solution from the government does not harm all parties (win win solution). Regarding the government policy on the establishment of religious tourism parks in Ex-kampong Texas field we should be escorted by involving all elements of society.

Associated with Makapetor alliance action that reap so many critics that can invite conflict between religious people should be a lesson and warning to the people of Manado city to be more vigilant and not easily instigated by issues that smell SARA. So that is harmed the people of Manado city itself.
Regarding the opinion that there are persons who take advantage of this case for personal gain. According to the authors, there should be further research because it is just an issue and opinion from individuals. It could be true, it could be just a misunderstanding between Makapetor and Muslim alliance parties against government policy

4. Conclusion

From the results of research and discussion above, the authors can conclude that so far, the city of Manado is still categorized tolerant of people of other faiths. This is evidenced by the statement of informants representing Christians and Muslims who only criticize the city government and suspect the existence of elements who try to take advantage of this situation, they do not blame other people. Furthermore, there are still interfaith activities routinely conducted in Manado city and the results of SETARA Institute's release on November 16, 2015 which empirically placed Manado as one of the 10 highest tolerant cities in Indonesia.

References

[1] Arifin, Samsul, (2015). Studi Islam Kontemporer Arus Radikalisasi dan Multikuturalisme di Indonesia, Malang: Kelompok Intrans Publishing.

[2] Artis, (2011). Kerukunan dan Toleransi Antar Umat Beragama, *TOLERANSI: Media Komunikasi Umat Beragama,* Vol 3, No. 1, h. 86-97.

[3] Aslati, (2012). Toleransi Antar Umat Beragama dalam Perspektif Islam (Suatu Tijauan Historis), *TOLERANSI: Media Komunikasi Umat Beragama*, Vol 4, No. 1, h. 52-58.

[4] Baidi, (2010). Agama dan Multikulturalisme: Pengembangan Kerukunan Masyarakat Melalui

Pendekatan Agama, *Millah: Jurnal Studi Agama,* Vol 10, h. 1-29.

[5] Bakar, Abu, (2015). Konsep Toleransi dan Kebebasan Beragama, *TOLERANSI: Media Komunikasi Umat Beragama,* Vol 7, No. 2, h. 123-131.

[6] Hertina, (2010). Toleransi Upaya Untuk mewujudkan Kerukunan umat Beragama, *TOLERANSI: Media Komunikasi Umat Beragama*, Vol 1, No. 2, h. 207-217.

[7] Indeks Kota Toleran (IKT) 2015, http://setara-institute.org/indeks-kota-toleran-ikt-2015/, di akses pada tanggal 20 Januari 2017 pukul 23.21 WITA

[8] Masykur, Said, (2016). Pluralisme Dalam Konteks Studi Agama-Agama, *TOLERANSI: Media Komunikasi umat Beragama* , Vol. 8, No. 1, h. 61-77.

[9] Maulana, (2016). Mempertegas Semangat Toleransi dalam Islam, *TOLERANSI: Media Komunikasi Umat Beragama,* Vol 8, No. 2, h. 117-133.

IOP Conf. Series: Earth and Environmental Science **156** (2018) 012047 doi:10.1088/1755-1315/156/1/012047

Improving Students' Sense to Learn Language in Islamic Institution of Coastal Area Indonesia

***St Kuraedah,[1] Nur Azaliah Mar,[2] Fahmi Gunawan[3]**

[1] Institut Agama Islam Negeri Kendari, Jl. Sultan Qaimuddin No 17 Baruga, Kendari, Sulawesi Tenggara, 93563, Indonesia
Email : st.kuraedahshahib@yahoo.co.id
[2] Universitas Islam Negeri Sultan Alauddin Makassar, Jl. Sultan Alauddin No 36 Kab. Gowa, Sulawesi Selatan, 92113, Indonesia
Email : nurazalia.na@gmail.com
[3] Institut Agama Islam Negeri Kendari, Jl. Sultan Qaimuddin No 17 Baruga, Kendari, Sulawesi Tenggara, 93563, Indonesia
Email : fgunawanp@gmail.com

Abstract. This research aims to examine the ways to develop a sense of love for learning Arabic among students in Islamic Higher education of Indonesia. This study is essential to do because Arabic should be the favourite subject by the students. In addition, Arabic is also the language of Al-Qur'an. As the language of Al-Qur'an, Arabic for Indonesian is not a foreign language as other foreign languages. In fact, the Arabic becomes one of the dreaded subjects by most students, especially at State Islamic Institute of Kendari. Therefore, it takes the tips and efforts by the Arabic teachers to make Arabic more interesting for the students. The results show that one way to increase the motivation to learn Arabic for students is to develop students' sense of love to Arabic. The teachers can do it by showing how easy Arabic is and how important it is since it is a language of religion and science, and providing the tips to learn the language. Besides, they also can explain some borrowing words from Arabic adopted into Indonesian and to be used in daily conversations without realizing it, and show the form of word derivation in Arabic that can help to enrich the Arabic vocabulary. The teacher should tell the students that knowing one word in Arabic can develop into some vocabulary with different meanings.

1. Introduction

According to historians and linguists, Arabic comes from a human race and a language family that has a significant role in the history of the ancient civilization of the Semites [1]. Then their descendants moved away from their homeland and settled in the Tigris and Euphrates rivers forming new language and nation, such as Babylonia, Assyria, Hebrew, Armia, Tunisia and others. The struggle between languages takes place in mutual influence and defeats, the dominant languages used and influential that comes out as winners, then the language of the tribe becomes the standard language, such as Arabic [2].

The Arabic language used by the Arabs now or contained in the Al-Qur'an and the hadiths of the Prophet originally grows and is developed in the region of Nejed and Hijaz and then spread to various regions [3]. It is because of the existence of Islam which exerts a far-reaching influence especially after holding the expansion of territory [4]. Arabic Baqiyah is used in common association, commerce, societies and government [5]. This Arabic language can survive and not disappear like any other, read: a related word is a word that cannot be separated from the influence and role of Islam at that time, where the main teachings of Islam that written in the Al-Qur' a using Arabic Baqiyah. Muslims seek to know Arabic by themselves, for non-Arabic speakers of Baqiyah make it as Arabic in their daily associations which have resulted in the previous languages being no longer abused, and gradually other languages die by themselves because of no more speakers. In the theory of language as Mahmud, Kamil al-Naqah says that a language can live if it is turned on by its speakers and otherwise the language will die when abandoned by its speakers (no longer used as a communication language). Subsequent developments indicate that Arabic influences appear to be increasingly widespread in international relations, and since 1973 this language has been officially recognized as a legitimate language for use in the United Nations environment. The Arabic language is also considered the language of Muslims [6], caused by the existence of the Al-Qur'an and the hadith of the Prophet who serves as two primary sources of Islamic teachings written in Arabic.

In fact, Arabic as the language of the Al-Qur'an should be the second language for Muslims, especially in Indonesia [7]. However, it is not the case; ironically it is not to be a second language, even a lot of Muslim communities who cannot read Al-Qur'an fluently. They only put the holy book as sacred items stored neatly in the cupboard. They use it only in the month of Ramadan or when there is a family *tahlilan*. Even lately in various regions, there is quite ironic spotlight related to religion teachers who cannot read Al-Qur'an well. Indeed, this phenomenon becomes a very unfortunate thing that should not happen. The condition is not going to happen when Arabic as the language of Al-Qur'an is studied in earnest since childhood, so that children can read the Al-Qur'an. By learning Arabic, it can help us to read the Al-Qur'an, or vice versa with knowing the Al-Qur'an reading will be easier to learn Arabic.

Research on learning Arabic is easy and interesting to do has been conducted by previous researchers. They examine Arabic language through online vocabulary games [8–11], virtual tutor [12], song lyrics modification [13], multicultural and islamic values approach [14-15], and active learning [16]. However, research on word derivation and borrowing words from Arabic adopted into Indonesian in Islamic Higher Education is still rarely done. Hence**,** this research aims to examine the ways to develop a sense of love for learning Arabic among students in Islamic Higher education of Indonesia.

2. Method

This research is qualitative descriptive research. To collect data, unstructured interview was conducted [17] to some students of Arabic study program that come from Madrasah and Public High School (SMU) whose fundamental stench of Arabic is still below of standard. The purpose of the interview is to find out how far their interest in Arabic learning materials. Thus, this research can reveal the result by explaining the forms of derivation and deciphering the word Arabic uptake into the Indonesian language used in everyday practice which can develop the students' interest and sense of love to learn Arabic.

3. Results

Arabic is included in a semitic language family that is widely used on the Earth [18]. Arabic is the primary language of 22 countries, studied not only by residents of Saudi Arabia but also in other countries [19]. It is because Arabic is a religious language used by more than one billion Muslims worldwide and is spoken during daily worship. Arabic is the most significant language regarding the number of speakers of the Semitic language family, namely the SAM family of languages derived from the language of the Prophet Nuh's son named SAM bin Nuh.

Arabic is part of Islam religion. In the extreme view, Ibnu Taymiyyah explains that "learning and teaching Arabic is *fardhu kifayah*" and he also points out, "The Arabic is part of religion and knows it is obligatory, because to understand Al-Qur'an and *Sunnah* is obligatory [20]. Both cannot be understood using other language than Arabic. Arabic is the noblest and oldest language still being used today. This language is used by the Prophets and will later be used by the inhabitants of Heaven [20]. Learning Arabic is very important, especially for Muslims because of Al-Qur'an and Hadiths of the Prophet that become a reference for Muslims in worshipping in daily life. The strength of Arabic is part of the glory of Islam and the Muslims [20]. Musthafa Shadiq Ar Rafi said that *"The lowness of a language will affect its speakers and its glory will be the power that never leaves its speakers* [21]*."* Therefore, some foreign invaders require their language to be learned by the people that they occupy. It can be seen how the English almost dominates the entire world especially the former colonies because they are aware of the existence of people that need to be supported by the strength and popularity of the language. No doubt for the existence and glory of Arabic that will disappear because Arabic as the language of Al-Qur'an will still exist throughout the ages. This has been described in the verses of Al-Qur'an which means *"We lowered Al-Qur'an by Arabic, so that you may understand.* "(Qs Yusuf: 2) and in (Qs Az-Zukhruf 43: 3) which means *"We have made the Qur'an in Arabic so that you may understand".*

Arabic is one of the fixed languages in the United Nations (UN). It is also one of the native languages of many nations, and some even claim that "Arabic is the origin of languages", and those who adopt this theory believe that Arabs can pronounce any sound in any language of the world quickly. In contrast, many non-Arabs find it difficult to pronounce some Arabic letters that are not in their native language. Thousands of years, Arabic still show its appropriateness in every age and have penetrated various civilizations. It is because Arabic has a wealth of vocabulary and a lot of derivation [22]. Derivation makes learning Arabic is easy as knowing one word can evolve into many new vocabulary, e.g., كتب (*kataba*) which means to write, if we change to كاتب (*Kaatibun*), it means secretary. The other examples are a writer, كتاب (*kitaabun*) implies book or book, and when it turns into مكتب(*maktabun*), it can mean writing desk for writing, office and after changing into مكتبة (*maktabatun*), it can be libraries, bookstores/place for gatherings of many papers.

Other data also indicate that Arabic has a significant influence on most languages present today. Perhaps, many Arabic words are used by many languages other than Indonesian, for example, sugar سكر (sukkarun) means sand sugar, cut قطع (Qatha'a) means Cut, credit قرض (Qorodha) means debt, cotton قطن (qotnun), cover كفر (*kafara) means close or* shut down, etc. This list continues to grow along with the growth of language development translated freely from the book *al-'Arabiyah Baina Yadaika*. The exposure of Arabic materials in the classroom with an explanation of the abundance of Arabic borrowed words into the international language and through the form of word derivation widely used in conversations such as the explanation above can help to improve the learning motivation and sense of love to Arabic. Based on the data that the writers found as many as 19 students from public high school (SMU) from the number of 61 students in the Arabic study program for the first semester in 2017 stated that they are happy to know how easy learning Arabic is.

The entry Arabic into the archipelago along with the entry of Islam between the 7th century until the 8th century through the Muslim traders from Arabia and Persia, that is when Islam began to develop in the archipelago around the 11th century until the 12th M [23]. It means the age of Arabic in the archipelago reaches ± 12-13 centuries. In the long span of time, the Arabic has become a vital part of the cultural expression of ethnic groups in the archipelago [24]. The existence of a synergistic relationship between Arabic and Indonesian is no other than because the majority of Indonesia's population is Muslim. Through Islam religion, the Arabic entry is as a language in the worship of Muslims in Indonesia, the language used in da'wah, the language used in Islamic studies, the language used in science, art, and Islamic literature in addition to Arabic as a language the holy book of Al-Qur'an that guides Muslims. Thus, there is no doubt for significant Arabic contribution to the development of language and culture of Indonesian. As the language of revelation and the language of religion, Arabic has long entered into the soul of the Indonesian nation so that some Arabic vocabulary has moved into Indonesian vocabulary. It is not only limited in vocabulary that runs into Indonesian from Arabic but Arabic grammatical system also moves into Indonesian grammatical system, in social aspect of social culture, laws, politics and the state of Arabic also has a small contribution to Indonesian,

The Arabic grammar system that moves into the Indonesian grammar system is male and female gender differentiation system [22]. Almost all words in Arabic are distinguished between men and women (masculine and feminine). The feminine type in the noun or adjective is marked with suffix (ah). "ah" is attached to the word such as *ustadz* 'guru/teacher' becomes *ustadzah* 'female teacher', *haji* becomes *hajah* 'female of *haji'*, preacher *'muballigh'* becomes *mubalighah* 'female of *muballigh'*, *qari* 'reader of Al-Qur'an for male' becomes *qari'ah* 'reader of Al-Qur'an for female' and so on. Another example is the system of forming the meaning of nature in the field of grammar in Indonesia. For instance, ordering of suffix - wi, - ani, and - i to show the nature, such as '*dunia*/world' into '*duniawi'*, '*akhirat'* becomes '*ukhrawi'*, '*roh'* becomes '*rohani'*, '*rab'* becomes '*rabbani'*, the human being of the word '*insan'* becomes '*insani'*; [22].

From the aspect of social culture, politics, laws, and the state, Arabic also has a contribution in addition to Indonesian vocabulary such as (1) Name of day, Sunday/*Ahad* "احد" *Ahad* means one, Monday/*Senin* from the word "اثنين" *Itsnain* means two, Tuesday/*Selasa* from the word "ثلاثة" *Tsalaatsah* means three, Wednesday/*Rabu* of the word "اربعة" *Arbaah* means four, Thursday/*Kamis* of the word "خمسة" *Khamsah* means five and the word Friday/*Jum'at* of the word "جمع" *Jama'a* which means to gather, and Saturday/*Sabtu* of the word "سبعة" *Sab'ah* which means seven. Thus Indonesian people when mentioning the names of the day then automatically they already know the Arabic number from one till seven. (2) Using Arabic pronunciation by Indonesian Muslims to name their sons as the example of Nur Fadhilah (نورفضيلة) which is the light of primacy, Abdul Hakim (عبد الحكيم) means wise of the servant, the several names are believed by Muslims as a form of prayer to Allah SWT when calling the name that in Islamic name. Just because the lack of understanding of Muslim society so found the naming of Arabic without regard to the aspect of meaning, which is important in Arabic, such cases as a sign that Indonesian Muslim society has an interest in Arabic. Then, it can be concluded that if someone has a friend whose name is taken from Arabic as much as 1000 people, it will help to add some 1000 new vocabulary in Arabic. (3) In social life, to interact with society, we need a lot of vocabulary. Many words are absorbed from Arabic for example, *syukur, ikhlas, sabar, tobat, laknat, azab, isyarat, alamat, umi, abi, om* (gratitude, sincerity, patience, repentance, curse, punishment, gesture, address, mother, father, uncle absorbed from the word "ammun" which means uncle, *ads, credit,* even the word *Alhamdulillah, lahaula Wala Quwata illa billah, subhanalllah, amiin, masya Allah* and *astagfirullah*. Those words are attached in oral communication of most Indonesian society without knowing the boundaries of belief. Only sometimes among Muslims there is an influence of unsupported interactions with a deep awareness of the word meaning. For example the word *Astaghfirullah* which is sometimes called *Astaga* which means in Arabic does not give specific meaning as a rosary, *tasbih* or prayer and etc, so it is highly recommended to mention *astaghfirullah* entirely. That is the development of Indonesian as the unifying language of the Indonesian nation which cannot be separated from the influence of assimilation of Arabic. The influence is not only in the lexical aspects but also in morphology and syntax. In the state institution, as expressed by Tajuddin Nur that the Arabic vocabulary used in socio-political institutions, formed political, and constitutional concepts, such as laws, judges, tribunals, chapter, verse, fair, civilized, prosperous, people, wisdom, representatives, deliberation, consensus, civilization and so on, has enriched and solidified the Indonesian national concepts in the field of administration [24].

Quantitatively, the Arabic contribution to the Indonesian vocabulary has been found by several researchers. Soedarno found 2,336 Arabic words absorbed into Indonesian language [25]. Ghazali also found 2,750 Arabic vocabulary in Indonesian [26]. In addition, he found 2,178 Arabic vocabulary in Indonesian in the Dictionary General Indonesian (KBBI) [26]. Arabic vocabulary will increase in line with the progress of development and the demands of global communications such as *tajir* word which means traders who come from Indonesian interpret the bourgeoisie, *risywah* be (*rasuah*) for the meaning of *suap/sogokan*, *ta'jil* which means accelerate absorbed into the Indonesian language and given the meaning of *panganan* prepared specifically for *iftar* (eating cake after fasting). Even several of these words have not been found containing in the Great Indonesian Dictionary (KBBI) till now [26].

The results of the data analysis explain that the process of changing the sound of the vocabulary of Arabic to Indonesia, according to the results of the research it is found that many symptomatic changes in Arabic absorption sounds change after passing some form of type, namely weakening the sound, the addition of sound, sound reinforcement, loudening sounds to monopolization. Several types previously explained before explaining the identification of symptoms of change and even the removal of phonemes to form sounds are different from the original transliteration.

When the writers convey the facts on the sidelines of the delivery of Arabic material, it is then followed by requests for responses from students about the positive attitude and love of Arabic as the language of religion and also the language of Science. The data that collected by the writers from 61 students, all give positive responses and initial assumptions stored in their memory of the difficulty of learning Arabic that has gained new insight and made them eager to learn Arabic. Thus, State Islamic Institute of Kendari that expands the vision to become a centre of Islamic studies transdisciplinary, studying Arabic cannot be an option. It should be a compulsory subject and certainly expected not only be the subject that fills the list in the transcript but importantly the students realize the importance of Arabic for them. Therefore, the interest and passion for learning Arabic, which in turn no longer a force for students. They eventually can think that Arabic is easy to learn.

4. Discussion

Arabic as one of international languages has also become a religious language which in turn becomes a unifying language for Muslims through a long process [7]. Arabic also has a particular function than other languages. Not only Arabic has high-grade literary values for those who know and deepen it, but even Arabic is destined to be the language of Al-Qur'an that's communicating the word of God which contains *uslub* language which is amazing, no one can match it. It is because Arabic destined as a language of religion derived from the holy book of Al-Qur'an. Al-Qur'an and Hadiths and some other books of religions classic are the main references of Muslims who all use Arabic in communicating the religious messages. It has been through the process of evolution within the history of the Arabic language as used now. It is not the same as old Arabic at the beginning of its emergence, but it has gone through a long journey in thousands of years of continuous processing to perfection Ali Abd al-Wahid Wafi as seen in the language of al-Qur'an and *Fusha* that used as a means of communication [27].

The changes occurs when a language user moves to another place as well as settling and reproducing with other languages, where there is a process of *step-by-step* changes without felt by the user and so on. Similarly, the entry and movement of Arabic absorption vocabulary into the Indonesian that sound change is a common condition occurs in the rapid dissolution of the language use of a nation. Moreover, the sound changes that come from foreign alien wording words in Indonesia, the appearance of different absorption in Indonesia is not because of the absence of vocabulary counterpart, but precisely because of the long history factor in Indonesia. Trade contacts between nations like Arab and Indonesia and the presence of the invaders, like the Dutch, Japanese and Portuguese also introduced their language to the ancestors of Indonesia. As a result, some Arabic vocabulary and the other foreign language also enrich the Indonesian language, even recorded by the Great Indonesian Dictionary (KBBI). In the process of assimilating languages with other languages, changes occasionally occur. Hadi mentions several types of sound changes, namely (a) a lenition that consisting of cluster reduction, apocopate, syncope (synaptology, haplology, and compression), (b) sound addition consisting of anaptyxis, epenthesis, and prosthesis, (c) metathesis, (d) fusion, (e) unpacking, (f) vowel breaking, (g) assimilation, (h) dissimilation, (i) abnormal sound change [28].

Sound changes as described above consist of 1) Sound's Weakening. According to Kridalaksana *"The weakening of sound is a change from a strong sound to a weak sound"* [29]. There is a relatively stronger sound, and there is a sound that is relatively weaker than the other sounds. 2) Sound's strength. The Sound's strengthening is a change from a relatively weak sound to a relatively stronger sound. This type is the opposite of the sound's weakening. The example of the word 'fahm' that absorbed into *paham*/understanding, there is a reinforcement of sound /f/ to /p/ caused by /f/ sound, is not a native Indonesian phoneme, phoneme /f/ is a loan phoneme, while phoneme /p/ is native Indonesian phoneme. 3) Loosen Sound. Loosen sound is an original single Arabic sound, evolving into a sequence of sounds, each with its original characteristics. An example of the word ‘adzan' supported by the phoneme /dz/ there is a relaxation of the phonetic features of the Arabic phoneme /dz/ later transforms into a phoneme /d / and phoneme /z/. Arabic /dz/ phonemic phrases are containing apiko dental shear sounding features. 4) Sound’s Addition. The sound’s addition has several types, the first insertion of sound or letters into a word also called epenthesis. Symptoms included in the changes caused by the addition of consonants between two consonants and consonants plus vowels. An Example word ‘Fahm’ فهم which is absorbed into ‘*paham*/understanding' has occurred the insertion of the vowel /a/. Secondly, the *paragog* is the sound recording at the end of the word for ease of pronunciation. The addition of this sound usually occurs at the end of a word that ends with a consonant, by the addition of the vowel. 5) Monopolization is a change due to the joining of two distinct sounds into a single sound and then containing some phonetic features of both sounds called monopolization. If we look at the examples ‘taubah’ توبة changed into ‘*tobat*/repentance' there has been a process of monopolization occurring in absorption words containing diphthongs /ai/ and /au/.

Arabic contributes many vocabularies to the languages of the Islamic world like the Latin role for most European languages. So, it can be found that several Arabic vocabulary through the process of sound changes, the sound’s addition. In the middle Ages, Arabic was the main language of science and politics. Interaction at that time made some European languages borrowed the hundreds of Arabic vocabulary. The most profound influence of Arabic is in predominantly Muslim-majority countries. Call it, in Kurdish, Persian, Swahili, Urdu, Hindi, Turkish, Malay, and Indonesian. From the description, it can be concluded that Arabic is not only a massive contribution to the Indonesian language but also in other languages in developed countries are like the conquest of Muslim in 711 M (started 8-15 M). In Spain, Arabic language becomes an important historical legacy that continues to live in the cultural treasures of Spanish society. Arabic adoption into the Spanish language lasted for approximately eight centuries of Muslim rule there. It contributes significantly to the modern Spanish vocabulary.

5. Conclusion

From the description above, the writers conclude that (1) developing the students’ sense of love in the Arabic courses can be done by describing the form of Arabic word derivation that can be changed to various types of words with different meanings. In addition, the Arabic language teachers are advised to provide reinforcement to the students by showing a number of vocabulary Arabic which are absorbed into the Indonesian and used in daily conversations unnoticed and understood, that they are absorption words from Arabic. (2) showing how easy is to learn Arabic for Indonesian students because Arabic is the language of Al-Qur'an and all at once the language of religion for Muslims who are the majority in Indonesia, and some words in Indonesia are actually derived from Arabic language.

References

[1] Chejne AG 1969 *The Arabic language: Its role in history.* U of Minnesota Press.
[2] Comrie B (Ed) 2009 *The world's major languages.* Routledge.
[3] Antonius G 2015 *The Arab awakening: The story of the Arab national movement.* Pickle Partners Publishing.
[4] Maisel S (Ed) 2009 *Saudi Arabia and the Gulf Arab states today: an encyclopedia of life in the Arab states* (Vol. 1). Greenwood Publishing Group.
[5] Al-Sulaiti L, Atwell ES 2006 The design of a corpus of contemporary Arabic. *Int. J. of Corpus Linguistics*. **11**(2) 135-171.
[6] Jaspal R, Coyle A 2010 “Arabic is the language of the Muslims–that's how it was supposed to be”: exploring language and religious identity through reflective accounts from young British-born South Asians. *Mental Health, Religion and Culture*. **13**(1) 17-36.

[7] Efrizal D 2012 Improving students' speaking through communicative language teaching method at MTS Ja-Alhaq, Sentot Ali Basa islamic boarding school of Bengkulu, Indonesia. *Int. J. of Humanities and Social Science*. **2**(20) 127-134.
[8] Sahrir MS, Yusri G 2012 Online Vocabulary Games For Teaching And Learning Arabic. *GEMA Online J. of Language Studies*. **12**(3).
[9] Sahrir MS, Alias NA, Ismail Z, Osman N 2012 Employing design and development research (DDR): Approaches in the design and development of online Arabic vocabulary learning games prototype. *TOJET: The Turkish Online J. of Educational Technology*. **11**(2).
[10] Ashraf H, Motlagh FG, Salami M 2014 The impact of online games on learning English vocabulary by Iranian (low-intermediate) EFL learners. *Procedia-Social and Behavioral Sciences*. **98** 286-291.
[11] Huang YM, Huang YM 2015 A scaffolding strategy to develop handheld sensor-based vocabulary games for improving students' learning motivation and performance. *Educational Technology Research and Development*. **63**(5) 691-708.
[12] Mahmoud A, Ragheb A, Sahrir MS 2013 Integration of an Interactive Program in Learning Arabic Language for Non-Native Speakers via Virtual Tutor. *GEMA Online Journal of Language Studies*. **13**(3).
[13] Gunawan F, Zulaeha 2017 Analysing Character Education Values at SDIT al-Qalam through Song Lyrics. Advances in Social Science, Education and Humanities Research. **84** 43-47.
[14] Wekke IS, Lubis MA 2016 A Multicultural approach in Arabic language teaching: creating equality at Indonesian pesantren classroom life. *Sosiohumanika*. **1**(2).
[15] Gunawan F, Syukri S, Humaera I 2016 Enhancing students' insights in constructing report texts with Islamic values by scaffolding. In *Proceedings of English Education International Conference*. **1** (2) 387-390.
[16] Yusuf M, Wekke IS 2015 Active learning on teaching Arabic for special purpose in Indonesian Pesantren. *Procedia-Social and Behavioral Sciences*. **191** 137-141.
[17] Doody O, Noonan M 2013 Preparing and conducting interviews to collect data. *Nurse researcher*. **20**(5) 28-32.
[18] Faber A 1997 Genetic subgrouping of the Semitic languages. *The Semitic Languages*. 3-15.
[19] Krauss M 1992 The world's languages in crisis. *Language*. **68**(1) 4-10.
[20] As' ad AbuKhalil 1994 The incoherence of Islamic fundamentalism: Arab Islamic thought at the end of the 20th century. *The Middle East Journal*. 677-694.
[21] al-Rafi'i MS 2010 The Grooming of Men. *For Better, For Worse: The Marriage Crisis That Made Modern Egypt*. 23.
[22] Gunawan F 2009 Perubahan Bahasa (Interaksi antarkomponen Tatabahasa Bahasa Arab), Language Change: Interactions between the components of The Arabic Grammar. *Adabiyyāt: Jurnal Bahasa dan Sastra*. **8**(2) 211-232.
[23] Hadi S 1995 Bahasa Arab dan Khazanah Sastra Keagamaan di Indonesia. *Humaniora*. (2).
[24] Madjid N 1988 Bahasa Arab dan Perkembangan Indonesia Modern. *Jurnal NADI Edisi September*.
[25] Soedarno 1992 *Kata Serapan dari Bahasa Arab*. Arikha Media. Jakarta.
[26] Ghazali AS 1999 "*Sumbangan Bahasa Arab terhadap Pembentukan Istilah Wacana Ilmiah Berbahasa Indonesia*". Makalah dalam Seminar PINBA I, 24-26 September, Malang.
[27] Wafi, Ali Abd al-Wahid TT *Fiqhi al-Lughah*. Dar al-Nahda. Kairo. 15-30.
[28] Hadi S, Chamamah S, Ramlan M, Wijana, IDP 2003 Perubahan Fonologis Kata-kata Serapan dari Bahasa Arab dalam Bahasa Indonesia. *Humaniora*. **15**(2) 121-132.
[29] Kridalaksana H 1993 Kamus Linguistik. *PT Gramedia Pustaka Utama*. Jakarta.

IOP Conf. Series: Earth and Environmental Science **156** (2018) 012048 doi:10.1088/1755-1315/156/1/012048

Democracy and responsiveness in Maritime Country: Interrelations between 'blusukan' of jokowi and the development of instrastructure of ocean toll

Suyatno Ladiqi,[1]Mohd Afandi Salleh[2]

[1]Faculty of Law & International Relations,Universiti Sultan Zainal Abidin, Gong Badak Campus, 21300 Kuala Terengganu, Malaysia
Email : yatno.ladiqi@gmail.com

[2]Faculty of Law & International Relations,Universiti Sultan Zainal Abidin, Gong Badak Campus, 21300 Kuala Terengganu, Malaysia
Email : afandisalleh@gmail.com

Abstract: In the context of the Asian financial crisis of 1997, a dictatorship in Indonesia that had lasted more than 30 years finally came to an end. Indonesia's transition process featured, among other aspects, a general overhaul and liberalization of the political system, including the establishment of a multiparty system, freedom of the press and the first free and fair elections in the country since 1956. The important thing is capturing relations between the state and the people as change phenomenon from authoritarianism toward transition democracy. Putting the people as a pivotal aspect of the context state relations is crucial democracy discussion not only reflected by people participation but also government responsiveness through pro poor policy or welfare policy. It should be discussing democracy in a complete definition. Implementation of democracy should reflection not only of people such as participation and representativeness measure but also for the people such as responsiveness measure.Understanding the relationship between the state and the people in the context of responsiveness through infrastructure development will certainly be more interesting when the 'Blusukan' method is put into the frame of development program implementation. Implementation of democracy such as "for the people" is evident in the government's efforts to meet the needs of the people (responsive) through pro people policy making. In addition to the existing institutional supervision, with the Blusukan method, Jokowi wants to get accurate, fast and direct get information from the people. This activity not only wants to bring the state closer to the people but also at the same time wants to control the implementation of infrastructure development that is and will be run.

1. Introduction

In the context of the Asian financial crisis of 1997, a dictatorship in Indonesia that had lasted more than 30 years finally came to an end. Indonesia's transition process featured, among other aspects, a general overhaul and liberalization of the political system, including the establishment of a multiparty system, freedom of the press and the first free and fair elections in the country since 1956 [4]. But the other hand, transition to democracy in Indonesia also has opened Pandora's box, fomented ethnic and religious conflict, and even encouraged regional separatism, thereby creating political and economic uncertainties [2].

The important thing is capturing relations between the state and the people as change phenomenon from authoritarianism toward transition democracy. Putting the people as a pivotal aspect of the context state relations is crucial democracy discussion not only reflected by people participation but also government responsiveness through pro poor policy or welfare policy. It should be discussing democracy in a complete definition. Implementation of democracy should reflection not only of people such as participation and representativeness measure but also for the people such as responsiveness measure. It is important since to put government responsibility to increase life standard through produce welfare policies.

Jokowi has a big commitment to implement democratic responsiveness. There are many policies that his administration produced reflects how is big government concern to increase people life standard. One of important is massive infrastructure development in across the region in Indonesia. It is very interesting if Joko Widodo (Jokowi) uses a unique approach to succeed his policy through 'Blusukan' methods. This approach is also reflecting how is closer relations between state and society in democracy context.

2. Method

The research literature-based, which data are obtained from many available sources from the internet, books, and journal. Considering new topics, the availability of resources are limited and related research is not widely observed in Indonesia. The research analyzes how relations between 'Blusukan' of Jokowi and development of infrastructure as refelection of responsiveness democractic implementation under Jokowi administration.

3. Result

This paper will use indicators that refer to implementing democracy that called Democracy Implementation Model [18] as a tool to elaborate democracy development. The model incorporates the philosophy of democratic, system and institution. The values of the philosophy of Abraham Lincoln, such as "of the people", "by the people" and "for the people" can be implemented if combined with Systems Theory that created by David Easton (1957). In Systems Theory consists three important components, such as "input", "process" and "output". The process of elaboration of the value of philosophy into the system will raise institutions which can be seen more clearly in shape as indicators of the democracy implementation.

Putting the value philosophy "of the people" to "input" will raise *representativeness* and *participation* institutions. Meanwhile, put the value philosophy "by the people" to "process" will embody institutions, such as *partnership*, *transparency,* and *accountability*. Put the value philosophy "for the people" will bring institutional *responsiveness*. Sixth institutions of democracy implementation are indicators that can be seen more clearly. The question of democracy implementation is not just a problem of power management of the people and government, such as participation and representation of the people, but also the issue of how the power is managed in transparency and accountability context, and in partnership with the people's position firmly. Equally important, the power should also for the people welfare in the form of responsiveness. Indicators of responsiveness that is the measure of the success of democracy in the context of power management that generates prosperity.

Responsiveness refers to the way in which a given decision-maker – public or private – perceives the needs and responds to the demands of particular groups, such as the poor [3]. Because it is the result of the implementation "for the people", it is understood responsiviness closer to the result of the public service through a series of policies pursued. In this regard, responsiviness defined as the ability of the bureaucracy to recognize the community needs, to sets the agenda and priorities of service, and to develop service plans according to the needs and aspirations of society [18].

'Blusukan' - Jokowi has a unique method to implement responsiveness. Comparing with previously leaders, Jokowi has different aprroach to the people since using 'blusukan' way as policy implementation control. 'Blusukan' in other words is inspection visit or impromptu visit. Jokowi explained that his 'blusukan' activity to the remote areas of the capital is not without purpose, but is based on a calculated management principle. In the early days of his leadership, blusukan had a function of knowing the society's conditions and needs (en.tempo.co, 31 July 2013). 'Blusukan' is also the latter's preferred method of overseeing developments and communicate with the people. Actually, 'Blusukan' method is also a habit Jokowi acquired as mayor of Solo, a small city in central Java, is a novelty in a country where politicians are usually glimpsed, if at all, on a distant rostrum. One of the reasons he is so popular is that he seems humble and approachable and to care about the views of those affected by his decisions (www.economist.com, 21 Janury 2014).

Jokowi's administration is determined to expand the development area not only to the Java island area but also other remote regions. For three years in his administration, Jokowi has conducted a series of 'Blusukan' activities that are almost evenly distributed throughout Indonesia. The picture below shows the map of 'Blusukan' of Jokowi during three years of his reign. The digital map issued by the official website of the presidential office of the Republic of Indonesia contains very comprehensive information about Jokowi's 'Blusukan' activities. By clicking one point will appear complete activities of President Jokowi in conducting a series of development activities in the area.

For example if clicking one of Blusukan point in Papua, it will inform about Jokowi Blusukan activity: "President Joko Widodo accompanied by First Lady Iriana reviewing road construction in Kenyam Village area, Nduga District, Papua Province (31/12). This path will connect Nduga and Wamena. President Jokowi hopes that the road will be built so that the Nduga District, including the red zone or unstable, will be more open" (presidenri.go.id).

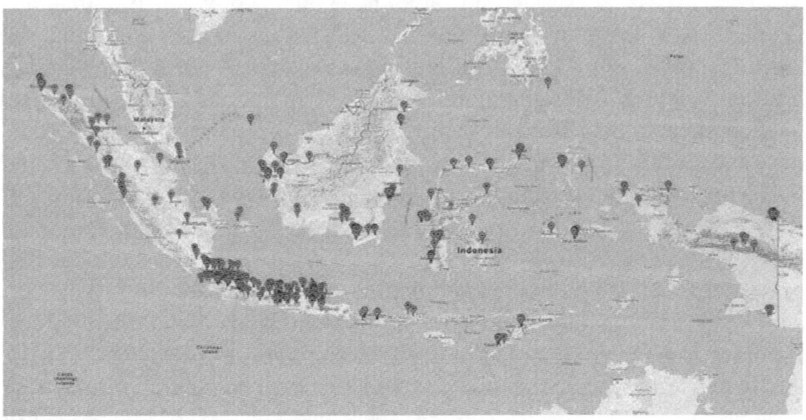

Figure 1. 'Blusukan' of Jokowi (Source: http://presidenri.go.id/blusukan.1.html. Downloaded 1 June 2017).

The Development of Instrastructure – One of important from responsiveness implementation is the development of insfrastructure that reflection governement responsibility to recognize community needs. In particular to encourage the economic activity of the people through the provision of infrastucture facilities. In World Economic Forum (WEF)'s Global Competitiveness Report 2015-2016, Indonesia ranks 62nd out of 140 economies in terms of infrastructure development, a mediocre ranking, and one that causes major problems. Since the era of reformation replaced Suharto's authoritarian New Order government in the late 1990s, expansion of Indonesia's infrastructure has not been able to keep up with robust economic growth that occurred after the recovery from the Asian Financial Crisis amid the lucrative commodities boom. As a consequence, Indonesia's economic growth fails to reach its full potential (indonesia-investments.com).

The development of infrastructure is very improtant because if the state of infrastructure is weak it implies that the economy runs in a highly inefficient manner as logistics costs are high, businesses lack competitiveness (because the costs of doing business are high), and there exists a high degree of social injustice when, for example, it is difficult for part of the population to reach healthcare facilities, or, for children to reach a school. In fact, infrastructure development and macroeconomic development go hand-in-hand (have a reciprocal relationship) because infrastructure development gives rise to economic expansion through the multiplier effect, while economic expansion gives rise to the need to enlarge existing infrastructure to absorb the larger flow of goods and people that travel across the economy. However, when existing infrastructure cannot absorb rising economic activity (and not enough new infrastructure is developed) then there occur problems, similar to blocked arteries in the human body that cause life threatening conditions.

As a basic fondation to develop infrastructure, Jokowi produces regulatory reform in economic sectors as a priority. Between September 2015 and February 2016, his government unveiled 10 economic reform packages, with the intention of promoting deregulation and supporting investment in key sectors. Jokowi has recently made a number of announcements about the urgent need for regulatory reform, including, in December, setting the highly ambitious goal of removing half of Indonesia's 42,000-odd business regulations by the end of 2016 (*Jakarta Globe*, 8 Dec. 2015). Two months later, in February, he unveiled plans for a 'big bang' loosening of restrictions on foreign investment in nearly 50 sectors, as part of the 10th reform package (Reuters.com, 12 Feb. 2016).

It is uncertain whether these packages and announcements will lead to actual reform. Foreign investors, at whom many of these measures are directed, have viewed them with a mix of cautious optimism and healthy scepticism. During the president's overseas visits, investors have frequently heard that Indonesia welcomes foreign capital, only to then see this commitment countermanded by the actions of ministers and officials wanting more restrictive regulations and practices. As Manning (2015) notes, this government has proven to be stronger on announcements than implementation, having often buckled in the face of opposition and reversed out of, or changed, its commitments [16].

Demonstrating success in delivering infrastructure, in particular, is pivotal to the Jokowi presidency, and requires a well-organised and coordinated effort in communicating the government's policies and accomplishments. His administration has struggled to sell its messages, both internationally and domestically, owing partly to inexperience in the president's office and, as Adam Schwarz suggests, partly to the president's own 'unstructured, ad hoc style' (*Jakarta Post*, 4 Jan. 2016). Being able to show that new infrastructure and better service delivery improve the lives of Indonesians will be the most critical test. This explains the urgency behind the government's push to accelerate disbursements [16].

The Indonesian government is well aware of the importance to offer a more attractive investment and business climate by improving the nation's infrastructure. Currently, there are not enough roads, harbors, airports, and bridges in Indonesia (Southeast Asia's largest economy), while - not unoften - the quality of existing infrastructure is weak. However, developing Indonesia's infrastructure (both hard and soft) is not an easy task. The Archipelago consists of about 17,000 islands (although it is fair to say that many of these islands are not inhabited and show no economic activity) which make it more complex (thus more expensive) to enhance connectivity and implies there exists a need to focus on maritime infrastructure. Currently, sea transport is more expensive than land transport as the country's sea transport is yet to be developed substantially. This also explains why - despite being the world's largest archipelago and, as such, having large quantities of waters and seas at its disposal - Indonesia's seafood business is far from being a flourishing business.

In this difficult context, Indonesian President Joko Widodo seeks a new approach to achieve much-needed breakthroughs to boost infrastructure development. For example, the government budget for infrastructure development has risen sharply since Widodo entered the office (see the table below). Something which is partly made possible by his decision to scrap a large chunk of energy subsidies. Furthermore, Widodo appointed a number of state-owned companies as developers of key infrastructure projects. The state-owned companies usually have bigger assets compared to privately-held companies and are also able to raise additional funds from (state-owned) banks more easily. There has also been an increase in capital injections from the state budget into several key state-owned construction firms. Another new tactic is to organize tenders in the year before the infrastructure project is expected to see groundbreaking (indonesia-investments.com).

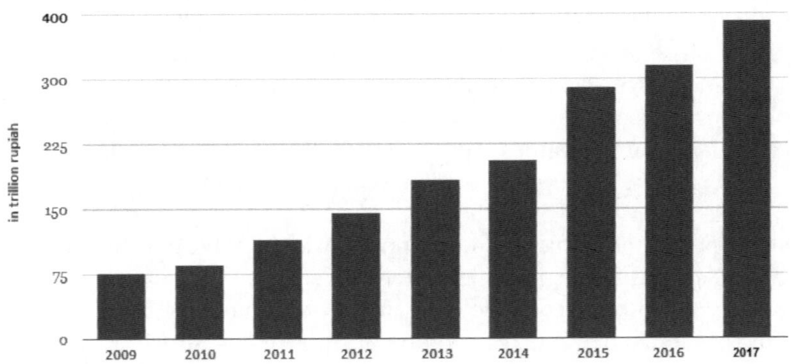

Figure 2. Funds Allocated to Infrastructure Spending in the

The Jokowi administration has made progress on a number of other fronts. At first glance, this progress looks to be quite recent. In mid-2015 there was much pessimism about the government's push on infrastructure. The largest government procurer of infrastructure, the Ministry of Public Works and Housing, had disbursed only 28% of its budget by the end of August (well below the 44% level achieved by the same point in 2014), mainly because of delays in finalising senior positions and institutional structures under the new administration (*Jakarta Post*, 3 Sept. 2015). The stalled infrastructure program had also contributed to weaker domestic prices for cement, while shares of the main cement producers, including Semen Indonesia and Holcim Indonesia, plunged by more than 55% between January and August. Construction stocks saw a similar sell-off. As noted in the previous section, the government's 2016 budget, announced in October 2015, including an increase in infrastructure allocations, from Rp 290 trillion to Rp 313 trillion. Importantly, the government committed to continuing its agenda of fiscal reform. It plans to boost capital spending from 2.2% of GDP to 5.3% of GDP by 2019, by shifting budget allocations away from subsidies and towards more productive spending (such as on infrastructure). The government also hopes to raise subnational infrastructure spending, by way of substantial increases in the DAK and in village funds. Questions remain, however, about the absorptive capability of subnational governments, and villages, to disburse these funds (World Bank 2015).

IOP Conf. Series: Earth and Environmental Science **156** (2018) 012048 doi:10.1088/1755-1315/156/1/012048

Public-sector infrastructure disbursements, after a slow start, accelerated sharply in the later months of the year. This is always the case, but the fourthquarter acceleration was much higher in 2015 than in previous years.6 Capital expenditure by line ministries reached Rp 209 trillion by the end of 2015, up from Rp 147 trillion in 2014. The Ministry of Public Works and Housing, which provided just under 50% of realised total government capital expenditure in 2015, managed to disburse almost 95% of its budget by year-end, a record for the ministry, which is also leading the 2016 spending program by having begun the procurement process in October 2015, three months before the start of the fiscal year [16].

Progress can also be seen in a number of projects that had been stuck in the planning or preparation stages for some time, in some cases for decades. The Makassar–Parepare segment of the Trans-Sulawesi railway has recently been completed, for example, while preparations in other segments, including land clearing, feasibility plans, and environmental impact assessments, are progressing quickly (*Media Indonesia*, 6 Jan. 2016). Likewise, viability-gap funding has been approved for the long-awaited Umbulan Springs project, which will supply water to Surabaya and surrounding areas (*Jakarta Post*, 8 Dec. 2015), and the winning bidder of the project was announced on 25 January. Also in this period, a $5.5 billion high-speed-train project to connect Jakarta to Bandung was awarded to a joint Indonesian–Chinese state-owned consortium. Construction commenced in late January, although the manner in which the project was awarded, and subsequently advanced, has invited controversy (box 1). The fourth quarter of 2015 also saw the completion of the first stage (Container Terminal 1) of the New Priok Port. This new development, designed to provide much-needed additional capacity at the country's premier port, in Tanjung Priok, North Jakarta, has progressed at an impressive rate since construction commenced in late 2012. Trial operations were successfully carried out in January 2016, and the new terminal is scheduled to be fully operational before the end of July [16]

The figure below shows that infrastructure development was built by government reaching across the regions. From the National Strategic Projects, the Government has selected a list of 30 Priority Project to be the focus of infrastructure delivery in the beginning of 2016. These projects are managed directly by the Committee for Acceleration of Priority Infrastructure Delivery (KPPIP).

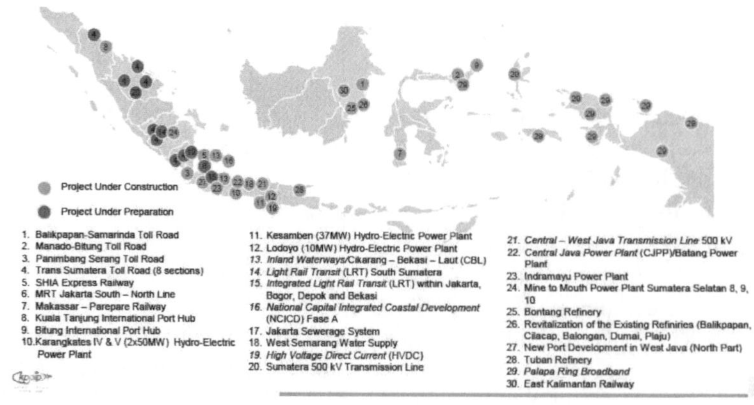

Figure 2. The Development of Infrastructure in the Beginning of 2016

4. Conclusion

Massive infrastructure development in Indonesia is indeed the best perfomance of Jokowi administration. However, the implementation requires close supervision. Jokowi has a unique approach to escort the development. In addition to the existing institutional supervision, with the 'Blusukan' method, Jokowi wants to get accurate, fast and direct information from the people. This activity not only wants to bring the state closer to the people but also at the same time wants to control the implementation of infrastructure development that is and will be run.

Understanding the relationship between the state and the people in the context of responsiveness through infrastructure development will certainly be more interesting when the Blusukan method is put into the frame of development program implementation. Implementation of democracy such as "for the people" is evident in the government's efforts to meet the needs of the people (responsive) through pro people policy making. A blusukan method as a typical method of Jokowi's administration can encourage implementation of democratic responsiveness in a more controlled manner because it can bridge and simultaneously bring the state to the people. No more local elite cheating is usually done to cover the weakness of development in the region. Jokowi can come so suddenly that control of development can be an effective shock therapy for other areas.

References

[1] Easton, David. 1957. "An Approach to the Analysis of Political Systems", *World Politics IX*, No. 3, April.
[2] Ghoshal, Baladas. 2004. "Democratic Transition and Political Development in Post-Soeharto Indonesia". *Contemporary Southeast Asia.* December: 26, 3; Social Science Database pg. 506
[3] Gloppen, Siri, Lise Rakner and Arne Tostensen. 2003. "Responsiveness to the concerns of the poor and accountability to the commitment to poverty reduction". *https://www.cmi.no/publications/1559-responsiveness-to-the-concerns-of-the-poor-and.* Downloaded 10 June 2017.
[4] Heiduk, Felix. 2014. "State Disintegration and Power Politics in post-Suharto Indonesia". *Third World Quarterly*. Vol. 35, No. 2, 300–315.
[5] *http://presidenri.go.id/blusukan.*1.html. Downloaded 1 June 2017
[6] *http://www.economist.com/news/asia/21594263-jakartas-governor-favourite-be-indonesias-next-president-though-he-not-candidate-no. No ordinary Jokowi: Jakarta's governor is favourite to be Indonesia's next president, though he is not a candidate.* 21 January 2013. Downloaded 1 June 2017-
[7] *https://en.tempo.co/read/news/2013/07/31/057501194/Jokowi-Laughs-at-Officials-Definition-of-Blusukan.* Dowloaded 1 June 2017. Jokowi Laughs at Official's Definition of 'Blusukan', 31 July 2013.
[8] *https://www.indonesia-investments.com/business/risks/infrastructure/item381*?. Infrastructure Development in Indonesia. Downloaded 25 May 2017.
[9] *Jakarta Globe*, 8 Dec. 2015
[10] *Jakarta Post*, 3 Sept. 2015
[11] *Jakarta Post*, 4 Jan. 2016
[12] *Jakarta Post*, 8 Dec. 2015
[13] *Jakarta Post*, 8 Dec. 2015
[14] Manning, Chris. 2015. "Jokowi Takes His First Shot at Economic Reform". *East Asia Forum*, 13 September. *http://www.eastasiaforum.org/2015/09/13/jokowi-takes-hisfirst-shot-at-economic-reform/.* Downloaded 15 May 2017.
[15] *Media Indonesia*, 6 Jan. 2016
[16] Ray, David & Lili Yan Ing. 2016. "Addressing Indonesia's Infrastructure Deficit". *Bulletin of Indonesian Economic Studies*, 52:1, 1-25, DOI: 10.1080/00074918.2016.1162266. To link to this article: http://dx.doi.org/10.1080/00074918.2016.1162266. Downloaded 5 June 2017.
[17] Reuters.com, 12 Feb. 2016

[18] Suyatno. 2011. *Dinamik Demokrasi Tempatan di Indonesia: Kajian Demokrasi Tempatan di Kota Yogyakarta dan Kabupaten Jembrana, Bali (The Dynamic of Local Democracy in Indonesia: A Study of Local Democracy in Yogyakarta City and Jembrana District, Bali).* Ph.D. Thesis. Penang: Universiti Sains Malaysia (USM).

[19] Wahyu Utomo and Rainier Haryanto. "Infrastructure Developments in Indonesia –Key Improvements and Priority Projects: Roundtable Discussion on Infrastructure Investment Opportunity in Indonesia". London, 15th July 2016. *http://www.ukabc.org.uk/wp-content/uploads/2016/08/160711_BKPM-London_Infrastructure-Investment-Opportunity-in-Indonesia_v11_Final-003.pdf.* Downloaded 5 June 2017.

[20] World Bank. 2015. *Indonesia Economic Quarterly: Reforming Amid Uncertainty*. December. Jakarta: World Bank.

[21] Government's Annual State Budgets *https://www.indonesia-investments.com/business/risks/infrastructure/item381?*. Downloaded 5 Jun 2017

[22] Wahyu Utomo and Rainier Haryanto. "Infrastructure Developments in Indonesia –Key Improvements and Priority Projects: Roundtable Discussion on Infrastructure Investment Opportunity in Indonesia". London, 15th July 2016. *http://www.ukabc.org.uk/wp-content/uploads/2016/08/160711_BKPM-London_Infrastructure-Investment-Opportunity-in-Indonesia_v11_Final-003.pdf* Downloaded 5 June 2017

IOP Conf. Series: Earth and Environmental Science **156** (2018) 012049 doi:10.1088/1755-1315/156/1/012049

Developing *Computer-Assisted Instruction* Multimedia For Educational Technology Course of Coastal Area Students

Husni Idris[1], Nurhayati Nurhayati[2] & Satriani Satriani[3]

[1]Faculty of Islamic Education Learning and Teaching, IAIN Manado, Sulawesi Utara, Indonesia.
E-mail *:* husniidris@iain-manado.ac.id
[2]Faculty of Islamic Education Learning and Teaching, IAIN Manado, Sulawesi Utara, Indonesia.
Email : nnurhayati@iain-manado.ac.id
[3]Faculty of Islamic Education Learning and Teaching, IAIN Manado, Sulawesi Utara, Indonesia.
Email : ssatriani@iain-manado.ac.id

Abstract. This research aims to a) identify instructional software (interactive multimedia CDs) by developing Computer-Assisted Instruction (CAI) multimedia that is eligible to be used in the instruction of the Educational Technology course; b) analysis the role of instructional software (interactive multimedia CDs) on the Educational Technology course through the development of Computer-Assisted Instruction (CAI) multimedia to improve the quality of education and instructional activities. This is Research and Development (R&D). It employed the descriptive procedural model of development, which outlines the steps to be taken to develop a product, which is instructional multimedia. The number of subjects of the research trial or respondents for each stage was 20 people. To maintain development quality, an expert in materials outside the materials under study, an expert in materials who is also a Educational Technology lecturer, a small groupof 3 students, a medium-sized group of 10 students, and 20 students to participate in the field testing took part in this research. Then, data collection instruments were developed in two stages, namely: a) developing the instruments; and b) trying out instruments. Data on students' responses were collected using questionnaires and analyzed using descriptive statistics with percentage and categorization techniques. Based on data analysis results, it is revealed that the Computer-Assisted Instruction (CAI) multimedia developed and tried out among students during the preliminary field testing falls into the "Good" category, with the aspects of instruction, materials, and media falling into the "Good" category. Subsequently, results of the main field testing among students also suggest that it falls into the "Good" category, with the aspects of instruction, materials, and media falling into the "Good" category. Similarly, results of the operational field testing among students also suggest that it falls into the "Good" category. Thus, it can be concluded that quality of the Computer-Assisted Instruction (CAI) multimedia developed in this research falls into the "Good" category viewed from the aspects of instruction, materials, and media. In other words, overall, the quality of this multimedia belongs to the "Good" category.

1. Introduction

Nowadays, there are two challenges instructional activities are dealing with, the first one comes from a shift in the perception about the word *learning* itself and the second one comes from the existence of information and telecommunications technology that show tremendous growth. Basically, the first challenge has been answered by the theory of Constructivism, which redefine learning as a constructive process in which information is transformed into knowledge through the

IOP Conf. Series: Earth and Environmental Science **156** (2018) 012049 doi:10.1088/1755-1315/156/1/012049

process of interpretation, correspondence, representation, and elaboration. Meanwhile, the rapid advancement of information and telecommunications technology to facilitate instructional activities might lead to a shift in the learning orientation from *outside-guided* to *self-guided* and from *knowledge-as-possession* to *knowledge-as-construction*. Moreover, this technology also plays an important role in redefining the concept of justification that initially focuses on learning as merely a presentation of various knowledge into learning as a form of guidance in order to be able to carry out the socio-cultural exploration which is rich in knowledge.

In the education and training sphere, an instructional process tends to be interpreted as the process of delivering information or communication. In this case, the instructional media is an integral part of an educational institution. Utilization of instructional media is a creative and systematic effort to create experiences that can make students learn so that in the end, educational institutions will be able to produce quality graduates. There is one thing to be taken into account, i.e. the use media, especially the Computer-Assisted Instruction (CAI), among students.

Some researchers and lecturers reported improvements in students' learning quality in the instruction that employed Computer-Assisted Instruction (CAI), namely: students were better prepared during classroom instruction (Cameron, 18-21), wrote down their papers more effectively, did better in the exams (Garham & Kaleta, 8(6)), and had deeper understanding and more meaningful discussion related to the course taught (King, KP, 2002: 231-246). Research conducted by Thomson Learning shows tasks tend to be completed faster and more accurately in the blended-learning environment (Martyn, M, 2003: 18-23).

So far, there are two problems related to the use of instructional media at IAIN (State Institute of Islamic Studies) Manado, namely availability and utilization. The availability of media is so limited that the utilization of media among lecturers during instruction is minimal. They usually use media in the form of printouts (lecture notes, modules, handouts, textbooks, magazines, newspapers, and so on) as the instructional media, coupled with simple equipment such as black/white boards and chalks/markers. As for audio-visual media such as videos/films, audio cassettes, and TV/Radio broadcasts, and electronic media (computer, LCD, and the internet), they have not been used intensively. The second problem concerns the use of media. Lecturers often use media in the form of printouts because such media can be easily developed and obtained from various sources. However, the majority of such media depend heavily on verbal symbols (words) which are abstract, thus requiring that students have a very high level of abstraction skills and this can be demanding. This is why the use of such media requires that lecturers be creative, in addition to careful thought with regard to instructional activities. In fact, many lecturers do not carefully select the instructional media they use by taking into account instructional consideration, they simply take pot luck. There are also those lecturers who use sophisticated media even though they do not need them.

Another equally important problem is lack of interest among lecturers to use instructional software (interactive multimedia CDs) to deliver course materials, even though at IAIN Manado, there have been facilities that are quite representative such as a computer laboratory which is quite conducive consisting of 20 units of PCs (Personal Computers) located at a separate building from the lecture building. Moreover, there is also a multimedia-based language laboratory which has been equipped with up-to date-software, making it have a dual function, i.e. to be used as a computer laboratory as well.

Consequently, it is necessary to develop computer-assisted instruction multimedia, especially to facilitate the instruction of the Educational Technology course as one of the learning sources and it is expected provide a stimulus to computer-assisted instruction of other courses. Moreover, lecturers have to be able to adopt this computer-assisted instruction as well.

Educational Technology as one of the courses students majoring in Islamic Education at the Faculty of Tarbiyah have to take is expected to facilitate the development of instructional media which constitute an integral part of learning in universities. So, in the near future, all courses taught at IAIN Manado in general and at the Faculty of Tarbiyah in particular can be delivered by adopting Computer-Assisted Instruction (CAI).

Based on the background above, the researcher believes that the time has come for IAIN Manado, which has just transformed from a STAIN (the State School of Islamic Studies) into an IAIN (the State Institute of Islamic Studies) and as the only state Islamic education institution in Manado to make the most of its facilities given rapid development of ICT, both in daily life and in the

IOP Conf. Series: Earth and Environmental Science **156** (2018) 012049 doi:10.1088/1755-1315/156/1/012049

educational sphere, thus requiring that the academicians of IAIN Manado be familiar with the use of technology to deal with rapid advances in ICT. Based on the foregoing, this research is entitled "Developing *Computer-Assisted Instruction* (Cai) Multimedia for the Educational Technology Course of Students Majoring in *Tarbiyah* at IAIN Manado".

2. Method

This is Research and Development (R&D). Research and Development (R&D) is product-oriented research. It employed the descriptive procedural model of development, which outlines the steps to be taken to develop a product, which is instructional multimedia. There were three basic stages developers had to followed, namely conceptualizing problems, developing a product, and trying out the product developed.

This research adopted the model proposed by Borg & Gall (Borg, W.R., & Gall, M.D, 1983: 775) based on a number of considerations: 1) This model has a clear prescriptive theoretical foundation because it sets optimal learning methods. this is consistent with the objectives of this activity to develop Computer-Assisted Instruction; 2) the development steps proposed in this model are detailed yet simple and flexible; and 3) this model can be used as a way to solve problems relating to instructional activities operationally.

Furthermore, subjects of the research trial or respondents for each stage was 20 people. To maintain development quality, an expert in materials outside the materials under study, an expert in materials who is also a Educational Technology lecturer, a small groupof 3 students, a medium-sized group of 10 students, and 20 students to participate in the field testing took part in this research.

The data collection instruments were developed in two stages, namely: a) developing the instruments, where at this stage, research instruments both questionnaires and tests were developed by the researchers. The questionnaires were developed based on the guidelines developed mostly based on instruments of the previous research into multimedia development, with necessary modification; and b) trying out instruments, where at this stage, the instruments that had been drawn up were tested to gain feedback for betterment of the instruments in order that they could be used as valid research instruments. In this research, the criterion for determining instrument quality was validity, especially logical validity.

As for data collection methods and instruments, they are described as follows: 1) To collect data whether the concept proves true or not, the researchers held discussion and handed in the product developed along with the evaluation sheet to be reviewed by experts and asked for their feedback relating to the product developed; 2) To examine the depth of the materials viewed from the indicators of display quality and instructional quality, questionnaires using a Likert scale were used, in addition to having discussion with a small group of responders and observation for the field testing; 3) Data on the effectiveness and efficiency of the product were collected by looking at students' work and time they spent during the field testing; 4) To examine the aspect of attractiveness, the researchers directly observed the students and from the observation results; 5) Students' improved learning outcome was seen by comparing results of the test and those of the previous one.

Techniques used to determine criteria for the score gained by the quality of the product developed were:

1. Data obtained from questionnaires on students' responses were converted first into interval data as follows:

Very Good	= 5	(100% consistent with the aspect contained on the questions)
Good	= 4	(80% consistent with the aspect contained on the questions)
Fairly Good	= 3	(60% consistent with the aspect contained on the questions)
Poor	= 2	(40% consistent with the aspect contained on the questions)
Very Poor	= 1	(20% consistent with the aspect contained on the questions)

2. Then, the obtained score was converted into a grade using a 5-point scale according to the table cited from Sukardjo (2005: 55) below:

IOP Conf. Series: Earth and Environmental Science **156** (2018) 012049 doi:10.1088/1755-1315/156/1/012049

Table 1. Score-to-Grade Conversion using a 5-Point Scale

Score Interval	**Grade**	**Category**
$\mathbf{X} > \overline{X}_i + 1.80$ SBi	A	Very Good
$\overline{X}_i + 0.60$ SBi $< \mathbf{X} \leq \overline{X}_i + 1.80$ SBi	B	Good
$\overline{X}_i - 0.60$ SBi $< \mathbf{X} \leq \overline{X}_i + 0.60$ SBi	C	Fairly Good
$\overline{X}_i - 1.80$ SBi $< \mathbf{X} \leq \overline{X}_i - 0.60$ SBi	D	Poor
$\mathbf{X} \leq \overline{X}_i - 1.80$ SBi	E	Very Poor

Description:

$\overline{X}_i$ = Ideal mean = ½ (maximum ideal score + minimum ideal score)

SBi= Ideal standard deviation = 1/6 (maximum ideal score – minimum ideal score)

X = Actual score

Based on the conversion formula in Table 1 above, clear description on how to convert quantitative data into qualitative data is obtained. Guidelines for converting quantitative data into qualitative data is presented in the following table:

Table 2. Pedoman Pengubahan Data Kuantitatif Menjadi Data Kualitatif

Score Interval	**Grade**	**Category**
$\mathbf{X} > 4.21$	A	Very Good
$3.40 < \mathbf{X} \leq 4.21$	B	Good
$2.60 < \mathbf{X} \leq 3.40$	C	Fairly Good
$1.79 < \mathbf{X} \leq 2.60$	D	Poor
$\mathbf{X} \leq 1.79$	E	Very Poor

Description:

Maximum score = 5 $\quad \overline{X}_i$ = ½ (5 + 1) = 3

Minimum score = 1 $\quad$ SBi = 1/6 (5-1) = 0.67

X = Actual score

3. To determine the mean of the scores obtained by the product developed, the following formula was used:

$$\overline{X} = \frac{\sum X}{n}$$

Description:

$\overline{X}$ = Mean

$\sum X$ = Total Score

N = Total Number of Respondents

Based on the conversion formula presented in Table 2 above, it is clearly described how to convert data on the percentage of learning mastery into qualitative data.

3. Results of Research and Discussion

A. Data from Field Testing and Suggestions to Revise the Product

1. Data on Students' Responses in the Preliminary Field Testing

There were a total of 3 respondents in the preliminary field testing, allof whom were students. Each student gave their responses separately. The purpose of the preliminary field testing was to determine eligibility of the initial quality of the product developed. At this stage, the evaluation focused more on the process factor rather than the factor of learning outcomes. The comparison of the means obtained in the Preliminary Field Testing can be seen in the following bar chart:

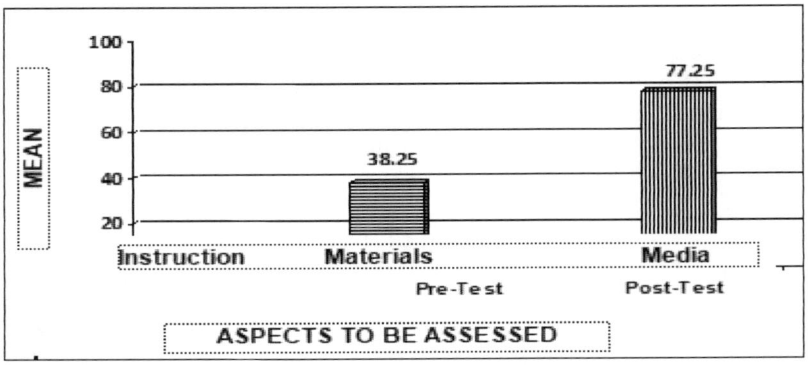

Figure 1. The Bar Chart Illustrating Students' Responses in the Preliminary Field Testing

In general, suggestions from the students to revise the Computer-Assisted Instruction (CAI) Multimedia product that had been developed in the preliminary field testing are: a) that the materials were too long; b) that the background should be made more attractive.

2. Data on Students' Responses in the Main Field Testing

In the main field testing, the respondents consisted of 10 students. Each student gave their responses separately. The purpose of the main field testing was the same as that of the preliminary field testing, what made them different was the number of research objects. In the preliminary field testing, the number of the objects under study was 3 students, while in the main field testing, the number of the research objects was 10 students. The comparison of the means obtained in the Main Field Testing can be seen in the following bar chart:

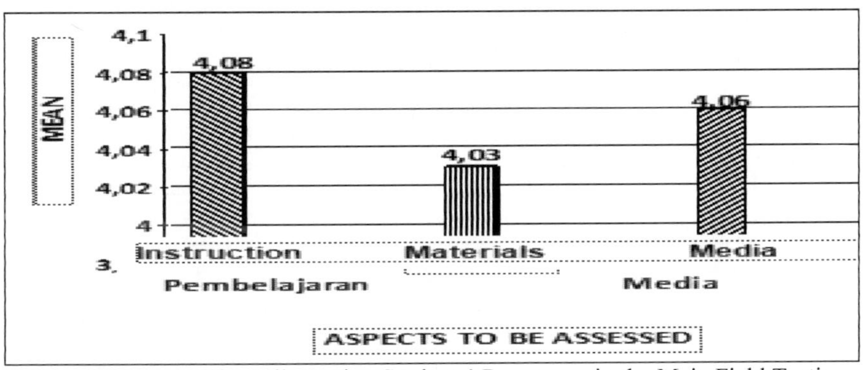

Figure 2 The Bar Chart Illustrating Students' Responses in the Main Field Testing

3. Data on Students' Responses in the Operational Field Testing
   a) Data on Quality of the Multimedia Product
      1) The Instruction Aspect

The percentages of the students' responses to the computer-assisted instruction multimedia viewed from the instruction aspect are presented in the following figure:

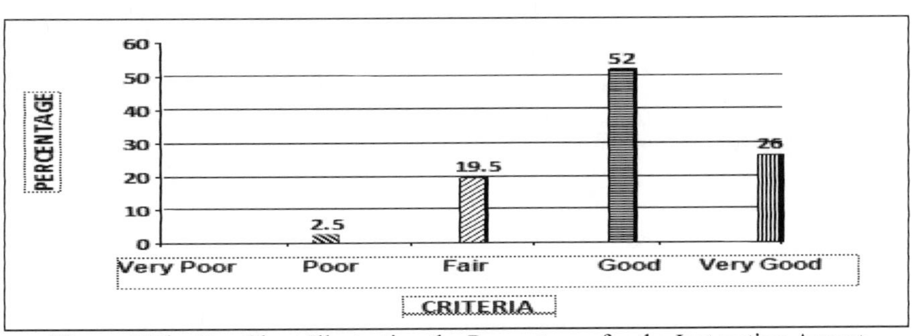

Figure 3 The Bar Chart Illustrating the Percentages for the Instruction Aspect

2) The Materials Aspect

The percentages of the students' responses to the computer-assisted instruction multimedia viewed from the materials aspect are presented in the following figure:

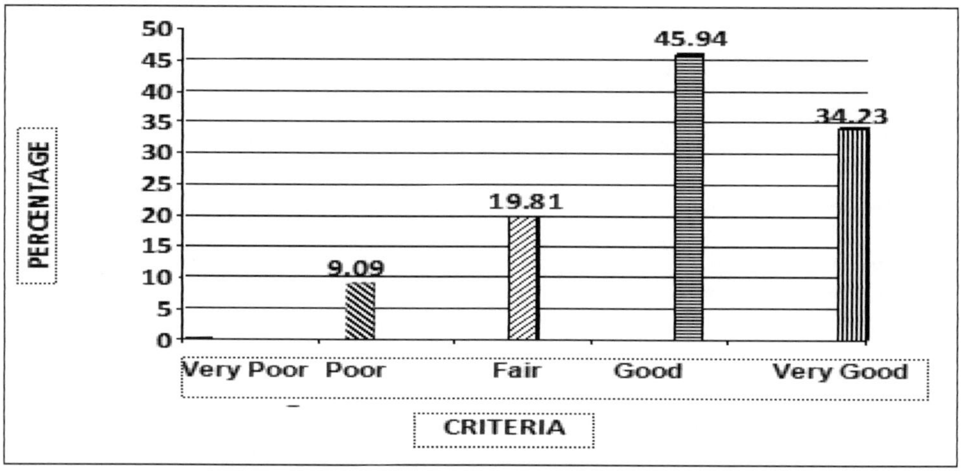

Figure 4 The Bar Chart Illustrating the Percentages for the Materials Aspect in the Operational Field

3) The Media Aspect

The percentages of the students' responses to the computer-assisted instruction multimedia viewed from the media aspect are presented in the following figure:

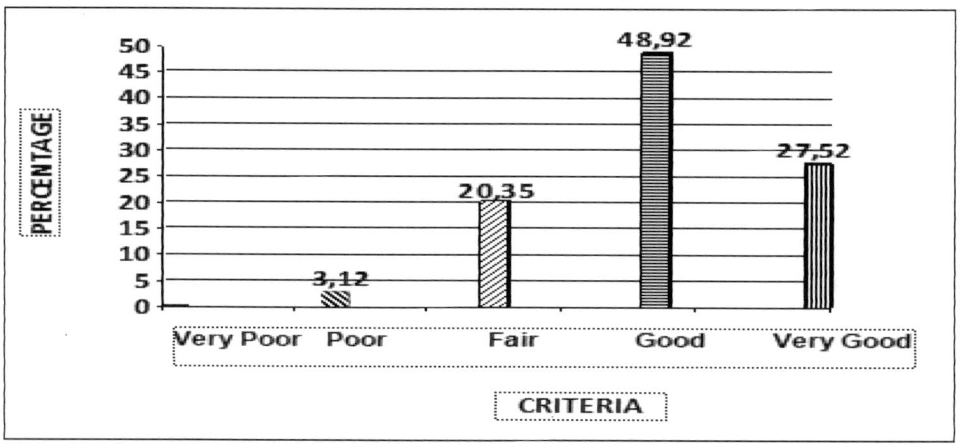

Figure 5 The Bar Chart Illustrating the Percentages for the Media Aspect in the Operational Field Testing

b) Data on the Motivation and Benefit Aspects

The field testing also generated data on the aspects of motivation and benefits perceived by the students with regard to the use of multimedia in the instructional activities of the Educational Technology coursee. The data are presented as follows:

Table 3. Percentages of Students' Responses

No.	Indicator	Yes	No
1	Multimedia helps me understand materials about Educational Technology.	20	0
2	For me, the concept presented is easy to understand.	17	3
3	The materials about Educational Technology presented challenge me to study better.	18	2
4	I want to learn about many things using this instructional multimedia.	20	0
5	This instructional multimedia gives me an opportunity to study at my own pace.	20	0
6	This instructional multimedia helps me study Educational Technology.	20	0
7	Answering the evaluation section helps me know whether I understand the materials about Educational Technology or not.	20	0
8	The materials about Educational Technology presented using multimedia allow me to gain understanding more quickly.	17	3
9	I love studying using multimedia.	20	0
10	I hope that the instructional acivities of the other courses use multimedia as well.	18	2
11	Without assistance from the lecturer, I will be confused and have no confidence.	5	15
12	The work sheet given helps me to study more actively and easily.	17	3
13	The materials are presented attractively using various media (texts, images/tables, audio, animation, and videos).	20	0
14	I have problems understanding some of the instructions.	3	17
15	I join the instructional activities continuously.	18	2

Based on the table above, in general, it can be concluded that there are a number of benefits students can get from studying using multimedia. This helps increase students' enthusiasm to study the materials presented. In conclusion, this multimedia positively affects the learning process.

c) *Pre-test* dan *Post-test*

The number of respondents in the Operational Field Testing was 20 people. A pre-test was held previously with a total of 20 questions, then a post-test was conducted afterwards with the same number of questions. The purpose of Operational Field Testing was to determine whether the use of the product that had been developed has a positive effect on the expected learning outcome. The comparison of the mean between the Pre-test and the Post-Test in the Operational Field Testing can be seen in the following bar chart:

Figure 6. The Bar Chart Illustrating the Comparison of the Mean Between the Pre-test and the Post-test in the Operational Field Testing

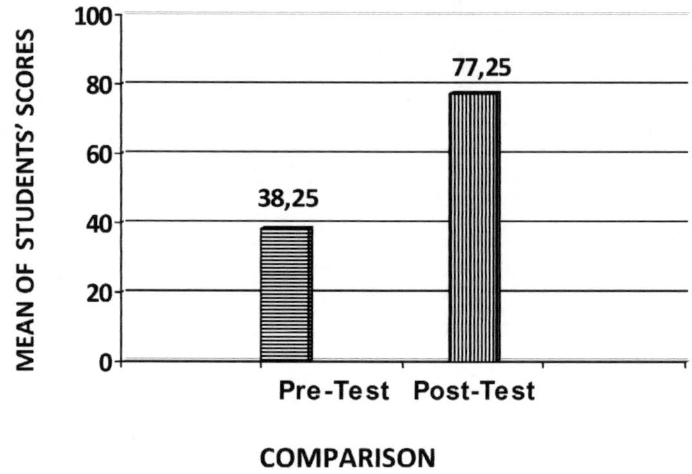

B. Final Product Analysis and Dissemination

After following a number of testing stages, both the limited feasibility test from an expert in materials and an expert in media, and field testing that included preliminary field testing, main field testing, and operational field testing, the computer-based multimedia product resulting from the development is deemed eligible to be a final product that can be disseminated and implemented among users. This eligibility is evident from the overall assessment results of almost all stages where the product gets a grade "B" and thus, falls into the "Good" category. This is consistent with the researchers' expectation who expect that in terms of eligibility, the product developed by the researchers should at least get a grade "C" and falls into the "Fairly Good" category.

In addition to its eligibility to be used, the computer-based multimedia product presented in the form of interactive multimedia CDs has a number of advantages and disadvantages as well. The first advantage is that the making of these interactive multimedia CDs using program *Macromedia Flash 8* proves much more practical and offers other advantages such as more attractive display, easy to operate, and so on and thus, students do not necessarily have to master this program in advance. The second one is that these CDs will run automatically once it has been inserted into the CD ROM on the computer because these interactive multimedia CDs have been published in the form of an "exe" file and adopts the "autorun" system. This computer-based multimedia also constitutes a major contribution as so far no Computer-Assisted Instruction (CAI) multimedia has been developed previously for the Educational Technology course, especially at IAIN Manado, and therefore both lecturers and students expect that this interactive multimedia that has been developed can be disseminated immediately in the form of a CD.

The use of these interactive multimedia CDs also has a significantly positive effect on students' learning outcome, especially for materials relating to Educational Technology as proved in the main field testing, where the percentage of students reaching the level of learning mastery after the implementation of the computer-based multimedia developed is equal to 72.25%.

In addition to having several advantages, this Computer-Assisted Instruction (CAI) multimedia also has several disadvantages, namely: (1) Program *Macromedia Flash 8* used to create interactive multimedia CDs requires numerous supporting files thus taking time; (2) The making of these interactive multimedia CDs needs meticulousness as the program *Macromedia Flash 8* is very prone to error if the developer does not carefully create these interactive multimedia CDs; and (3) The selection of the type of music fails to accommodate the whole students in terms of their taste in music because they usually have different learning styles and music preferences.

IOP Conf. Series: Earth and Environmental Science **156** (2018) 012049 doi:10.1088/1755-1315/156/1/012049

4. Conclusions

Based on findings of the research and development of this multimedia product, the following conclusions can be drawn: Findings of the research into the development of this Computer-Assisted Instruction (CAI) multimedia syggest that: in the preliminary field testing, the developed Computer-Assisted Instruction (CAI) multimedia tried out to 3 students falls into the "Good" category, with the aspects of instruction, materials, and media falling into the "Good" category. Subsequently, results of the main field testing where the product developed was tried out to 10 students also suggest that it falls into the "Good" category, with the aspects of instruction, materials, and media falling into the "Good" category. Similarly, results of the operational field testing where the product developed was tried out to 20 students also suggest that it falls into the "Good" category. The quality of the Computer-Assisted Instruction (CAI) multimedia developed in this research falls into the "Good" category viewed from the aspects of instruction, materials, and media. Overall, the quality of this multimedia belongs to the "Good" category.

5. References

[1] Ade Koesnandar, 2004, *Unsur-Unsur Pokok Dalam Penilaian Kualitas Program Multimedia.* (Powerpoint). Jakarta: Pustekkom.

[2] Borg, W.R., & Gall, M.D, 1983, *Educational Research: An introduction (4th ed.).* New York: Longman.

[3] Bower, G. H. & Hilgard, E. R, 1986, *Theories of learning (5th ed.),* New Delhi: Prentice Hall of India Private Limited.

[4] Cameron, B. The Effectiveness Of Simulation In A Hybrid And Onlinemetworking Course. *Techtrends*, 47(5), 18-21

[5] Garham & Kaleta, Introduction to hybrid courses, *Teaching with Technology Today,* 8(6)

[6] Kennedy, G., Petrovic, T., & Keppell, M, 1998, *The Development Of Multimedia Evaluation Criteria And A Program Of Evaluation For Computer Aided Learning. ASCILITE'98.*

[7] King, KP. *Identifyng Succes In Online Teacher Education And Development Professional. Internet And Higher Education.* 5, 231-246 2002

[8] Kumar, M. *A Critical Discourse In Multimedia Design: A Pedagogical Perspective To Creating Engaging Online Courseware.* Singapore: Nanyang Technological University. Diambil pada tanggal 10 Mei 2007, dari http://www.usq.edu.au/electpub/e-jist/docs/Vol7_no2/FullPapers/CriticalDisc_MM.htm.

[9] Martyn, M. The hybrid online Model: Good Practice. *Educause* Quarterly, 1, 18-23.2003

[10] Walker, D.F., & Hess, R.D, 1984, *Instructional Software: Principles And Perspectives For Design And Use.* Belmont: Wadsworth Publishing Company.

Improving the reading comprehension through cognitive reading strategies in language class of coastal area in indonesia

Abdul Gafur Marzuki[1] , Nur Alim[2], Ismail Suardi Wekke[3]

[1]English Education Department, Palu State Institute for Islamic Studies (IAIN) Palu, Diponegoro Street 94221, Palu city, Central Sulawesi, Indonesia
Email : gbudiperwira@gmail.com
3 Institut Agama Islam Negeri (IAIN) Kendari, Southeast Sulawesi. Indonesia
E-mail: nalim@iainkendari.ac.id

[2]Sekolah Tinggi Agama Islam Negeri (STAIN) Sorong, West Papua. Indonesia

E-mail: ismail@stain-sorong.ac.id

Abstract. The aim of the study is to solve reading comprehension problems of students in an EFL class through cognitive reading strategies. The study employed Classroom Action Research design in a reading class at Islamic Education Department at an Islamic university in Indonesia. The study consisted of two cycles. Each cycle consisted of planning, implementation, observation and reflection. The criteria of success were that ≥75% of the students must gain reading comprehension score of ≥75 and the number of students who participated actively during the reading class must achieve ≥75%. The results of cycle 1 revealed that 56.7% or less than 75% of the students gained score and 63.3% or less than 75% of the students who participated actively; consequently the revised plan for cycle 2 was implemented. The results of cycle 2 revealed that 83.3% or more than 75% of the students gained score ≥75 and 86.7% or more than 75% of the students who participated actively during the reading class and therefore the cycle was stopped.

1. Introduction

Reading is an important skill in learning a language. The requirement of being knowledgeable in this global era can be fulfilled by reading. As stated by Wallace (1992), reading serves the wider role of extending our general knowledge of the world. This s also supported by Hood et al. (2005), the ability to read well in English will influence learning potential in all other areas. This can be fulfilled by the use of learning strategies which can also be applied in reading.

There are two kinds of learning strategies which can also be applied in reading, i.e. direct strategies and indirect strategies [17]. Direct strategies consist of memory strategies, cognitive strategies and compensation strategies. Meanwhile, indirect strategies are composed of metacognitive strategies, affective strategies and social strategies. Among these strategies, cognitive strategies and metacognitive strategies are the most popular strategies used in reading [20] [18] [6] [8].

Metacognitive strategies in reading already known material, paying attention, finding out about language learning, organizing, setting goals and objectives, identifying the purpose of a language task, planning for a language task, seeking practice opportunities, self-monitoring, and self-evaluating. Meanwhile, cognitive strategies in reading identified are repeating, getting the idea quickly (skimming and scanning), using resources for receiving and sending messages, reasoning deductively, analyzing expressions, analyzing contrastively, translating, transferring, taking notes, summarizing and highlighting [17].

The cognitive strategies mentioned above are also in line with those identified by O'Malley & Chamot which include resourcing, repetition, grouping, deduction, imagery, keyword method, elaboration, transfer, inferencing, note-taking, and summarizing. The use of cognitive strategies in reading can help someone to be a proficient reader [16]. This is supported by the theories of cognitive reading strategies and the results of the study conducted [6] [8][18][10].

Since the number of research on cognitive reading strategies is still limited in Indonesia and a study on cognitive reading strategies to solve the students' problems in reading comprehension is not yet investigated in the previous studies, this study attempts to fill the gap. The previous studies become parts of theoretical foundations the researcher uses in this study which is focused on employing cognitive reading strategies to improve the reading comprehension in an EFL class in Indonesia.

The reading comprehension ability of the students at the Islamic Education Department of the Faculty of Tarbiyah and Teacher Science at Palu State Institute for Islamic Studies needs to be improved. It is observed that the achievement of the students in the reading classes has not been satisfying. Many of them still find difficulty to identify the topic and the main idea of the passage, to find the meaning of the unknown words, to identify the tone of the passage and the author's purpose, and to understand what implies in the text or inferring from the text. It is also noted that the students are not active in the process of teaching and learning in the classroom.

2. Method

This research employed classroom action research (CAR). The steps of the CAR; planning, implementation, observation and reflection were implemented. The research was conducted in two cycles where each cycles consisted of four meetings. The setting of the research was at the Islamic Education Department Faculty of Tarbiyah and Teacher Science, Palu State Institute for Islamic Studies in Indonesia. The subjects were 30 students who enrolled in the English 1 class 2016/2017 academic year. The data were collected by scoring individual reading comprehension and students' active participation in each cycle. Field notes and observation were also conducted. The field notes were based on the classroom situation and interaction and the observation was based on the way individual students speak and respond during the class. The criteria of success were used in this study, namely: ≥75% of the students must gain reading comprehension score of ≥75 and the number of students who participated actively during the reading class must achieve ≥75%.

3. Results

3.1 Cycle 1

Cycle 1 consisted of four meetings. In the planning phase, the lesson plan was prepared based on the syllabus and the previously identified problems. The identified problems are; difficulty to identify the topic and the main idea of the passage, difficulty to find the meaning of the unknown words, difficulty to identify the tone of the passage and the author's purpose, and difficulty to understand what implies in the text or inferring from the text. The instructional objectives were used as the basic of the selection of the possible instructional material which related on the Islamic topic. The students were arranged in pairs. Each pair was assigned to discuss their problems in understanding the content of the text during their English study period. The researcher prepared a reading text which related to Islamic topic then distributed to each pair, there was one text for each student to be discussed.

It was decided that the scoring system was based on the individual score gained by the students in answering the questions based on the reading text correctly. In the implementation phase, the three phase technique, pre-reading, while-reading, and post-reading was implemented. During the pre-reading, the purpose of the study and the instructional objectives were introduced. The students were divided in pairs so there were fifteen pairs. The placement of the students in pairs was done based on their tablemate. Then each pair was requested to read and discuss their reading text. In this phase, the students used dictionary to find the meaning of important and unknown words such as the title and the first line of the text. In the while-reading phase, the students were demanded to employ skimming-scanning, guessing the meaning of a word from the context, considering other sentences to figure out the meaning of a sentence, and repetition strategy. Skimming involves searching for the main ideas the speaker wants to get across, while scanning means searching for specific details of interest to the learner, and repetition is the strategy of reading a passage more than once to understand it more completely. Then the students were asked to answer the questions based on the text. Next the students' pairs were requested to present their answer from the questions based on the text in front of the class. There were four pairs in meeting 1-3 and three pairs in meeting 4 because there were fifteen pairs from the total number. After pairs' presentation, the others pairs were given opportunity to comment, argue, or even ask questions to the presenter. This interaction was observed by the researcher, individual active participation was check-listed in the column that has been previously prepared and notes were jotted down based on the classroom situation during the four meetings implementation phase. In the post-reading phase, the students were asked to employ summarizing strategy. Summarizing is making a written summary of new information gained through reading. It enables the students to grasp the original text better. The class was ended up by highlighting how to employ cognitive reading strategies. The achievement of the set criteria in cycle 1 is graphed in figure 1.

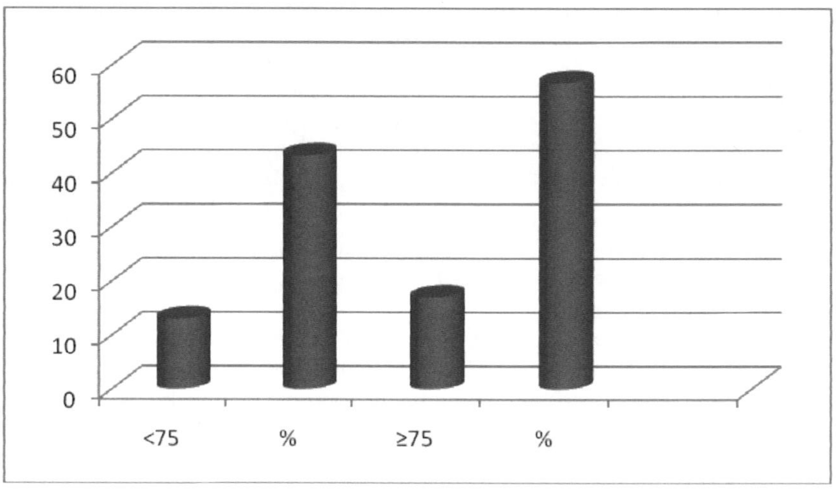

Figure 1. The results of cycle 1

The figure indicates that there are still quite a few students who have not passed the set criteria. The results are also presented in the percentage table as can be described in Table 1.

	<75	%	>75	%
Cycle 1	13	43.3	17	56.7

Table 1. Results of cycle 1

The data indicate that there are still 13 or 43.3% of the subjects who have not passed the set criteria.

Since the criteria of success were planned based on the individual gained score and active participation, the results of the observation on the students' active participation are graphed in figure 2.

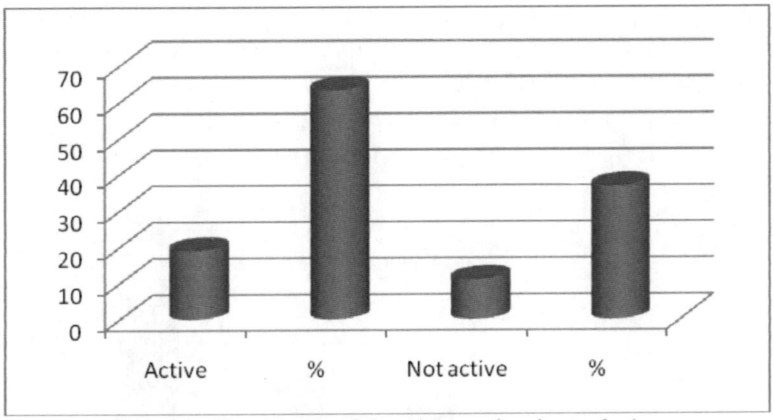

Figure 2. The results of the observation in cycle 1

The figure indicates that there are many students who are not active during the discussion and meetings in cycle 1. The number and the percentage of the students who are not active are presented in table 2.

	Active	%	Not Active	%
Cycle 1	19	63.3	11	36.7

Table 2. The results of the observation

In the reflection phase, it is concluded that the criteria of success has not been achieved and that the revised plan must be prepared. Consequently the research has to continue to cycle 2. The results of the reflection indicate that there must be improvement in the process of the teaching and learning procedures.

3.2 Cycle 2

Cycle 2 consisted of four meetings. The pairs in cycle 2 were the same students as the pairs in cycle 1. In the planning phase, the revised plan, the way of topic given for each pair was revised. Each pair freely chose and decided one topic of reading text based on their interest. The reading texts were provided by the researcher from many sources. There were five reading texts provided by the researcher in this cycle. Before starting the teaching and learning process, every member of the pair was assigned to read the chosen topic individually and then read and discuss the reading text in pair. The implementation phase in cycle 2 was the same as the one in cycle 1; pre-while-post reading. However, the employment of cognitive reading strategies in pre and while-reading were revised. In the pre-reading phase, the students were guided and requested to employ inferencing strategy. It is represented by the strategy of using the title to predict the content of the text and paying attention to words or phrases that show how text is organized. In while-reading phase, the students were demanded to employ inferencing, elaboration, and note-taking strategy. Inferencing strategy was used in this phase was guessing the meaning of a word from the context and skipping the unknown words. And elaboration strategy used in this phase was thinking about previous knowledge on the topic of the text or associating to the students' background knowledge, while note-taking is writing down key words and concepts in abbreviated verbal, graphic, or numerical form while reading. It makes students active participants in their learning, helps them organize important concepts, remember information, and becomes one of their study aids. In post-reading phase, the students were guided and demanded to summarize the main ideas of the paragraphs and the researcher focusing ended up the class by highlighting how to employ inferencing, elaboration, and note-taking strategy which are parts of cognitive reading strategies. The results of cycle 2 are graphed in figure 3.

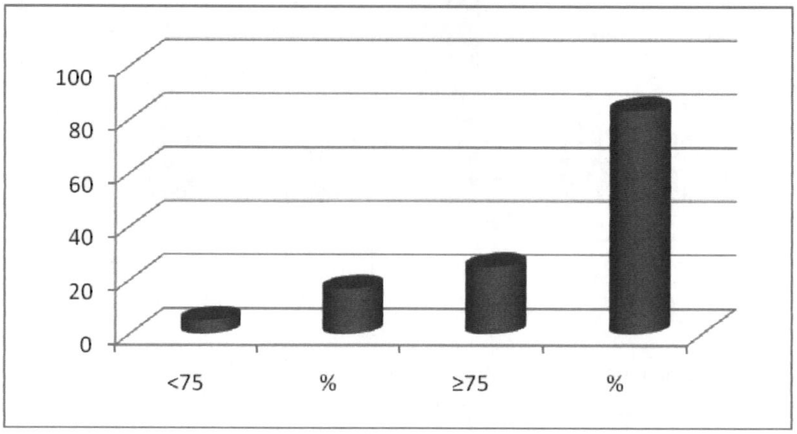

Figure 3. The results of cycle 2

The data in figure 3 indicate that only few students who get lower than the set criteria and the ones who achieved the set criteria increased. The number and the percentage of the students who have achieved the set criterion are presented in table 3.

	<75	%	≥75	%
Cycle 2	5	16.7	25	83.3

Table 3. Results of cycle 2

The data in table 3 indicate that 25 out of 30 or 83.3% have achieved the criteria of success and only 5 out of 30 or 16.7% who have not passed the set criteria. The active participation of the students in cycle two was also observed. The data of the active participation are graphed in figure 4.

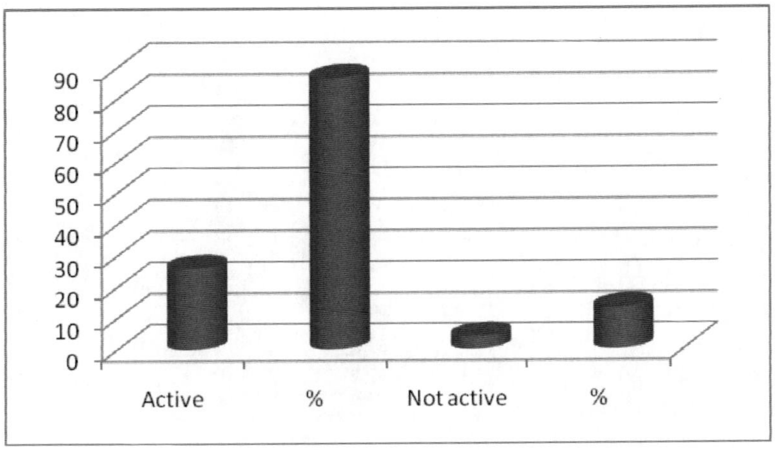

Figure 4. Participation of the students in cycle 2

The data in figure 4 indicate that not all of the students are active. There are still 4 students are not active. It can be seen also in Table 4 that 26 or 86.7% of the students are active and 4 or 13.3% of the students are not active.

	Active	%	Not Active	%
Cycle 2	26	86.7	4	13.3

Table 4. Active participation

The data in table 4 indicate that not all of the students participate actively during the classroom meetings in cycle 2 however; the students' active participation has reached 86.7% or more than ≥75. Since the criteria of success have been achieved, cycle was stopped.

In the reflection of cycle 2, it is noted that the students' achievement increase from cycle 1 to cycle 2. The improvement is graphed in figure 5.

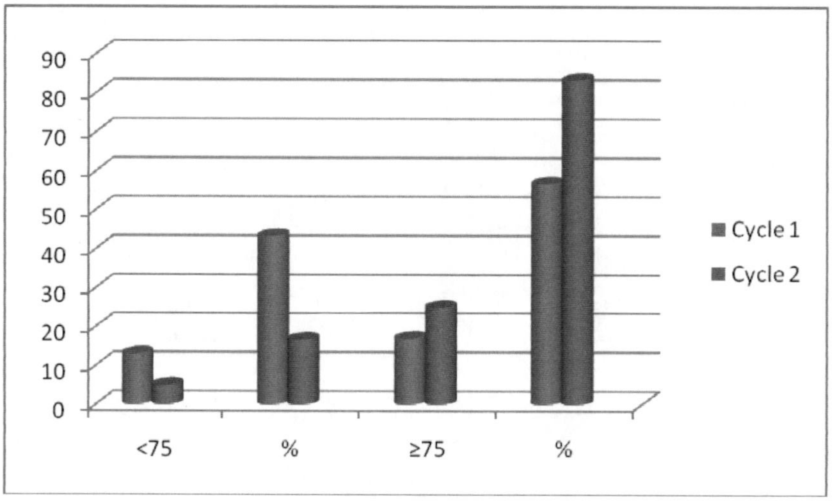

Figure 5. The comparison of the achievement of cycle 1 and 2

The data in figure 5 indicate that the students who passed the criteria increase significantly from cycle 1 to cycle 2. The same case applies to the improvement in the student's active participation during the class meeting. The improvement in the student's active participation between cycle 1 and cycle 2 is graphed in figure 6.

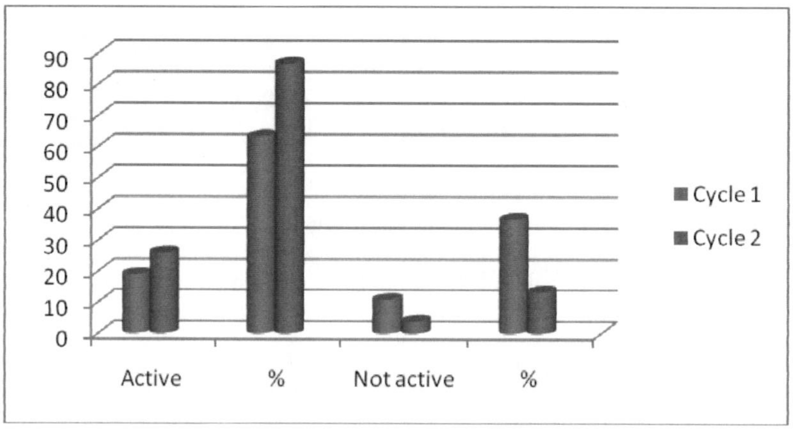

Figure 6. The comparison of the active participation in cycle 1 and cycle 2

The data in figure 6 indicate that most of the students who are not active in cycle 1 become active in the classroom meeting cycle 2. The improvement of the students' reading comprehension and the active participation during the classroom meeting is believed to be affected by the implementation of the revised plan.

4. Discussion

The research question has been answered during the implementation of cycle 1 and cycle 2. The reading comprehension of the students gets improved through the employment of cognitive reading strategy in the reading class in two cycles. There are at least two potential factors that have been implemented during the cycles to overcome those problems, the first is the way to choose instructional materials (reading text) for each pair and the second is the instructional procedures which employed cognitive reading strategies.

4.1 The reading text selection

Referring back to the previous identified problems in the reading class difficulty to identify the topic and the main idea of the passage, difficulty to find the meaning of the unknown words, difficulty to identify the tone of the passage and the author's purpose, and difficulty to understand what implies in the text or inferring from the text, It is noted that the way of selecting the topic of the reading text in cycle 1 holds important role. The change of the way of selecting topic of the reading text in cycle 1 and in cycle 2 provides the students with opportunities to choose topic of the reading text in which they are interested in. The opportunity to freely choose topics of the reading text given opens up wide possibility for the students to choose topic of the reading text which is contextual with their prior knowledge and based on their interest. This technique seems potential to create efficient and joyful learning to increase the students active participation in reading class and to improve their reading comprehension ability in the following ways; Firstly, the topic of the reading text that the students is interested in motivates them to read and discuss with others; Secondly, the students should have got previous knowledge and experiences related to the topics and consequently there have been at least stock of vocabularies related to the topics of the reading text; Third, when the topic is contextual and related to their prior knowledge, the students get opportunities to hear or even to listen to the topics in their daily encounter and hence help them broaden their understanding of the topics.

4.2 The instructional procedures

The second potential factor that has contributed to the way to solve the problems in reading comprehension and the improvement of the reading comprehension ability is the instructional procedures. The instructional procedures help solve the reading problems of the students in the following ways; firstly, the pre-reading phase, the students were guided and requested to use the title to predict the content of the text and paying attention to words or phrases that show how text is organized. This strategy helped the students to comprehend the content of the text in general case and in a very limited time. Secondly, in while-reading phase, the students were demanded to guess the meaning of a word from the context, skipping the unknown words, associating the text to the students' background knowledge, and writing down key words and concepts in abbreviated verbal, graphic, or numerical form while reading. It makes students active participants in their learning, helps them organize important concepts, remember information, and becomes one of their study aids. Third, in post-reading phase, the students were guided and demanded to summarize the main ideas of the paragraphs and researcher focusing ended up the class by highlighting to guide the students in employing cognitive reading strategies more effectively. Finally, the scoring system which was based on the reading test during the classroom could train the students in employing cognitive reading strategies more effective and efficiently. In short, the implemented teaching and learning procedures in this CAR (Classroom Action Research) proposes three main phase, pre-reading, while-reading, and post-reading. The steps can be briefly proposed as follows: 1) pre-reading: freely choose a topic of reading text for a pair, use the title to predict the content of the text and paying attention to words or phrases that show how text is organized; 2) while-reading: guess the meaning of a word from the context, skip the unknown words, associating the text to the students' background knowledge, and write down key words and concepts in abbreviated verbal, graphic, or numerical form while reading; 3) post-reading: summarizing the main idea of the text.

5. Conclusion

The implementation of cognitive reading strategies improves the reading comprehension ability of the students. The improvement in the reading comprehension is seen in the individual student in answering the questions based on the text correctly, the ability to use the title to predict the content of the text, to guess the meaning of a word from context, and to consider other sentences in the paragraph to figure out the meaning. There are two ways on how the improvement is achieved using the cognitive reading strategies; the first is the reading text selection provides the students with opportunities to choose topic of the reading text in which they are interested in. The opportunity to freely choose topics of the reading text given opens up wide possibility for the students to choose topic of the reading text which is contextual with their prior knowledge and based on their interest. This technique seems potential to create efficient and joyful learning to increase the students active participation in reading class and to improve their reading comprehension ability; the second is the instructional procedures of cognitive reading strategies which cover various strategies such as resourcing, repetition, deduction, skimming-scanning, elaboration, inferencing, note-taking, and summarizing help the students to understand the text and motivates students to actively participate in the classroom teaching and learning process. These procedures allow the students to actively participate in the classroom teaching and learning process such as asking questions and discussing in pairs. The implementation of the cognitive reading strategies in the instructional process allow the students to read the text individually and then in pairs prior to classroom teaching and learning. The findings imply that cognitive reading strategies potentially improve the reading comprehension of the students in the EFL class teaching at the university level.

6. References

[1] Adler, C.R. (2004). *Seven strategies to teach students text comprehension*. Retrieved September 6, 2016, from http://www.readingrockets.org/article/3479.

[2] Aebersold, J.A. & Field, L.M. (1997). *From reader to reading teacher: Issues and strategies for second language classrooms.* New York: Cambridge University Press.

[3] Bell & Woodard. (1998). *Reading: strategies for parents*. Cullowhee: The reading center. Retrieved September 6, 2016, from http://www.paec.org/david/reading/parents.pdf.

[4] Chorney, (2005). *Interactive reading, early modern texts and hypertext: A lesson from the past.* Retrieved September 4, 2016, from http://www.academiccommons.org/commons/essay/early-modern-texts-and-hypertext

[5] Frase, L. (2008). *Reading strategies good readers use*. Retrieved September 4, 2016, from http://ezinearticles.com/?Reading-Strategies-Good-Readers-Use&id=1824654.

[6] Ghonsooly, B. & Eghtesadee, A. (2006). *Role of cognitive style of field-dependence/independence in using metacognitive and cognitive reading strategies by a group of skilled and novice iranian students of english literature*. Retrieved September 4, 2016, from http://www.asian-efl-journal.com/Dec_06_bg&are.php.

[7] Goodwin, W. L. & Goodwin, L. D. (1996). *Understanding quantitative and qualitative research in early childhood education.* New York: Teachers College Press.

[8] Hamdan, A. R., Ghafar, M. N., Sihes, A. J. & Atan, S. B. (2010). The cognitive and metacognitive reading strategies of foundation course students in teacher education institute in malaysia, *European Journal of Social Sciences*, 13 (1).

[9] Hernandez, J.M., Rodriguez, M., & Bulnes, M.A. (2009). *Reading comprehension strategies: A case study in a bilingual high school*. Revista Electronica Matices en Lenguas Extranjeras. Retrieved 24 August, 2016, from www.revistamatices.unal.edu.co.

[10] Hood. (2005). *Focus on reading*. Sydney: NCELTR.

[11] Keshavarz, M.H., and Mobarra, M.K. (2003). *The effects of simplification and elaboration on reading comprehension of iranian EFL students*. JAL, 6(1), 171-184.

[12] McMillan, J.H. and Schumacher, S. (2001). *Research in education: A conceptual introduction* (5th Ed.). New York: Addison Wesley Longman.

[13] McNamara, D. S. (2007). *Reading comprehension strategies*. New Jersey: Lawrence Erlbaum Associates.

[14] Ming Xu, L. (2007). *A comparative study of reading strategies among college students.* Retrieved August 23, 2016, from http://www.celea.org.cn/pastversion/pdf/liumingxu.pdf.

[15] National Capital Language Resource Center (NCLRC). *The essentials of language teaching.* Retrieved August 23, 2016, from http://nclrc.org/essentials.

[16] O'Malley, J. and Chamot, A. (1990). *Learning strategies in second language acquisition.* New York: Cambridge University Press.

[17] Oxford, R.L. (1990). *Language learning strategies: What every teacher should know*. New York: Newbury House Publishers.
[18] Ozek, Y & Civelek, M. (2006). *A study on the use of cognitive reading strategies by ELT Students*. Retrieved September 2, 2016, from http://www.asian-efl-journal.com.http://www.asian-efl-journal.com/PTA_August_06_ozec%26civelek.pdf
[19] Pressley, M. (2000). *A focus on reading comprehension strategy*. Retrieved September 12, 2016, from http://www.readingonline.org/articles/handbook/pressley/index.html.
[20] Richard, J.C. (1997). *From reader to reading teacher: issues and strategies for SL classroom*. New York: Cambridge Language Education.
[21] Spielvogel, J. (2005'). *Journey across time-the early ages*. New York: McGraw Companies.
[22] Wallace, C. (1992). *Reading*. Oxford: Oxford University Press.

IOP Conf. Series: Earth and Environmental Science **156** (2018) 012051 doi:10.1088/1755-1315/156/1/012051

Local Wisdom of Sasak Islamic to Enhance Unity and Togetherness Among Communities on the Lombok Island

I Wayan Wirata[1]

[1]Sekolah Tinggi Agama Hindu Gede Pudja Mataram, Jl. Pancaka No.7b, Mataram Barat, Selaparang, Kota Mataram, Nusa Tenggara Bar. 83126, Indonesia
Email: iwayanwiratastah@yahoo.co.id

Abstract: The Islamic society of Sasak Lombok as part of the Muslims archipelago has had a tradition that is maintained by the community. The cultural tradition is a recorder of social solidity in improving the integrity of the nation to increase the sense of brotherhood and kinship as part of human moral and character. The strength of the local culture is a culture of the nation's power. Besides, the cultural forces have the potential to destroy the existence of violence, radicalism, and terrorism. This can be achieved by increasing the solidity and harmony, to create comfort and safety which emphasises the values of brotherhood and togetherness. Violence, terrorism, and radicalism have been worrying and causing fear, anxiety, and disrupt the stability of comfort. Violence and terrorism in the name of religion as part of the nature of arrogance and radicalism lead to social tensions that could disrupt public order and public convenience Lombok to be a safe community, peace, harmony, and peace. Thus it is necessary to anticipate the collective culture disharmony and instability of Lombok. The local knowledge from Lombok Sasak Muslims that can enhance human values, pluralism, and togetherness involves: 1) The celebration together *mauludan*; 2) local cultural performances; 3) The meeting between religious leaders and community leaders; and 4) implementation of mutual assistance; and 5) Attend and witness traditional ceremonies.

1. Introduction

In a free nation, local knowledge is a capital of unity and togetherness which creates peace among religious groups in Indonesia. Thus, simultaneous and coordinated efforts are required by the various components (i.e. race, ethnicity, and nation) to achieve religious, harmonious, and peaceful society. The government of Indonesia has established an agency to maintain the conducive conditions through the National Counter Terrorism Agency (BNPT). The organisation mediates dialogues across components of society. It prevents the idea of separatism including radicalism, terrorism, and nonviolence to be happening in the community. The example is a dialogue called: "Community Involvement in Terrorism Prevention Through the Perspective of Social Culture" with the intention of creating a sense of religious, harmonious, and peaceful community. In the context of community empowerment especially in Lombok, cultural power maintains the harmony among the society. It has created integrative understanding, tolerance, harmony, and peace within society.

IOP Conf. Series: Earth and Environmental Science **156** (2018) 012051 doi:10.1088/1755-1315/156/1/012051

On the other hand, the power of local wisdom of Sasak Islamic society has a high potential to block violence, radicalism, and terrorism by increasing the solidity and harmony through the emphasize of fraternity, solidarity, and religious values. Violence, terrorism, and radicalism have been worrying and causing fear, anxiety, and disrupt the stability of the society. Violence and terrorism, especially in the name of religion, might lead to disharmonies for the communities in Lombok. With local wisdom, Lombok community appears to have had internal tolerance and inter-religious harmony necessary to facilitate unity and brotherhood among Muslims in Lombok.

The Islamic tradition of Lombok has been part of local wisdom combined with ancestral traditions. The local wisdom is on the basis of mutual respect, family values and a sense of brotherhood in the Islamic civilisation of Sasak Lombok. These values are upheld by not eliminating the traditional value of brotherhood *(pasemetonan).* It indicates that the Islamic community of Sasak Lombok has had intensive and conducive communication media to the creation of the nuances of prosperous, harmonious, and peaceful Islamic society.

2. Findings and Discussion

The local knowledge of Sasak Islamic in enhancing unity and togetherness among the people on the island of Lombok is as the following:

2.1 Joint Celebration *Mauludan*

Increasing the brotherhood and kinship among the Sasak Islamic community of Lombok is done through a joint celebration called mauludan ceremony. The ceremony improves the solidity and unity *a*mong the local Islamic community in Lombok. It becomes the media for social gathering and communication for Muslims with the family and other communities in society. It facilitates respect between the Islamic family and their relatives as they built join *mauludan* celebration together. This event is conducted by the family to increase a sense of family and brotherhood. This is similar to the explanation of Koentjaraningrat that activities pattern social system and structure [1]. Besides, social interaction occurs for mutual appreciation and respect among people praying together in the rituals. The pray is addressed to create a balance and awareness of the universe and the environment. Beside seeking guidance and directions from *Allah* (God Almighty) and direction to his people so that, there is a balance outwardly and inwardly. Sanderson explains that the social structure always refers to the actual patterns of behaviour in the domain of hierarchical social structure [2]. Eating together is conducted after finishing the prayer together with family and friends along with other invitees. This is done jointly to get closer to family, so as to create the sense of belonging and respect, in addition to the exchange of experiences and exchange of information about the condition of the family, in addition to the development and advancement of the family.

2.2 Performance of Local Culture

Sasak Islamic Society of Lombok has established interaction with various components in both the internal and external environment on the island of Lombok. With right and continues communicationd, a sense of family and togetherness are created on the island of Lombok, describing in the ceremony of War *Topat* (*Perang Topat*) and Pujawali Pura Lingsar (*Lingsar Ceremony*). During the ceremony, the Islamic community of Sasak (Islam) performs the ritual together to establish kindship and brotherhood among the people on Lombok island. The event is an ancestral cultural heritage which needs to be preserved to create a sense of togetherness and brotherhood among the people on the island of Lombok. It grows a sense of belonging, respect, and mutual respect among ethnics and religious communities. Local Government also supports the event to improve adhesiveness and favourable spiritual climate in the territory of the Lombok island.

IOP Conf. Series: Earth and Environmental Science **156** (2018) 012051 doi:10.1088/1755-1315/156/1/012051

2.3 Meetings between Religious Leaders and Community Leaders

Religious and community leaders are public figures to provide role models and examples of the social values and religion in society Lombok. Religious and community leaders are respected in the culture of Lombok. As for the various religious and community leaders meeting has been carried out by Muslims and the surrounding communities, especially communities Sasak (Islam) to create unity and togetherness on the island of Lombok. Efforts undertaken in mediating the integrity and brotherhood in the region of the island of Lombok in the meeting is mutually to convey social conditions of the issues and their actual phenomenon. This is done to look for strategic measures for the realisation of Lombok advanced society, religious, and cultural. The same thing is confirmed by Mulyana who explains that through conversations with others, we understand the messages and ourselves better [3]. Edkins and Willliams describe that everyone can understand each other through a process of dialogue [4]. During the meeting sought the synergy of understanding and views on the importance of fraternity to create a sense of tolerance as well as the anticipation of violence and radicalism that leads to divisions and social tensions Lombok Island. Thus, it results harmonious, serene, and peaceful society.

2.4 Implementation of the Mutual Cooperation

Cooperation is the most critical part of Muslims, especially for Muslims Sasak of Lombok. This attitude as part alleviate, in addition to helping provide power for activities carried out can be organised and carried out well. Kembarawan explains that attitude is part of the sub-culture learned through the process of socialization [5]. Cooperation reflects mutual affection, mutual trust, and mutual understanding between the people of Lombok. This is in line with the statement of May who explains that attitude cannot stand alone but always related to something or object; or in other words no attitude without the object [6]. Attitudes have a significant influence in determining the behavior of a person or group of people. In addition, it was confirmed by Sarnoff that the attitude as willingness to react positively or negatively to certain objects [7]. Attitude has never happened in vain, but formed in social interaction and is associated with specific objects. With the attitude of cooperation creates unity and togetherness with high cohesiveness towards harmonious, peaceful, and prosperous Islamic society.

2.5 Attending and Witness Ceremony

A religious ceremony is a ritual process undertaken by community religious community adherents. The Islamic community in improving the fraternity and concern for other people is by inviting other people to visit (relationship) in large-scale activities such as ceremonies Lebaran (Eid Haji) and other ceremonies. This is in line with the statement Carey who explains that in a ritual activity does everyone and fellowship come together [8]. Ritual activities happen to respect religious values that have been embedded from ancestors. For example, when fasting ceremony *Eid al-Adha*, the Hindus man (*Pecalang*) may join and help to create conducive climate between religious communities in the island of Lombok. With the cooperation conducive integrative way to create more harmonious feeling of familiarity among religions in the island of Lombok. Also, during the ceremony of Eid, and *mauludan* Islamic society helped create temporary gazebo (*tetaring /terob*) to increase the sense of awareness and tolerance among the religious communities in the island of Lombok. It is maintained in order to increase the sense of brotherhood and emotional cohesion as a reflection of culture harmonious, safe, serene, and peaceful. Couldry explains that the ritual associated with the patterned behaviour voluntarily by the community through hereditary (by habit) concerning the behaviour pattern [9].

3. Conclusion

The Sasak Islamic Society of Lombok as part of the Muslims archipelago has had a particular tradition maintained by the community. It is to preserve social solidity to improve the integrity of the nation and to increase the sense of brotherhood and kinship as part of human moral and character. The local wisdom of Lombok appears to enhance human values, pluralism and community togetherness Lombok through: 1) The celebration together *mauludan*, 2) Performing the local culture, 3) The meeting between religious leaders and community leaders, 4) Implementation of mutual assistance, and 5) Attend and witness traditional ceremonies.

Reference

[1] Koenjaraningrat 2003 *Pengantar Antropologi*. Jakarta, Rineke Cipta.
[2] Sanderson S K 2003 *Makro Sosiologi* (Terjemahan Farid Wajidi S Menno). Jakarta, *PT Raja Grafindo Persada*.
[3] Mulyana D 2013 *Metodologi Penelitian Kualitatif: Paradigma Baru Ilmu Komunikasi dan Ilmu Sosial Lainnnya.* Bandung, PT Remaja Rosdakarya
[4] Edkins J and William N V 2013 Teori-Teori Kritis; Menantang Pandangan Ulama Studi Politik Internasional, Yogyakarta, Pustaka Pelajar.
[5] Kembarawan IGK 2014 *Wacana Simplikasi Ritual Pada Masyarakat Hindu di Lombok Nusa Tenggara Barat (Hasil Penelitian).* Mataram, *Sekolah Tinggi Agama Hindu Negeri Gde Pudja.*
[6] May A 2006 Persepsi dan Sikap Tokoh Agama Terhadap Eksistensi Pariwisata Lombok (Kajian Perspektif Kajian Budaya). *Tesis Program Magister Kajian Budaya,* Denpasar, Universitas Udayana.
[7] Sarwono S W 2003 *Teori-Teori Psikologi Sosial.* Jakarta, PT Raja Grafindo Persada.
[8] Care J W 1992 *Communication as Culture: Essays on Media and Society.* New York, Routledge.
[9] Couldry N 2005 *Media Rtual Beyond Functionalism dalam Media Antropology.* SAGE Publications Thousand Oaks.

IOP Conf. Series: Earth and Environmental Science **156** (2018) 012052 doi:10.1088/1755-1315/156/1/012052

Community Empowerment Strategy Based on Social and Cultural Capital of Coastal Communities at Makassar Island

Tanzil Tanzil[1]
[1]Universitas Haluoleo Kendari, Kampus Hijau Bumi Tridharma, Anduonou, Kambu, Kendari, Kambu, Kota Kendari, Sulawesi Tenggara 93132, Indonesia
Email: tanzilsosio@gmail.com

Abstract: The study aims to (1) analyze the form and the function of social capital in fisher communities of Makasar Island (2) formulate a strategy of empowerment through reinforcement of social capital in fisher communities in Makasar Island. The methodological design used is a case study. The data is then analysed through descriptive-qualitative, an analysis carried out continuously from the collection of data until the investigation is completed. The study findings showed that the fisher communities in Makasar Island have social and cultural capital that can be identified on the forms of trust, values/norms, and networks in which each social and cultural capital has become a power for fishermen to survive and adapt to its environment. The results of this study also identify various problems related to the process of empowering fishing communities on the island of Makassar. The problems also become a constraint for fishermen to improve their business so that they obtain relatively low incomes. The problems are: firstly, the weakness on the use of technology so that the productivity of fishermen is relatively low and the small business scale, the second, the difficulty in business development because of the limited access to capital and the third, the poor of business management as the result of limited ability in business management. These factors then lead the fishing communities on the island of Makasar powerless to exploit the rich potential of marine resources.

1. Introduction

Nationally, as proposed by Kusnadi, the policy of Economic Empowerment of Coastal Communities (PEMP) undertaken since 2000 has not been able to improve the life of coastal communities [1]. According to him, the lack of pro-development policies by both national and local government to the coastal area and communities, increasingly making coastal communities are in a state of powerlessness. The same condition also occurs in fisher communities in Bau-Bau City, without exception to the fisher communities on the island of Makasar. This fact is in line with the statement of the Head of Department of Fisheries and Marine of Bau-Bau, Amiruddin that the existence of business in the field of fisheries in the region is still faced with the lack of supporting facilities, fisheries management is only supported with modest equipment and technology resulting in insufficient production to meet basic needs of the family.s

To overcome the problems faced by fisher communities on the island of Makasar, it seems very important to pay attention to the dimensions of social capital. Interest in social and cultural capital dimensions in alleviating poverty issues relating to the development paradigm changes is not only emphasized on the material aspect but also on the human aspect. This approach is done to see the extent to which these problems can be solved based on the potential and resources of the fishing communities themselves.

IOP Conf. Series: Earth and Environmental Science **156** (2018) 012052 doi:10.1088/1755-1315/156/1/012052

The study on social and cultural capital owned by fisher communities on the island of Makasar refers to the ideas of Coleman [2], Putnam [3], Fukuyama [4] where there are three important elements of social capital: trust, social norms, and social networks. The idea at the core of the theory of social capital is extremely simple: Social networks matter. Networks have value. As Putnam [3] describes that social networks and the associated norms of reciprocity as social capital, because like physical and human capital (tools and training), social networks create value, both individual and collective, and because we can investǁ in networking. Social networks are, however, not merely investment goods, for they often provide direct consumption value.

According to Piere Boudieu [5], social capital is a capital of social relationships which will provide, if necessary, useful supports': a capital of honourability and respectability which is often indispensable if one desires to attract clients in socially important positions, and which may serve as currency, for instance in a political career . Social capital is an attribute of an individual in a social context. One can acquire social capital through purposeful actions and can transform that capital into conventional economic gains. The ability to do so, however, depends on the nature of the social obligations, connections, and networks available to you. Otherwise, according to Fukuyama [4], an instantiated informal norm that promotes co-operation between two or more individuals. By this definition, trust, networks, civil society, and the like, which have been associated with social capital, are all epiphenomenal, arising as a result of social capital but not constituting social capital itself. While social capital has been given some different definitions, most of them refer to manifestations of social capital rather than to social capital itself.

Concerning that, according to Dasgupta Partha [6], Social capital is most usefully viewed as a system of interpersonal networks. There is no single object called social capital; there is a multitude of bits that together can be called social capital. Each bit reflects a set of interpersonal connections. Just as the productivity of manufactured or natural capital goods depends upon the use to which they are put, the worth of social capital depends upon the kinds of activities in which members of networks are engaged. This is why writings on social capital so frequently have been studied by institutions as it holds societies together to be recognized as a necessary functional social order, along with a certain degree of common cultural identifications, a sense of belonging and shared behavioural norms. This internal coherence helps to define social capital. Without it, society at large would collapse, and there could be no talk of economic growth, environmental sustainability or human well-being [6].

2. Method

Design methodology used is a case study, which attempts to analyze forms of social capital of *Pobhinci-bhinciki Kuli* principle practiced by fisher communities in their life and then developed to formulate a model of empowerment based on the social and cultural capital, which is a novelty in this study. To obtain reliable findings, this study uses a variety of data sources including field notes, interviews, conversations, photography, documents or records existing in the field. Each of these data sources is collected together to obtain data matched to the need (reliable). The data is then analyzed through descriptive-qualitative method; an analysis carried out continuously from the collection of data until the study is completed

3. Findings and Discussion

The results of the observation showed that fisher communities on the island of Makassar (ethnic Buton in general) still believe and practice a variety of values and norms in their life. The values and norms which still live in the island are known as the philosophy "*pobhinci-bhinciki kuli'*. The philosophy is outlined in four basic behaviours, i.e. first *pomae-maeka* (fear/shy around each other), second, *pomaa-maasiaka* (love each other), third, *popia-piara* (care each other), the fourth, *poangka-angkataka* (respect each other). In general, the value that guides the fisher communities in a social interaction is not only as cognitive structure (concept) but also as conduct which is implemented in the daily life of fisher communities.

As a consequence, at a macro level, fisher community on the island of Makassar is known as an open-minded community that easy to interact with other people from any region. At mezzo level, trusting each other among the members of the community is also strongly established which encourage the existing of social-economic institutions, so that if they need assistance among others, it is usually easily obtained. The trust raises social awareness, where people with sufficient economic resources will help needy neighbours and relatives. Usually, the assistance includes various household needs. The helped people would reciprocate the help with the readiness to provide any services according to their ability. It then forms patron-client relationships as occur in other fisher communities in Indonesia.

Moreover, these habits also become one of the teachings of their religion, namely Islam which teaches that men must establish a relationship with others because men are created in a weakened state. Islam teaches human beings to establish cooperation in goodness and piety and to create friendship relationship with others, as stated by one of the fishermen on the island "If anyone comes to us and mean well, we will respond with goodness and trust. We are used to establishing a relationship of friendship and that we have done since long time ago. Religion also teaches us to trust each other and creating friendship"

Thus, the trust, in the fishing community on the island of Makassar, rooted in the teachings of Islam and local culture. This phenomenon is consistent with the view of Nahapit and Ghosal in Hasbullah [7] that the source of trust is derived from professed religion, competence and openness of someone that has become the norm in society. It has encouraged the fisher communities to maintain his livelihood as fisherman even under challenging conditions. The position of powerlessness of fisher communities in Makassar Island is adapting ideas of Woolcock [8] as the implementation of the empowerment of fisher communities on the island tend to be partial and unsustainable. The role of business and society, in general, is not optimal and there is no synergy among the various parties involved in the empowerment of fisher communities, of course with this condition, it will be tough to realise the goal of empowerment.

The empowerment of fisher communities on the island of Makasar tends to be inconsistent in organising cycles of poverty reduction and seems to be just negating obligations. Empowerment approach with attention to the social and cultural capital as a people-centred development has not been realised in the implementation. The forms of social and economic organisation in fishing communities on the island of Makassar, internally, can be said not well developed. During this time the primary support of the fisheries industry is the institution of family and kinship. Cooperative as a very important institution in improving the welfare of fishermen encounters fluctuation and has not been able to support the fishers' business.

Figure 01. [The Koli-koli, The Traditional Makasar Island Fishing Boat]

Fishers on the island of Makasar are still categorised as peasant fisher namely traditional fisher in which the orientation to meet their own needs.

4. Conclusion

IOP Conf. Series: Earth and Environmental Science **156** (2018) 012052 doi:10.1088/1755-1315/156/1/012052

The empowerment efforts on fishing communities in Makassar Island need to be directed not only to address the cultural issues but also structural problems. As the approach of empowerment is still partial, mainly just related to funding and technical training, it is recommended that the empowerment needs the application of concept and multidimensional efforts to all aspects such as social, economic and culture. Therefore, it should use an approach called two-track approach, i.e. top-down approach through an active role in the decision-making process related to the policy of fisher communities and bottom-up approach by working to strengthen communities at the grassroots level by promoting the principle of involvement/participation of fisher communities.

References

[1] Kusnadi KSN 2006 Kemiskinan dan Perebutan Sumber Daya Alam. *Yogyakarta, LKiS*
[2] Voss T 2017 James S. Coleman: Foundations of Social Theory. In *Schlüsselwerke der Wirtschaftssoziologie*, Springer.
[3] Putnam RD 1995 Bowling alone: America's declining social capital. *J. Democr.* **6**(1) 65–78.
[4] Fukuyama F 1999 *Social Capital and Civil Society.* The Institute of Public Policy, George Mason University.
[5] Bourdieu P 1986 *The forms of capital Handbook of theory and research for the sociology of education.*
[6] Dasgupta P 2001 *Social capital and economic performance: Analytics.* Beijer international institute of ecological economics.
[7] Hasbullah J 2006 *Social Capital: menuju keunggulan budaya manusia Indonesia*. MR-United Press.
[8] Woolcock M 2002 Social Scientist, Development and Research. *Soc. Cap. Particip.*

IOP Conf. Series: Earth and Environmental Science **156** (2018) 012053 doi:10.1088/1755-1315/156/1/012053

Directive Speech Act of *Imamu* in *Katoba* Discourse of Muna Ethnic

Ardianto Ardianto[1] Hardiman Hadirman[2]

[1]Institut Agama Islam Negeri Manado, Jl. Manguni, Malendeng, Tikala, Kota Manado, Sulawesi Utara

Email: ardianto@iain-manado.ac.id

[2]Institut Agama Islam Negeri Manado, Jl. Manguni, Malendeng, Tikala, Kota Manado, Sulawesi Utara

Email: hhardiman@iain-manado.ac.id

Abstract: One of the traditions of Muna ethnic is *katoba* ritual. *Katoba* ritual is one tradition that values local knowledge maintained its existence for generations until today. *Katoba* ritual is a ritual to be Islamic person, repentance, and the formation of a child's character (male/female) who will enter adulthood (6-11 years) using directive speech. In *katoba* ritual, a child who is in-*katoba* introduced to the teaching of the Islamic religion, customs, manners to parents and his brother and behaviour towards others which is expected to be implemented in daily life. This study aims to describe and explain the directive speech acts of the *imamu* in the *katoba* discourse of Muna ethnic. This research uses a qualitative approach. Data are collected from a natural setting, namely *katoba* speech discourses. The data consist of two types, namely: (a) speech data, and (b) field note data. Data are analyzed using an interactive model with four stages: (1) data collection, (2) data reduction, (3) data display, and (4) conclusion and verification. The result shows, firstly, the form of directive speech acts includes declarative and imperative form; secondly, the function of directive speech acts includes functions of teaching, explaining, suggesting, and expecting; and thirdly, the strategy of directive speech acts includes both direct and indirect strategy. The results of this study could be implied in the development of character learning materials at schools. It also can be one of the contents of local content (mulok) at school.

1. Introduction

The directive act is one of the classifications of speech acts according to Searle. Taxonomy proposed alternative Searle about the classification of illocutionary acts, such as assertive, directive, commissive, expressive, and declarations [1]. The directive speech acts are intended for speakers to create the effect that hearer will do the instruction mentioned in the speech. This speech acts included in the competitive function, which competes with social goals.

The representation of this directive speech act can be found in a variety of discourse practice, which one of them is in *katoba* ritual discourse on the speech community of Muna ethnic. In the process of implementation, *katoba* ritual formed a variety of speech acts. One is the dominant speech acts is directive speech acts.

The word "*katoba*" consists of two morphemes are *ka-* and *toba* which means "tobat/repentance''. Thus, *katoba* implies "the particulars of repentance" or "prepent" [2]. *Katoba* derived from the *toba* which means repentance or conviction. It said that *katoba* was a party at the time of Muslim children to be at the age of about eleven or nearly reached adulthood. *Katoba* ritual is a ritual performed after the child was circumcised or uncircumcised [3]. *Katoba* derived from the word 'taubat'. The point is that children who have been circumcised or pledge led to pronounce the sentence of 'taubat/ repentance'. The children will be given the explaination about the repentance terms, after greeting accompanied by two sentences *syahadat.*

Also, *katoba* ritual is the ritual of "repentance" with Islamic culture nuance. *Katoba*, besides showing tradition, also showed characteristic traits of orality. Ritual of *katoba* is cultural products of Muna act as acculturation local values and Islamic values. Ritual of *katoba* disclose essential matters relating to character education within the family and religious education for children who grow up to live in the future. Speech about *katoba* has become a tradition and cultural practices that still exist in Muna society. The tradition of the *katoba ritual* has a profound meaning in moral essential in Muna ethnic who always maintained from generation to generation until today.

The purpose of the *katoba's* ritual is to teach the appeals to the children which are aged approximately 8-12 years to be good, moral and ethical children in treading life within family and society. This ritual is followed by children, *katoba's* ritual leaders, parents, and audience witnesses. *Katoba* ritual in the implementation process will also create a sense of togetherness and solidarity within the family. Ritual *katoba* regarded as Islamic ritual because in this ritual child would be named as a Muslim in a way guided *syahadat.*

The use of language is the reality of the communication interaction of Imamu and children who repentance that takes place in the ritual of *katoba. Imamu* must be able to communicate well to create a sacred, fun and meaningful atmosphere. That communication was realised as one of the directive speech acts. Directive speech is a speech that provokes the response of children to have better and depth understanding and interpretation the speech of *katoba* which full of moral teachings which are useful to shape the Islamic character and personality of the children from an early age.

In the activity of the conversation, the *imamu* utilize specific speech acts as an initiation to stimulate the child's response. *Imamu* affects the child for ritual execution process *katoba* to give direction and stimulus through speech acts to participate in the context of the conversation.

In the context of that conversation, the unit or units of speech acts imamu's expression appears in verbal communication (and nonverbal) in ritual *katoba*. One type of speech acts used by *imamu is* directive speech acts; namely speaker speech acts committed with the aim of producing the effect of an act committed speakers [4].

During interaction formed in *katoba* ritual, the *imamu* has the power and control of children who is being repentance. Children who repentance as partners in a ritual *imamu* said *katoba* has lower status so that children can be set within the limited. In this context, directive speech acts of *imamu* are used to deliver certain communications functions in *katoba* ritual. During directive speech acts based on local wisdom is speech or expressions sounds of language which means complete and contains instructions or advise that needs to be nurtured, taught, practised in everyday life to the life of the community that is safe, serene, peaceful, and prosperous.

Based on the understanding of Katoba discourse *above,* the overall is one of the expressions of local wisdom nuanced. This is because the directive speech acts expressed in katoba discourse *imamu* contains noble advice life. The directive speech acts in katoba discourse *imamu* belonging to local wisdom discourse.

The problem of this research is how a form of directive speech acts *imamu* in discourse *katoba*. The issue is directed at three main focuses, namely (1) the use of form, (2) function and (3) strategy directive speech acts *imamu* in discourse *katoba*.

2. Theory and Methods

Definition theories used in this research is the theory of discourse, pragmatics and speech act. Those three theories were used to explain the issues which examined in this study.

Discourse is the cultural construction of the building as the actualization of texts, and texts as well as a practice of the use of signs, both verbal and nonverbal [5]. Discourse can also be used to refer to particular contexts of language use, and in this sense, it becomes similar to concepts like genre or text type [6].

At the level of discourse meaning phenomenon is often considered to have sufficient understanding of the complex. This is not only due to the terms of discourse is the area of language, but also caused directly related to the practice of social signals and daily [5]. As area of language, Bloor and Bloor define discourse as sometimes used in contrast with 'text' where 'text' refers to actual written or spoken data, and 'discourse' refers to the whole act of communication involving production and comprehension, not necessarily entirely verbal [7]. The discourse is not only a way of speaking, but more importantly is directly related to the practice of speaking.

Discourse in this paper is a record of language intact or narrative verbal discourse of story or narrative that builds unity of meaning as a result of acts of communication between the characters involved in the ceremony who communicated by *imamu* (leader of ritual *katoba*) to participants (children repentance).

This study used a qualitative approach. Data were collected from natural background, i.e. katoba discourse speech. Researcher's acts as a key instrument prioritize process rather than results. Data of this study consists of two types: (a) utterances, and (b) field notes. Data speech contains the form, function and strategy of *imamu* directive speech acts. Data analysis was performed through several stages, namely (a) the classification of the data, (b) the data that has been grouped analyzed using interactive model [8] with four stages, namely: (1) data collection, (2) data reduction, (3) data display, and (4) conclusion and verification.

3. Results and Discussion

3.1. Form of Directive Speech Acts Imamu

From the results of research conducted, the directive speech acts in katoba discourse *imamu* embodied in speech mode declarative and imperative. As the shape of the directive, the speech with declarative and imperative mode can use the language and certain linguistic variation.

3.1.1. Forms of Directive Speech Act Imamu with Declarative Mode

In the discourse of *katoba*, the findings of this study indicated that the directive speech act of *imamu* was manifested in a declarative which consists of four categories, namely in modus of the statement of the willingness, statement of necessity, statement of prohibition, and the statement of allowances. Here is an example of speech that contains a directive speech acts in a declarative mode in the form of an expression of interest that represents the demand function that is characterized by the use of modalities to be explicit.

[1]	*Imamu:*	(a) Atumobakomu ini, (b) tamaka bharie-bharie sonipogaughooku omobhaloane uumbe (I will repentance you, but all I'm saying, you said yes)
	Children:	(c) Uumbe (Yes)

Citations [1] above is a form of directive speech acts in a declarative mode in the form of an expression of interest. At speech, it uses the marker insertion *imamu* -um- 'means to be' on the word *atumobakomu* (I will repentance you). In the speech, *imamu* informs the child that the *katoba child* to begin. Because of this information, followed by a desire of *imamu* for children to follow what he said and the child must answer *uumbe*. Apparently, *imamu* expect verbal reactions of children (b). Shaped child's responses verbal answer, as in speech (c).

3.1.2. Forms of Directive Speech of Imamu with Modus Imperative

In the discourse of *katoba*, using the imperative mode utterances are potentially used by *imamu*. In the katoba ritual *imamu* has much higher position than children. Such situation allows the *imamu* to use

IOP Conf. Series: Earth and Environmental Science **156** (2018) 012053 doi:10.1088/1755-1315/156/1/012053

speech with a direct imperative modus to express the orders, advice, or prohibition in the implementation of *katoba's* ritual process.

Data findings indicate that in the katoba discourse, directive speech acts manifested in the imperative mood *imamu* consists of four categories, namely the modus imperative command, imperative errands, imperative prohibition, and imperative advice. The use of the imperative errand and prohibitions as exemplified below.

[2]	*Imamu:*	*(a)Ane ameenako, mobhalo elae* (I asked, you answer yes)
	Children:	(b) *Uumbe* (Yes)
	Imamu:	*(c) Ane omobhalo fekaghosa-ghosa suaramu. (d) Peda dua nasebantara itu datumolako komobhaloghoo ohaeno, fokoumbe. (e) Natumolako ahae-ahae dofokoumbe mokesahano.* (If the answer should be rather loud voice. Similarly, a minute later, if called upon do not say? Answering yes. Called anyone should be answered yes, that's a good answer.
	Children:	*(f) Uumbe (Yes)*
	Imamu:	*(g) Aforatoangko mokesahano ini, (h) okabhalo miina naembali dobhalogho ohaeno...* (I'll tell you this good, answered question should not be answered "what" ...?)
	Children:	*(i) Umbee (Yes)*

Directive speech act on the data [2] (a) and (b) is realized with a command in imperative mode[6]. *Imamu* commands action punctuated with the phrase *mobhalo fekaghosa-ghosa* (replied half aloud). Speech (b) *imamu* ordered the children to say yes.

The use of ban directive imperatives that are commonly used by *imamu* as exemplified as follows.

[3]	*Imamu:*	*(a) Pedamu dua datumolako, kofoko aa..., panaembalia.* (Similarly, when called, do not answer aa ..., not to be)
	Children	*(b) Uumbe* (Yes)
	Imamu:	*(c) Pedamu dua natumolako inamu, koe mobhaloghooa ohaeno..gara ina, mobhaloghoo uumbe..* (Similarly, when you are called by mother, do not answer "what, mother? Please answer yes)
	Imamu:	*(d) Pedanagha elae*? (Like that)
	Children	*(e) Uumbe* (Yes)a

Directive speech act on the data [3] realized the imperative mode restrictions. *Imamu* commands action punctuated with the phrase *oohaeno*. Speech of (b) *imamu* prohibits a child not to answer the call (the parents) with *aa...* or *ohaeno*, because according to community of Muna ethnic such speech was an impertinent answer. Parents who hear answers such children will be hurt because his child did not appreciate them. Directive speech act on the data [4] realized the imperative mood prohibition *koe* "do not'. Speech (b) *imamu* forbids children to answer *ohaeno*. It shaped child's responses verbal form of answers as in speech (b).

Pronunciation sound *koe mobhaloghoo aa…* or *ohaeno* ... the imperative prohibition from flat to high pitch tone elongated, the phoneme /o/ lengthened and given the pressure on the phoneme /o/. *Ohaeno* sound is long spoken utterance forms unethical and prohibited use of a child to answer the call of parents as contained in the speech (b).

3.2. Function of Directives Speech Act of Imamu

The data of research's findings indicated that the function of *imamu's* directive speech act in discourse of *katoba* was represented by different modes of speech which are classified into six

IOP Conf. Series: Earth and Environmental Science **156** (2018) 012053 doi:10.1088/1755-1315/156/1/012053

categories, i.e. the functions of ordering, prohibiting, permitting, advicing, expecting, and inviting. Those six-directive function of speech act is commonly used by *imamu* in *katoba's* verbal interaction, both delivery with directly or indirectly speech. The directive functions which represented in various modes of speech are related to the activities of *imamu* to directing the children, directing the children to affirm what is delivered associated with the speech of *katoba*, encourages the children to remember and implement the counsels of *katoba*. The following quote was a function directive sent to syahadat in Muna and Arabic language associated with the coronation as true Muslims.

[4]	*Imamu:*	*(a) Angkafi kanau elae* (Follow what I say)
	Children:	*(b) Uumbe (Yes)*
	Imamu:	*(c) Asyhadu Allah ilaha Ilallah*
	Children:	*(d) Asyhadu Allah ilaha Ilallah*
	Imamu:	*(e) Wa asyhadu anna Muhammadar Rasulullah*
	Children:	*(f) wa asyhadu anna Muhammadar Rasulullah*

Directive speech manner function as indicated by the speech quote [4] is command, i.e. *imamu* through utterances that ordered the child to follow what he had told. Speech with the imperative mood subtly is done. Angkafi kanau use of the word 'follow me'. In *katoba* discourse, following his words is one of the strategies from *imamu* to teach of syahadat to children. This way is *imamu's* strategy to teach syahadat to children so that children can easily utter such speech. Besides of for ease to understand of children for syahadat in The Arabic language, on data [4][7], and *imamu* says the meaning of Muna language. This carried out *imamu* for easy understanding of syahadat meaning for children, in the data [5].

[5]	*Imamu:*	*(a) Asumakusiimo, sakotu-kotughuno miinahi bhe Ompu soni somba sapaeno Ompu Allah Taala* " (I bear witness that in truth there is no God but Allaah Ta'ala)
	Children:	(b) Uumbe (Yes)
	Imamu:	(c) *Maka asumakusiimo tora, sakotu-kotughuno omputo anab Muhammadi kantudu-ntudno Allah Ta'ala* " (Then I testified as well, truthfully tell-Muhammad was the messenger of Allah Ta'ala)
	Children:	(d) *Uumbe* (Yes)

3.3. Strategy of Directives Speech Act Imamu

In the discourse of *katoba*, the findings indicated that the realization of directive speech of *imamu* was conducted by using a strategy of direct and indirect speech. The *imamu* speech strategy options related to the communications objectives to be achieved in the context of the implementation of *katoba's* ritual. The research findings indicate that *imamu* uses direct and indirect strategy in revealing the various its directive acts intent on implementing *katoba* tradition. Relating to the usage of direct strategy, it is used by *imamu* to express the function of order, prohibit, advising, and expecting. In other respects, the indirect strategy is also used to declare the function directive govern, advise, prohibit, and expect.

The data findings in this research indicated that the usage of the directive form which represents a specific directive functions realized in diverse according to the context of the purpose and goals to be achieved in oral communication. With says no *katoba* ritual discourse. Examples of the *imamu's* directive act usage with direct strategy in the discourse of *katoba* by using the mandatory statement when ordering the children to respect the elderly.

IOP Conf. Series: Earth and Environmental Science **156** (2018) 012053 doi:10.1088/1755-1315/156/1/012053

[6] *Imamu:* (a) *Motehie amamu, kapae amamu itu lansaringino kabolosino ompu Allah Taala* (Fear to father, because the father is a substitute for Allah)

Children: (b) *Uumbe* 'Yes''

Imamu: (c) *Inando itu suano kaawu ina motobusaangko ne dunia ini, taaka lahae-lahae membalino kamokula robhine, inamuo dua itu, **tabeano** dotehie itu* (The so-called people old woman was not just mothers who gave birth to us in this world, but anyone who has an old woman, your mother, it should be feared of "

Children: *(d) Uumbe (Yes)*

Imamu *(e)Omoghondohi malaikati omaiane nehamai; isamuoitu kabolosino malaikati mentaleano. Nikonando isando itu suanokaawukapokakutahando ghule, taaka lahae-lahae foliuno umuru isamuo dua itu, **tabeano** dotehie itu* (Look for the angels will come where your brother that is a substitute for a real angel. The so-called big brother, not the only sister sibling us, but anyone which exceeds the old is your brother too, but feared too)

Children: *(f) Uumbe (Yes)*

Imamu: *(g) Okoasigho o ne ai maitu bea dapotooanebe kaasigho ne mie bhari. Nikonando ai maitu suano kaawu ai kapokakutaha ghule, taaka lahae-lahae niliumu umuru, aimuo itu, **tabeano** doasiane itu* (Affection of younger siblings was equated with affection by the people. The so-called sister was not only younger your birth, but anyone under old you are, it's your sister, it should be cherished it)

Children: *(h) Uumbe (Yes)*

Speech [6] above is a form of directive rule with an immediate imperative mode. The speech command was marked by the use of the word *tabeano* (should, must) on the data (c, e, and g) indicating that the advice is delivered *imamu* become a necessity. It should emphasis the command based on the teachings of Islam. This is the emphasis of this advice for children who repentance, namely the formation of the individual child on respecting their parents. For that purpose, *imamu* ordered the children to respect the elderly.

The directive act is also expressed by using indirect speech, or the intended meaning is not the same as what appears in the mode of speech (metaphoric/figurative). The following excerpt shows an example of the use of strategies prohibits the meaning nonlateral (figuratively) with the imperative mode.

[7] *Imamu:* *(a) Ane gholeno labu nopolobhighoo kangkaha fopalie gholeno, **koe** findahiea. (b) Ane owora kokarawuno ghofano koe seliea atawa **koe** meuta ne kobhakeno palolano atawa kogholeno labuno koe findahiea. (c) Karawuno ghofa, bhakeno palola, gholeno labu ainiini maanano hakuno manusia bhe nehakuno Allah Taala (*If the shoots pumpkins stretched across the road, you move the bud, do not you stomp '.' If you look at the pile of potato taro others, do you dig or not to pluck eggplant of others or do not step on a leaf gourd of others'. Heaps of potatoes Tatas, eggplant, pumpkin leaves, shoots before, the meaning of human rights are the rights of Allah)

Children: : (d) *Uumbe* (Yes)

Speech [7] is a form of directive bans realized using indirect speech with ban directive *'koe'* 'not' in speech (a) and (b). Advisingadvice by directive indirect considered ineffective because it tends to be metaphorical (figurative meaning).

It is such in data (b) is phrase *karawuno ghof*a phrase 'pile yam taro', *bhakeno palola* 'eggplant', and *gholeno labu* 'leaf pumpkin' explained human behaviour. Heaps of potatoes taro, eggplant and leaf pumpkin is a plant. Plants that belong to others should not be disturbed or taken. *Karawuno ghofa* phrase 'pile potato taro' in speech (b) refers to humans, in particular, that is the woman (wife) others. *Bhakeno palola* phrase 'eggplant' and *gholeno labu* 'leaf pumpkin' in speech (b) refers to humans, especially men (husbands) of others. *Katoba imamu* language options above teach our children (male/female) as adults and are married are not allowed to interfere with other people's spouses.

Based on the description of the research findings that have been described, it can be stated that the directive speech acts in katoba discourse *imamu* represented variously through the form, function, and specific strategies. Form, function, and strategy of directive speech acts are realized by *imamu* based on the diversity of the context and intent of delivery. A discussion of the findings of the form, function, and the strategy outlined as follows *imamu* directive.

First, regarding to mode of directive speech acts, *imamu* realized with speech declarative and imperative mode. Both categories form the shape of the directive speech act represents directive functions according to the needs of communication in context of *imamu* and children. In the discourse, *katoba* use directive forms associated with certain types of directive speech acts. Concerning the specific type of following directive, the uses of directive forms have a pattern or specific traits. Or characteristic patterns are influenced by the context of their use in discourse or conversation took place during the implementation of *katoba*. According to Mey [8] and Cummings [9], the context is more than just a problem of reference and understanding of something that meant, context utterances make us more depth. This is confirmed by Sperber and Wilson (in Ardianto) that to obtain the maximum relevance, language activities should involve contextual impact [4]. The greater the impact of contextual conversation, the greater the relevance will be.

In general, the findings of this study suggest that context is an inseparable part of speech. Every act of directive spoken by *imamu*, the inferences is always based on the context of the speech. This suggests that a very dominant role in the context of directive speech acts of *imamu*.

Secondly, regarding function realization directive, based on the description of the research findings can be stated that the function of directive speech acts of *imamu* in *katoba* discourse includes functionality ordering, commanding, forbidding, suggesting, expecting and reprimands. Directive functions expressed the diverse types and their use, as well as having certain patterns and characteristics. The pattern and characteristics influenced by the context of use in discourse *katoba*. The use of the function follows the directive rule and reign; for example, there is a speech delivered by direct imperative mode. In addition to using lateral meaningful speech acts, *imamu* also use non-lateral meaningful expression by using figurative (metaphorical).

4. Conclusion

Based on the findings of this research, researchers noted that the results as follows: First, the form of directive speech acts in *katoba* discourse of *imamu* realized child-Islam to include speech act in declarative and imperative modes. Declarative speech act is embodied in a directive speech statement mode of intent, orders, errands, prohibitions, and licensing statement. Directive speech acts in directive-imperative mood embodied in the speech of directive-imperative command, imperative errands, imperatively demand, imperative prohibition, imperative invitation, imperative suggestions, and in other words, the directive speech act of *imamu* in *katoba* discourse manifested in various ways through the use of forms of speech with the variant marker in accordance with certain linguistic context underlying the discourse of conversations and the needs and goals to be achieved in the ritual *katoba* communication.

Second, the communications functions contain through teacher directive speech acts in katoba discourse include directive function of enjoining, prohibit, permitting, advice, expecting, and inviting. In other words, the follow directives embodied in the katoba discourse of *imamu* represents diverse directive functions by the underlying context of *katoba* ritual conversation discourse and the needs and goals to be achieved in ritual communication. Third, the realization strategy in directive speech acts of katoba discourse by *imamu* includes direct strategies and indirect strategies. The direct strategy carried out by *imamu* explicitly conveys the message. In other respects, the indirect strategy is also used to declare the function of directive speech acts. The message conveyed is metaphoric. In other words, the shape of the directive which represents a specific directive function by *imamu* realized in various ways according to the context of the underlying discourse of conversations in *katoba* ritual.

References

[1] Sadock J 2004 Speech acts. *Handb. Pragmat.* 53–73.
[2] Huberman A, dan Miles M 2009 Manajemen Data dan Metode Analisis, Handbook of Qualitative Research. Yogyakarta: Penerbit Pustaka Pelajar.
[3] Berg R van Den La Ode Sidu Marafad. 2013. *Kamus Muna-Indonesia*
[4] Ardianto A 2013 Tindak tutur direktif guru dalam wacana interaksi kelas anak tunarungu. *LITERA* **12** (1)
[5] Khanizar K 2010 Seni Pertunjukan Wacana-Semiotika Sawirman's E135. *Linguist. Kult.* **4** (2)
[6] Baker P, Ellece S 2011 Key terms in discourse analysis. A&C Black.
[7] Bloor M, Bloor T 2013 The practice of critical discourse analysis: An introduction. Routledge.
[8] Mey JL 2001 Pragmatics: an introduction. Blackwell Publishing Company.
[9] Cummings L, Setiawati E 2007 Pragmatik: Sebuah Perspektif Multidisipliner. Pustaka Pelajar.

IOP Conf. Series: Earth and Environmental Science **156** (2018) 012054 doi:10.1088/1755-1315/156/1/012054

Philosophy of Poroomu Yinda Saangu, Pogaa Yinda Koolota: Management of Government System of Buton Sultanate

Abdul Kadir[1], Mahruddin Mahruddin[2], Erdianti[3] Mardlina Gazali[4]
[1]Institut Agama Islam Negeri Kendari, Jl. Sultan Qaimuddin No 17 Baruga, Kendari, Southeast Sulawesi, 93563, Indonesia
Email: akadir@yahoo.co.id
[2]Institut Agama Islam Negeri Kendari, Jl. Sultan Qaimuddin No 17 Baruga, Kendari, Southeast Sulawesi, 93563, Indonesia
Email: udinmahru@yahoo.co.id
[3]Institut Agama Islam Negeri Kendari, Jl. Sultan Qaimuddin No 17 Baruga, Kendari, Southeast Sulawesi, 93563, Indonesia
Email: eerdianti@yahoo.co.id
[4]Institut Agama Islam Negeri Kendari, Jl. Sultan Qaimuddin No 17 Baruga, Kendari, Southeast Sulawesi, 93563, Indonesia
Email: mgazali@yahoo.co.id

Abstract: This article examines the management of the government system of Buton Sultanate in one philosophy: "*Poromu Yinda Saangu, Pogaa Yinda Koolota*". This philosophy is believed to be the unifying of the Buton Sultanate for hundreds of years. This paper is a qualitative descriptive with ethnographic approaches. The results indicate that this philosophy provides insight to the citizens of the Buton sultanate on the importance of relations among elements of government such as *Kaomu, Walaka,* and *Papara.* These three elements are likened to a human body consisting of *Kaomu* as Head, the place of the good brain, brightly lit because it always receives the light of Divine *Nur. Walaka* is as a body; the heart contains a sacred heart because there is a place where the balance of justice is published. *Papara* as an accomplice has the power of *lahiriyah.* The relationship of three elements is bound and intertwined in one philosophy: "It is near but untouchable, it is far away, but no space between them". This implies that all three must unite to uphold justice and common interests, but they should not violate other rights. Therefore, this philosophy can be used as the central management system to build good governance.

1. Introduction

The position of various local histories in the archipelago is an integral part of the long journey of Indonesian today. Therefore, the historical inheritance brought by some kingdoms in Indonesia and known by a lot of people nowadays needs to be appreciated. Furthermore, there is a need to explore its roles in human's life. The position is highly appreciated, and Indonesian people are required to study it at all education levels. However, in the implementation process, the history is centralized on Javanese history among others. The implications have such an enormous impact on the history inherited to the next generation. The community is more familiar with the history of other regions than the scope of history around them. The historicity and culture of Indonesian society are not placed in equal treatment in academic nuances; rather it seems to glorify certain ethnocentrism of history as a motivation to participate in the development of nation and state. This phenomenon is called "the forgotten history of the country".

The assessment and values of the historical wisdom of local government that has been implemented in the Sultanate of Buton is still an interesting and ongoing study in the field of political science, historical science, other scientists, national development, and especially for the sake of Buton government. In fact, here have been some values of the Sultanate government system implemented and will be applied to local government. Historical experience relating to the traditional management of the government, based on existing sources, shows that in the Sultanate of Buton, the government structure has been regulated according to the constitutional provisions and the implementation of democratic government, as well as the power relations of the central and local government. Moreover, the governance of the people leads to define the characteristics of an initial state. This shows at least that in the period of empire/colonial, colonial and post-independence is an interesting historical phenomenon to be discussed because past historical events have causality with the present.

Reviewing the power-sharing system in the era of the Sultanate of Buton shows that the state is not a dominant factor in carrying out its authorized functions. Before the birth of Montesquieu who popularized the power-sharing system within the state government, the Sultanate of Buton had long embraced the decentralized system of government in an atmosphere of regional autonomy for approximately five centuries [1]. In governing the power-sharing system, the government is governed according to the customary provisions for Buton society contained in the text consisting of three elements of *kaomu* (king), *walaka* (*Syara/Dewan Dewan Harian*), and *Papara* (citizen). The relationship of the three elements is closely related as seen from the words of Buton's philosophy "*Poromu Yinda Saangu, Pogaa Inda Koolota*" (It is near but untouchable, it is far away, but no space between them) [2].

Research on the Sultanate of Buton is ubiquitously done by many experts. Some talk about *Kaomu*, *Papara* and *Walaka*: a study on social structure and powerful ideology in the Wolio Sultanate. The study reveals that the classification of society on the Sultanate of Buton is based on the origin of the region. There are two classifications within this region; Wolio and non-Wolio people. The first two social categories, namely *Kaomu* and *Walaka*, fall into the Wolio category. The last two social categories, *Papara* and *Batua*, belong to the non-Wolio category [3].

The subsequent research done by Isaac Bagea is about the implementation of *sarapatanguna* cultural values in government leadership in Bau-Bau City. This study reveals that *Sarapatanguna* cultural values have been implemented in everyday life and the government leadership of Baubau [4].

Among all the previous studies, it is known that the philosophy research about *poroomu yinda saangu pogaa yinda koolota*: the B management of the government system of the Buton Sultanate has not been studied and discussed in depth by the researchers. Therefore, this study aims to discuss the importance of the relationship among the regulatory elements such as *Kaomu*, *Walaka*, and *Papara* in the government system.

2. Methods

This study uses historical method. The historical method consists of four stages, namely (a) heuristics, which is the activity of collecting traces of the past, (b) the criticism refers to the investigation whether the traces are true to both the form and the content, (c) the interpretation which means to decide the interconnected meanings from the facts that have been criticized, (d) the writing which means to convey the synthesis obtained in the form of the story [5]; [6].

3. Findings

Sultanate Buton has adequate authority over territorial and its surroundings. Since the arrival of the Miapatamiana which led to the government birth became the forerunner of a country called Butuni [7]; [8]. Buton had ever achieved its glory of its time. The celebration began when Sultan Dayanu Iksanuddin became the fourth Sultan to reform the government by establishing the Basic Law of the

IOP Conf. Series: Earth and Environmental Science **156** (2018) 012054 doi:10.1088/1755-1315/156/1/012054

Buton Sultanate (*Murtabat Tujuh*). This historical process continues in its development to experience ups and downs and then reaches the height of glory occurred to the twenty-ninth sultan of Buton Muhammad Idrus Qaim ad-Din who ruled Buton in 1824-1851. In his time, Islamic ideology called "Wahdatul Wujud" was implemented with the implementation of Islamic Shari'a. Buton ended in the reign of Sultan La Ode Muhammad Falihi in 1960, which coincided with his disappearance (death) in the same year [9]. The end of self-government, the former territory of Buton Sultanate had previously been integrated into the Unitary State of the Republic of Indonesia.

In regulating the execution of power-sharing government arranged according to the customary provisions of Buton society, there is three level of group namely *Kaomu*, *Walaka* and *Papara* which are collected in a philosophy called *Poromu Yinda Saangu*, *Pogaa Inda Koolota*. The position of the three classes can be described as follows.

3.1. Kaomu (*Sultan*, *Sapati*, *Kenepulu*, *Kapitalau*)

As stated according to the custom, the head of Buton people should be *Sultan*-blood person who has the law of power, vast influence, and nobility in the area of his empire. Therefore according to the custom and the people's belief, the one who is considered as *sultan* should become the people's guide as *khalifatullah* (successor of God). Since the name of the sultan is exemplified by the nature of *khalifatullah*, it is obligatory for the *Sultan* to receive the shadow of a perfect form of God of his divine nature. Therefore, the one who becomes *sultan*, he should have perfect attitudes as having been set based on the custom. He should have (1) a sense of obligatory perfection, his descendants should be recognized (2) the perfection of his being, that is, the nature of the perfection of his birth body (a perfect man of his senses, his limbs have no blemish or deficiency) In addition to the nature of perfection, a *sultan* must have perfect inner beauty (intelligence of wisdom and wisdom).

Inner beauty becomes the most important perfection of a *sultan* because it will lead to the body perfection. The intellectual perfection based on the intelligence of reasoning or deciding wisdom is generally some properties of excess and the nature of the glory or the nature of human virtues, for example, temperament or *tabiaat*, the main character, polite behavior, his sociable friendship, sweet speech, love to the crowd and loved by the crowd. Because of his fairly perfect qualities of the inner self-glorious nature of the *sultan*, the *sultan* is worshipped by his servants not because of the right of the perfection of his being but because of his soul and the perfection of his inner self which receives the shadow of its inner self that reaches the name *ismurrahmani*.

On the determination of his monotheism unity, there arise four qualities of his inner glory and become the main duty of the *Sultan*: (1) The wisdom of intelligence of the mind to investigate the state of the true inner self in order also to consider and investigate others. (2) the fair intelligence of feeling that weighs equally weight in the measure of life (3) love, refinement so that the compassion to human beings as well as pitying the true self because all human beings do not stand alone but *Qadim* with the form of God. (4) Inner or soulless kanaa (wealth) which does not base his thought on excessive wealth because soul or inner self is more important.

3.2. Walaka (Religious Means / Legal Facilities)

This independent institution is the institutional system of the sultanate which remains functioning with all its order in our life. In general, the scope of duty and its function is to enforce Islamic and religious social law in the Sultanate, so it is also called *Sara* Religion. The Position division system is as follows: (1) *Lakina* Religion (*Khadi*) is one person, included in *kaomu* class (royalty). The position of *Lakina* religion is a religious leader. (2) One priest, the position is *Kaomu* (Bangsawan). The position of Imam other than as a leader of a prayer is also a leader of *Sara kidi*. (3) *Khatib* is four people, included in *Kaomu* class (nobility). (4) *Moji* (Bilal) is for twelve people, included in *Walaka* Group (customs' Stakeholder). *Moji* circle consists of several positions namely, *Tungguna Tobha* (*Tungguna* (Wolio language, from the root word, means to guard, while *Tobha* is a container to save money and or the like, so the task and its function as Treasurer), *Tungguna Aabha* (*Aabha* means ask, *Tungguna Aabha* means a place to ask, but in carrying out his duties, he is the communicator in *sara kidi*) and *Tungguna Bula* (*bula* means moon, guardian of the moon. One of the implementations of his

IOP Conf. Series: Earth and Environmental Science **156** (2018) 012054 doi:10.1088/1755-1315/156/1/012054

task is to determine the day of the beginning of fasting in Ramadhan). (5) *Mokimu* (mosque management) takes up forty people including in *walaka* group. Mokimu is associated with mosque because in this position comes from young age and serves as a container of religious leadership.

In carrying out his duties and functions one of his guided according to "Makhafani", among others, he understands the matters about the marriage; *talak*, and reconcile. In addition, serving the society is related partially to the death. The system of division of office above is equipped with four experts of chemistry called "Can Patamiana" consisting of (1) *Mojina Silae*, (2) *Mojina Peropa*, (3) *Mojina Kalau*, (4) *Mojina Waberongalu*. According to *bisa patamiana*, a *sultan* cannot be dismissed. The one who deserves this position is only the one who has the same lineage.

3.3. Papara (People)

Papara groups or free societies are scattered throughout the country that inhabits *Kadie* (local Government). The term *Kadie* (Wolio language) comes from the root word "die" which means iris and *kadie* lexically mean slice. The use of the term *kadie* in the context of government is associated with the village-level governmental area. In the Sultanate of Buton consists of 72 *Kadie* commonly referred to as "*pitu pulu Rua Kaomuna*" (72 local governments).

The Government of the Buton Sultanate established a system of autonomy to 72 *Kadie* by managing his government. The central government has no right to interfere directly with the implementation of governance in the territory of the *kadie*. The position of the central government only exercises a supervisory system that is divided into two regions: *Pale Matenaeyo* (Eastern Region) and *Pale Sukanaeyo* (Western Region). In the division of supervision, areas of the *kadie* or class of *papara* conducted by *Kaomu* and *Walaka* groups. The *Kaomu* (Nobles) group oversees 40 *Kadie* territories under the title of *Lakina* (*Bobato*) while the *Walaka* group oversees 30 *Kadie* Regions by the name of *Bonto* (Minister).

The governance structure of Buton is governed by a vision of mission as a maritime country symbolized as a curate boat by focusing its attention on shipping and trade. The country is likened to a boat that must be kept in balance so as not to sink and reverse. The point is the balance of power-sharing between *Kaomu* and *Walaka*, the balance between the central and regional portions and the balance of the division of defence tasks.

From the above description, it is known that the elements of Buton society are the pillars of Buton society which consists of three elements, namely: King / *Sultan* (*Kaomu*), Syara Board (*Walaka*), People (*Papara*). The relationship of the three pillars of society to each other is very closely related as seen from the words of the philosophy of Buton: "*If Only One Wasted, it Became Three Perish*". These words of philosophy mean that if the three elements are incomplete, for example only one element alone without any other, it will be futile him. What is the meaning of a King or *Sultan* without *Syara* and the People? Furthermore, when the three elements of society are separated into three parts stand alone and apart from each other, there is no unity between them, there will be destroyed because each may feel equally strong, equally courageous, equally capable and others.

In arranging the common life of the Sultanate of Buton, the relationship between the *Sultan*, the Syrian Council and the People are manifested in two kinds of parables: First, the three elements are likened to one human body as follows: (a). *Sultan* as head, where the great brain, brightly lit because it always receives the light of Divine *Nur*. (b). Syara Council as the body, where the heart contains the sacred heart because there is a place where the balance sheet of justice. (c). People as members (Foot of Hands) who have the power of the outer. Second, the three elements are likened to a household (family), namely: (a). Sultan as the head of the families who is responsible and always think about all the interests of family members, namely his wife and children. (b). *Syara* Council as a housewife always tries to foster members (c). The people as children must be nurtured with the love of their parents to be virtuous people, devoted to their parents, fearing God Almighty, and useful for the nation and nation. The relationship of the three elements is bound and intertwined in a Buton philosophy: *Poroomu Yinda Saangu, Pogaa Yinda Koolota*. This implies that all three must unite in custom, and the distance between them is governed by custom. All three must unite to uphold justice and common interests, but should not touch or offend let alone violate the rights of others. The position of the three has a distance that is limited by custom, in such a way as if not intermediate. If the separation distance is too far, it is likely to cause the initial dispute between them, the distance that is the point of contact between contacts and separate that is the duty and responsibility of the board. If the intersection between touch and separation is off balance, then it will likely bring undesirable consequences, i.e. the words touching will become clashes, and the distant words will mean hostile.

4. Discussion

The government system of the Buton sultanate is compiled in one body. *Sultan* as Head, where the place for great brain takes place, brightly lit because it always receives the light of Divine *Nur*. The *Syara* Council as the Body; the heart contains the sacred heart because there is a place where the balance of justice is recognized. People as members (accomplices) have the external power. The relationship between these three Buton poles is closely related to each other as can be seen from the words of Buton's philosophy: "*If Only One Wasted, it Became Three Perish*". These words mean that if all three elements are incomplete, it will be useless. What is the meaning of a king or *sultan* without people? And vice versa, when the three elements of society are split up into three parts and stand alone, and apart from each other, there is no unity between them, there will be destruction because each may feel equally strong, equally brave, the same equally capable and others.

Research on the management system based on the philosophy of local wisdom in Indonesia has been done by many experts. In Banjar and the structure in the Sultanate of Banjar in the XIX century are more centred on the Sultan. The power of Sultan Banjar has full authority in determining royal policy although in the structure of the Banjar kingdom there is the Crown Council as a deliberation institution in the Sultanate of Banjar whose members consist of the slaughter (relatives) near the king and the palace [10]. All of these studies have similarities and differences because they are both sultanate-based. If in Buton, the government management system is divided into 3, namely *kaomo*, *walaka*, and *papara*, then Banjar consists of kin close relatives. Thus, it can be generalized that the Sultanate-based governmental management system in Indonesia is divided into three.

It has similarities and differences with the imperial system in Britain, Brunei and the Middle East. In Brunei, the Sultan of Brunei considers the position of Head of State, Head of Government, and Leader of Religion of Islam (Brunei majority Shafi'i school). All state affairs are run by the Sultan. But in the process, execution of the Sultan's government does not work alone, but the sultan is assisted by counsellors and ministers of state. The country's advisers and ministers are incorporated into a Brunei Darussalam state cabinet. The authority of all the counsellors and ministers of the state joined in this cabinet is mere to provide advice and suggestion to the Sultan in determining the direction of policies and decisions that the Sultan will take in the life of the state. The decisions and policies advised by advisers and ministers of the state shall have no legal force and shall not apply in Brunei without the consent of a Sultan of Brunei Darussalam [11].

While in England, the Government is run by the Prime Minister as head of government assisted by ministers. The Queen and the King of England are only heads of state that function as a symbol of state (symbol of sovereignty, majesty and unity of the state) [12]. As with the kingdom of Saudi Arabia other than as head of state, he is also prime minister, supreme commander of the army, guardian of two shrines (Mecca and Medina), appoints and dismiss the Council of Ministers, and interpret the law. The highest political authority under the king is the crown prince. The crown prince is determined by the king, provided that it is still taken from the descendants of Abdul Aziz. The crown prince can even rule on behalf of the king, even before the crown is consecrated. The Council of Ministers acts as the legislator and executive executor of the king. Both roles are based on the king's blessing. The law established by the council of ministers will be applicative law within 30 days unless the king uses his veto's rights. Members of the council of ministers are the descendants of Abdul Aziz. As Shura board is a consultative council [13].

Finally, this study confirms that the governance system of the Buton sultanate philosophy of *poromu yinda saangu pogaa yinda koolota* is divided into three parts and this is in line with Montesquieu's theory of politics that divides the system of governing into three, namely the executive, the legislature, and the judiciary. So, the question of statehood in Indonesia today is practiced 15 centuries ago in the land of Buton. Not only that, the governmental structure in the Buton sultanate of these three levels can controls each other or supervise, if one violates the rules, it will be punished not to mention the sultan and get punishment. This proves to be one of the sovereigns of Buton who is found to have violated the customary provisions of the hanging law by the council of Syara and is not present in any sultanate in the world.

5. Conclusion

The philosophy of *Poromu yinda saangu pogaa yinda koolota* confirms that in the management of the government system in the Sultanate of Buton is divided into three parts that are interconnected and cannot be separated because if separated there will be imbalances. In addition, in the management of government between the three levels which are *Kaomu*, *Walaka* and *Papara* each can control each other when one of them there is a violation of customary provisions or abuse of authority, there will be punishment in accordance with the rules of law in the Sultanate, even it applies to sultan.

Broader research related to the management of the imperial government can be done in order to refine this research. Government management in some forestry in Southeast Sulawesi such as the Muna, Moronene and Mekongga sultanates can be done in subsequent research. In addition, further research can also be focused on good governance that is applied in the Sultanate of Buton.

References

[1] Rukiyah R 2017 Sistem Pemerintahan Kerajaan Butun dalam Naskah Istiadat Tanah Negeri Butun. *Anuva* **1** (2) 111–118.

[2] Mansyur FA Onina Manga Mancuana Mangenge: Ekspresi Kebahasaan Orang Wolio (Kajian Linguistik Antropologis).

[3] Rudyansjah T 2014 Kaomu, Papara dan Walaka: Satu Kajian mengenai Struktur Sosial dan Ideologi Kekuasaan di Kesultanan Wolio. *Antropol. Indones.*

[4] Bagea I 2017 Implementasi Nilai Budaya Sarapatanguna Dalam Kepemimpinan Pemerintahan Di Kota Baubau, Provinsi Sulawesi Tenggara. *Kandai* **12** (2) 297–308.

[5] Daliman A 2012 Metode Penelitian Sejarah. Penerbit Ombak.

[6] Abdurahman D, Safa A 2007 Metodologi Penelitian Sejarah. Ar-Ruzz Media.

[7] Zahari AM 1977 Sejarah dan Adat Fiy Darul Butuni.

[8] Rosdin A 2015 Buton, Islamization, and this Manuscript Tradition. *Int. J. Nusant. Islam* **2** (2) 101–116.

[9] Iwan M Pasang-Surut Hubungan Buton–VOC: Studi Masa Sultan Himayatuddin Muhammad Saidi (1751-1752, dan 1760-1763)".

[10] Suriadi A 2014 Syekh Muhammad Arsyad Al-Banjari dalam Dinamika Politik Kerajaan Banjar Abad XIX. Pusat Penelitian dan penerbitan LP2M IAIN Antasari.

[11] Ghofur A 2016 ISLAM DAN POLITIK DI BRUNEI DARUSSALAM (Suatu Tinjauan Sosio-Historis). *TOLERANSI* **7** (1) 53–69.

[12] Braddick MJ 2000 State formation in early modern England, c. 1550-1700. Cambridge University Press.

[13] Maisel S 2011 Kingdom of Saudi Arabia. *Gov. Polit. Middle East North Africa* 161–205.

IOP Conf. Series: Earth and Environmental Science **156** (2018) 012055 doi:10.1088/1755-1315/156/1/012055

Fuzzy Multi Criteria Decision Making To Determine The Best Rice Quality in Riau Island

Zamzami[1,2], Didik Siswanto[1], Yogi Yunefri[1], Bayu Febriadi[1] and Sutejo Sutejo[1]

[1]Universitas Lancang Kuning, Pekanbaru, Indonesia 28265
[2]Pangkalan Data Perguruan Tinggi (PDPT) Unilak, Pekanbaru, Indonesia 28265
Tel/Fax : +6271 52324
E-mail : zamzami@unilak.ac.id,

Abstract: Indonesia as maritime country has designed national food security plan not only in sea food resources it also in mainland food resources such as rice crops that will produce rice (staple food source) which is intended as consumption of staple food source of the people of Indonesia. This research aimed to help the national food security plan which is to determines the superior quality of various types of rice and rice varieties that exist in this maritime country through Fuzzy Multi Criteria Decision Making method (FMCDM) as a applied method to determineof high quality rice. In this method researchers have been able to determine some types of seeded rice that has been through the process of calculating the fuzzy logic system which later can be utilized in the needs of the wider community, in the hope that the government can used this method to the selection of superior rice quality.

Keywords: Fuzzy Multi Criteria Decision Making (FMCDM)

1. Introduction

Farmers are one of the most important factors in the region of Indonesia relying on its life as a rice farmer. But often people have a lot of difficulty in determining what rice they will use in farming. Rice crops basically have a way of determining their own characteristics or characters to determine whether the rice is good or good and deserves to be consumed. Rice is the main food crop in Indonesia because most of Indonesia's population is mainly rice. Many farmers or rice enthusiasts are difficult to determine the best rice. The best selection of the best rice is one of the efforts made by the farmers or rice enthusiasts to solve the problem, the lack of interest of farmers or people who consume to choose the best rice available. Decision-making refers to the selection or ranking of available alternatives by using several criteria for selection consideration. As for rice farmers basically have a way to determine the characteristics of its own or the character to determine the good or good rice and deserve to be consumed. Many sometimes farmers or rice enthusiasts are hard to determine the best rice. With the existing alternative on decision support system with FMCDM method to choose the best rice.

The Decision Support System (DSS) is a system intended to support managerial decision-makers in semi-structured decision situations, in order to be a tool for decision-makers to expand their capabilities, but not to replace their assessment of these currencies in (Muqtadir A. and Purdianto I., 2013). Problem solving not only refers to the solution of the problem area / difficulties but also includes an investigation of the opportunities that exist. Definition of ideal decision support system is:

a. DSS is a computer-based system with an interface between machine / computer and user.
b. DSS is intended to assist decision makers in solving a problem at various levels of management and not to change the position of humans as decision makers.
c. DSS is able to provide an alternative solution for semi / unstructured problems for both individuals and groups and in a variety of processes and decision-making styles.
d. DSS uses data, databases and analysis of decision models (Marto, 2012).

IOP Conf. Series: Earth and Environmental Science **156** (2018) 012055 doi:10.1088/1755-1315/156/1/012055

The system is a collection of different components that are interconnected, work together and support each other to achieve a certain goal (Munir, Usman M. and Angtriswono, 2013), while according to this rate states the System is a unity of component procedures that are related to each the other, work together in accordance with the rules applied so as to form a common goal, where in a system if there is one part that is not working or damaged then a goal can occur error results or output, the system is a collection of elements that mutually (Ardiansyah, 2013), While the system is an order or alignment consisting of a number of functional components (with units of functions or special tasks) that are interconnected and together aims to fulfill a process or work certain. Fuzzy Multi Criteria Decision Making (MCDM) is a decision-making method that aims to define the best alternative decision of the number of alternatives based on certain criteria to be considered (Kahar N. and Fitri N., 2011). And Fuzzy Multi Criteria Decision Making (MCDM) is a decision making method to establish the best alternative of a number of alternatives based on certain criteria. Criteria are usually in the form of measures, rules or standards used in decision making (Puspitorini S. and Sitohang S.A., 2011).

2. Methodology

The research methodology is a description of the steps that will be implemented in conducting the research. The steps of this research will be further illustrated in the form of a research framework. This methodology needs to be established so that research can be conducted in a structured way. Steps that will be done must include starting from studying the problem to the creation of a system that can solve the problems that exist. Using Fuzzy Multi Criteria Decision Making (MCDM) is one of the methods that can help decision makers on some decision alternatives to be taken with some criteria and the degree of compatibility of each alternative. In the best rice selection process using Fuzzy Multi Criteria Decision Making, criteria are needed, the importance of each criterion and the rating of an alternative match against the criteria for calculation so that the best alternative will be obtained.

3. Results and Discussion

In this case the data source can be from the Office of Agriculture Asahan, therefore the researchers wrote that A1 is the same as alternative 1, researchers also grouped between A1 with existing data sources can be seen in table 4.1 below :

Table 1. Alternative

Alternatif	Information
A_1	Rice Situ Bagendit
A_2	Inpari Rice
A_3	Ciherang rice
A_4	Cibogo Rice
A_5	Cigeulis Rice
A_6	Mekongga Rice

While the criteria used are shown in the following table:

Table 2. Criteria

Kriteria	Information
C_1	Plant Age
C_2	Forms of Plants
C_3	Plant height
C_4	Productive Seedlings
C_5	Texture of Rice
C_6	Weight of Grain
C_7	Average Results
C_8	Potential Result
C_9	Resistance to pests
C_{10}	Illness
C_{11}	Advice Planted

Where Table 4.2 explains that the best rice is rice that has been decided as the best rice among various types of rice (alternative) along with these criteria comes from the Department of Agriculture Kab. Asahan. Whereas UT = Planting Age (C1), BT = Plant Form (C2), TT = Plant height (C3), AP = Productive tiller (C4), TN = Texture Rice (C5), BB = Weir Grain (C6), RRH = Average Results (C7), PH = Potential Result (C8), H = Pest (C9), P = Disease (C10), AT = Cultivation (C11), are part of predetermined criteria. Rice Situ Bagendit (A1), Inpari Rice (A2), Rice Ciherang (A3), Cibogo Rice (A4), Cigeulis Rice (A5) and Mekongga Rice (A6) is an alternative to determine the best rice to be determined. Multi Criteria Decision Making fuzzi model can be seen in Figure 4.2 below:

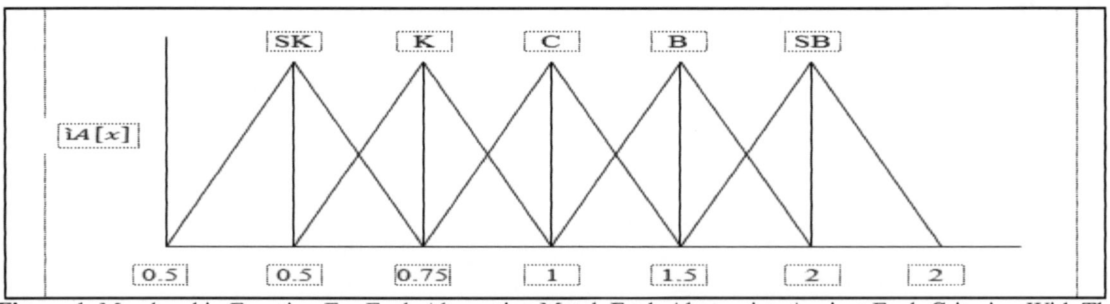

Figure 1. Membership Function For Each Alternative Match Each Alternative Against Each Criterion With The Fuzzy Triangle Set of Numbers.

Then the degree of matching alternatives with the decision criteria are: T (match) S = {SK, K, C, B, SB}, by SK = Very Less; K = Less; C = Enough; B = Good; and SB = Very Good; which are each represented by the triangular

fuzzy numbers as follows:

1. SK = (0.5, 0.5, 0.75)
2. K = (0.5, 0.75,1)
3. C = (0.75, 1, 1.5)
4. B = (1, 1.5, 2)
5. SB = (1.5, 2, 2)

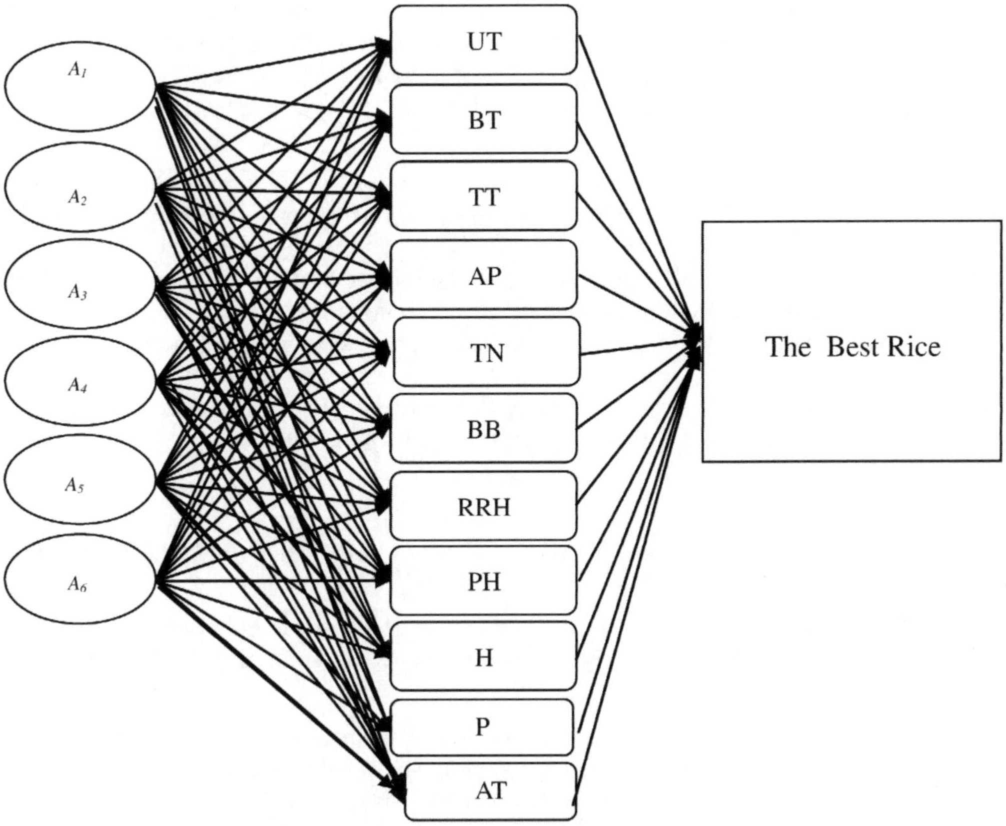

Figure 2. Model Fuzzy Multi Kriteria Decision Making

Table 3. Index Matches For Any Alternative

Alternatif	Fuzzy Matching Index		
	Y_1	Q_1	Z_1
A_1	0,857954545	1,670454545	2,534090909
A_2	0,869318182	1,625	2,534090909
A_3	1,386363636	2,659090909	3,636363636
A_4	0,971590909	1,852272727	2,840909091
A_5	1,011363636	1,954545455	3,04545445
A_6	1,113636364	2,159090909	3,363636364

In completing the results of interviews in rice research obtained several alternatives as well as criteria or characteristics of rice that has various types of variation, apart from that in the determination of the assessment from the lowest level to the highest level. Basically the result of the value of table 3 above is a problem of confused farmers in choosing rice crops and the most sought after by consumers. To solve the problem the first point is to collect the next alternative to collect the criteria, then get the membership results.

To get the final result like table 3 above do multiplication of value between interest rating and match rating value, then all the result of multiplication are mutually added and multiply with result of alternative division and criteria. To produce the final result in the need to distribute the fuzzy match index, and by taking the optimum degree (α) = 0 (not optimistic), α = 0.5 and α = 1 (very optimistic), it will obtain the integral value for each alternative. In the column α = 0 in use all match results are added and shared to ½ and to find the results in columns 0.5 and 1 using the same way.

4. Conclusions

Based on the results of the discussion from the previous chapter it can be concluded as follows:

a. The application of the Fuzzy Multi Criteria Decision Making Method can help determine the best alternative in a decision-making system for the best rice selection.
b. Fuzzy Multi Criteria Decision Making using decision support system can generate criteria and weights on each criteria.
c. Fuzzy Multi Criteria Decision Making by utilizing Software Tools Web Browser this system can we apply only as a tool for decision makers in determining the best rice.

5. References

[1] Agustina S 2012. *Sitem Pendukung Keputusan Penentuan Prioritas Pelangan Dialer Suzuki Soekarno-Hatta Malang Mengunakan Metode AHP dan SAW.*

[2] Dey *et al* 2012. *A Moora based fuzzy multi-criteria decision making approach for supply chain strategy selection,* Seminar Nasional Informatika 2011 (semnasIF 2011), ISSN: 1979-2328

[3] Damanik M.A 2013. *Sistem Pendukung Keputusan Pemindahan Tugas Karyawan dengan Menggunakan Metode Profie Matching(Studi Kasus: PT. Perkebunan Nusantara III Medan).* Pelita Informatika Budi Darma, Volume : Nomor: 2, Agustus 201. ISSN 2301-9425

[4] Edi S.W.M., 2014. *Optimization of Group Decision Support System in Updated Pranata Mangsa System Using Fuzzy Multi Kriteria Making (FMCDM) Method. IJCSI International Journal of Computer Science Issues,* Vol. 11, Issue 1, No 1, January 2014. ISSN (Print): 1694-0814 | ISSN (Online): 1694-0784. www.IJCSI.org

[5] Hasugian P.M. 2012. *fuzzy Multiple Attribute Decision Making Untuk Menentukan Tepisang kerja dengan Metode Simple Additive Weighitng (Studi Kasus : PT Cahaya Bintang Medan).* Pelita Informatika Budi Darma, Volume II, Desember 2012 ISSN : 2301-9425.

[6] Hafsah *et al*, 2011 *Sistem Pendukung Keputusan Penentuan Hotel Dengan Menggunakan Metode Promitee dan AHP,* Seminar Nasional Informatika 2011 (semnasIF 2011), ISSN: 1979-2328

[7] Kahar N. dan Fitri N., 2011. *Aplikasi Metode Fuzzy Multi Kriteria Dicision Making (FMCDM) Untuk Optimalisasi Promosi Produk. Seminar nasional Aplikasi Teknologi Informasi 2011 (SNATI 2011)* Yogyakarta, 17-18 Juni 2011 ISSN: 1907-5022

[8] Muktadir A. and Purdianto I., 2013. *Sistem Pendukung Keputusan Kenaikan Jabatan Menggunakan Metode Profile Mitching (Studi Kasus di PT. Industri Kemasan Semen Gersik). Seminar Nasional Aplikasi Teknologi Informasi (SNATI)* Yogyakarta, 15 juni 2013, ISSN:1907 – 5022

[9] Munir A., 2013. *Sistem Informasi Perawatan pada Peraktik Gigi Ingnawati. Makasar.*

IOP Conf. Series: Earth and Environmental Science **156** (2018) 012055 doi:10.1088/1755-1315/156/1/012055

[10] Sałabun W., 2014. *Application of the Fuzzy Multi-criteria Decision-Making Method to Identify Nonlinear Decision* Model, International Journal of Computer Applications (0975 8887), Volume 89 - No. 15, March 2014

[11] Sri Yulianto J.P *et al,* 2008. *Aplikasi Pendukung Keputusan Dengan Menggunakan Logika Fuzzy (Studi Kasus : Penentuan Spesifikasi Komputer Untuk Suatu Paket Komputer Lengkap). Jurnal Informatika,* Vol.4, No. 2, Desember 2008: 159 – 173.

[12] Sagala A., 2011. *Sistem Pendukung Keputusan Penerimaan Jabatan Dengan Metode Fuzzy Multi Criteria Decision Making,* Volume: V, Nomor 1, ISSN: 2301-9425

[13] Valverde R., 2012. *A Risk Management Decision Support System for the Real Estate Industry. 2011 ISSN 2223-4985,* Vol. 1, No 3, Juli 2011.

Rising Company's Performance through Leadership Role: Culture, Strategies, and Management System as a Marine State

Jeni Wardi[1] and Alexsander Yandra[1]

[1]University of Lancang Kuning, Pekanbaru, Riau, Indonesia
Phone: +6281270059669
Email: jeniwardi@unilak.ac.id

Abstract: This research aims to learn the direct influence of transformational and transactional leaderships on Indonesian company's performance through company's culture, strategy, management accounting and control system as a marine state. This research involves descriptive and inferential designs in solving the research problem. To test the model and the hypothesis, SEM analysis is used. The populations of this research are companies registered in Indonesian stock exchange in 2012. The sampling technique uses purposive sampling. The data of the research are obtained from questionnaires distributed to respondents. The respondents are companies' managers represented by accounting and finance managers with the positions 1 and 2 levels below top management team who have direct communication with the top management. The results of the research show that transformational leadership influences company's performance directly, but not the transactional leadership. The company's culture is not the mediation variable in indirect influence on the company's performance, either in transformational or transactional leadership. On the other hand, management control system proves to be the mediation in transactional leadership on the performance but not for transformational leadership. Meanwhile, management accounting system proves to be the mediation variable in the influence of transformational and transactional leaderships. Except the variables of company's culture, strategy, management accounting system and management control system, each directly influences the performance.

Keywords: leadership, transformational, transactional, and system.

1. Introduction

It is known that one of the ways to find out a company's performance comprehensively is using CGPI measurement (Corporate Governance Perception Index). CGPI is the ranking of good corporate governance (GCG) in Indonesia's companies through research designed to support companies to increase quality in applying corporate governance (CG) through continuous improvement by evaluating and benchmarking (The Indonesian Institute for Corporate Governance, 2012:2). In reality, the performance of a company increases and decreases from time to time The below graph shows the CGPI score of some influential companies in Indonesia which are not stable.

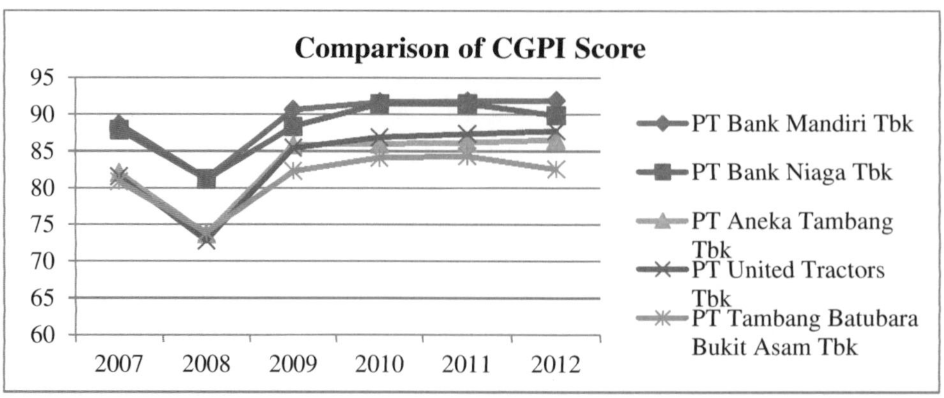

IOP Conf. Series: Earth and Environmental Science **156** (2018) 012056 doi:10.1088/1755-1315/156/1/012056

Figure 1: Graph CGPI Some Companies Year 2007-2012.

Therefore, related to the explanation above, some problems are identified that Company's performance is not stable, seen from the instability of CGPI score of some companies. Nevertheless, The unstable CGPI score shows the instability of some aspects judged by CGPI including leadership. Moreover, Leadership is an important part dealing with human resource related to company's performance, but here's no conclusive research about the influence of leadership on a company's performance. Therefore, in this research, the problem is limited to a direct relation between leadership and company's performance. Besides, it involves contextual variables (culture, strategy, management accounting system, management control system). This study seen several problems that needs to underlie.

However, the main objective of this research is to analyze whether transformational and transactional leadership with contextual variables influence company's performance. Therefore this study consider to propose the *Contingency Theory* states that strategy in line with external business environment determines the company's sustainable performance to overcome the problem. Mangkunegara (2016) urges that performance can be defined as the working results qualitatively and quantitatively achieved by an employee in doing his/her job parallel to responsibility given to him/her. Nevertheless, leadership is a process to influence others to understand and to agree with what should be done and how the job is done effectively and process to facilitate individual or collective to reach the aim together. Changing business environment needs a leader who functions as changing catalysator instead of manager who administrative works and control it. Moreover this study proposed the following framework to overcome the situations.

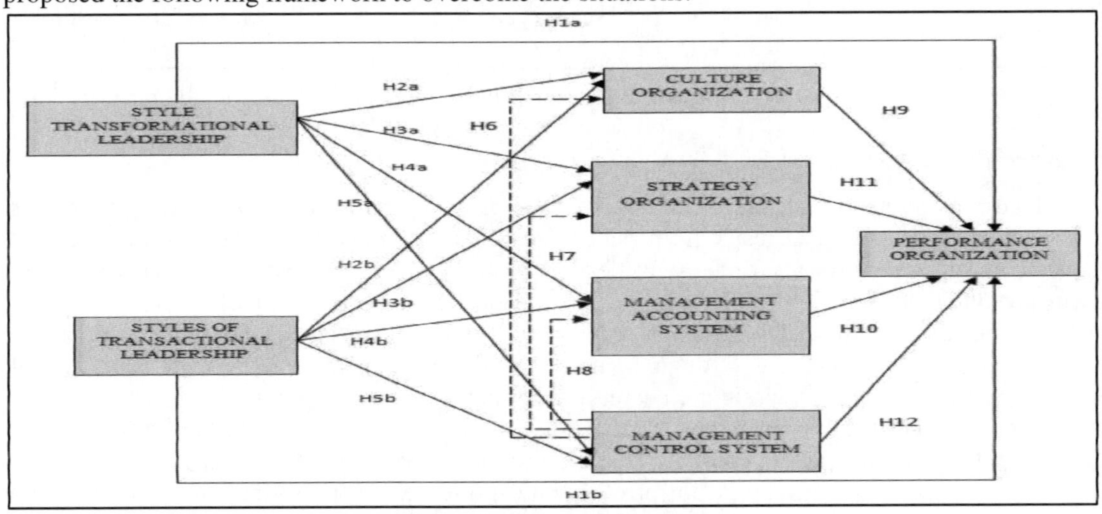

Figure 2: Research Framework

2. Research Methodology

First, description, i.e., research which aims to get a description of the characteristics of research variables. Second, inferential, i.e., to test the causal hypothesis on research model through data collecting. The research data are obtained from questionnaires distributed to respondents. The respondents of this research are companies' managers represented by accounting and finance managers with level 1 and two below the top the top management team who have direct communication with top management. Data collecting is done by using research instrument, i.e., questionnaires which use Likert scale 1-5. This research is done by using primary data, collected through a survey by sending them via post, facsimile, email or giving them direction.

IOP Conf. Series: Earth and Environmental Science **156** (2018) 012056 doi:10.1088/1755-1315/156/1/012056

The research hypothesis is tested by using Structural Equation Model (SEM) which develops and has similar function with double regression with multiple latent independents in which each is measured by some indicators. This SEM uses program or LISTREL software. The research data are obtained from questionnaires distributed to respondents.

The respondents of this research are companies' managers represented by accounting and finance managers with level 1 and two below the top the top management team who have direct communication with top management. The numbers of respondents are 400 managers (from 80 companies) who work in companies registered in Indonesian stock exchange.

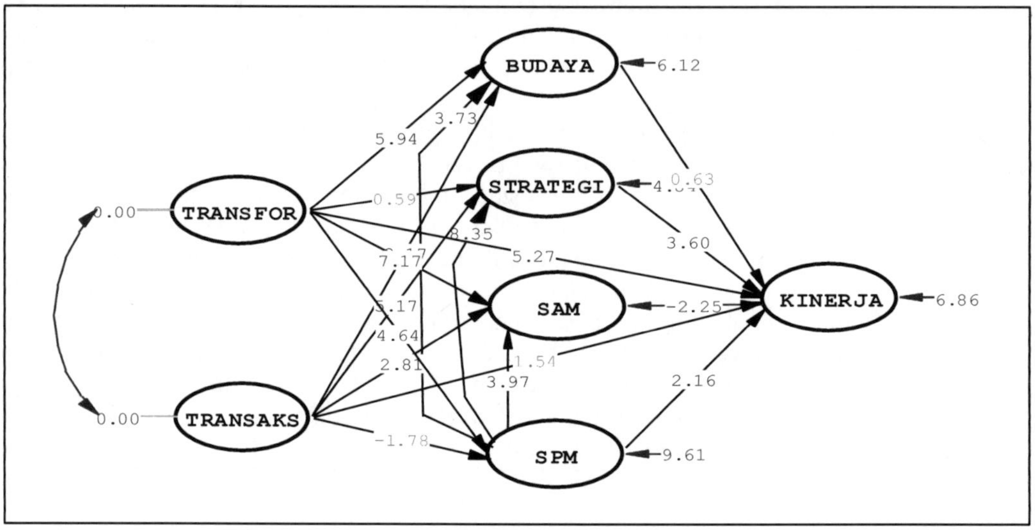

Figure 3: Lisrel Output Model

3. Research Result And Discussion

Transformational leadership influences company's performance directly. It means transformational leadership can result from better performance. Transactional leadership deoes not influence company's performance directly. It means transactional leadership does not motivate better performance. In this research, the indirect influence of leadership to company's performance cannot be seen from company's culture. The research results show that transformational leadership does not influence company's performance through company's culture. It means there is no influence of company's perception on the increasing or decreasing of company's performance.

The research done proves that transactional leadership does not influence company's performance through company's culture. That's why the hypothesis that states that leadership influences company's performance is not proved. The influence of transformational leadership on company's performance through company's strategy is rejected. So, the company's strategy does not strengthen or weaken the influence of transformational leadership on company's performance.

These research results show that the indirect influence of transformational leadership on company's performance occurs through management control system. This result opposes the previous research done by Muchlis (2009) whose research proved that management control system, on one side, can give negative influence on performance. In this case, Muchlish's research showed that socialization process influenced negatively by controlling power, not as intervening variable.Research hypothesis which states that there is an influence of management control system on company's performance through company's culture is rejected. This research shows that management control system can not indirectly influence company's performance through company's culture. Research conducted by Chow et, al., (1999) showed that culture influenced company's design and employees' performance on management control theoretically. Chow et.al.,(1999)

IOP Conf. Series: Earth and Environmental Science **156** (2018) 012056 doi:10.1088/1755-1315/156/1/012056

Different from this research, Lekatompessy (2012) found out that company's culture strengthened the relationship between management control system and company's capability. From this idea, it can be concluded that management control system indirectly influences company's performance through company's culture and this is supported by some previous researchers.

4. Conclusion,

The result of the research can be used as the reference in giving the suggestion, particularly for company's leaders to pay attention to mediation variables to optimize the influence of leadership on company's performance. Related to that, the implications of this research are as follows:

1) This research finds out that there is the direct influence of transformational leadership on performance, but not transactional leadership. So, to increase company's performance it is needed to apply the elements of transformational leadership which can give positive impact optimally to the company's performance.
2) This research also finds out that the most dominant indirect influence of leadership to performance is transformational leadership through management control system. It means that management control system is the most active mediation variable to increase production. So, the companies' leaders can pay more attention to management control system to increase the influence of transformational leadership on the performance optimally.

Managerial ideas that can be recommended related to the result of the research are as follows:

1) Hypothesis test result proves that indirect influence of transactional leadership on performance through management control system is the smallest influence. So, the effort is needed to increase this in the order it can give bigger influence to increase performance.
2) For Hypothesis test result also proves that direct influence of management control system on company's performance is the smallest. So, the effort is needed to increase the influence of management control system on performance, particularly those related to controlling the action management to be parallel with the goal or for the sake of the owner.

5. References

[1] Abdullah dan Herlin Arisanti. 2010. Pengaruh Budaya Organisasi, Komitmen Organisasi dan Akuntanbilitas Publik terhadap Kinerja Organisasi pada Kantor Wilayah Direktorat Jenderal Perbendaharaan Bengkulu. *Jurnal Ekonomi dan Bisnis Fakultas Ekonomi Universitas Syiah Kuala Banda Aceh,* Volume 9 Nomor 2, Agustus 2010.

[2] Akram, M., H. M. Alam, L. Ali, dan M. M. Mughal, (2012), "How Leadership Behaviors Affect Organizational Performance in Pakistan," *Journal of Economics and Behavioral Studies,* Vol. 4 (6), 354-363.

[3] Ho, J.L., C.J. HuangIsmail, T., (2013), "Formatting Strategy and Management Control System: Evidence from Indonesia", *International Journal of Business and Social Science,* Vol. 4 (1), 196-205.

[4] Lay, T. A., dan Ruzita Jusoh, (2012), "Business Strategy, Strategic Role of Accountant, Strategic Management Accounting, and Their Links to Firm Performance: an Exploratory Study of Manufacturing Companies in Malaysia," *Asia-Pacific Management Accounting Journal,* Vol.7 (1), 59-94.

[5] Lekatompessy, J, E., (2012), "Peran Sistem Pengendalian Manajemen dalam Meningkatkan Kinerja Perusahaan: Analisis Kontinjensi dan *Resource-Based View*", *Disertasi,* Semarang: Program Studi Doktor Ilmu Ekonomi Universitas Diponegoro.

[6] Muclish, M., (2009), *Pengaruh Sistem Pengendalian Manajemen terhadap Kinerja dalam Strategic Supply Relationship dengan Kerjasama sebagai Variabel Intervenin",* Semarang: Program Pascasarjana Universitas Diponegoro.

[7] Satyawati, N.M.R., dan I.W. Suartana, (2014), "Pengaruh Gaya Kepemimpinan dan Budaya Organisasi terhadap Kepuasan Kerja yang Berdampak pada Kinerja Keuangan", *Jurnal Akuntansi Universitas Udayana,* 6 (1), 17-32.

Needs Analysis of Employee in Riau Main Island: A Reflection from Universitas Lancang Kuning Context

Hasnati, Bunga Chintia Utami, and Trio Saputra

Universitas Lancang Kuning, Pekanbaru, Indonesia, 28265
Telp: (0761) 53108-53236; Fax (0761) 52 248
E-mail: hasnati_tati@yahoo.co.id

Abstract: Performance of employees is a major concern for all organization. It's depend on workload. The availability of staff in Lancang Kuning University which established in Riau Main island is not adjusted to the changing needs of the Organizational Structure and the workload of the employees. One preparation to Achieve the Strategic HR Planning Is that ripe with a set amount of human resources that is ideal to get *UNILAK UNGGUL 2030*. But, it will not happen if the distribution of employees have not Referred to the real needs of the organization. The distribution has not been based on the workload of the organization. Stacking employees in one unit without clear work and lack of staff in other units is a fact of the underlying problem at Lancang Kuning University. So in order to Achieve the performance of employees, we must to to be done structuring. In the implementation, Lancang Kuning University conducted an analysis of positions that include positions produce information Position Information Analysis. The purpose of this study is to projection of the Needed Employees with qualitative method. The research informants were selected from three institutions / units in the University of Lancang Kuning. The results of the workload analysis of staff at the University of Lancang Kuning can be done if there is clarity of guidelines or work instructions, a good understanding of the work guidelines to everyone, the suitability between work guidelines and knowledge of educational personnel. Based on job information and workloads, it is found that only two institutions are LPPM and BPM roomates is suitable of needs but in other side there are excess and lack, library excess 1 employee.

Keywords: Need Analysis, Employee, workload

1. Introduction

Availability Lancang Kuning University officials today considered not adapted to the changing needs of organizational structure and workload of employees. Inventories of employees is the number of employees who belong to a work unit at this time. In order to support the smooth implementation of the tasks of educators and educators in a faculty, the necessary number and quality of human resources is appropriate to be able to provide services for all stakeholders in an optimal (Nindi Aristi, 2014).By knowing the employee supplies this year will make it easy forclickinghcounted inventory forecasts an employee at a period of years to come. Estimates on the number of employees consist of existing employees, reduced by the amount of the pension during the year. Reduction of employees outside of retirement such as transfer, promotion, stop, and dies hard to predict, so that these reductions are not included in the estimates. In addition, the problem of availability in terms of quantity, it is necessary to note also the quality of employees (performance). Employee performance under the spotlight of various groups and institutions. The problems of employee performance is an important issue because the employee is considered less productive and low work ethic will be a source of waste and be a bad influence on other employees who have a good performance. Stigma was generally addressed to all employees, including the University of Lancang kuning.

Employee culture at the University of Lancang kuning similar to the work culture in other government agencies. It is shown from the cultural and environmental systems in government offices are generally more oriented toward public service and not to the product. As a result, the performance and discipline will always relate to the duties (roles, responsibilities, and functions) as well as the rules governing bureaucracy that is standard. To be able to carry out the work, the employee is always based on the technical and operational guidelines, SK, the work order, and the like. Therefore natural that in each unit and the work environment of government and universities many employees seem less productive. However, it does not mean there are no employees who have potential. Currently just a lot of potential employees, but less chance to be empowered. The ability or the potential of the employees is dependent upon the respective superiors -masing in each work unit. If the boss in each unit to intelligently take policy in empowering subordinates, then there is a chance that employees will work according to their potential. So, the bad stigma on employee performance among employees will be lost if the potential employee was empowered suit its domain expertise and has managed by the leadership of the right.

One of the main causes of low performance of employees is the occurrence of a mismatch between the competence of the employees to the position it occupies. Mismatches caused by the composition of the membership or employee skills are not proportional. Similarly, the distribution of employees still refer to the real needs of the organization, which means that the distribution is not based on the workload of the organization. The backlog of employees in one unit without any specific job and the lack of employees in other units is the reality of the problem. On the other hand the formation of the organization tend to be based on real needs, in terms of organization formed too large while the workload is small, so as not to organizational objectives efficiently and effectively. So in order to achieve the expected gains of employees who need to do the arrangement, in the implementation of the arrangement, Lancang Kuning University conduct job analysis provides information covering positions Position Description, Terms Position, Map Position and Strength Employees. If the information office is available, the Lancang Kuning University undertake a reconsideration of the job information. To facilitate in preparing or reviewing information office, Lancang Kuning University uses the example of office information that has been compiled by BKD agencies that have adapted to the legislation and the need Lancang Kuning University. Further, structuring employees include aspects of quantity, quality, composition, and distribution. The first phase is set arrangement Aspects existing quantity in order to obtain the number of employees in accordance with the needs of the organization, while the quality aspect, composition and distribution will be regulated in further Rector. Implementation of the Arrangement include the following steps:

1) Count the number of required staff conducted in accordance with the statutes and other laws.
2) Elaborate structural and functional positions (job analysis).
3) Develop a map of positions by analyzing the gap between employee profiles with terms of office.
4) Analysis of the workload.
5) Determinutesing Category Number of Employees by comparing the calculation results with the employee needs any office of the existing number of employees, such as employee number Less categories (K), Match (S), and More (L).
6) Projected needs of employees for 5 years
7) an employee redistribution plan report.
8) Report on the evaluation and recommend follow-up measures

The research focuses on the calculation of the workload of employees based on their respective work units. Before calculating the work load, job analysis is done first. Job analysis aims to look at employee understanding of the job description, job specification and job conditions.

2. Method

The research method used Concurrent Mixed Method Triangulation, ie in-depth research by analyzing both quantitative and qualitative methods well then both in comparison (Creswell, 2009). Comparative analysis of positions is expected to be obvious, then the results of research carried out can be used to determinutese the predictive kuning sassy needs of university employees, the study's respondents kuning sassy university employees 60 people and ordinary 9, 1 key informants and informant. Data was collected by using a questionnaire for the quantitative method for qualitative methods while using interviews and comes with documentation.

2.1. Job Analysis

Job analysis is a process of collecting, research, the data decomposition stage position includes the step of preparation and planning, phase of data collection and data processing stage. The implementation stage is described as follows:

1) Planning and preparation phase At this stage some of the activities carried out are:
   a. Reaffirmation of the organizational structure that will be a guide for subsequent processes, including the names of title and place.
   b. Inventory existing positions in each work units and collated based on hierarchy and given an identification code.
   c. Establish data collection methods to be used and set up tools and equally necessary (form etc.).
   d. Forminutesg fielded the analysis and describes the methods to be used.
   e. Communication / explanation by the leaders of work units and all employees of the intent and purpose of job analysis to be carried out.

In order to obtain qualified employees expected, the best way to be done by the leader of an organization is to conduct job analysis or job analysis so that it can see the type and competence of employees required for an organization (Muhammad rizky ramadhan, yoanita yunianti, 2014) government regulations on office management Hours of work Ideal employee that is 72,000 minutes / year (Kepala BKN, 2011) the effectiveness and efficiency of work is a comparison between the weight of work with the normal conditions of the stakeholders to complete the tasks and functions of the organization (Negeri, 2008).

2) Data Collection phase

The data collectionpositions can be done in several ways:
   a. Methods of Observation and Interview. An observation method means executing a job analysis to directly observe on the spot how job tasks performed and recorded to process into information. While the method of interviewing officers interviewed directly incumbent analysis by asking questions in the first set and record the answer to be processed into information necessary.
   b. questionnaire Methods (List of Questions). Data collected through the deployment of a questionnaire to all employees to fill them. The list of questions could be "open" (Open ended) means, answering should give an answer according to his own will in its own way, is not restricted.

3) Phase Data Processing. After the data collection process was completed, the data processing is:
   a. Determinutesing factors of appraisal office
   b. Determinutesing the weight value of each factor
   c. Analysis of the interviews and questionnaires that have been in the content

d. Analysis of job requirements
e. Develop job descriptions
f. Doing patterns ratings workload as the basis of the determinutesation of the system

3. Results and Discussion

Distribution of employees based on work unit can be seen in Table 1 below.

Table 1 : Distribution of Employees Under the Work Unit

No	Unit	Number of Employees		
		Permanent	Contracts University	Contract Institution
1	LPPM			3-1
2	BPM	1	-	-
3	BHE	1	-	-
4	Libraries			6-2

Based on table 1 above it can be seen that the educational staff with the status of education personnel in institutions / unit environment Lancang Kuning University. This study focuses on the permanent staff in 4 units / institutions.

3.1. Library

Optimization of work to do when the number of employees in accordance with the amount of workload. Based on the results of form filling information office, it is known that the workload associated with job descriptions include:

1) Providing service borrowing and returning books
2) Manage Feasibility computer device
3) Caters to college students looking for theses, books, journals and other
4) Inventory of theses, books, journals and other
5) Recapitulate and data input to the system of OPAC, such as data theses, books, journals and other incominutesg
6) doing checking and repairs to the thesis that damaged
7) Making newspaper clippings
8) Processing fines student who fails to return the book
9) doing routine tasks adminutesistrative offices such as sorting the incominutesg and outgoing mail, manage finances and making financial reports, monthly coordination meeting
10) of library materials procurement
11) Print and CopyService

Services borrowing and returning books given to both students and lecturers environment Lancang Kuning University and outside the University Lancang Kuning. Services are provided to the 50 to 100 people every day. Everyone is served about 5 minutesutes. Thus, in the first day of service spent about 500 minutesutes / day or 120,000 / year. Details of the workload based on the task can be seen in the table below.

Table 2 : Details of Output and Turnaround Time Job

No.	Job Description	Output	Time (1 Year)
1	Providing service borrowing and returning books	100 students / day	120 000 minutes
2	Manage Feasibility computer device	12 Activity /week	1440 minutes
3	Caters to college students looking for theses, books, journals and the other on the rack provided	50 students / day	60.000 minutes
4	inventory of theses, books, journals and other	4 activities / week	2880 minutes
5	recapitulate and input data to the OPAC system, such as data theses, books, journals and other entry	8 activities / week	5760 minutes
6	perform checking and repairs to the thesis that damaged	four activities / week	5760 minutes
7	Creating a newspaper clipping	of 4 activities / week	5760 minutes
8	Processing fines student who fails to return the book	10 students / day	12,000 minutes
9	Doing routine tasks such as sorting incominutesg and outgoing mail, managing finances and making financial reports, monthly coordination meeting	1 activity / day	7,200 minutes
10	Acquisition of library materials	4 events /week	480 minutes
11	Print and Copy Service	1 activity / day	43,200 minutes
	Total Time		264 480 minutes / year
	Workload		3.67
	Employee Needs		4

Time completion of the work in the first year in which the Library 264 480 minutes / year divided by the effective working time 72.000 minutes / year so, the number of employees to efficiently this section is four (4) out of the five (5) existing employees. This means that this unit has an excess number of employees as much as 1 (one). Based on the results of form filling information office, it is known that the workload associated with job descriptions include:

1. Providing service borrowing and returning books
2. Manage Feasibility computing devices
3. Serving students seeking theses, books, journals and other racks available
4. inventory of theses, books, journals, and more
5. recapitulate and input the data into the system of OPAC, such as data theses, books, journals and other incominutesg
6. doing checking and repairs to the thesis that damaged
7. Making newspaper clippings
8. Processing fines student who fails to return the book
9. doing routine tasks adminutesistrative offices such as sorting the incominutesg and outgoing mail , managing finances and making financial reports, monthly coordination meeting
10. of library materials procurement
11. Print and CopyService

Services borrowing and returning books given to both students and lecturers environment Lancang Kuning University or from outside environment Universit as Lancang Kuning. Services are provided to the 50 to 100 people every day. Everyone is served about 5 minutesutes. Thus, in the first day of service spent about 500 minutesutes / day or 120,000 / year. Details of the workload based on the task can be seen in the table below.

Table 3 : Details of Output and Turnaround Time Job

No	Job Description	Output	Time (1 Year)
1	Providing service borrowing and returning books	100 students / day	120 000 minutes / year
2	Manage Feasibility computer device	12 Activity /week	1440minutes / year
3	Caters to college students looking for thesis , books, journals and others on the shelf that provided	50 students / day	30,000 minutes / year
4	inventory of theses, books, journals and other	4 activities / week	2880 minutes / year
5	recapitulate and input the data into the system of OPAC, such as data thesis, books, journals and other incominutesg	8 activity / week	5760 minutes / year
6	Perform checking and repairs to the thesis that damaged	four activities / week	5.760 minutes / year
7	Creating a newspaper clipping	four activities / week	5760 minutes / year
8	processing fines student who fails to return the book	10 students / day	12,000 minutes / year
9	doing routine tasks such as sorting incominutesg and outgoing mail, managing finances and making lapo financial ran, the monthly coordination meeting	1 activity / day	7,200 minutes / year
10	Acquisition of library materials	4 events /	480minutes / year
11	Print and Copy Service	1 activity / day	43,200 minutes / year
	Total Time		234 480 minutes/ year
	Workload		3.25
	needs employees		4

Time of completion of work in one year in the Library ie 264 480 minutes / year divided by the effective working time 72.000 minutes / year so, the number of employees who are efficient for this section is four (4) out of the five (5) employees who already available. This means that this unit has an excess number of employees as much as 1 (one)

3.2. Institute for Research and Community Service (LPPM)

Based on information form filling positions, it is known that the workload associated with LPPM job descriptions include:

1. Doing Assistance improvement proposals APBF , APBU and Higher Education
2. Conducting Research and Services APBF
3. Conducting Research and Services APBU
4. Conducting Research and Service of Higher Education
5. trainings to
6. enter the performance
7. inventory of incominutesg and outgoing mail
8. Making financial reports, meeting monthly coordination

Service Assistance improvement proposals APBF, APBU and Higher Education awarded to lecturers environment of the University of Lancang Kuning. Funded research and service with as many as 70 proposals APBU scheme, APBf as many as 100 proposals and 40 proposals of Higher Education scheme. Each mentoring, LPPM do the checking against the proposal format. Usually one proposal could be improved as much as 2-3 times. Each correction process takes about 15 minutesutes. So that 218 proposals with a repair requires 6540 minutesutes / year. Details of the workload based on the task can be seen in the table below.

Table 4 : Details of Output and Turnaround Time Job

No.	Job Description	Output	Time (1 Year)
1	Perform on improvement research proposal	218 documents / year	6540 minutes / year
2	Data Recapitulations and research exam	1 events / year	48,000 minutes / year
3	data recap. and examinutese on APBU and DIKTI	1 events / year	48,000 minutes / year
4	Melakaksanakan Preparation Research and Services APBF, APBU and DIKTI like to contact the reviewer and prepare for the adminutesistration of	3 events / year	24 000 minutes / year
5	training in	5 events / year	3000 minutes / year
6	input performance	2 events /	1200 minutes / year
7	Make activity Report	5 events / year	3000 minutes / year
8	Make Financial Statements	12 events / year	7200 minutes / year
9	inventory of incominutesg and outgoing mail	1 events / year	2400 minutes / year
10	Meeting monthly coordination	12 activities / year	2160 minutes / year
11	Dropping Letter	1 Activity / year	7200 minutes / year
12	Make SOP	1 Activity / year	6000 minutes / year
13	Assisting and m engontrol information on information systems RISTEK DIKTI	1 Activity / Year 30X20X12	7200 minutes / year
	Total Time		165 900 minutes / year
	Workload		2.3
	Needs Employees		3

Time of completion of work in one year in LPPM ie 165 900 minutes / year divided by the effective working time of 72,000 minutesutes / years so, the number of employees to efficiently this section is three (3) persons. It shows that in this unit already has a number of employees in accordance with the workload.

4. References

[1] Creswell, J. W. (2009). *Research Design Pendekatan Penelitian Kualitatif, Kuantitatif, Dan Mixed.* Yogyakarta: Pustaka Pelajar.

[2] Kepala BKN. (2011). *Pedoman Pelaksanaan. Analisis Jabatan.* Jakarta.

[3] Muhammad Rizky Ramadhan, Yoanita Yunianti, Sugih Arijanto. (2014). Analisis Beban Kerja Dan Pengukuran Gap Kompetensi Teknisi Laboratorium Umum Dan Fakultas Z PTS XYZ. *Jurnal Insitutr Teknologi Nasional*, *2*(3), 384–394.

[4] Negeri, M. D. (2008). *Perhitungan Beban Kerja Dalam Rangka Menyususn Formasi Pegawai.* Jakarta.

[5] Nindi Aristi, Hanny Hafiar. (2014). Analisis Beban Kerjatenaga Pendidik Dan Kependidikan Di Fakultas Y Universitas X. *Jurnal Kajian Komunikasi*, *2*(1), 53–60.

[6]Yandra, A. (2017). Fisibilitas Pilkada Serentak Tahap II Kota Pekanbaru Pasca Permendagri No 18 Tahun 2015. *Jurnal Niara*, *9*(2), 62-74.

IOP Conf. Series: Earth and Environmental Science **156** (2018) 012058 doi:10.1088/1755-1315/156/1/012058

A Closer look on Ineffectiveness in Riau Mainland Expenditure: Local Government Budget Case

Alexsander Yandra, Sri Roserdevi Nasution, Harsini and Jeni Wardi
Universitas Lancang Kuning, Pekanbaru, Indonesia, 28 265
Tel / Fax: +62761 53 353
E-mail: alexsy@unilak.ac.id

Abstract: this study discussed about the issues on ineffectiveness of expenditure by one Indonesia local Government in Riau. This Provence were amounted Rp. 10.7 trillion through Local Government Budget (APBD) in 2015. According to data from Financial Management Board and Regions Assets (BPKAD) APBD Riau in 2015 stood at approximately 37.58% until October 2015,another data taken from the Ministry of Home Affairs, Riau regional budget, from January to December 2015, it shows the lowest in Indonesia which amounted to 59.6%. The percentage described the lacking implementation of the budget, this can be interpreted that Riau government is less optimal and irrelevant in spending the budget in 2015. Through a theoretical approach to government spending, the implementation of public policies showed the ineffectiveness of the budget that have implicated towards regional development. As regional budget is only the draft in achieving the targets. Budget management in 2015 by the provincial administration through the Local Government Unit (SKPD) shows unsynchronized between the Medium Term Development Plan with the work program from SKPD.

Keywords: ineffectiveness, expenditure, Riau, Policy.

1. Introduction

Local government Budget (APBD) prepared in accordance with the requirements of governance and capacity of local revenue. Budget preparation was guided by local government work plan (RKPD) in order to actualize the public service to achieve the state purpose. Based on Government Regulation No. 58 Year 2005 on Regional Financial Management in Chapter IV Preparation of the draft budget Article 29 through Article 42 explained that: "The *process of preparing the proposed budget based on the Local government medium term development plan (RPJMD), Local Government Work Plan (RKPD), Policy general local government Budgets, interim Budget priorities and funding ceilings (PPAS) Local government department budget and work plan (RKA-SKPD)*".

The regulations above supposed to be a strict guidance for management budget in Riau. Yet, in 2015, the draft budget sets revenue and expenditure amount to Rp10.7 trillion. The Riau Government and the Regional representative assembly (DPRD) approving the draft to become a Regional Regulation No. 4 of 2014 and Governor Regulation No. 27 Year 2014 concerning the budget in 2015 as a reference for government work units in the use of local finance. [1]

But the tendency of budget implementation in Riau Province in 2015 is ineffective. The data from Financial Management Board and Asset (BPKAD) is budget realization in 2015 only in 37.58% until October 2015. Uptake of APBD Riau in 2015 got bad predicate from the central government. Until 30 July 2015, the realization of APBD Riau 2015 is the fourth lowest in Indonesia [2].

Furthermore, the Ministry of Home Affairs noted that the calculation of Riau local government budget in January-December 2015 is the lowest in Indonesia by 59.6% and including in top 5 provinces with lowest realizations. Issues of APBD Riau Province in 2015 is interesting to be examined and assessed by conducting a comprehensive and in-depth research to find an area of the financial governance model that is effective as a manifestation of intensive regional development.

This study refers to the implementation of development public policy research focusing on the problems of implementation constraints Public Policy Regional Regulation No. 2 2005 on transparency and participation. The problem is comparable to the budget implementation of Riau in 2015, the policies that made by government are not relevant to the needs of the public and making some ineffectiveness. For that reason, we need the analysis that refers to the grand theories that approach the concept of regional spending which describes about standardization expenses and integrated with theory of public policy implementation.

2. Methods

The methodology used in this research is descriptive qualitative phenomenological [7]. The unit of analysis is the Riau provincial government, namely the Board of Finance and Asset Management (BPKAD), Regional Planning Agency (Bappeda), Riau DPRD Budget Committee. The researcher is using observation technique with a pattern of investigation, interview and documentation. Information obtained being paired through Focus Group Discussion (FGD) with the Indonesian Forum for Budget Transparency [5]. Then there will be qualitative interpretations by using emic and ethics interpretation. Emic interpretation is interpretation of the data subject problem on the environment and the surrounding world. While ethics interpretation is interpreted the view of the researcher that based on the study of relevant literature [2].

3. Results and Discussion

Riau province has high revenue (PAD), however it is not supported yet by the local government capabilities in realizing the targets and public expectation. The low budget uptake in 2015 that is set in Regional Regulation No. 4 of 2014 and Governor Regulation No. 27 Year 2014 for 2015 budget describe that the provincial administration has ignored the rules they had created with Parliament (DPRD). Therefore, Riau APBD in 2015 is not optimal in the context of the financial budget use. Even the relevance is slightly missing the target when viewed from the infrastructure development program because most of the financial outlay was used for personal expenditure. Some of the findings indicated that the most budget constraints preference in the implementation of Riau APBD in 2015, whose realization has not yet reached the optimal target, is due to the political conditions and the condition of public administration Riau Province in 2015.

3.1 Budget Gap APBD Riau Province

Revenues and expenditures from Riau Provincial Government from 2011-2015, tend to overestimate (underestimate) expenditure and underestimate (underestimate) revenue. The following table illustrates the disparity in Riau APBD expenditure if reviewed in the last five years.

Table. 1: Riau APBD expenditure 2011- 2015

	Total Revenue (Miliar)		Total Expenditure (Miliar)		Surplus/Defisit (Miliar)		Variance Surplus/Defisit (Miliar)	%	Variance Revenue Surplus/Defisit (Miliar)	%	Variance Expenditure Surplus/Defisit (Miliar)	%
	OB	AB	OB	AB	OB	AB						
2011	4.625,00	5.440,00	4.798,00	4.265,00	-173,00	1.175,00	-1.348,00	-679,19	815,00	117,62	-533,00	88,89
2012	5.488,00	6.847,00	6.367,00	8.737,00	-879,00	-1.890,00	1.011,00	215,02	1.359,00	124,76	2.370,00	137,22
2013	6.938,00	6.995,00	8.915,00	7.525,00	-1.977,00	-530,00	-1.447,00	26,81	57,00	100,82	-1.390,00	84,41
2014	7.127,00	8.132,00	8.277,00	5.602,00	-1.150,00	2.530,00	-3.680,00	-220,00	1.005,00	114,10	-2.675,00	67,68
2015	7.407,00	6.911,00	11.388,00	7.761,00	-3.981,00	-850,00	-3.131,00	21,35	-496,00	93,30	-3.627,00	68,15

Keterangan:

TR : Total *Revenue*
TE : Total *Expenditure*
OB :*Original Budget* ; AB : *Actual Budget*; VE: *Variance Expenditure*
(-) : minimum realization
(+) : Above standard realization

The table above shows the imbalance or budget gap when viewed from the side of high revenue is not proportional to the low expenditure. The pattern becomes the mainstream of APBD 2015 Riau which impact on the performance of Riau Province government in the eyes of the public, because the commitment in the acceleration of welfare and development feels sluggish away from public expectations. The expenditure of APBD Riau shows an ineffective tendency due to the orientation of budgetary usage based on difficult work programs achieved within the span of a year. Governments in Riau Province tend to perform positive variants on the receiving side and negative variants on the expenditure side systematically. These conditions illustrate that the consistency of Riau Provincial Government in realizing the 2015 APBD is still considered very minimal. When viewed from the regional revenue from 2014-2017 fluctuation of regional income in Riau Province is relatively good and does not interfere with the course of the economy and development.

3.2 Political Implications vs Riau APBD 2015

Political support is essential in the process of formulation, implementation, discussion and evaluation of the budget. This means that political stability will affect the use of funds in the implementation of public administration, economic development and infrastructure. Reviewing the determination of the budget in 2015 there was no discussion with the Legislative Lists 2014 to 2019, because of APBD Riau in 2015 held before the inauguration of Parliament Members in September 2014. There was an error when the 2015 budget set but did not involve the members elected legislature. It describes that the local government, in this case the governor, tends to enforce APBD in 2015 before the inauguration of the new council, the assumption is to reduce the budget and the political dynamics, and also Annas Maamun as the Governor have more flexibility in determining the appropriate budgets for political will. The high desire (the importance) of political governor in making some programs and activities in the budget 2015 creating a negative impact because it does not consider a possible future political and budget items. A governor is supposed to be able to become a right leader in determining the direction of short term, medium term, and long term regional development policy, and not based on the political. Elected legislators considered that the inconsistency of APBD Riau in 2015 with Draft Medium Term Development (RPJMD) is impact on some programs in APBD Riau 2015 doesn't run well (not effective), according to the Law No. 23 of 2014 on local governments states that everything that is using APBD should refer to RPJMD.

The regulations directing the organization as well as local government bureaucracy that resulted in a change of regional work units (SKPD) and the changing in government. The role of the bureaucracy is still an important political instrument in realizing political goals [3]. Instead, bureaucracy failed miserably as a political force that is professional and value-free society partnership. The effectiveness of this regulation is two years after its enactment; therefore, APBD Riau in 2015 is still in a transitional period of local government regulations. The enactment of Law No. 23 of 2014 on Government Regional and at the same time Governor of Riau suspected to diversion of protected forest case in Kuantan Sengingi. The series of events had implicated for shifting the regional head (governor) Riau replaced by the deputy governor as head of the region with temporary status (Plt) or not yet definitive. This is the political factor which lead to weak *political will* of governments to implement the pure-APBD and there is changing in revised-APBD.

3.3 Administrative Governance APBD Riau

Administration process is an important step in the implementation of the Local Government Budgets Riau Province. Political instability can be controlled if the administrative is able to be the best partnership between the political elite (Governor), bureaucracy and society (Civil Society). The commitment in administrative process must be supported by the atmosphere of governance that oriented towards achieving the vision by short, medium, and long term Development Plan. Looking the administration of the budget in 2015, there is some problems of ineffectiveness due to APBD poor absorption. One of the reason is the result of bureaucratic performance because twice rotational leadership in bureaucratic organizations is actually slow down the APBD progress due to an adaptation and a slow synchronization program. The change of leadership in public administration institutions (bureaucracy) still held hostage by political interests, these interests are not progressing the budget but rather slowing down the progress of the budget. The reason is APBD Riau in 2015 is a political agreement and political consensus of the non-active governor and the bureaucracy leadership worried to be dragged in the corruption case.

The basic APBD Riau Province in 2015 was 10.7 trillion then revised-APBD and reaching 11.3 Trillion with the realization only 7.7 trillion or 68%. If seen from the implementation data of APBD Riau in 2015, 82% over personnel expenditure, 86% for profit sharing shopping and shopping hibah is 102% that exceeding the expectation. The visible lowest uptake is 35 % from financial assistance spending, 32% from social assistance expenditure and 0% from expenditure unexpected. From the total uptake of the APBD Riau in 2015 is 59.76% capital expenditure and 6.19% on goods services. The last four years (2014-2017) there was a trend that shows the decision taken by the local government in the APBD implementation is not effective comparing to the increase in revenue and the budget realization.

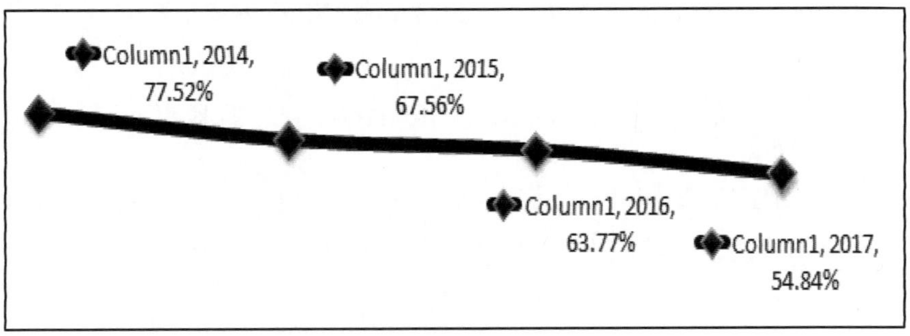

Chart. 1: Graphic description of slow performance

The graphic description is a form of slow performance of the bureaucracy and the fiscal discretion in Riau province, reviewed from the growing number of PAD from pure-APBD to revision-APBD. The local fiscal strength relies heavily on its revenue (PAD), especially in taxes. The larger the fiscal space available, the more flexible the local governments develop the regional development. Unfortunately, the statement is not in line with the expectations, as the facts show the fiscal strength of Riau Province in the last four years (2014-2017) continued to decline despite the revenue sector (PAD) increasing annually.

4. Conclusion

The ineffectiveness of local government budget Riau in 2015 is due to the high political desire governor that did not consider the possibility of political and financial budgets eye area, such as no synchronization between program and budget activities with the Medium Term Development Plan (RPJMD) Riau. Regulations changes on local government also have implications for changes in the bureaucracy organization structure that slows down the performance because the adaptation to the new leader towards adjusting the previously planned program, the implication make discretion in fiscal policy in Riau Province in 2015.

5. Acknowledgment

The authors would like to express their sincerely gratitude to the Indonesia Ministry of Higher Education for granted with the PDP Scheme 2017.

6. References

[1] Antara and Riau Pos, September 5, 2015

[2] Riau Setdaprov Assistant II, Masperi. Personal files of APBD-2015.

[3] Gaffar, Afan. "Politik Indonesia Transisi Menuju Demokrasi." Pustaka Pelajar, Yogyakarta: 2006.

[4] J. Moleong, Lexi. *Metode Penelitian Kualitatif Edisi Revisi*. PT. Remaja Rosdakarya, Bandung: 2011.

[5] Forum Indonesia Transparansi Anggaran (FITRA), Riau

[6] Mangkosoebroto, Guritno. Ekonomi Publik. Yogyakarta: BPFE.

[7] Nugroho, D Riant. 2009. *Public Policy*. PT Elex Media Komputindo Kelompok Gramedia, Jakarta: 2009.

[8] Yandra, A. (2016). E-goverment dengan Memanfaatkan Teknologi Informasi. *POLITIK*, *12*(1), 1769.

[9] Yandra, A. (2017). Fisibilitas Pilkada Serentak Tahap II Kota Pekanbaru Pasca Permendagri No 18 Tahun 2015. *Jurnal Niara*, *9*(2), 62-74.

IOP Conf. Series: Earth and Environmental Science **156** (2018) 012059 doi:10.1088/1755-1315/156/1/012059

Riots in Indonesian Main Island: Peace Action of Defend Islam Impact on Economic Environment in Local Media Discourse

Junaidi Junaidi[1], Alexsander Yandra[2] and Budianto Hamuddin[3]

[1]Dept. of Indonesian Studies, Universitas Lancang Kuning, Pekanbaru. Indonesia
Email: junaidi@unilak.ac.id
[2]Dept. of Politic and administration, Universitas Lancang Kuning, Pekanbaru 28265-Riau, Indonesia
Email: alexsy@unilak.ac.id
[3]Dept. of English, FKIP. Universitas Lancang Kuning. Pekanbaru 28265-Riau, Indonesia
E mail: budihamuddin@unilak.ac.id

Abstract: Indonesia is a maritime country with the largest numbers of islands in the world which covering more than seventeen thousands islands. There are thousand tribes and ethnics with their cultures suppose to be enriching the diversity of Indonesia. However a series of riots happening in Indonesia including *Aksi Bela Islam (ABI)* recently challenging the *unity in diversity* of Indonesia. This present study tries to describe *Aksi Bela Islam (ABI) (*The Peace Action of Defend Islam) rally that run peacefully brings impacts on Indonesia economic sector as stable social and political condition will bring significant impact on Indonesia economic sector. The rally is a long journey of Indonesia of Muslim majority to seek justice as the Republic of Indonesia State Police seems not to be serious to handle the Islamic blasphemy case. Through the framing approach, it will be described how ABI brings impacts on the economic sector focusing on the study on the media perspectives. From the perspective of one leading economic magazines in Indonesia *Bisnis Indonesia.*The framing from the media evidently showed that the rallyfollowed by millions of Indonesian Muslims does not bring negative impacts on the economic sector of Indonesia sinceit run peacefully, well manage and safe so the market gives a positive response and appreciationto the action.

Keyword: Media Discourse, Indonesian Economic and Islamic Movement

1. Introduction

It is interesting to read an article tulisan "*MUI, Gerakan Islamis, dan Umat Mengambang" (*MUI: the Indonesian Ulama Council), the Islamist Movement and Swing Muslims' by Nur Ichwan (2016) regarding the different perspective on the Peace Action of Defend Islamthat takes place as an attempt to demand justice by the majority of Muslims of Indonesia. In his presentation, Nur Ichwan explained that by the end of 2016, Indonesia was slanted by *Aksi Bela Islam (henceforth: ABI)* rallies carried outon October 14, 2016, November 4 and December 2 - the two latter rallies are known as the '411' and '212'Actions. These actions were triggered by the statement of the Governor of Jakarta, Basuki Tjahaya Purnama (a.k.a: Ahok), in Kepulauan Seribu on September 27, 2016 that is convicted of blaspheming against the Koran, the Ulama and even Muslims around the world. The proponents of these rallies demand that Ahok be immediately arrested and imprisoned.

The rallies were followed by various elements of Muslim society mobilized from various regions in Indonesia. Some estimated that the number of Muslim taking a part in these rallies was approximately seven million. However, based on the mathematical calculations on Google map,it is estimated the number was lower reaching about 2.3 million. Whatever the exact number is, there has been no protest of Indonesian Muslims involving a very huge number of participants so far.

The actions were taken along with the high tension of Jakarta gubernatorial election campaign of 2017, followed by three candidates and their running mates, Agus Harimurti Yudhoyono-Sylviana Murni, Basuki Tjahaya Purnama (Ahok)-Djarot Saiful Hidayat, and Anies Baswedan-Sandiaga Uno. The first and the third candidates and their running mates are all Muslims, while the second candidate and his running mate are non-Muslim and Muslim. The second candidate, the incumbent had stirred up a controversy after Joko Widodo was elected as the president of Indonesia and Ahok, previously deputy governor substituted Joko Widodo as the governor. From the start, even since Joko Widodo-Ahok won in the election, the Islam Defenders Front (FPI) had become the opposition and taken the resistance movement.

When Ahok started to place his gubernatorial position, FPI appointed KH. Fakhrurrozi Ishaq a shadow governor, Ahok is known for his uncompromising style of leadership and his frank style of speech. Some of his policies are considered controversial. In one of his speeches, he said that he would dismiss FPI. The speech made FPI that claims the movement of '*nahi mungkar'* (abandon bad ones) increasingly inflamed. Anti-non-Muslim and anti-Chinese sentiments are echoed through various forums and media. The peak of the controversy was that when the non-Muslim Chinese figure commented on'*dibohongi pakai al-Maidah 51*'(lied with al-Maidah 51) that is assumed to be edited by Buni Yani with "lied with al-Maidah 51" on his Facebook wall on October 6, 2016 (Nur Ichawan 2016). Nur Ichwan (2016) affirmed clearly that all the boisterousmovement of ABI started from there. Islamist movements were then mobilized maximizing networking and social media usage and cyber army by framing the 'blasphemy' against the Koran and an insult to Ulama and Muslims. The anti-non-Muslim and anti-Chinese were combined with the political agenda of gubernatorial election of DKI Jakarta. Under such circumstances, "religious opinions and attitudes" emerged, which were then popularized as 'fatwa' (order) issued by MUI. The council assured that Ahok did commit blasphemy against the Koran, Ulama and Muslims. Ahok-Rizieq hostility then found its momentum to be blown up as Ahok-Muslim hostility, and even hostility between Christian (religious mission) and Chinese (business, even in certain cases between Chinese communism versus Muslim. The blown hostility was successfully used to convince a huge number of Muslims to participate in the rally of ABI 1, 2 and 3.

Businessmen, of course, require peaceful and stable political atmosphere as a guarantee for them to invest. In the writer's point of view, the super peaceful action of ABI certainly bringsan impact on the economic development of Indonesia nowadays regardless whether the impact is significant or not. The impact can be seen from the way entrepreneurs to respond the ABI actions in Indonesia, such as (1). Before ABI run, a lot of businessmen went abroad. (2). Investors postponedinvesting their capitals in Indonesia (3) Businessmen strongly rejects the rallies resulted in unfavorable business atmosphere and made economic activity hampered. From the perspective of Culture and Media Studies, the phenomenon of ABI rallies is certainly interesting to be studied, especially on how the mass media in Indonesia do the construction on informing of the demand of the majority of the Muslims. Considering the politics and economy of the media, the owners of the media, of course, believed that this case is differentand the way to reconstruct it is different too. The above fact raised a research question: *What are the impacts of the Peace Actions of Defend Islam on the Indonesian economic sector from the mass media perspective?*. From the research question, the purpose of this study is designed. The purpose of this study is to find out patterns of mass media framing about the Impact of the Peace Actions of Defend Islam on the Indonesian Economic Sector.

IOP Conf. Series: Earth and Environmental Science **156** (2018) 012059 doi:10.1088/1755-1315/156/1/012059

2. Methodology

There are a number of studies that seek to examine the relationship between the growth of Islamic movement in post-reform Indonesia and the political-economicalcondition that surrounds it. Hadiz (2014) argues that the post-reformist Islamic movement in Indonesia is characterized by the expansion of the social base of the movements. The Islamic movement that is often supported by the merchant class, little bourgeoisie, or educated middle class is now supported by a large alliance deriving from cross-clas.

Further, the Islamic movement, in this case, uses the rhetoric of the Muslims to influence and to change the existing political-economic condition in such a way that resources can flow more for the movement. Meanwhile, in his study of the Islamic vigilantism group in Jakarta, Wilson (2011) argues that along with the fall of Suharto, the political elite approached the vigilantism groups in order to gain support, especially from conservative Muslims. It made the vigilantism groups no longer merely a client of the political elite as they experienced during the New Order. The vigilantism groups grew to be an established actor in the Indonesian political arena and their existence is required by the political elite to keep their power stable. At this point, the rhetoric of Islamic morality is present to legitimize the existence of the vigilantism groups in the political arena. At this point, the vigilantism groups have freedom to channel resources for themselves as well as the group they represent.

Rizky Alif Alvian affirmed that the strengthening of Islamic narrative in the political arena of Indonesia reflects not only efforts to change the distribution of resources within society but also the fragility of pluralism issue where economical gap and uncertainty grow in society. The inability of the narrative of pluralism to integrate the difficulties of community life into its systematic explanation of the world created so-called *Laclau and Mouffe*as a 'dislocation'. At this point, if a gap between the daily experience of society and the narrativethat seeks to confine the experience, other narratives can penetrate and fill the gap left by the narrative of pluralism. The right-wing populism primarily supported by Front *Pembela Islam* (FPI) (Islamic Defenders Front), *Gerakan Nasional Pengawal Fatwa Majelis Ulama Indonesia* (GNPF-MUI) (National Movement of the Fatwa Guards of the Indonesian Ulama Council), Hizbut Tahrir Indonesia (HTI), and other elements of Nahdlatul Ulama (NU) and Muhammadiyah gain a lot of support and the climax was in ABI because the promotion to defend Islam was strongly echoedat the moment. The existence of the right-wing populism, therefore,is not merely instrumental to the interests of the group that carries it. The strengthening of Defend Islam issue also enabled a transformation process that preceded it. ABI provided means for community to voice their unrest while at the same time the right-wing populism tried to understand the daily experiences of the community.

Robert N. Entman is one of the experts who laid the groundwork for framing analysis for media content studies. The concept of framing is written in an article for the Journal of Political Communication and other journals that implanted the concept in a media case study. The concept of framing by Entman is used to describe the selection process and to highlight certain aspects of reality by the media (Eriyanto, 2001). In Entman's point of view, media framing is implemented in four stages. The first stage isDefining Problems. This stage is the master frame / the most important frame. The second stage is Diagnosing causes (estimates the cause of the problem). It is a framing element used to frame who is considered an actor of an event. The cause here can refer to what, but may refer towho. The third stage is Making moral judgment (making moral choice). Framing Element is used to justify / to argue the definition of problems that have been made. The four stage is Treatment recommendation (Emphasizingproblem solving).

Therefore this descriptive qualitative study used the used the framework inline with the constructivist paradigm proposed by Schwandit, 1994: 128. Therefore this study considers that there is no objective reality, because reality is created through a certain construction process and view. This study aims to describe and summarize various conditions, situations or various phenomena of social reality that exist in the society that became the object of research, and attempt to pull that reality to the surface as a characteristic, character, model, sign or picture of condition, certain situation or phenomenon. Thus, the qualitative descriptive format is more appropriate when it is used to examine problems that require in-depth study such as the study of consumer behavior, media effects, and implementation of a policy (Bungin, 2008: 28-29), namely by using sentences that correspond to theories contained in the presentation of data.

Table 1.
Research Unit (*Bisnis Indonesia*)

No	Edition	Headline
1	5 Nov. 2016	*Jakarta Terkendali* (Jakarta under Control)
2	3 Dec. 2016	*Damai Tertib Aman* (Rallies are Peaceful)

3. Findings and Discussion

How the Peaceful Actions of Defend Islam as an attempt to seek justice for the majority of Indonesian Muslims on the blasphemy against the religion by the Governor of DKI Jakarta, Ahok bring impacts on the economic sector in the media frame. In the discussion, the framing will be analyzed based on Robert N Enman model which represented by the headline.

3.1 The Headlines

The Peaceful Action of Defend Islam held on November 4, 2016, went so peaceful that it did not bring negative impact on economic sector of Indonesia is *defining problem* of *Bisnis Indonesia* headline, "*Jakarta Terkendali" (*Jakarta under Control), Saturday, November 5, 2016 Edition. Even, the President of the Republic of Indonesia, Joko Widodo expressed to the community leaders, the Ulama who coordinated their people and control the action so that it ended peacefully. The factual text is written:

> *Presiden Joko Widodo menghargai aksi unjuk rasa yang tertib dan damai. Kepala Negara juga menyampaikan terimakasih kepada tokoh dan ulama yang memimpin ummatnya mengawal aksi unjuk rasa sampai maghrib bisa berlangsung tertib dan damai (*Bisnis Indonesia, ***Jakarta Terkendali***. Edisi, 5 November 2016, Hal. 1*)*
>
> (President Joko Widodo appreciated the peaceful and orderly rally. The President also expressed his gratitude to the prominent figures and Ulama who coordinated their people in controlling the rally until the maghrib that can run orderly and peacefully (*Bisnis Indonesia*, Jakarta under Control. Edition, 5 November 2016, p.1)

The Peaceful Action of Defend Islam (ABI) is framed by the media on two stressingpoints that basically do not bring negatively impact on economic sector of Indonesia: *Firstly*, that the Peace Action of Defend Islam (ABI) that run in Jakarta is positively responded by market makers and businessmen. Secondly, ABI that run peacefully became an indicator of the maturation of the democracy in Indonesia. The two factors mentioned above are *diagnose course*of the headline. The fact is written as in the following:

*Para Pelaku Pasar dan kalangan usaha merespons positif demo yang berjalan damai dan disebut mencerminkan semakin matangnya iklim demokrasi di tanah air"(*Bisnis Indonesia, ***Jakarta Terkendali***. Edisi, 5 November 2016, Hal. 1*)*

(Market markers and businessmen respond the peaceful rally positively and the rally reflectsthat democracy in Indonesia is getting better and better (*Bisnis Indonesia*, ***Jakarta under Control***. 5 November 2016Edition, p.1)

The Chairman of Kamar Dagang dan Industri (Kadin) (the Chamber of Commerce and Industry of DKI Jakarta, Salman Simanjorang appreciated and strongly affirmed that this is a strong signal for potential investors and economic actors not to hesitate to invest in Indonesia. The fact is written as in the following:

*Ini membuktikan kita sudah dewasa dalam berdemokrasi dan menghormati perbedaan yang ada.Ini juga sinyal yang kuat bagi calon investor dan pelaku ekonomi untuk tidak ragu menanamkan modalnya di Indonesia(*Bisnis Indonesia, ***Jakarta Terkendali***. Edisi, 5 November 2016, Hal. 1*).*

(This proves that we have grown up in democracy and respect the diversity. This is also a strong signal for potential investors and economic actors not to hesitate to invest their capital in Indonesia (*Bisnis Indonesia*, ***Jakarta under Control***. 5 November 2016Edition, Page 1)

That ABI does not affect the decline of Indonesian economy can be seen from various perspectives. Fundamentally, it can be seen when ABI run, Composite Stock Price Index (IHSG) and Rupiah currency strengthened on the day despite weakening in the morning session. It certainly makes moral judgment of the 'Jakarta under Control' headline.The fact is written as in the following:

*Dari pelaku pasar keuangan, pada sesi perdagangan kemarin, Indeks Harga Saham Gabungan (IHSG) dan mata Uang Rupiah kompak menguat setelah dibuka melemah pada sesi perdagangan pagi. IHSG ditutup menguat 0,62% atau 33,16 poin menuju 5.362,66. Indeks berhasil bergerak dalam rentang 5.303,36 – 5.362,66 dan berbalik menhijau pada sesi II setelah dibekukan melemah pada posisi 5.310,78 (*Bisnis Indonesia, ***Jakarta Terkendali***. Edisi5 November 2016, Hal. 1*).*

(From the financial market markers, at the previous day trading session, both the Composite Stock Price Index (IHSG) and Rupiah currency strengthened after opening weakened in morning trading session. IHSG closed up 0.62% or 33.16 points to 5,362.66. The index moved within the range of 5,303.36 - 5,362.66 and turned in session II after it was recorded weakened to 5,310.78 (*Bisnis Indonesia*, ***Jakarta under Control***. 5 November 2016Edition, Page 1).

Meanwhile, the*treatment recommendation*is that the Republic of Indonesia State Police will investigate the allegation of blasphemy against Islam in which the reported doer is the Governor of DKI Jakarta, Basuki Tjahaja Purnama. The fact is written as in the following:

IOP Conf. Series: Earth and Environmental Science **156** (2018) 012059 doi:10.1088/1755-1315/156/1/012059

> *Sebelumnya, Polri menyatakan akan mengusut dugaan penistaan agama dengan terlapor Gubernur DKI non aktif Basuki Tjahaja Purnama. Bareskrim Polri telah mengagendakan pemeriksaan Ahok pada Senin (7/11) dan mempersilahkan Ahok menghadirkan saksi lain(*Bisnis Indonesia, ***Jakarta Terkendali***. Edisi, 5 November 2016, Hal. 1*).*
>
> (Police had affirmed before that it would investigate the alleged blasphemy in which the reported doer is the non-active Governor, Basuki Tjahaja Purnama. Criminal Investigation Board of the Republic of Indonesia State Policehas scheduled a preliminary investigation on Ahok on Monday (7/11) and ordered Ahok to bring another witness (*Bisnis Indonesia*, ***Jakarta under Control***, 5 November 2016 Edition, Page 1).

The peaceful, orderly and safe rally inolving millions of people got the President of the Republic of Indonesia, Joko Widodo decide to attend and do the Friday prayer amid millions of mass action. Jokowi passed through the south gate of *Istana Merdeka* at about 11:30 pm, accompanied by Vice President Jusuf Kalla and a number of *Kabinet Kerja* ministers. Some colleagues of *Istana Merdeka* suggested that The First Person of the Repulic of Indonesia met the crowd. However, some others prohibited Jokowi to attend the historic moment. That the president came after he had obtained security assurance from the Commander of the Presidential Security Guards, Brigadier General Bambang Suswanto who affirmed that the atmospher was fairly safe, and eventually Jokowi came is the *dignose course* of *Bisnis Indonesia* headline written **"*Damai Tertib Aman*"** (Peaceful, Orderly and Safe, Saturday, December 3, 2016 Edition. The fact is written:

> *Bahkan Presiden Joko Widodo dan Wapres Jusuf Kalla turut bergabung salat Jumat bersama lautan massa di Lapangan Monumen Nasional(*Bisnis Indonesia, ***Damai Tertib Aman***. Edisi, 3 Desember 2016, Hal. 1*).*
>
> (President Joko Widodo and Wapres Jusuf Kalla even joined the Friday prayers together with millions of the rally participants at the National Monument Square (Bisnis Indonesia, Peaceful, Orderly and Safe. December 3, 2016 Edition, P. 1)

Then, in the next analysis that becomes *make moral judgment* is that ABI 212 that went peacefully made the market and entrepreneur enthusiatic and show positive trends. More specifically, the Rupiah exchange rate moved in the green territory along the trading that coincides with ABI 212 which closed Rp.13,512/US dollar. In addition, IHSG(CompositeStock Price Index) climbed and closed 0.91% by 47.2 points to 5245.96. The composite index even climbed 2.61% from 5122.1 of the previous week. This condition is affirmed by the statement of Chairman of Apindo (Indonesian Enterpreneurs Association), Hariyadi B. Sukamdi. Hesaid "*berbagai ancaman yang menakutkan berhasil dibalik oleh Presiden.Ini akan direspons positif oleh pelaku usaha*" ('various threats that are frightened are successfully settled by the President. It will be positively responded by the business actors'). Furthermore, Chairman of Kadin (Indonesia Chamber of Commerce and Trade), Rosan Roeslani said"*Semua damai, semua positif*"('It is peaceful, it is possitive'). The fact is written:

> *Nilai tukar Rupiah bergerak di teritori hijau sepanjang perdagan kemarin hingga ditutup Rp. 13.512 per dolar AS. IHSG ditutup naik 0,91% sebesar 47,2 poin ke level 5.245,96 pekan ini, IHSG melejit 2,61% dari 5.122,1 pekan sebelumnya(*Bisnis Indonesia, ***Damai Tertib Aman***. Edisi, 3 Desember 2016, Hal. 1*).*
>
> (Rupiah exchange rate that was in the green territory in the yesterday's trading session closed Rp.13,512/US dollar. Composite Stock Price Index recorded increased 0.91% or 47.2 points to 5245.96 this week, the stock index climbed 2.61% or 5122.1 of the previous week session (Bisnis Indonesia, Peaceful, Orderly and Safe. Edition, December 3, 2016 Edition, P. 1)

Then, the *treatment recommendation* is referring to the statement of the Minister of Religious Affairs, Lukman Hakim Saifuddin "*Ya tentunya kita sudah tidak mengharapkan lagi ada aksi. Ini kan tuntutan mereka sudah, agar (Ahok) diproses hukum.* Hal senada juga disampaikan oleh Kapolri Tito Karnavian "*Tidak ada satu pohon pun yang patah.Yang melaksanakan kegiatan ibadah bentul-betul menjaga agar pohon tidak rusak* ('Yes, we certainly do not expect any action anymore. The participants of the rally had voiced their demand that Ahok be investigated. The Chief of the Indonesia State Police, Tito Karnavian also affirmed that 'There is no even single tree damaged. Those who did the worship truly keep the tree is not damaged (Bisnis Indonesia, ***Damai Tertib Aman***. Desember 3, 2016 Edition, P. 1).

3.2 Impact of ABI: Economic Sector

Indeed, the social political atmosphere is very influential on the economic development of a nation. Stable economics and security is an ideal condition expected by the market and business actors. ABI which is a place to express freedom of speech to demand justice will generally be judged differently by the market. Different from other rallies, ABI 212, which runs very peacefully is very positively responded by the stock market and business actors. In this paper, the writer will attempt to explore the impact of ABI on the economic sector in a different perspective that is in the view of the mass media. The two major national media (*Bisnis Indonesia*and*Kontan*) which write ABI as headlinesaffirmed in almost all of their headlines that ABI has no negative impact on economic sector, meaning that only one of the four headlines that stated that ABI brings negative impact on economy but the impact is even not significant.The following table illustrates more detailed about the headlines written in the media.

4. Conclusions

The objective of a reasearch commonly is to answer the research problem and prove the research objectives. Therefore, based on the research findings and data interpretation in the previous chapters, the writer draws following conclusions:

First, ABI is framed by the media on two stressingpoints that basically do not bring negative impact on Indonesia's economic sector: *(1)* ABI running in Jakarta is positively responded by market makers and businessmen as it runs peacefully, orderly and safe. *(2)* The peaceful rally is an indicator of the maturity of the democratic climate in Indonesia. *Second*, the angle of *Kontan* is different from that of *Bisnis Indonesia* in the construction of the reality of ABI. An action that ends unrest must be stopped, if not, it will impact on crippling economy of Indonesia. *Third*, ABI 212 that went peacefully made the market and businessmen enthusiatic andpositive market response can be seen. In addition, Rupiah exchange rate that was in the green territory in the yesterday's trading session that coincided with ABI 212 closed Rp.13,512/US dollar.In addition, IHSG (Composite Stock Price Index) climbed and closed 0.91% by 47.2 points to 5245.96. The composite index even soared 2.61% from 5122.1 of the previous week trading session. *Fourth*, ABI 212 brings no impact on the Composite Stock Price Index. The stock price index is even influenced more by global sentiment such as the OPEC decision in which Indonesia withdrew temporarily from the policy of Organization of Petroleum Exporting Countries. The stock exchange price index strengthened supported by the strengthening of the exchange rate to US dollar. Rupiah perched at Rp. 13,463/US dollar.

A research is not 100 percent accomplished if it does not close with recommendations or suggestions. The recommendations of the study are as follows: *First*, this media research uses Robert N Entman framing model and an opportunity to carry out similar studiesusing other framing models is possible. *Second*, in addition to framing, to obtain a comprehensive research findings, the writer recommend to study the same object in the perspective of discourse analysis, content analysis and other research approaches.

5. Reference

[1] Al-Turabi, Hasan. 2003. *Al-Islam wa al-Hukm*. London: Al-Saqi

[2] Alif Alvian, Rizky. *Ekonomi-Politik Aksi Bela Islam: Pluralisme dalam Krisis?*.MAARIF Vol. 11, No. 2 — Desember 2016.ISSN : 1907-8161

[3] Bayat, Asef "*Islamism and Social Movement Theory*," Third World Quarterly, 26: 6 (2005), 891-908.

[4] Berger, Peter & Thomas. 1967. *The Social Construction of Reality: A Treatise in the Sociological of Knowledge*.NY, A Double Day Anchor Book

[5] Berger, Peter & Thomas. 1967. *The Social Construction of Reality: A Treatise in the Sociological of Knowledge*.NY, A Double Day Anchor Book.

[6] Moelong, Lexy.2007. *Metodelogi Penelitian Kualitatif*, Bandung , P.T Remaja Rosdakarya

[7] Burhan Bungin.2008.*Penelitian Kualitatif, Komunikasi, Ekonomi, Kebijakan Publik, dan Ilmu Sosial lainnya*.Kencana. Jakarta.

[8] Eriyanto.2001. *Analrsis Wacana, Pengantar Analisis Teks Media*.Yogyakarta: LKiS

[9] Hadiz, Vedi R. 2014. "*The Organizational Vehicles of Islamic Political Dissent: Social Bases, Genealogies and Strategies*." Between Dissent and Power. Ed. Khoo Boo Teik, Vedi R. Hadiz, and Yoshihiro Nakanishi. London: Palgrave Macmillan UK

[10] Hamuddin, Budianto. 2015. DISCOURSE ON MEDIA: Bringing Hot News into ELT's Classroom Discussion. *Proceedings of ISELT FBS Universitas Negeri Padang*, *3*, pp.87-95.

[11] Nur Ichwan, Moch. *MUI, Gerakan Islamis, dan Umat Mengambang*. MAARIF Vol. 11, No. 2 Desember 2016. ISSN: 1907-8161

[12] Telle, Kari. 2013"*Vigilante Citizenship: Sovereign Practices and the Politics of Insult in Indonesia*." Bijdragen tot de taal-, land- en volkenkunde / Journal of the Humanities and Social Sciences of Southeast Asia 169.2–3: 183–212

[13] Wilson, Ian. 2011 "*Reconfiguring Rackets: Racket Regimes, Protection and the State in Post-New Order Jakarta*." The State and Illegality in Indonesia. Brill

Blogs as Powerful Learning Tools: The Perception from EFL Students in Riau Main Island Indonesia

Budianto Hamuddin[1] and Dahler[2]

[1]Dept. of English, FKIP. Universitas Lancang Kuning, Pekanbaru, 28265. Indonesia
E-mail: budihamuddin@unilak.ac.id [2]
[1]Dept. of English, FKIP. Universitas Lancang Kuning, Pekanbaru, 28265. Indonesia
E-mail: dahler@unilak.ac.id

Abstract: Promoting the potential value of blogs in English Language Teaching (ELT) for English as a Foreign Language (EFL) students is the focus of this present study. It's aimed to explain in a quantitative way the perceptions from English as a Foreign Language (EFL) students in using blogs in Riau main island context. The students from PBIG FKIP Universitas Lancang Kuning in Riau mainland, Indonesia was selected based on the increasing trends of the blog used as alternative media to learn English during these two years. The analysis of this present study based its data from online questionnaire as the main instrument to collect the data. The online questionnaire was open from August 2016-February 2017 (6 months) and filled up by 161 students from English Department of FKIP Unilak who enroll in 6 different blog-based classes at PBIG FKIP Unilak. The analysis showing that the majority of the students had a positive perception that blog can be used as powerful tools to promote English language skills especially reading and writing. The data also reveals that using the blog as a learning journal can raise students' motivation in learning English and publishing article. These positive results somehow showing that blog can be promoted as an alternative media of learning English for English as Foreign Language Students at PBIG FKIP Unilak or in other institutions around the globe that teach English as a Foreign language.

Keywords: Blog, ELT, EFL, Unilak, Riau Mainland.

1. Introduction

A mammoth alike studies rise in the past few years discussed the use of the blog in educational institutions to teach English as a Foreign language (henceforth: EFL). Blog In recent years has been widely used as an effective tool to reach the global result related to teaching EFL. In many studies, blogs were found increased students' knowledge as well as their critically thinking. This trend happened around the globe because blog used, not only limited as media to improve students skills in English but also strategies to raise students motivation in EFL classrooms. There are many studies can be taken as references in teaching EFL context around the globe related to this matter such as Aydan: 2014; Aydin: 2014; Gutler: 2011; Noytim: 2010; Hamuddin: 2016, which have proven the effectiveness of blog to improve skills, motivation and critical thinking in many educational institutions. There are so many beneficial reasons in using the blog in the learning process. The study above underlined Blog can increase students' motivation and ability to think critically. It also develops students' motivation as well as to train the students more independent in learning. For that reason, the study above having a similar thought to use the blog as learning to raise the classroom's creative atmosphere.

In many institutions blog usage has changed education point of view to be more comprehensive and varied in contributing to ease English language teaching including in ESL or EFL context. Through the use of the blog, accessing information will be easy and allowed the students can learn anywhere and anytime, as well as the lecturer can upload their material with no limitation of time and place. The development of education which used a little touch from IT has to contribute the development and renews the conventional learning media. It becomes more creative and innovative as this is what is happened in Universitas Lancang Kuning (henceforth: Unilak) as one of the renowned private University in Indonesia expected so far. Realizing these benefits, English Department of FKIP Unilak (henceforth: PBIG) starts to used blog in teaching English for EFL students since two years ago in six different blog based classes. In the past two years the number of blog writers (blogger) in PBIG Unilak, Pekanbaru Indonesia increasing significantly [5].

Based on pre-observation in this study in May 2016, some lectures and hundreds of students in PBIG Unilak was an active blogger, this is in line with the department policy that recommended the use of IT in the ELT classroom. The trends of blogs usage increasing in PBIG FKIP Unilak somehow because blog was used as learning media as well as a strategy in some courses i.e. Reading, Writing, Intro to Linguistic, Semantic, Discourse Analysis, and Prose [5]. Moreover, blogging club also founded and becomes one of the favorite clubs in co-curricular activities in PBIG Unilak (Hamuddin: 2016). The pre-observation also found blogs in these classes were used to post an assignment such as article review, presentation slide, and discussion forum. Nevertheless, the pre-observation also found that in the learning process in some of these courses is somehow showing that blog is not only for writing and reading exercise in ELT context but also becoming tools for students to up-to-date with the recent study regarding their assignment and stay connected with peers and their lecturer [5].

2. Method

The study sees perception as cognitive processes with the aims of giving interpretation, brand, judgment, opinion, regarding on something, someone or situation, based on the experience, background, and the individual's knowledge to form a comprehensive idea about it. Therefore, this study employs quantitative descriptive to interpret the students' perceptions on the use of the blog as a learning tool to study English in PBIG. This study was started in August 2016 focusing on two different semesters (2 and 4) students that enroll in 6 blog-based classes namely, Reading 2, Writing 2, Structure 2, Introduction to Linguistic, Semantic dan Morpho-syntax at PBIG FKIP Unilak.The data collection in this study was based on the questionnaire to maintain the quantitative perspective.

The online questionnaire designed to describe the majority of students' perception on the blog usage as well as to see the major description for this topic regarding the blog as learning media, student's journal or strategy in some courses at PBIG FKIP Unilak. The questionnaire in this study starts to distributed since the beginning of August 2016-February 2017 (6 months) and analyzed at the end of March 2017. The main data of this study comes from an online questionnaire. The questions in the questionnaire were designed to covers; demography, and the use of the blog as a learning journal at the English Department Faculty of PBIG FKIP Unilak. It consists of with multiple choice options to ease the students in answering the questionnaire. The students' demography described in the next section below:

3. Result

The data recorded 161 out of 168 of the total students of 2nd and 4th semester who participated in the survey. This study calculated 89.8 % of the respondent was female with 136 students, and the remaining 10.2 % was 23 was male students as seen in the figure below.

IOP Conf. Series: Earth and Environmental Science **156** (2018) 012060 doi:10.1088/1755-1315/156/1/012060

The data showed that the majority who took part in this study somehow was female, this is clearly reflecting the quantity which indicates female students is more dominant in term of quantity at PBIG Unilak.

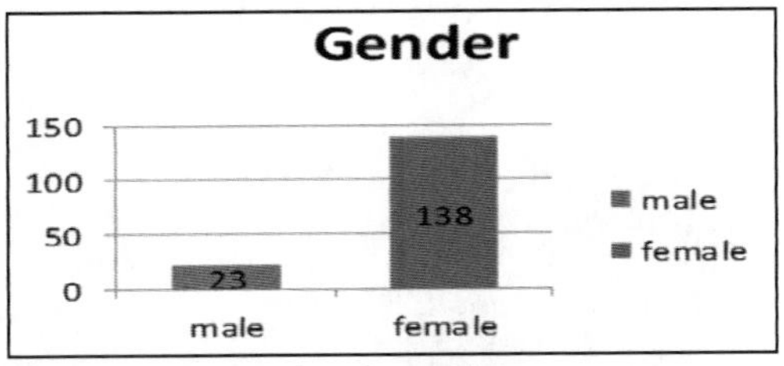

Figure 1: respondents based on gender

The questionnaire was softly distributed to the 2nd and 4th-semester students which consist of demographic questions, perceptions, and their problems. However only this paper only discusses and describe the students' perception on blog usage at PBIG FKIP Unilak. The result described based on the items below.

On the question number 1-2 merely on demographic questions such as "*Saya sudah memiliki blog selama?...*" (I have been blogging since?) The data showed that majority of the students (84%) use blog for 1 month – 6 months, meanwhile 6 % have the blog for 6 month – 1 year. and as much as 5 % from students user blog has been blogging for less than 1 month. This is means that the students were newbie blogger or just blogging during the semester. The data somehow calculating 98% of the students enroll in the blog-based classes have positive experiences regarding blogging activities or blog based resources in PBIG Unilak

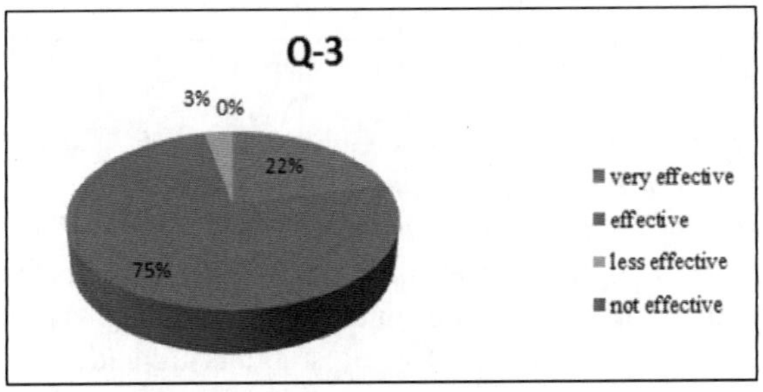

Figure 2: Blog as an Effective media

On the question number. 3 "*menurut saya media blog sangat... sebagai media pembelajaran bahasa Inggris di PBIG FKIP Unilak*". (In my mind blog is . . . as learning media in learning English at PBIG) Data on the diagram above showed that 75% PBIG students declared that blog was effective as learning media then followed 22 % with very effective and only 3 % students mention less effective. This data means that majority (97%) of students at PBIG giving a positive response about the effectiveness blog as media of learning. Therefore, this showing that blog was effective in the learning process at PBIG Unilak.

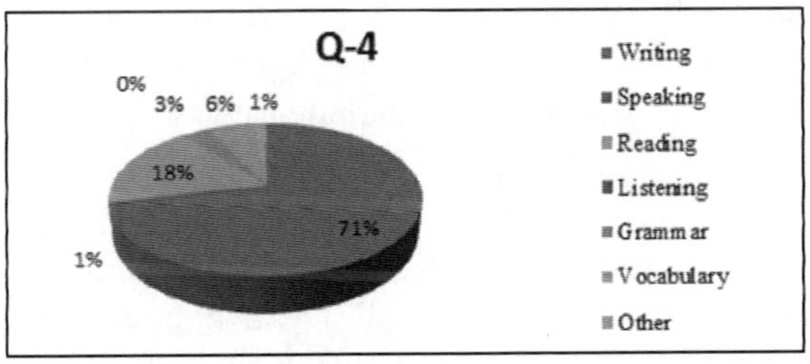

Figure 3: Blog usage in ELT context

On the question number. 4 "*Menurut saya, menggunakan blog sebagai media pembelajaran lebih tepat digunakan dalam membantu meningkatkan kemampuan /pengetahuan .... saya*" (In my mind using Blog as media to learn English is useful to develop my...skill" Data on the diagram above showed most of the students (71 %) were prefer writing as one of the skills which can be increased significantly using blog as the media. While 18 % the students chose to improve the reading skill than 6 % from the students chose vocabulary skill followed with 3 % choosing grammar and 1 % answered speaking skill and 1 % choosing the other skills. This data showed that based on their experience, writing and reading skill were the 2 most improved skills regarding the use of blog at PBIG.

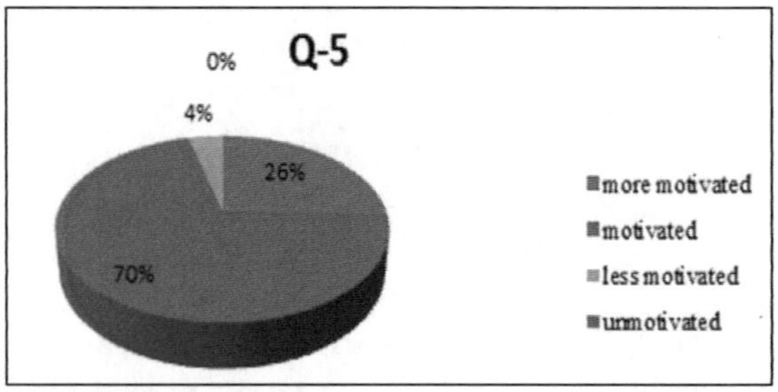

Figure 4: Rising motivation

On the question number 5 "*Dengan memiliki blog, saya....... dalam mengerjakan tugas yang diberikan dosen*".(by having a blog I feel … to do my task) the diagram above showed most of the students (70 %) feeling motivated in doing the task, while 26 % of the students mentioned more motivated and only 4 % of the students were less motivated in doing the task. This data indicated that the majority of the students at PBIG feel the motivation was rising regarding the use of a blog.

On the question number 6 "*Saya percaya penggunaan blog sebagai media pembelajaran akan ..... skill saya dalam Bahasa Inggris*" (I believe my English skill will be… with the use of blog as a learning) The data on the diagram above reveals that the majority students (81 %) at PBIG experiencing improvement after studying using blog and only 2 % of students were less improving with the use of blog.

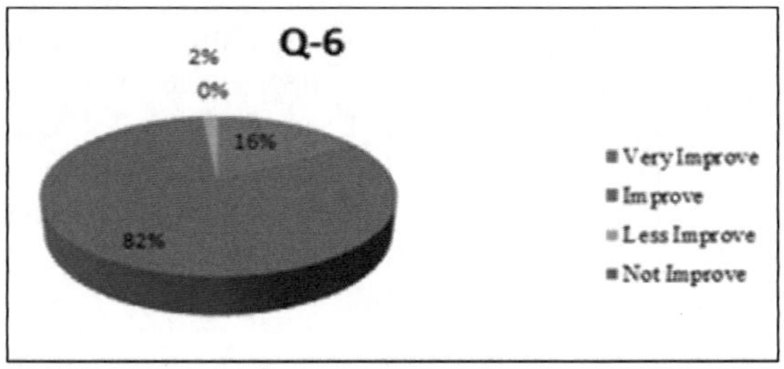

Figure 5: Students' believe on improvement

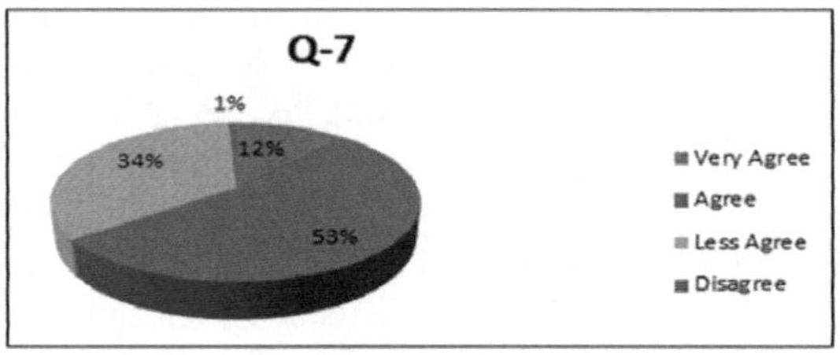

Figure 6: Preferring to use a blog than other media

On the question Number. 7, "*Saya lebih suka belajar bahasa menggunakan blog dibandingkan belajar dengan menggunakan media lainnya*" (I prefer studying using blog that other media in learning English) The data above showed that more than a half (53%) of students agreed that they would rather be using blog than other media, and 12% was very agree. However, 34% of students answered less agree to use blog than to learn using other media. However, 12% of students answered strongly disagree in learning using the blog, and 1% of the students mentioned disagree. From the data above showing the majority of students (67%) prefer to use a blog than to learn by using other media. This may be due to media blog that was easy to apply in learning English.

On question number 8, "*Saya berfikir penggunaan media blog bisa diterapkan dihampir semua mata kuliah di jurusan Bahasa Inggris*"(I think the use of media like blog as learning journal can be implemented in almost every courses in English department) This study showed 57% of students agree and 15 % very agree. This is means the majority of the students (72%) having a positive perception on the implementation of blog in many courses, even though there were 15% of students answered strongly disagree and 4% of students' answered did not agree with the use of blogs as a media can be applied in almost all courses in the department of English. From the data above this study believed blog can be used and implemented in many courses

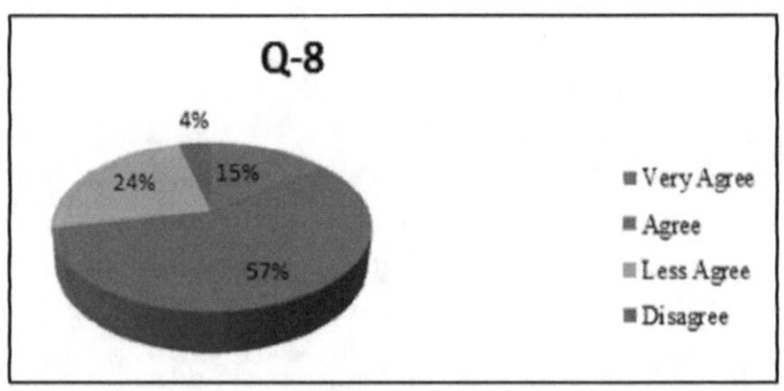

Figure 7: Perception on Blog implementation in ELT context

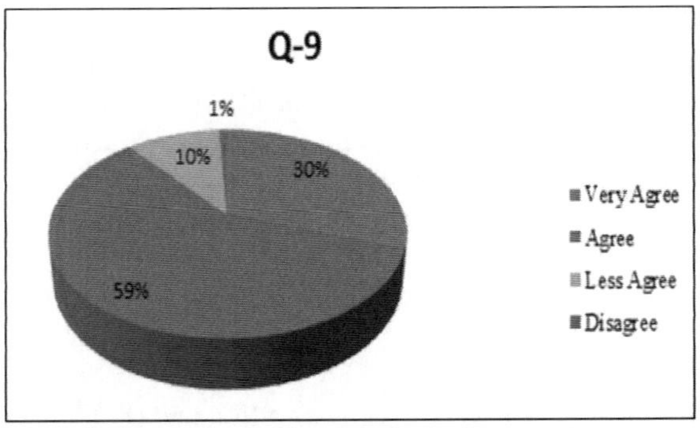

Figure 8: Expectation on blog usage in the department

On the question number. 9, "*Saya berharap penggunaan strategi pengajaran menggunakan media blog bisa terus di terapkan di jurusan Bahasa Inggris FKIP Unilak*" (I am expecting that blog can be used consistently as a teaching strategy in my department) the diagram above showed that 59% of students agree and 30% said very agree and expecting that blog can be used again in the next semester. Since only 11% of students answered less agree and disagree that blogs can be continued to be applied in PBIG Unilak.

The data above reveals that the majority students PBIG Unilak expecting that blog can be used continuously and consistently as a learning media and strategy at PBIG FKIP Unilak as it somehow effective and efficient to be applied.

On the question number 10, *Menurut saya aplikasi maupun feature di blog itu... dalam belaja bahasa Inggris di PBIG FKIP Unilak*" (in my opinion apps and features on blog ...... to learn English). Data on the diagram above showing the majority of students (61 %) mentioned that apps and features on blog support to learn English and 34 % students mentioned very support in learning English, meanwhile only 5 % students felt less support. This data showed a positive perception that the apps and features on blog can help as well as facilitating the learning of English in PBIG Unilak.

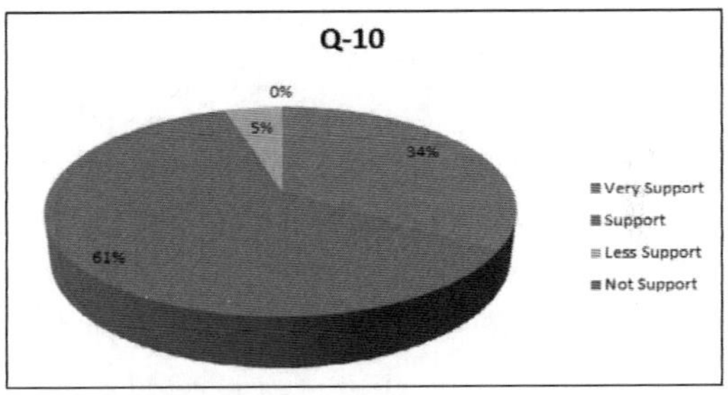

Figure 9: Blog usage to support ELT

Blog as learning journal somehow connecting students with many parties. It showing clearly On the question number 11, *"Menurut saya, penggunaan blog sangat memudahkan saya dalam ....."* (In my mind blog helping me a lot in ...) The data showed that most of students (57 %) were felt blog can be used as source of learning and 43% can share knowledge and getting feedback instantly. This data somehow showing a positive perception that blog is a powerfull tools to help students in ELT context.

4. Conclusion

This descriptive quantitative study based its data from 161 students who enroll 6 different blog-based classes. This study tries to help academia in ELT context by sharing PBIG students' experiences on what they have been through in the last few semesters regarding the use of the blog in the class. PBIG starts using a blog in the past two years to broadening their ELT perspectives. This present study conducted as to review the use of blog whether as a media of learning, strategy and learning journals at PBIG FKIP Unilak. The idea is to try to raise students' English skills and rising their motivation in their ELT classes. Pre-observation and online questionnaire were used to find out students perspective on the use of the blog. This study starts with collecting data soon after the semester end, August 2016 until February 2017. The study can reveal some points based on the data collected.

The data revealed that PBIG students had a positive perception on the implementation of the blog in many courses. They believed their English skill especially writing, reading skills, as well as grammar and vocabulary developed significantly with the use of the blog as learning journals, media, and strategy at PBIG FKIP Unilak. Nevertheless, their positive perception also described in the data on how they believed its effective as learning media due as it facilitating them in learning of English at PBIG Unilak.

Nevertheless, it caused the majority of the students experiencing an improvement in learning English. The majority expressing that blog is a suitable strategy and effective media to help EFL students at PBIG Unilak. Through the blog, the students experiencing a positive academic atmosphere and help them to train their communication skill whether among peers as well as with their lecturers. This present study believed through the data lies in the analysis that blog somehow perceives as powerful tools for English as foreign students in PBIG FKIP Unilak. Therefore, this study recommended the blog as an effective tool and should be able to broaden ELT goals and objectives in a higher education.

5. Acknowledgement

The authors would like to express their sincerely gratitude to the Indonesia Ministry of Higher Education for granted with the PDP Scheme 2017.

6. References

[1] Aljumah, F.H.2011. Saudi Learner Perceptions and Attitudes towards the Use of Blogs in Teaching English Writing Course for EFL Majors at Qassim University. *The Journal of English Language Teaching*. 4(3): 100-116.

[2] Aydan, Lütfiye Seda. 2014. Student and Teacher Perceptions on Benefits and Challenges of Using Blogs in English in Foreign Language Reading Classes. Master *Thesis*. Ankara: Midle East Technical University Press.

[3] Aydin, Selami. 2014. The Use of Blogs in Learning English as a Foreign Language. *Mevlana International Journal of Education (MIJE)*. 4(1): 244-259.

[4] Fageeh, A.I. 2011. EFL Learners' Use of Blogging for Developing Writing Skills and Enhancing Attitudes Towards English Learning: An exploratory study. *Journal of Language and Literature*. 2 (1): 31-48.

[5] Hamuddin, Budianto. 2016. "Using blog to Promote English Skills for EFL Students: The Students' Perception." ELT-Lectura 3 (2): 22-28

[6] Noytim, Usa. 2010. Weblogs Enhancing EFL Students' English Language Learning. *Procedia - Social and Behavioral Sciences*. Volume 2, (2): 1127–1132.

[7] Özdemir, Emrah, Selami Aydın. 2015. The Effects of Bl ogging on EFL Writing Achievement. The Proceedings of the 1st *GlobELT* Conference on Teaching and Learning English as an Additional Language., Balikesir: Balikesir University

[8] Wekke, Ismail Suardi, Alexander Yandra, and Budianto Hamuddin. 2017."Learning Strategy in Class Management: A Reflection from Manado Case." *IOP Conference Series: Earth and Environmental Science*. Vol. 97. No. 1. IOP Publishing.

Tourism and Environmental Policy Strategies: Promoting Local Destination in Riau Province

Prihati Prihati[1], Surya Dailiati[2], Hernimawati Hernimawati[3], Alexsander Yandra[4]

[1,2,3,4]Dept. of Public administration, Universitas Lancang Kuning, Pekanbaru, Indonesia
Telp: +620761 53108, Fax: +620761 52248
Email: alexsy@unilak.ac.id

Abstract: This study is based on the principal issues concerning the policy implementation of Pekanbaru tourism promotion which has not been optimally being able to develop the tourism potential in Riau Province. It was estimated to be caused the Pekanbaru City Government strategy c.q Pekanbaru City Tourism Office has not been able to develop the tourism potential in Pekanbaru City. The method used in this research is descriptive qualitative, which aims to formulate strategies that can be applied to tourism promotion policy implementation of Pekanbaru City can support the development of the tourism potential of Riau Province. The collected data were analyzed with an interactive model of Miles and Huberman. The research result indicated that strategy can be applied so that the Pekanbaru City tourism promotion policy implementation be able to support the development of tourism potential of Riau Province create based on reality that tourism development policy of Pekanbaru City which has failed and benchmarking or comparison both with Batam City and Padang City, then the strategy which can be applied is strategy for making the Pekanbaru City as a MICE (Meeting, Incentive, Conference, Exhibition) City.

Keywords: Tourism, Development, Policy Implementation.

1. Introduction

Pekanbaru City is not main tourism destination cities nationally, but the city has a variety of tourism potential which includes 14 leading tourism attractiveness, i.e.: Nature tourism attractiveness including *Kaca Mayang* Fishing Park, Princess *Kaca Mayang* Park, Rumbai Lake Recreational Park (3 tourism attractiveness); Cultural tourism attractiveness including Indigenous House of Melayu Riau, *Sang Nila Utama* Museum, An-Nur Great Mosque, Seri Harbor, Great Mosque and *Marhum Bukit* Cemetery as well as *Marhum Makam* Cemetery (5 tourism attractiveness). Shopping tourism attractiveness including Dekranasda Riau, Lower Market Shopping Tourism in Pekanbaru City, Senapelan Plaza, Sukarani Plaza, Pekanbaru Mall and SKA Mall (6 tourism attractiveness). That potential should be developed to attract tourists for a visit so that the tourism of Pekanbaru City will align with other regions in Indonesia. The policy of Tourism of Pekanbaru City starting with the enactment Regional Regulation of Pekanbaru City No. 13 of 2000 on Regional Tourism Promotion. This policy based on the ideas that each region has its peculiar attraction to motivate tourists to visit.

This research is based on the principal issues concerning the policy implementation of Pekanbaru tourism promotion which has not been optimally being able to develop the tourism potential in Riau Province. It was estimated to be caused the Pekanbaru City Government strategy c.q Pekanbaru City Tourism Office has not been able to develop the tourism potential in Pekanbaru City. Related to this Head of Pekanbaru City Tourism Office said that for the development of a leading tourist attraction still has constrained in funding because investors who visited the locations are always asking various requests and can not be fulfilled. This research was inspired by previous studies such as research by Krutwaysho (2013) and research by Wirudchawong (2012) who found the model and development strategy to successfully develop Phuket and Koh Yao Island as a tourist destination. Besides this study also refers to the study by Sukmana (2010) who developed a model of tourism environment in Batu City and the study by Kartodihardjo *et al.,* (2010) who succeeded in formulating the strategy to develop tourism in the Yapen Islands Regency of Papua Province.

The excess of this study lies in the formulation of strategy so that the policy implementation of tourism promotion in a region can promote the development of the tourism potential in the regions concerned. This strategy formulated by comparing the existing policy with 2 (two) regions with a background of the same customs or almost the same which has advanced in tourism and combining with the theory/model of policy implementation of Mazmanian and Sabatier which actualized by Gustama (2013: 125-126). Referring the above conditions, This study intends to examine strategies that can be applied to tourism promotion policy implementation of Pekanbaru City can support the development of the tourism potential of Riau Province.

2. Method

In this study, the authors used qualitative research methods. Qualitative research is essentially observing people in their environment, interact with them, trying to understand the language and their interpretation of the world around it (Sugiyono, 2012: 64). With the use of qualitative methods, the data obtained will be more complete, more in-depth, credible and meaningful, so that the research objectives can be achieved. The use of this method is done considering the study is related to many factors. When using quantitative methods will not find data that is the work process, the development of an activity, a description of a broad and deep, feelings, norms, beliefs, mental attitude, work ethic and culture adopted person or group of people in their environment (Sugiyono, 2012: 65).

The reason for using qualitative methods because in essence is to observe people in their environment, interact with them, trying to understand the language and their interpretation of the world around him. With the use of qualitative methods, the data obtained will be more complete, more in-depth, credible and meaningful, so that the research objectives can be achieved. The use of this method is done considering the study is related to many factors. When using quantitative methods will not find data that is the work process, the development of an activity, a description of a broad and deep, feelings, norms, beliefs, mental attitude, work ethic and culture adopted person or group of people in their environment. This study used data collection techniques of observation / observation and in-depth interviews /*in-depth interviews* and documentation. The third technique is described as follows:

1) Observations / Observations in question is the observation that systematically about the incident and behavior in *settings* socialselected for the study.
2) In-depth interviews*(in-depth interviews)*in-depth interviews are data collection techniques that are based on an intensive conversation with a specific purpose. Interviews were conducted to obtain information concerning the matters raised in the study. Interviews were conducted with the informant who considered master research problem.
3) Documentation of data collection techniques by studying documents, reports, pictures, electronic documents, encyclopedias, articles, regulations and other literature relevant to the research problems.

This study uses the *triangulation* of data sources for data validity checking, meaning that the researchers collected similar data from various data sources are different. The accuracy of data obtained from one of the informants will be confirmed by the data obtained from informants. Thus, each informant will act as informants control one another. Also, in this qualitative study, the researchers also functions as the main instrument*(key instrument)* that went into the field, as well as trying to compare the accuracy of data obtained so that the data collected in this study will be guaranteed its validity. According Moleong (2011: 87) is a triangulation technique that utilizes data validity checking something else outside of the data to check or as a comparison of the data. In this study, triangulation is used the triangulation of data sources.

Triangulation of data sources to compare or check the degree of confidence in information obtained through time and different tools in qualitative methods. This can be done in the following way (Moleong, 2011: 88):

IOP Conf. Series: Earth and Environmental Science **156** (2018) 012061 doi:10.1088/1755-1315/156/1/012061

a. Comparing the observed data with the results of interviews.
b. Comparing the situation and perspective of someone with different opinions of others.
c. Comparing the results of interviews with the contents of the documents related.

In this way is expected to be obtained data proved its validity so that research results can be applied to determine the Pekanbaru city ecotourism policy implementation to develop the tourism potential in Riau Province. The validity of the data will be the starting point of the research results are accurate and reliable. Determinations based policy valid data will be quickly implemented for the improvement and progress to be achieved.

Data analysis is the process of arranging and combining the data into a pattern, theme, category, whereas interpretation is to give meaning to the analysis, explaining the pattern or category, and find the relationship between some of the concepts. Researchers describe the perspective of interpretation is not the truth. Analysis and interpretation of data in qualitative research is not the things that go together; both carried out since the beginning of the study (Nasution, 2010: 96). Data analysis was performed so that the data have been obtained will be more meaningful. Data analysis is the process of simplification of data into a form that is easier to read and interpret.

The data obtained in the field that is the result of interviews, observation, and documentation requires analysis and interpretation of data to meet the demands of the purpose of research and other information. To obtain accurate data, the researchers made field notes which further simplified or refined then coded the data and issues. Encoded data was based on the criticisms made; the corresponding data is separated by a certain code of data that does not correspond to the research problem. The data were analyzed qualitatively using repeated and continuous between the collection and analysis of data, both for data collection in the field and after the data are collected (Bogdan and Biklen, 2012: 72).

In the first phase should be done after obtaining the data is to: *(1) checking (2) organizing, and (3)*coding.The next phase of data that have been obtained is tested triangulation method, researchers, and data sources. After testing the next triangulation qualitative analysis was made using a model that matches the theme of the study. Furthermore, Bogdan and Biklen (2012: 73) explain the data analysis is a systematic process of searching for and set the record interviews, field notes, and *the riders* other gathered to escort understanding. The analysis involves working with the data, set it up, separating into units that can be managed, combine them, searching for patterns meet important matters and what was known and decide what will be passed on to others.

According to Taylor and Powell (2013: 87) process analysis data can be done by stages:*(1)understanding the data (2) focus the analysis, (3) categorizing information as coding the Data or indexing the data (4) identify patterns and connections within and between categories, (5) interpretation - bringing it all* together.The first stage researchers to find exactly the data regarding eligibility, information, and meaning. The second phase of research done to know focus on a particular case, the individual, group or organization with a statement within a certain period or a particular event. The third stage to organize the data by coding or give the index based on the theme, shape, or the desired category. The fourth stage saw the diversity of these data and linking various categories derived either narrowly or broadly, as well as examine the level of importance of the data. The fifth stage of doing an interpretation of data to sort, select and determine the significance, synthesis, and significance.

3. Findings and Discussion

Based on the data obtained by the first dialysis using interactive model analysis*(Interactive Model* Analysis) of Miles and Huberman (Yandra:2017). In technical analysis interactive model after data collection, the activities of the analysis carried out by following the pattern of interaction between data reduction, data display, and conclusion or verification, which is described as follows:

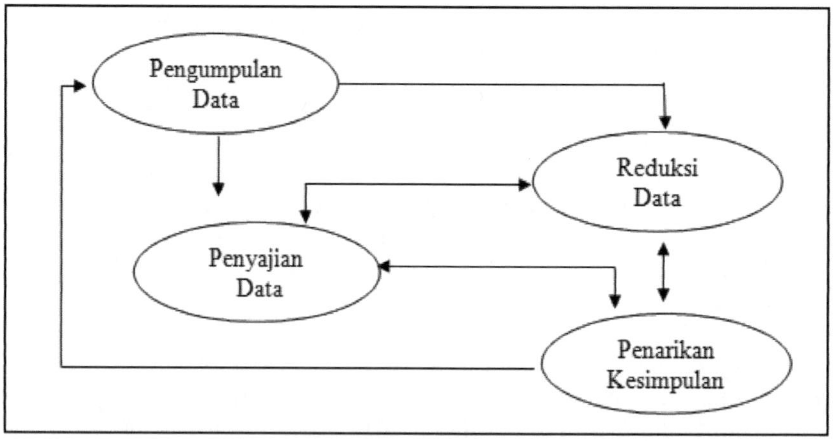

Figure 1: Analysis of Model Interactive (Miles and Huberman, 1992: 20)

Miles and Huberman (1992: 20) states that activity in the qualitative data analysis performed interactively and continues over time*(cyclical process)*at every stage of research, through to completion, and saturated. Activities conducted in this analysis is the collection of data*(data collection),*the reduction of the data*(data reduction),*exposure*(data display),*and the conclusion or verification*(conclusion/*verification).The collection of data*(data collection),* on these investigators, collect all the notes, photos of activities, recorded interviews, written documents, from the observation and interviews, then sorted according to the problem and arranged in order of time of data collection activities.

Reduction of data*(data reduction),*in this activity the researchers conducted a data reduction of the amount of data that is still common and complex to be chosen where the principal / relevant, focused on things that are important, look for a theme or a pattern that deserves to be presented (Nasution, 2010).

Data reduction can be done in various ways, namely by careful selection, with a summary, a description of the sentence, taking into account the importance of data and a moderate degree of data. Exposure to the data*(data display),* the activities carried out are arranging or composing, organizing data into new information that can be retrieved at the whole picture, as a conclusion or action. Exposure to good data and are often used in the form of narrative text, a wide variety of matrices, images graphics, *networks*, and *charts ;* Conclusion / verification of data *(conclusion /*verification) is the principal activity in the data analysis, namely the inductive process that does not ignore the validity of such principles, the ability for authenticated, reasonable, measurable, and firmness.

This study builds upon a theoretical framework concerning about analysis of policy implementation of Pekanbaru City tourism promotion in the development of tourism potential in Riau Province based on the logic thinking of the study of public administration theory. The theory used in this study is the Public Administration Theory of Starling (2011:42) as a Grand Theory, Public Policy Theory of Carl J. Friedrich actualized by Seibel (2015: 72-73) as the Middle Range Theory and Public Policy Implementation Theory of Mazmanian and Sabatier actualized by Gustama (2013: 125-126) as the *Applied Theory*, which can be illustrated as follows:

Figure 2: Theoretical Framework

Besides the above three theories, this study also used the tourism concept to become a knife analysis to answer the problem statement of the study. The strategy proposed in this study is the strategy which can be applied so that tourism policy implementation of Pekanbaru City be able to support the tourism potential in Riau Province. This strategy was drawn up following a review of the tourism policy implementation of Pekanbaru City currently ongoing and which has been considered failed as well as results of benchmarking or comparison of tourism development policy in Batam City and Padang City. According to these conditions, the strategy that can be applied so that the tourism policy implementation of Pekanbaru City be able to support the tourism potential in Riau Province is a strategy for making the Pekanbaru City as a MICE City. So, in the future strategy of tourism development in Pekanbaru City endeavored so that this city became one of the MICE City be reckoned with in Indonesia in general as well as in Sumatera Island in particular. The potential which supports Pekanbaru City as MICE City such as (1) Pekanbaru as the capital of Riau Province; (2) Strategic Location/Access; (3) Tourism attractions diversity; (4) Supporting facilities for tourism; (5) Facilities and infrastructures supporting MICE, become potential alternatives.

From the above description, it can be said that Pekanbaru City potential to be made as one of MICE destination in Indonesia. From the criteria for selecting a MICE destination, Pekanbaru City generally was able to meet the criteria required for MICE destinations, although there are still shortcomings on several things, such as (1) There are still limited international flights in Sultan Syarif Kasim II International Airport Pekanbaru City; (2) Opportunities additional tourism activities that still have not varied; (3) Lack of human resources to handle MICE in Pekanbaru; (4) Currently Pekanbaru City still intensively conducting meetings and incentive, whereas the convention and exhibition activities still untapped due to lack of supporting infrastructure; (5) Lack of promotion and marketing of MICE; (6) The presence of trauma as a result of tourism development policy that is not optimal or even fail.

4. Conclusions

Based on the discussion of the results of research that has been described above, it can be concluded the following:

1) factors affecting the implementation of the policy of promotion of tourism in Pekanbaru in the development of tourism potential in Riau Province are:

a. From the characteristics of the problem, it common problems faced by the City of Pekanbaru to develop tourism in the city of Pekanbaru is very difficult because it involves people's habit of Pekanbaru in meeting the needs of life. During this community, Pekanbaru City has been spoiled by the abundance of petroleum and palm oil. Moreover, it can also be concluded that the characteristics of Pekanbaru City tourism development issues are more difficult to be realized due to the implementation of this policy, policy implementers society actors fully Pekanbaru City.

b. Of the characteristics of the policy, the policy of tourism development of the city of Pekanbaru, this policy may be said to be poorly understood by policy implementers. This is because the policy is understood to pitch to bring in tourists from outside the area of Pekanbaru. Also, although the consistency and commitment of the implementing agency of the Department of Culture and Tourism of the city of Pekanbaru recognized many circles, the consistency and commitment are less obvious implementation is less able to create conditions Pekanbaru City tourism and growth.
c. From the environmental aspect, it can be concluded that some of the factors that cause the condition of Pekanbaru City tourism can not thrive either is (1) the unavailability of *the Master Plan for* the development of tourism as a reference in planning in a focused and sustainable keparawisataan; (2) the limited budget for the development of tourism in the city of Pekanbaru impact on the lack of infrastructure facilities and infrastructure as well as the conservation and management of objects and attractions; (3) non-optimal tourism promotion activities in order to attract tourists to visit tourist destinations; and (4) lack of public awareness of tourism as an opportunity to boost the economy.

2) The strategy can be applied to the implementation of policies promoting tourism Pekanbaru City can support the development of tourism potential of Riau Province were made based on the fact that the policy of tourism development Pekanbaru city that has failed and SWOT analysis, the strategy can be applied to the implementation of tourism policies Pekanbaru City to support potential travel in Riau Province is (a) fix the strategy attractions during this lie fallow; (b) the strategy to make the city of Pekanbaru as MICE*(Meeting,Incentive, Convention,*Exlibilition);and (c) make strategy Pekanbaru City as shopping and culinary tourism destination.

Based on the discussion that has been said above, it can be said that the *novelty* of this research finding weaknesses theory Mazmanian and Sabatier that can not be used to formulate a strategy that needs to be coupled with the SWOT analysis to formulate a strategy taking into account the strengths and weaknesses of the internal side and the opportunities and threats of the externally.

5. References

[1] Bogdan. R dan SK. Biklen. 2012. *Qualitative research for Education An Introduction to the Theory and Methods.* Boston: Allyn and Bacon.

[2] Gustama, Dielson B. S. 2013. *Exploring Policy Implementation Problems and Solutions.* Washington DC: Liberal Study Inc.

[3] Kartodihardjo, Hariadi, Rinekso, dan Karsudi, Soekmadi. 2010."Strategi Pengembangan Ekowisata di Kabupaten Kepulauan Yapen Provinsi Papua". *JMHT*, Vol. XVI, No. 3, hlm. 148-154.

[4] Krutwaysho, Oratai. 2013. "Obstacles to the Implementation of Tourism Policies and Regulations in Phuket, Thailand". *Journal of Tourism,* Vol. 8, No. 6. pp. 1-16.

[5] Miles, Matthew B dan A. Michael Huberman. 1992. *Analisis dan Kualitatif: Buku Sumber tentang Metode-Metode Baru.* Jakarta: UI Press.

[6] Moleong, Lexy. 2008. *Metode Penelitian Kualitatif.* Bandung: PT. Remaja Rosdakarya.

[7] Nasution, S. 2010. *Metode penelitian naturalistic-kualitatif.* Bandung: Tarsito

[8] Seibel, Wolfgang. 2015. *Hybridity and Responsible Leadership in Public Administration.* Duisburg: DVPW Press.

[9] Sugiyono. 2012. *Struktural Equation Modeling: Teori, Konsep dan Aplikasi.* Semarang: Badan Penerbit Universitas Diponegoro.

IOP Conf. Series: Earth and Environmental Science **156** (2018) 012061 doi:10.1088/1755-1315/156/1/012061

[10] Starling, G, 2011. *Managing the Public Sector.* 5th Edition, Florida; Harcourt Brace and Company.

[11] Sukmana, Oman. 2010. "Model Pengembangan Lingkungan Kota Ekowisata (Studi Di Wilayah Kota Batu)". *HUMANITY*, Vol. V, No. 1, hlm. 42-47.

[12] Taylor, Ellen dan Marcus Powell Renner. 2013. *Analyzing Qualitative Data*. Madison: Program Development and Evaluation

[13] Wirudchawong, Niti. 2012. "Policy on Community Tourism Development in Thailand." *Journal of Tourism,* Vol. 6, No. 2, pp. 13-26.

[14] Yandra, A. (2017). Penyelamatan Hak Pilih Warga Perbatasan Jelang Pilkada. *POLITIK*, *13*(1).

IOP Conf. Series: Earth and Environmental Science **156** (2018) 012062 doi:10.1088/1755-1315/156/1/012062

Selecting the Right Varieties in Riau Main Island: Sweet Corn Context

Surtinah[1] and Niken Nurwati[1]
[1]Faculty of Agriculture, Universitas Lancang Kuning (Unilak), Pekanbaru (28265), Indonesia
E-mail:surtinah@unilak.ac.id
Tel/Fax: +62761 7xxx

Abstract: The Riau Main island long way known has a unique geographic location and minerals contains in its soils. To help in selecting the best varieties that can help to sustain the food security plan in Riau main island is the aims of this study. Corn is a one of the local food that support the coastal and highland community in Riau and Sumatera Islands. Therefore corn was selected to be the object of this study. This experiment was conducted experimentally using a complete randomized environmental design and the treatment design with eight varieties of sweet corns. Data analysis uses Sidik variety and Duncan 5 % distance tes. The parameter observed is sweet corn which deals with the weight of cobs with cornhusk, the weight of cobs without cornhusk, the length of cobs without cornhusk and the sugar content of the seeds. The result shows that the best varieties are Sweet boy, sweet lady and bonanza.

Keywords: sweet corn, varieties, sweet boy, sweet lady, bonanza

1. Introduction

The cultivation of food crops in Pekanbaru always faces the constraints of soil condition. The soil in Pekanbaru is dominated by Red Yellow Podsolic Mineral Soil (RYPMS), a soil that is known to have low fertility and contains very low organic matter and some, even, do not contain organic matter. In order to take advantage of it, it is necessary to add organic materials and nutrients through fertilization to meet the needs of the plants.

The food crop that can be cultivated in RYPMS is sweet corn. Sweet corn is one of food commodities that has bright future because of its sweet taste, soft texture and high market demand so that sweet corn becomes of the commodity choices of food that is feasible to be cultivated. Moreover, sweet corn market is avaiable from traditional to modern market.

The potential yield and sugar content in sweet corn will greatly determine its quality. The quality is measured from the weight of the cobs without cornhusk and sugar content. The higher the cob weight and ther sugar content are, the better quality it gets. Sucrose and reducing sugar (glucose and fructose) of photosynthesis are transfered to various organs which are then used to maintan the integrity of the organ, partly converted to plant structure material and the rest is used as food reserves (Harini, 1993).

Sweet corn varieties have different sugar content. It is influenced by genetic characteristic of each variety, as reported by Siswono (2004) that super sweet variety has sugar content of 16-18 %, colse to sugar cane level 19 %, compared to local sweet corn which only has 9-11 % .Meanwhile, the sugar content of Bisi variety is between 15-16% , sweet boy variety reaches 16% and super honey variety reaches 28 %.

Sweet corn contains energy of 96 cal, 3,9 g protein, 1.0 g fat, 22,8 g carbohydrat, 3.0 mg calcium, 111 g phosphorus, 0.7 mg iron, 400 SI Vitamin A, 0,15 mg vitamin B, 12,0 mg vitamin C and 72,7 g water. (Iskandar, 2006).

Nuhayati and Siti (2002) reported that super sweet varieties harvested in 65 days resulted in the weight of cornhusk cob 355,82 g; the weight of cobs without cornhusk 271,81 g, and sugar content of the seeds is 5,4 %. In different research, Surtinah (2008) indicated that sweet boy harvested in 70 days results the weight the cobs 384,53 g, the weight of the cobs without cornhusk 288,89 g, and sugar content 15,78 %.

Surtinah (2012) reported that Bonanza variety planted in Pekanbaru results 14,82% of sugar content when it is harvested at 17.00 pm in 65 days. In other research (Surtinah, 2013) reported that Bonanza Variety is the best compared to other two varieties, with the weight of the cobs without cornhusk 295 g. Syafruddin, Nurhayati and Wati (2012) reported that Bonanza variety showed high growth and production with NPK liquid fertilizer treatment.

The soil condition of planting sweet corn is very significant in determining the growth and productivity of the plant. In RYPMS soil, sweet corn can grow well with appropriate organic material on the soil. (Lidar % Surtinah, 2012). Dinariani, Heidy dan Guritno (2014) reported that the application of goat manure 10 ton/ha results the highest sugar

IOP Conf. Series: Earth and Environmental Science **156** (2018) 012062 doi:10.1088/1755-1315/156/1/012062

content i.e 16,07 brix. The objective of this research is to get the sweet corn variety with the highest production with high sugar content in Pekanbaru.

2. Methodology

This research is conducted in experimental garden of Agriculture department, Lancang Kuning University. Then kind of soil is Yellow Podsolic mineral soil with the height 0f 20 m dpl.
The research is conducted experimentally by using complete randomized environmental design, done for 3 times, and treatment design with 8 varieties of sweet corns (V): v1= sweet boy variety, v2= sweet lady variety, v3= bonanza variety, v4= secada variety, v5= Bontang Asia variety, v6= valentino variety, v7= master sweet variety, v8= mulia variety.
The population of the plants in each experimental unit is 10 plants and 3 plants are the samples and the numbers of experimental units is 24 units.
Primary data is analyzed by using Sidik Variety with 5 % chance, continued by 5 % Duncan distance test.

3. Result And Discussion

Variety analysis used to conclude the weight of the cobs with cornhusk of each variety is different and is not real at the test level F 5 %, and Duncan distant test is as follows:

Table 1. The weight of the cobs with cornhusk (gram) 8 varieties of sweet corn.

Perlakuan Varietas	rata-rata perlk	uji Duncan
V1 = Sweet Boy	430,00	b
V2 = Sweet Lady	438,83	b
V3 = Bonanza	466,67	b
V4 = Secada	457,17	b
V5 = Bintang Asia	461,33	b
V6 = Valentino	465,00	b
V7 = Master Sweet	462,83	b
V8 = Mulia	375,83	a

The numbers followed by the same letters are different unreal based on Duncan test.

The low weight of the cobs with cornhusk of the sweet corns is only found in Mulia variety. Other varieties show the weights of the cobs with cornhusks which are different unreal. It means the seven varieties, they are: sweet boy, sweet lady, bonanza, secada, valentino, master sweet and Bintang Asia, have the same weight of the cobs. The best weight of the cobs is found in bonanza variety. In a research done by (Surtinah, Susi % Lestari, 2016) it was found that the weight of the cobs of sweet corn with cornhusk from Bonanza variety increased 14,82 %, in Bintang Asia variety, it decreased 9,83%, master sweet variety increased 22,22 %. In other research (Surtinah, 2008) reported that the weight of the cobs with cornhusk of sweet boy variety increased 6,99 % compared to the previous research.

Tabel 2. Berat tongkol tanpa kelobot (gram) delapan varietas jagung manis

Perlakuan Varietas	rata-rata perlk	uji Duncan
V1 = Sweet Boy	275,00	b
V2 = Sweet Lady	292,83	b
V3 = Bonanza	322,83	b
V4 = Secada	300,17	b
V5 = Bintang Asia	294,17	b
V6 = Valentino	298,17	b
V7 = Master Sweet	281,67	b
V8 = Mulia	212,83	a

The numbers followed by the same letters are different unreal based on Duncan test.

The lowest weight of the cobs without cornhusk is Mulia variety, while the other seven varieties show unreal different results. From the seven tested varieties, the best weight of the cob is Bonanza, followed by Secada and Bintang Asia. Surtinah (2013) reported that the weight of the cob withour cornhusk of Bonanza variety is 295 g. It means, in this research it occurs the increasing of the result of Bonanza variety 8,62 %. Meanwhile in Bintang Asia variety it decreases 7,29 %. From the research done by (Pradipta, Kruniawan Puji and Guritno, 2014), it showed that the best weight of the cob was 188,90 gram.

IOP Conf. Series: Earth and Environmental Science **156** (2018) 012062 doi:10.1088/1755-1315/156/1/012062

Table 3.The Test Result of Duncan's Multiple Distance for sugar content of Sweet corns (%).

Perlakuan Varietas	Kadar Gula	uji Duncan
V1 = Sweet Boy	13,05	d
V2 = Sweet Lady	13,80 .	d
V3 = Bonanza	13,73	d
V4 = Secada	10,65	a
V5 = Bintang Asia	11,02	abc
V6 = Valentino	12,55	bcd
V7 = Master Sweet	12,83	cd
V8 = Mulia	10,88	ab

The numbers followed by the same letters are different unreal based on Duncan test.

The best sugar contents of the seeds is Sweet Lady, Bonanza and Sweet Boy varieties. There is difference in sugar content compared to the previous research. It is caused by the climate at the moment the research conducted from year to year.

Table 4. the difference of sugar content sweet corn seeds between description and research result (%)

Perlakuan Varietas	Kadar Gula	Kadar Gula Deskripsi
V1 = Sweet Boy	13,05 %	12,1 %
V2 = Sweet Lady	13,8 %	13,8 %
V3 = Bonanza	13,73 %	13 %
V4 = Secada	10,65 %	11,19 %
V5 = Bintang Asia	11,02 %	12 %
V6 = Valentino	12,55 %	Tidak diketahui
V7 = Master Sweet	12,83 %	13,3 %
V8 = Mulia	10,88 %	Tidak diketahui

From the comparison of research results and description of sugar content of sweet corn seeds, it finds out that sugar contents of Sweet boy and Bonanza varieties are better in the research result, while the sugar content of Sweet lady variety is the same between the research result and the description. This can be reference in dertermining conformity grows of sweet corns in an area based on the quality of the seeds related to the sugar content. And Sweet boy, Sweet lady and Bonanza varieties are the chosen varieties to be tested for the next year in various planting locations in Pekanbaru.

4. Conclusion

The Riau Main island long way known has a unique geographic location and minerals contains in its soils. To help in selecting the best varieties that can help to sustain the food security plan in Riau main island is the aims of this study. Corn is a one of the local food that support the coastal and highland community in Riau and Sumatera Islands. From the result, it is concluded that the potential varieties to be developed in Pekanbaru are Sweet boy, Sweet lady and Bonanza varieties.

References

1] Dinariani,Y.B.S., Hedy,B., & Guritno. (2014) Kajian Penambahan Pupuk Kandang Kambing dan Kerapatan Tanaman yang Berbeda pada Pertumbuhan dan hasil Tanaman Jagung Manis (Zea mays saccharata Surt) *J. Produksi Tanaman, 2* (2): 128-136

2] Harini (1993) *Pengaruh Umur Panen dan Suhu Pendinginan Terhadap Perubahan Beberapa Sifat Fisika dan Kimia Jagung Segar Selama Periode Penyimpanan .* Universitas Gajah Mada

3] Iskandar, D. (2006) Pengaruh Dosis Pupuk N,P, dan K Terhadap Pertumbuhan dan Produksi Tanaman Jagung Manis di Lahan Kering. *J, Sains dan Teknologi. Balai Penelitian dan Teknologi:* 1-2.

4] Lidar, S., & Surtinah (2012) Respon Tanaman Jagung Manis (Zea mays saccharata Sturt) Akibat pemberian Tiens Golden harvest. J. Ilmiah Pertanian, 8 (2): 1-6.

5] Nurhayati dan Siti (2002) Pengaruh Takaran Pupuk Kandang dan Umur Panen Terhadap Hasil dan Kandungan Gula Jagung Manis.

6] Pradipta,R., Kurniawan Puji,W., & Guritno (2014) Pengaruh Umur Panen dan permberian Berbagai Dosis Pupuk Kalium Terhadap Pertumbuhan dan Kualitas Jagung Manis(Zea mays saccharata Sturt). *J. Produksi Tanaman,2 (7)*: 592-599.

7] Siswono,(2004) Jagung Manis Rendah Lemak dan Kolesterol. Retrieved from www.gizi net.

8] Surtinah (2008) Umur Panen yang Tepat Menentukan Kandungan Gula Biji Jagung Manis. (Zea mays saccharata Sturt).*J. Ilmiah Pertanian, 4(2)*: 1-6.

9] Surtinah (2012) Korelasi Antara Waktu Panen dan kadar Gula Biki Jagung Manis (Zea mays saccharata Sturt).*J. Ilmiah Pertanian, 9(1)*: 1-6.

10] Surtinah (2013) Menguji 3 Varietas jagung Mnias di Rumbai, Pekanbaru..*J. Ilmiah Pertanian, 11(1)*: 1-10.

11] Surtinah, Susi, N., & Lestari, S.U (2016) Komparasi Tampilan dan Hasil lima Varietas Jagung Manis (Zea mays saccharata Sturt)di Kota Pekanbaru..*J. Ilmiah Pertanian,13(1)*: 32-37.

12] Syafruddin, Nurhayati & R,W. (2012) Pengaruh Jenis Pupuk Terhadap Pertumbuhan dan Hasil Beberapa Varietas Jagung Manis. *J. Floratek, 7(1).*

IOP Conf. Series: Earth and Environmental Science **156** (2018) 012063 doi:10.1088/1755-1315/156/1/012063

Agroforestry Models in Riau Main Island Indonesia: Kampar Regency Context

Enny Insusanty, M. Ikhawan, and Emy Sadjati

Universitas Lancang Kuning, Pekanbaru, 28265, Indonesia
Email: annovisa@yahoo.com

Abstract: This paper focused on the identification of species composition dan standing stock of agroforestry model in District XIII Koto Kampar in Riau Main Island. The study aimed to show kinds of agroforestry model and estimate the standing stock. Data were collected from 12 sample plots with 60 sub sample plots. The research found that plants of agroforestry have three models that are agroforestry based rubber-durian-gaharu (*Hevea brasiliensis-Durio zibethinus-Aquilaria* sp,), agroforestry based rubber-durian and agroforestry based rubber-gaharu. Agroforestry based rubber-durian-gaharu has 4 sample plots with average potential standing stock 108,93 m^3/ha. Average the number of tree species per ha is 657 rubber, 214 gaharu, durian 15 and another tree 20. Agroforestry based rubber-durian has 2 sample plot with average potential standing stock 81,35 m^3/ha. Average the number of tree species per ha is 316 rubber, 41 durians, and another tree 71. Agroforestry based rubber-gaharu has 6 sample plot with average potential standing stock 59,48 m^3/ha. Average the number of tree species per ha is 545 rubber, 124 gaharu, and another trees 8.

Keywords: agroforestry model, species composition, standing stock

1. Introduction

Agroforestry is one of the options taken by upstream farmers for their land management. The principle of agroforestry that implements a land management system by combining different types of seasonal crops (food, vegetables, medicines), and annual crops (wood, fruits), has in many cases proven to have positive implications for farmer welfare and environmental conservation (Sambas, 2004; P3P Unram, 2003). The successful application of agroforestry is influenced by various factors, generally dependent on local characteristics, namely the biophysical, social, institutional, and external factors that support it. According to ICRAF (2000), there are many agroforestry patterns found in Indonesia, so each region in Indonesia has a distinctive agroforestry application pattern. Therefore, successful implementation of agroforestry in one area may not be generalized as an excuse to be applied in other areas.

These variations can be seen, among others, the variety and variety of vegetation, planting systems, patterns of individual, family, and group work relationships, as well as local institutional and policy support factors. The diversity of models of agroforestry application is a fact that needs to be explored more deeply. Of the various models that exist, of course, each has its advantages. In District XIII Koto Kampar which is one of the District in Kampar district found much agroforestry-based rubber planted with other plants such as trees and fruit trees. Plants that many become farmers' choice is gaharu planted with rubber. These agroforestry activities are undertaken by the community to gain benefits to improve the welfare of communities in and around the forest while maintaining the condition of the forest. According to De Foresta (1997) that complex agroforestry is a sedentary farming system that contains many types of plants with (tree-based) planted and treated with cropping patterns and forest-like ecosystems. Agroforestry activities carried out with a combination of crop types will provide by choice of farmers with various considerations such as conformity of the place of growth, economic factors, and others. The type composition of the constituent components of the agroforestry stand will make a difference to the pattern and model of agroforestry in each place.

The stand potential of each agroforestry model is based on the density and type that dominates a stand. High-density stands with large tree diameters will provide a larger value for the standing potential. The aims of this study to determine the potential of existing agroforestry model in Kecamatan XIII Koto Kampar.

2. Research Methods

The research was carried out agroforestry land in Desa Gadang Village, Koto Masjid Village and Tanjung Alai Village, XIII Koto Kampar District, Kampar Regency, Riau Province. The research was conducted in February - June 2017. The method of the research was survey method by taking agroforestry biophysical data, namely: composition of agroforestry, plant density, planting pattern, diameter, height, topography, soil type, rainfall. Also, interviews with agroforestry owners and measurements were made in the sample plots. In measuring the potential of agroforestry system that has wide plant spacing, plot 20 m x 100 m = 2000 m^2 where the total plot is 12 plots with 5 sub plots measuring 20 m x 20 m = 400 m^2 so that there are 60 sub plots with total sampling area is 2.4 ha.

Measurement standing stock are calculated by the formula

V = ¼ x 3.14 x d x t x cf
V = volume
d = diameter
t = high
cf = correction factor

3. Result and Discussion

The topography is flat and hilly. Soil type is yellow-red podzolic or commonly called FMD. Climatic conditions in the category of tropical climate with six months dry months while the number of wet months six months. The average rainfall in 2016 is about 2420 mm, and the average amount of rain in 2016 is about 15 days. The temperature of this research location temperature ranges from 20.2^0C - 22.8^0C. For air humidity in this study, location ranges between 97% - 99%. Geographically located in the lowlands. These three villages are located on the edge of Koto Panjang hydropower reservoir so that this village can be categorized into a tourist area. This area is dominated by rubber plantations and fish ponds

The agroforestry system located in Kecamatan XIII Koto Kampar is an agroforestry based rubber that is combined / intercropped with forest plants (Meranti, Sengon, teak, Tampui, etc.) or fruits (Durian, Petai, Jengkol, cocoa, mangosteen and others) other). The advantage of agroforestry system is diversification of plants so that it resembles the condition of the forest, it can be a house of millions of fauna and flora, and can prevent the occurrence of landslides and floods. Also, the most important is a very suitable agroforestry system selected to suppress and prevent global warming is becoming the hottest issue in the world. Agroforestry type composition grouped based on 12 agroforestry land located in XIII Koto Kampar based on observation and interview with land owner obtained three models of rubber based agroforestry based on the dominant tree species found in the field. These three models consist of 3 dominant tree species, ie, rubber, gaharu and durian are presented in Table 1

Table 1. Agroforestry model based on tree species composition

Agroforestry Model	Species composition	∑ Plot	N/ha	Vol/ha
Model 1 Rubber-Gaharu-Durian	Rubber,gaharu, meranti,bamboo,durian,jengkol, tampui, mangosteen, oil palm	4	907	108,93
Model 2 Rubber -Durian	Rubber, Durian, mangosteen, cocoa	2	429	81,35
Model 3 Rubber -Gaharu	Rubber, gaharu, jackfruit, sengon, candlenut, meranti, cocoa, teak, rambutan	6	678	59,48

Based on Table 1 it can be seen that the most dominant species that exist in each field are rubber, durian, and gaharu. Rubber plants grown initially is a government assistance activity in the form of rubber plants grown at plant spacing 3 x4, but over time some plants have died. The regeneration is only a natural regeneration of the scattered rubber seeds. The farmers also plant other crops among the rubber plants of durian, jengkol, petai, cocoa and others. The plants are planted with irregular patterns among the rubber plants with different times so that there are various structures and compositions. In general, fruit tree seedlings grown in rubber gardens are local seeds. There is no spacing or pattern, and even tillers are allowed to grow naturally by themselves. Seeds are dispersed on existing land and left without treatment such as treatment, thinning and even fertilization. This is done by farmers because it is considered to provide additional income other than rubber sap. Rubber based agroforestry has a variety of composition or plant composition and can be utilized by the owner (farmer) at any time by the wishes of the owner.

Durian one of the many plants found among rubber plants because durian tree has a high economic value of durian fruit. With the diameter of large and high trees then the durian becomes dominant among other plants. Wood from durian trees also has good quality to be used as raw materials for construction timber. The interest of landowners to grow gaharu arises when gaharu is considered beneficial because the price of gaharu has a very high value coupled with the existence of farmers groups who received help from gaharu seeds from the government. Gaharu plant is an intolerant plant which means need shade for its growth so that planted under rubber will give good growth. The composition of the type of each model is presented in Graph 1

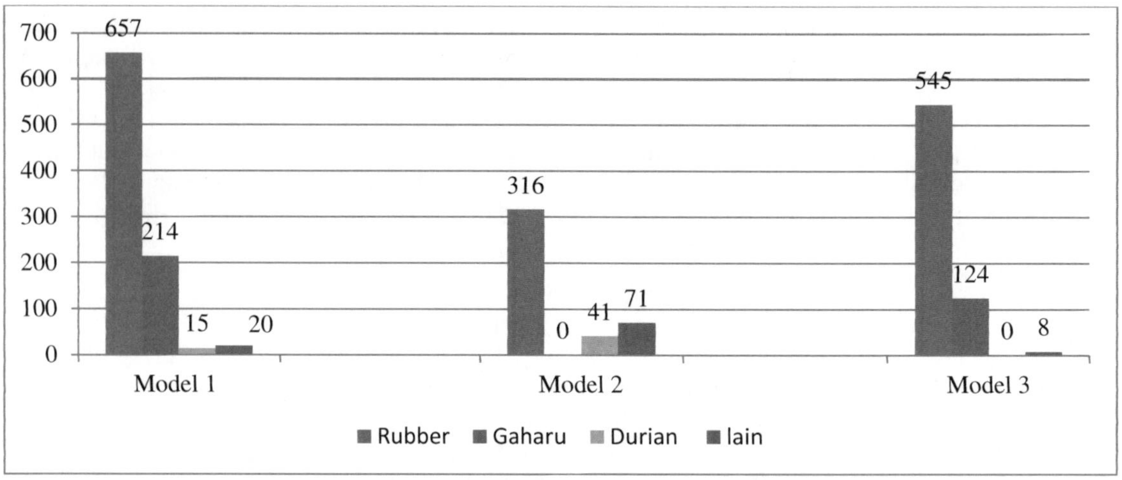

Figure 1. The composition of species and Number of trees/ha (N/ha) of Agroforestry models

Based on Figure 1 it can be seen that model 1 has some species dominated by rubber, then gaharu and durian. Other types that are encountered are jengkol, tamui, bamboo, meranti. Although regarding the number of durians is not too much but found in every observation plot and has the largest diameter of 64 cm. While on the 2nd model is not found gaharu species only durian and other plants. Other plants include mangosteen and cocoa. In model 3 the dominant plants are rubber and gaharu and also there are other plants that are jackfruit, sengon, candlenut, meranti cocoa, teak, rambutan. In terms of density or number of trees per ha (N/ha) model 1 has a larger number of trees than these two models because it has many tree species and gaharu planted is also young with a diameter between 4 cm- 9 cm while model 2 has N/ ha at least with only 429 trees this is due to the planted tree already has a large diameter and no additional of gaharu trees among the rubber plants.

The results of 12 plots of 20 mx 50 m in XIII Koto Kampar were recorded in 15 species of trees obtained in plots such as rubber (*Hevea brasiliensis*), gaharu (*Aquilaria sp*), durian (*Durio zibethinus*), cocoa (*Theobroma cacao*) Mangosteen (*Garcinia mangostana L*) and palm oil (*Elais guinensiss* Jacq). The staple crop is the rubber which is then between the oil palm planted another tree by planting applied by the farmer is a random system with irregular spacing.

In the research plot, rubber is found almost throughout the plot with the diameter range is 24-30 cm because the rubber plant is a staple crop. For gaharu plants are planted around the year so that the age of the plant is still young with diameter 5-10 cm. Gaharu plant aims to produce gaharu either naturally or through the treatment by introducing microorganisms after > 5 years old or diameter has reached 15 cm. Diameter range of durian is 31-64 cm, it shows that rubber plant and durian are superior plant species and old age and the average diameter of large stems in the research location, can be seen in Table 2

Table. 2 Description of plots

Plot	The number of trees.	Average diameter	Average height	Volume
1	49	26,7	14,2	103,88
2	32	34,9	17,8	116,40
3	56	27,8	16,2	167,86
4	52	25,5	14,8	99,82
5	37	26,2	11,1	62,89
6	43	25,1	12,6	25,44
7	26	23,4	13,9	38,25
8	27	24,6	14,8	47,58
9	22	22,0	12,4	29,54
10	53	22,5	14,0	83,76
11	38	29,1	14,6	95,77
12	38	27,5	14,2	84,14
average	39,42	26,275	14,22	79,61

The number of trees per observation plot varies between 22-56 stems with a plot area of 0.1 ha. Diameter tree ranges between 22.1 cm - 34.9 cm and 11.1 m high up to 17.8 m. This condition indicates that each plot of observation has a good diversity in the number of trees, diameter, and height. Agroforestry as one form of land utilization that has characteristics of diversity both regarding the number of plants, types of plants, diameter, and height.

The structure of rubber agroforests resembles secondary forests so that ecologically agroforests provide a good hydrological function and can prevent erosion. The composition becomes protective in the upstream area as it can serve as a buffer of the litter layer on the soil surface through the leaves of rubber and other fallen trees and prevent the formation of ditches due to erosion. Agroforestry is a multifunctional system of the landscape that is as a source of income for farmers, protection of the surrounding soil and water (Young, 1989), protection of biodiversity, controlling carbon emissions, and maintaining the aesthetic value of the landscape (Hairiah et al., 2001; Nair 2012)

4. Conclusion

Agroforestry based rubber models encountered are three forms: Rubber-Gaharu-Durian model, Durian-Rubber model and Rubber-Gaharu model with agroforestry plant composition comprise looking for rubber-gaharu-durian mixture and other varied plants in each plot._The number of trees per ha of the model Rubber-gaharu-durian is 657 rubber, 214 gaharu, 15 durian and 20 other species. For Durian, rubber model is 316 rubber, 41 durians, another type 71 while the model of Rubber-new species of plant is 545 Rubber, 124 gaharu, and 8 other Type. The average diameter of rubber is 25,5 cm durian 46,5 cm and gaharu 6.6 cm. Standing stock model Rubber-gaharu-durian is 108,93 m^3/ha, model Rubber-durian is 81,35 /ha and Rubber –gaharu is 59,48 m^3/ha

5. Acknowledgments

Many thanks to Kementerian Riset dan Teknologi Pendidikan Tinggi Republic of Indonesia, which has funded this research activities under the Penelitian Produk Terapan(PPT) scheme “Models of agroforestry as climate change mitigation efforts in District XIII Koto Kampar, Kampar Regency, Riau”.

6. REFERENCES

[1] De Foresta. 1997. When the Garden is like a Forest: Agroforest Typical of Indonesia. A Community Contribution. Bogor, Indonesia. International Center For Research in Agroforestry, SEA Regional Research Program. 249 p.

[2] Hairiah, K., Sitompul, S.M., Van Noordwijk, M., and Palm, C. 2001b. Methods For Sampling Carbon Stocks above and below ground. ASB Lecture Note 4B. ICRAF, Bogor, Indonesia.

[3] [ICRAF] International Center for Research in Agroforestry. 2000. When Forest Garden: Indonesian Agroforest A Community Contribution. Bogor: World Agroforestry Center

[4] Nair P K R and Garrity D, 2012. Agroforestry research and development: The way forward. In: Nair P K R and Garrity D (eds.). Agroforestry – the future of global land use. Adv. Agroforestry, 9:

[5] P3P Unram, 2003. Implementation of Agroforestry and Integrated Farming System through Participation of Forest Area Community and Dried Land. Cooperation MFP DFID-P3P Unram.

[6] Sambas Sabarnurdin, 2004. Agroforestry: Technical, Environmental, and Education Sides.

[7] Young A, 1989. Agroforestry for soil conservation. CABI-ICRAF. 276p.

IOP Conf. Series: Earth and Environmental Science **156** (2018) 012064 doi:10.1088/1755-1315/156/1/012064

Factors Which Influence The Fish Purchasing Decision: A study on Traditional Market in Riau Mainland

Latifa Siswati[1] and Asgami Putri[2]

[1] Universitas Lancang Kuning, Pekanbaru, 28265, Indonesia
E-mail: latifasiswati@unilak.ac.id
[2] Universitas Lancang Kuning, Pekanbaru, 28265, Indonesia
E-mail: asgami@unilak.ac.id

Abstract: The purposes of the research are to analyze and assess the factors which influence fish purchasing by the community at Tenayan Raya district Pekanbaru.Research methodology which used is survey method, especially interview and observation technique or direct supervision on the market which located at Tenayan Raya district. Determination technique of sampling location/region is done by purposive sampling. The sampling method is done by accidental sampling. Technique analysis of factors which used using the data that derived from the respondent opinion to various fish variable. The result of this research are the factors which influence fish purchasing decision done in a traditional market which located at Tenayan Raya district are product factor, price factors, social factor and individual factor. Product factor which influences fish purchasing decision as follows: the eyelets condition, the nutrition of fresh fish, the diversity of sold fish. Price factors influence the fish purchasing decision, such as: the price of fresh fish, the convincing price and the suitability price and benefits of the fresh fish. Individual factors which influence a fish purchasing decision, such as education and income levels. Social factors which influence a fish purchasing decision, such as family, colleagues and feeding habits of fish.

Keywords: Analysis factors, fish consumption, fish purchasing decision

1. Introduction

Fish is one of food that contains varieties of nutrition. According to the water where it lives, fish consist of freshwater fish and sea fish. Both of it is the food which rich in protein that really important for body growth. As food, fish is sources of protein, fat, vitamin and mineral which really good and perspective. The main advantage of fish protein rather that another product is a complete composition of amino acid and the simplicity of digest (Astawan, 2004). The behavior of consumer decision making is greatly influence whether or not consumers buy the products offered by the manufacturer / dealer. People consumption of fish products until the present time is still far below the average national consumption, and when it compared with consumption in other countries of our national consumption is far below the average. The scope of this research include: the factors that influence fish purchasing decisions at Tenayan Raya district Pekanbaru.

2. Research Methodology

This research will be conducting at Tenayan Raya district Pekanbaru. The research was planned from October until December 2016. Research method which used is survey method. Determination technique of sampling location/region is done by purposive sampling. While the sample of respondents was deciding as many as 50 participants. Sampling is done by accidental sampling. The analysis techniques to find the factors which influence the purchase by using weighting and scoring techniques. The weighting of criteria which related to the commodity set based on the importance level of fish purchasing activity. The analysis factor which used is using the data which come from the respondent opinion to various fish variable. Mathematically model of analysis factors is as follows:

$$Fi = Wi1X1 + Wi2X2 + ...+WinXn$$

Where:

Fi = 1st estimate factors, Wi = weight or coefficient score factor, X_{n} = mix variable marketing which observed

Mix variable marketing which observed were:

X1 = Fish smell, X2 = Fish nutrition, X_3 = Fish eye condition,
X4 =Various types of fish, X5= Price, X6= Price Suitability,
X_7 = Convincing price, X_8= Fish availability, X_9= Seles services
X_{10}= Fish eating habit, X_{11}= Income level, X_{12}=Educational level
X_{13}=Family, X_{14}= Colleagues, X_{15}= Friend

3. Result and Discussion

Testing variables which have been determining, incorporated into analysis factors to tested KMO value and Bartlett Test and MSA (measures of sampling adequacy). MSA value should be above 0,5. This is the table of KMO value and Bartlet Test.

Table 1: The Result of KMO and Bartlett's Test

Kaiser-Meyer-Olkin Measure of Sampling Adequacy.		.633
Bartlett's Test of Sphericity	Approx. Chi-Square	197.157
	Df	105
	Sig.	.000

Based on the table above can be seen that the number KMO and Bartlett Test is 0,633 with significant level 0.000 therefore, the variable and the sample can be analysis further. The number of MSA in the table anti image matrix, which contained in anti image correlation, shows typical aroma values (XI) is 0,409, friend (X2) is 0,448, fish eye condition (X3) is 0,582, various types (X4) is 0,593, price as consideration in purchasing fresh fish product (X5) is 0,791, the suitability of price product with benefit product (X6) is 0,655, convincing price (X7) is 0,497, fish availability (X8) is 0,593, sales services (X9) is 0,771, fish eating habit (X10) is 0,759, income level (X11) is 0,597, educational level (X12) is 0,600, family (X13) is 0,658, colleagues (X14) is 0,651, nutrition content (X15) is 0,752. From 15 variables that exist, can be seen the value of MSA. If there are MSA value under 0,5, then that variable cannot analyze further. From 15 variables, the smallest MSA values are typically aroma (X1) with MSA value 0,409, friend (X2) with MSA value 0,488.

Therefore, typically aroma and friend are removed from the factors because it has the smallest values of MSA. After typically aroma and friend variables removed from the factors, then the next step is to retest the 13 variables which remain. It can be seen in KMO value and Bartlett Test and also MSA.

Table 2: The Result of KMO and Bartlett's Test continued

Kaiser-Meyer-Olkin Measure of Sampling Adequacy.		.673
Bartlett's Test of Sphericity	Approx. Chi-Square	167.397
	Df	78
	Sig.	.000

*SPSS 17

From the table of output result above can be seen that KMO value and Bartlett test has increased from 0,633 become 0,673 with fixed significant level (0,000). KMO value and Bartlett test should be over 0,5, and the significant should be under 0,05. Based on KMO value and Bartlett test, then from 0,633, there are MSA values which less than 0,5, then those variables must be removed. Thus there is increasing become 0,673 KMO value and Bartlett Test. Therefore, with MSA value over 0,5, it can be analyzed further.

3.1 Conducting factor and rotation

In this extraction process, the method which used is Principal Component Analysis, after four-factor are formed to know the 13 variables will be included in which factor, then rotation process does by using varimax method (part of orthogonal). More details can be seen in table 3.

Table 3: Result of SPSS Communalities

	Initial	Extraction
Eye condition	1.000	.490
Various types	1.000	.644
Price	1.000	.588
Price suitability	1.000	.555
Convincing price	1.000	.680
Fish availability	1.000	.743
Sales services	1.000	.584
Fish-eating habit	1.000	.647
Income level	1.000	.691
Educational level	1.000	.615
Family	1.000	.492
Colleagues	1.000	.739

Communalities table is variant number (can be in percentage), at first, a variable can be explained by the factors that already exist. According to the values which include in communities table, it can be concluded that the variables which exist can be explained in formed factor, the more communalities value, the more closely with the formed factor. Total Variance Explained table, describe the number of formed factor. In looking the formed factors, it must be seen in the Eigen value.

To determine the factors that formed it must be seen the Eigen value is more than one (1), if it is less than one then it is not appropriate. More details can be seen at table 4 below

Table 4: Initial of Eigen value

Component	Initial Eigenvalues		
	Total	% of Variance	Cumulative %
1	3.251	25.007	25.007
2	2.228	17.138	42.145
3	1.428	10.984	53.128
4	1.269	9.760	62.888
5	.922	7.095	69.984
6	.788	6.063	76.046
7	.708	5.450	81.496
8	.587	4.515	86.012
9	.463	3.562	89.573
10	.411	3.163	92.736
11	.394	3.032	95.769
12	.319	2.453	98.221
13	.231	1.779	100.000

Extraction Method: Principal Component Analysis.

In the table above, can be seen that there are 13variables (component) which include in analysis factors. In the table above, shows that only four factors are formed. After known that the factors are the most optimal amount, then, *component matrix* table shows that the distribution of those 13 variables in four factors which formed. Can be seen in table 5:

Table 5: Output result SPSS Rotated Matrix Component

	Component			
	1	2	3	4
Nutrition content	.759	.311	.028	-.187
Eye condition	.756	.041	-.261	-.050
Fish availability	.748	.020	.370	-.217
Various types	.471	-.355	-.404	-.277
Price	-.140	.789	.303	.077
Price suitability	.138	.730	.225	.209
Convincing price	.322	.670	.350	.067
Fish eating habit	.026	-.101	.692	.256
Income level	.337	.284	.667	.229
Educational level	.197	-.031	.663	-.104
Family	.071	.274	-.003	.642
Colleagues	-.140	.077	.303	.789
Sales services	-.282	.397	.332	.487

Table 6 : The Result of SPSS Transformation Matrix Component

Component	1	2	3	4
1	.566	.653	.479	.157
2	.751	.692	.632	-.021
3	.222	.186	-.289	.912
4	.259	.709	-.536	.677

Sources : Output result SPSS

From the table can be explained that in diagonal factor (component) 1, 2, 3, 4 (0,566; 0,692; -0.289; 0,677). The number with a minus (-) that indicates the correlation. While in others diagonal that shows the value under 0,5 it indicates there are others components of each factor which has the higher correlation. And only three which the value is over 0,5: factor 1 (component 1),factor 2 (component 2), and factor 4 (component 4), each values are: 0,566; 0,692; 0,677. Based on the formation, there are only three factors in the diagonal which the value is over 0,5 it is enough to represent the four factors that formed.

3.2 Interpretation of Formed Factors

After conducting factor and rotation, the next step is to interpret the factors which already formed. Giving name and concept for each factor is deciding based on the general meaning of variables which include. The result of the research can be seen that psychological factor, product, social, distribution, price, and individual are influence the consumers. Psychological factors become an important part of consumers purchasing process, to the research object that is fresh fish, it seen that they have different reasons in purchasing of fresh fish. Product factors are the part that dominant in a product. Product attribute of fresh fish is seen in nutrition content, fish eye condition, typical aroma. From these attributes can judge that fresh fish product is qualified or not. Social factors are the factors that important in consumers purchasing decision.

Considering the research object is the consumer fresh fish purchasing, it can explain that the reference group which consists of a group of people with direct influence can come from colleagues, the close friend, community friends, and family can influence them to buy a product. Based on the research result can be determine the factors which most dominant by seeing at the total value of variant on the *Total Variance Explained* table based on the result of the research and the result of the analysis as whole, evidently, the most dominant factors that lead decision of fresh fish purchasing, in sequence are; product factors, which is the most dominant factors with the variant values 25.007%, price factors with variant value is 17.138%, individual factors with variant value 10.984%, social factors with variant value 9.760%.

4. Conclusion

Based on the analysis and the discussion there are several conclusion that can be taken as follows: the factors which influence fresh fish purchasing decision which conducted in traditional market at Tenayan Raya district are the product factor (eye fish condition, nutrition content, the various types of sold fish) Price factors (the price of that fresh fish), Social factors (family, colleagues, fish-eating habit), and Individual factor (the level of educational and Income)

5. Suggestion

The suggestion than can be given in this research is pushing people to like to eat fish should start to observe what the factors that cause the people to buy the fresh fish and for academicians to continue the research about consumers behavior for a fresh fish product in particularly and agriculture product in generally.

6. References

[1] BPS Kota Pekanbaru, 2014. *Pekanbaru Dalam Angka* 2014. BPS Pekanbaru

[2] BPS Provinsi Riau. 2016. *Riau DalamAngka 2016.*BPS Provinsi Riau.

[3] Destiana, Fransisca Tunjungsari. 2009. *Analisis Perilaku Konsumen Terhadap Pembelian Ikan Kakap Merah Di PasarTradisional Kota Surakarta*. Surakarta.

[4] Dinawan, M Rhendria 2010.*Tesis Program Studi Magister ManajemenUniversitasDiponegoro, Analisis Faktor-faktor yang Mempengaruhi Keputusan Pembelian, (*diaksestanggal 28 Agustus 2016)

[5] Ghozali, Imam 2005, *AplikasiAnalisis Multivariate dengan Program SPSS*, Badan penerbit Universitas Diponegoro, Semarang.

[6] Jayakusumah, Herdi. 2011. *Analisis Analisis Faktor-Faktor Yang Mempengaruhi Konsumen Dalam Keputusan Pembelian The Celup Sariwangi*. Jakarta

[7] Kotler, Philip. 2007. *Manajemen Pemasaran, Analisis, perencanaan, Implementasi, dan Pengendalian*. Salemba Empat, Jakarta, 2007.

[8] Nitisusastro, Mulyadi 2012, *Perilaku Konsumen dalam Perspektif Kewirausahaan,* Alfabeta, Bandung.

[9] Rianto K, 2007. *Gizi dan Pola Hidup Sehat*. CV Yarama Widya.

[10] Saladin Djaslim. 2003. *Manajemen Pemasaran, Analisis, Perencanaan, Pelaksanaan, dan Pengendalian*. Linda Karya, Bandung, 2003.

IOP Conf. Series: Earth and Environmental Science **156** (2018) 012065 doi:10.1088/1755-1315/156/1/012065

Developing Maritime Thematic Expressions on English Reading Materials Used In Indonesia: Junior High School Context

Syahdan[1] and Budi Lusandi[2]

[1]English Education Department, FKIP. Universitas of Lancang Kuning
E-mail: syahdan@unilak.ac.id
[2]English Education Department, FKIP. Universitas of Lancang Kuning
E-mail: bulus@unilak.ac.id

Abstract: Indonesia is a maritime country with the largest islands in the world covering more than seventeen thousands islands. This wealth should be wisely explored for all their advantages. Also, all parties should take responsibility to inherit values embedded in the wealth. This study is a Research and Development focusing at developing English worksheet materials in the theme of Maritime Thematic Expressions. Specifically, it aimed at identifying initial contents of the English worksheet, describing the development of worksheet materials and justifying contents of worksheet materials in facilitating the students' needs in reading English materials. The flow of this R&D refers to Sugiono's model that includes analyzing data, collecting data, designing materials, validating material, revising the material, piloting products, revising the material, piloting materials, revising the material and material development. The development of this material adjusted to the needs of syllabus and national strategic issues in maritime that were coined with students of Junior High School *Terbuka* Pekanbaru. The worsksheet development has been revised and validated by some experts of curriculum and linguistics. This material has also been piloted by the English teacher to determine of the students' ability in comprehending the materials. Based on the results of this development, it can be concluded that the materials development were relevant with the standard competence and basic competence in syllabus of the students' reading skill in Junior High School Terbuka Pekanbaru.

Keyword: Material Development, English Maritime Expressions, Worksheet Contents, Reading

1. Introductions

Teaching English Reading for students at schools requires authentic materials (Thakur, 2015) so that they will see the relevance between materials studied at school and issues discussed outside of it. As Indonesian government, currently, soars theme related to maritime, it is therefore important for teachers to consider this as a theme that is also studied by students at school, specifically Junior High School. It is teacher role to determine best teaching material for improving students' archipelagic awareness. Worksheet, as a teaching material used in Junior High Schol Terbuka Pekanbaru, might be a tool to be developed to raise the students' awareness in maritime issues. The development English materials should also consider the need of curriculum as the teacher's guideline. In the syllabus, it is stated that students should be able to identify information orally, read loudly and respond to short functional text clearly. Moreover, the material should cover instruction, greeting card, list of item and announcement. In fact, it was found that announcement, and instruction materials were not included.

Furthermore, the worksheet was circulated practically and presented instantly without having detailed explanation of its basic concept. Moreover, the worksheet has limited task for reading skill at each chapter. It might consequence to students' low achievement in reading, as Suyitno (2007) highlighted that worksheet has crucial influences in learning process and support students' development [4].

The development of English materials, like worksheet, also refers to cognitive level of Bloom's Taxonomy in the level of Understanding. It requires information and concepts that the students should be able to compare, describe, discuss, and restate activities in learning.

2. Method

This study is an R&D to develop a product, namely English worksheet. It is to cover the students' awareness toward national strategic issues in maritime. In developing material, this research combined the ADDIE's and Sugiono's model. It included of analyzing the potential problem, collecting the data, identifying the worksheet material, designing and developing worksheet material, validating worksheet material, revising the worksheet material, piloting product, revision worksheet material, validating worksheet material, revising worksheet material, piloting product, revising worksheet material, developing worksheet material. In the step, the researchers conducted two times validation and piloting product, namely worksheet material as data source. The researchers also used focus group discussion as instrument of this research. In analyzing the data, the researchers did pra-planning, identifying the product, identifying the content, identifying the social condition and strategy. While analyzing the data, the researchers conducted intensive discussion with the experts to get their judgments.

3. Finding and Discussion

In the description of initial contents, there were two materials that were not covered in the worksheet materials and they were not supported by some tasks that as stated in the syllabus. It can be seen on the table below:

Table 1. The Worksheet Material

No	Materials	Contents	Task	
			Essay	Multiple Choice
1	Announcement	Have not included in the worksheet book	-	-
2	Instruction	Have not included in the worksheet book	-	-
3	Greeting card	Definition Function Kinds of the greeting card	5	15
4	Shopping List	Have not discussed	5	2

Based on the table above, the students' worksheet covered four materials; they were instruction, greeting card, list items and announcement. However, it was found that announcement and instruction material were not included. In greeting card material, its presentation did not cover discussion of generic structure and kinds of tenses properly. Meanwhile, in shopping list materials, it presented procedure text without discussing the meaning and function of the materials. To add, the task material presented has limited and innovative questions. The contents of material worksheet presented not more than three essays (15 items) and fifteen multiple-choices for each task. To cover this limitation, the researchers developed the materials that are expected to cover the students' awareness and achievement in national strategic issues in maritime and learning objectives. The steps of development in the following chart:

IOP Conf. Series: Earth and Environmental Science **156** (2018) 012065 doi:10.1088/1755-1315/156/1/012065

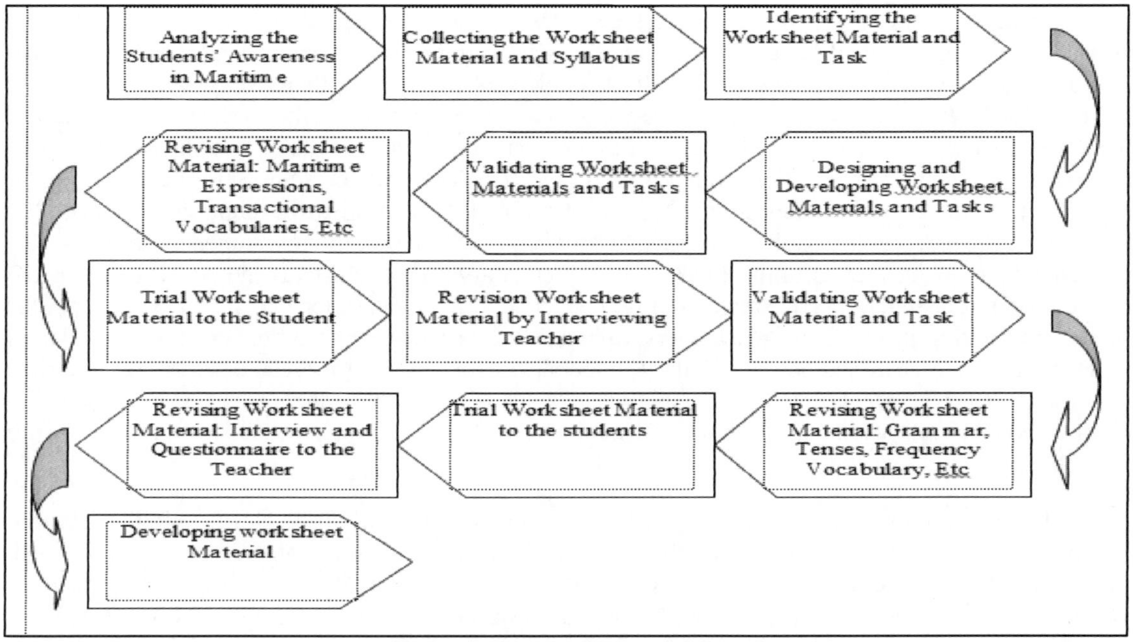

Chart 1. Stages of Material Development

In the stages of developing materials above, the researchers developed material more clearly referring to some weaknesses found in the worksheet material. In the first stage, the researchers analyzed the students' awareness in maritime expressions. In this step, the researchers found that they were lack of maritime expressions. According to the teacher's claim that the materials were never promoted in the class. Moreover, the teacher only took some related materials from other sources to cover the weakness of materials. Based on the teacher explanation above, in the second stage, the researchers checked the syllabus and worksheet material. At this stage, the researchers found that the syllabus covered four materials that might facilitate the required materials. The material covered announcement, greeting card, instruction and list items. However, there were only two materials included in the worksheet materials. The materials which were not included were announcement and instruction.

In the third stage, the researchers checked the materials and tasks in the worksheet. The researchers found that the worksheet material was circulated practically and presented instantly without any detail explanation of basic concept. Moreover, the worksheet material had some limited tasks for each material. It might consequence to the students' optimum achievement. Based on the weakness of the students' worksheet material above, the researcherss developed the worksheet material by following standard competence and basic competence in the syllabus. The researchers designed the materials by arranging each material to be quite simple and easily understood by the students. Considering of the students' achievements, the researchers presented some tasks that included 20 innovative questions in the essay and multiple-choice instrument form. Furthermore, the researcherss provided the materials for the validators to get some suggestion of the developing materials. In this validation, the researchers got some suggestions from validator to identify, add and correct the material into better level. The suggestion included useful expression in maritime, minor materials on tenses, transaction vocabularies, innovative question and word choices. After identifying the error above, the researchers re-checked, revised, and re-added the material into the expected level. After revising (checked, added and corrected) the material, the researchers returned the material to the validator for getting an approved to apply the material in the class.

In the seventh stage, the researchers gave the material to the teacher in order to pilot the product. The teacher applied the materials consisting of definitions, functions and etc. The teacher also gave one worksheet that consisted of 10 multiple choices. As the result, there were some weaknesses of the materials. In the revision of the materials, the teacher explained that the students felt difficult to finish the task because the material had difficult level so that the students did not understand very well to finish the task. This result indicated that the materials needed some revision. Before conducting the second trial product, the researcherss revised and explained the result of first piloting product to the validators to develop the materials in the second product. Furthermore, the researchers designed a new material by following the learning objective in the syllabus and level of taxonomy bloom, that the expected reading materials were categorized into understanding level. After making a new product for the second piloting product, the researchers validated the last material. In this stage, the validators suggested to use high frequency vocabulary, good distractors of multiple choice items, appropriate tenses and better word choices. Following some suggestions above, the researchers revised and returned the materials to the validator to get an approved in applying the materials as the last piloting product.

In the second piloting product, the teacher applied the material in the class as similar with the step in the first piloting product. The teacher explained the material and continued providing the task to the students. As a result in the second piloting product, the students were able to improve their vocabularies comprehension on maritime theme. Based on the teacher explanation, the students were able to improve their comprehension because the materials used were categorized into high frequency vocabularies in the maritime theme and the level of the task was moderate so that they understood very well in the learning process. The teacher suggested that the material should be developed into the second product materials. Continuing the teacher's suggestion above, the researchers discussed it with the validators to have some feedbacks before taking further development on the material.bFurther, the researchers described the materials development as shown in table 2. The development highlighted the constraints of the materials and tasks of the worksheet. The developing materials consisted of material explanation that was equipped with the tasks. The result of material development can be seen below:

Table 2. The Materials Development

No	Materials	Contents	Task	
			Essay	Multiple Choice
1	Announcement	– Definition – Function – Kinds of announcement – Generic structure – Examples – Tasks	20 Items	20 Items
2	Instruction	– Definition – Function – Generic structure – Examples – Tasks	20 Items	20 Items
3	Greeting card	– Definition – Function – Kinds of Greeting Card – Generic structure – Examples – Tasks	20 Items	20 Items
4	Shopping List	– Definition – Function – Generic structure – Examples – Tasks	20 Items	20 Items

Based on the table above, the product results was equipped with the tasks organized based on the findings of curriculum needs in learning English. This development product consisted of the definition, function, generic structure, and lexical grammatical by considering some factors like quite frequent vocabularies to facilitate and improve students' ability in comprehending the material easily. Moreover, the researchers also presented some tasks consisted of 20 multiple-choices and 20 essays for each material to support the students' reading ability. Furthermore, the materials were being validated by some experts before it was applied by the teacher in the class. It helped the researchers to ensure the product's conformity of the syllabus and to find the weakness of the new material before piloting product. The result of validation can be seen below:

Table 3. Result of Validating Product

No	Materials	Results
1	Announcement	– The task was difficult level – Grammar was not good – Many passive sentences
2	Greeting Card	– Tenses should be appropriated – Grammar was not good – High vocabulary – Word choice was not good – Choose high frequency – The task is not innovative – Check spelling – The task did not have good destructor
3	List of Items	– Tenses should be appropriated – Grammar was not good – Many passive sentences – The task was not innovative – The task did not have good destructor
4	Instruction	– Word Choice was not good

Based on the table above, the researchers found there were some suggestions from validators to review, add and correct the material into most suitable material. The materials that should be reviewed were announcement, greeting card, shopping list and instruction. The review included of the task was level of difficulty and some grammatical errors, the use of numerous passive sentences, high frequency vocabularies, word choices was not good, the need of innovative and various materials and some inappropriate distractors on multiple choice questions. Considering the review above, the researchers revised, re-checked and re-added the material into most suitable level. After revising (checking, adding and correcting) the material, the researchers returned the material to the validator to get an approved in order to apply the material in the class. Below is the table of the last validation:

Table 4. Draft of Revision

NO	Activities	Command and Discussion
1.	**Greeting Cards**	1. The theme : Greeting Card 2. The lesson plan: Good 3. Content suitability: – Good
2.	**List of Items**	1. The theme : Shopping List 2. The lesson plan: Good 3. Content suitability: – Good
3.	**Instruction**	1. The theme : Instruction 2. The lesson plan: Good 3. Content suitability: – Good
4.	**Announcement**	1. The theme :Announcement 2. The lesson plan: Good 3. Content suitability: – Good

Based on the result of the revision draft above, the experts argued that the material development was suitable and could be given to the teacher to apply in the class as trial product. The results of this research presented the answer of the third formulation of the research: how is piloting a model result relevant to the syllabus and user's needs. In this research, the researchers piloted the product two times. In the first piloting product, the material was used by the teacher to assess students in classroom quiz. The teacher gave a sheet of the task consisted of 10 multiple choices created by the researchers that has been validated by validators. As the result of the first piloting product, the researchers found the material was quite difficult for students accomplish. In the analysis, it showed that the average of the students' score was 66.7 or lower than passing score determined by the school, 72. This result indicated that the material needed some revisions and better development. Furthermore, the researchers conducted focus group discussion with the English teacher. As the result, the teacher explained that the material was quite difficult level required appropriate frequency vocabularies so that the students would get the better advantages of the development.

Before conducting the second piloting products, the researchers revised the result of the first trial product by explaining the results of focus group discussion to get some suggestions from the validator to develop the materials in the second piloting product. In the second piloting product, the teacher applied the new material that followed the learning objective in the syllabus and level of taxonomy bloom as similar with the step in the first trial product. The teacher explained the material and continued to provide the task for students. As the result of the second piloting product, the teacher gave another quiz to assess students' comprehension on the materials. It was found that the average of students' score was 83.5, which is higher than the passing score. According to the teacher's statement, this result was achieved for the reason of better and appropriate materials with moderate frequency vocabularies and the level of the task was relatively easy understood by students. Based on the result of the second piloting product above, the teacher gave some response of the result in the focus group discussion, the teacher argued that the material could be accepted and understood by the students easily.

Moreover, the teacher argued that the material had good level because her students did not feel difficult when finishing the task. Finally, the teacher concluded that the material development had supported learning objective in the syllabus and level of taxonomy bloom, since the students were categorized into understanding level. Based on the teacher's response above, the researchers concluded that the final materials had been already appropriate for the needs of curriculum with moderate level of difficulty for students. It has suitable with the purpose of R&D by Mulyatiningsih (2012:202) that the research aimed to make a new product empirically. This result was also supported by the previous research by Noor (2014) that the students can improve their average score by new product and have the feel of suitability with the new product. Although the material gave improvement, the material still adapted curriculum KTSP and and should have an update for curriculum 2013. The material developed, also, only covered one skill and discussed four materials of the functional text. It means that the materials should be supported by other sources of the worksheets to cover the students' skill and to make more innovative tasks needed by the students in learning English.

4. Conclusion

Based on the data analysis, the researchers described some conclusions as follows:

1) The students' worksheet book has not covered the material that stated in the syllabus. It was found that announcement, and instruction materials are not included. Furthermore, the worksheet was circulated practically and presented instantly without accompanied by detailed explanation of basic concept.

2) The material development is designed by following the standard competence, basic competence and learning objective in the syllabus also following the level of taxonomy bloom, that the students were categorized into understanding level.
3) The material development reaches the average score is 83.5. It means that the students can improve their reading comprehension by using these materials.

5. Bibliography

[1] Acevedo. 2009. *Defining Reading.* Lima, Peru: Instituto Cultural Peruono Norteamericano.

[2] Alyousef. 2006. *Teaching Reading Comprehension to ESL/EFL Learners.* Vol. 5, The Reading Matrix. Retrieved on May 2011. From http://www.scribd.com.

[3] Faridi, Ahmad. 2008. *Pengembangan model Ajar Muatan Lokal Bahasa Inggris di Sekolah Dasar Jawa Tengah yang Berwawasan Sociokultural.* PhD Dissertation of Semarang State of University.

[4] Goodman, et. al. 2005. *Reading miscue inventory* . (2005). *Reading miscue inventory.* Katonah, New York: Richard C. Katonah, New York: Richard C. Owen Publishers, INC. Owen Publishers, INC.

[5] Herlina. 2012. *Developing Reading Narrative Text Material for the Eighth Grades of Junior High School Implemented with Character Building.* Conservation University. /http/journal.unnes.ac.id./sju/index.php.eej

[6] Hutchins. 2003. *7 keys to comprehension.* New york: Three Rivers Institute for Curriculum Development *Introduction to Educational Research.* Enschede, Netherland : National

[7] Klingner et.al. 2007. *Teaching Reading Comprehension to Students with Learning Difficulties.* New york: The Guilford press.

[8] Mulyatiningsih. 2012. Metode Penelitian Terapan Bidang Pendidikan, Yogyakarta.

[9] Mustofa. 2013. *Pengambangan Lembar Kerja Siswa Berbasis Observasi Pada Taman Sekolah sebagai Sumber Belajar Sains di SDN 01 Tinjamoya.*Semarang University.

[10] Noor. 2014. *Pengambangan Lembar Kerja Siswa (LKS) Berbasis Inkuiri Terbimbing untuk Memfasilitasi Pemahanan Konsep Materi Sistem Persamaan Linear Dua Variabel*: Islamic University of Yogyakarta.

[11] Plomp. 2007. "Educational Design Research : An Introduction", dalam *An* Qureshi (2010) *The Important of Speaking Skill for EFL Learners*, Pakistan University.

[12] Ruslan et.al 2015. *Developing supplementary Speaking Material for the Seventh Grades.* Univesity of Malang.

[13] Salmah et.al. 2012 *The Development of Students worksheet Using PMRI Approach on Material of Rectangle and Square for the Seventh Grades of Junior High School.* University of Surabaya.

[14] Snow, Catherine. 2000. *Toward an R & D Program in Reading Comprehension.* New York:Science & Technology Policy Institute.

[15] Sugiono. 2014. Metode Penetian Kuantitatif, Kualitatif dan Research and Development.Bandung.

[16] Thakur, Vijay Singh. 2015. Using Supplementary Materials in the Teaching of English: Pedagogic Scope and Applications. English Language Teaching; Vol. 8, No. 12; 2015. Retrieved from https://files.eric.ed.gov/fulltext/EJ1084305.pdf

[17] Westwood, 2001. *Reading and Learning Difficulties Approaches to teachingg and assessment.* Victoria: The Australian council for Educational Research Ltd press.

[18] Wekke, I.S., Yandra, A. and Hamuddin, B., 2017, December. Learning Strategy in Class Management: A Reflection from Manado Case. In *IOP Conference Series: Earth and Environmental Science* (Vol. 97, No. 1, p. 012053). IOP Publishing.

Institute of Physics
Dirac House, Temple Back
Bristol BS1 6BE UK

ISSN: 1755-1307
ISBN 978-1-5108-6396-5

CURRAN ASSOCIATES INC.
proceedings.com

5th International Conference on Advanced Materials, Mechanics and Structural Engineering (5th AMMSE 2018)

IOP Conference Series: Materials Science and Engineering
Volume 473

Seoul, South Korea
19 – 21 October 2018

IOP
Institute of Physics